Collision Processes of Ion, Positron, Electron and Photon Beams with Matter

Collision Processes of Ion, Positron, Electron and Photon Beams with Matter

Proceedings of the Latin American School of Physics

Caxambu, MG, Brazil 4 – 24 August 1991

Editors
Ana Cecilia de Azevedo e Souza *(IQ -UFRJ)*
Enio Frota da Silveira *(PUC – Rio)*
José Carlos Nogueira *(DQ – UFSCar)*
Marco Antonio Chaer do Nascimento *(IQ – UFRJ)*
Danilo de Paiva Almeida *(IF–UFRJ)*

World Scientific
Singapore • New Jersey • London • Hong Kong

Published by

World Scientific Publishing Co Pte Ltd
5 Toh Tuck Link, Singapore 596224
USA office: 27 Warren Street, Suite 401-402, Hackensack, NJ 07601
UK office: 57 Shelton Street, Covent Garden, London WC2H 9HE

British Library Cataloguing-in-Publication Data
A catalogue record for this book is available from the British Library.

ELAF 91
Collision Processes of Ion, Positron, Electron and Photon Beams with Matter

ISBN-13 978-981-02-0863-9
ISBN-10 981-02-0863-4

PREFACE

*The LATIN AMERICAN SCHOOL OF PHYSICS (Escola Latino Americana de Fisica - **ELAF**) has been held since 1959 with the aim of stimulating and improving the interaction between latin american physicists.*

*The founders and first organizers of this school were the physicists M. Moshinsky (Mexico), J. Leite Lopes (Brasil) and J. J. Giambiagi (Argentina). They made the school possible each year, in their respective home countries. Later, some other countries such as Venezuela, Chile and Colombia also joined **ELAF**. This was an important point for the enlargement of scientific colaboration within Latin America.*

Initially, the main subjects covered by the School were Nuclear Physics and Elementary Particle Physics. Gradually, some other areas were included such as Solid State Physics, Statistical Mechanics and Gravitation.

*For political reasons, the last **ELAF** in Brazil was held in 1963. Now, after many years, **ELAF** returns to our Country reflecting the democratic stability very much desired by all of us.*

*The **ELAF91** happened in Caxambu, a charming brazilian resort town, under the title* **"Collision Processes of Ion, Positron, Electron and Photon Beams with Matter"**. *We joined together 23 invited lecturers coming from many countries in Europe, North and Latin America. They gave 6 courses and 17 invited talks in different topics related to Surface, Atomic and Molecular Physics. The School was dedicated to Prof. Jose Leite Lopes for his valuable contributions towards the development of Physics in Brazil.*

We gathered 94 latin american students who had the opportunity of learning and discussing, during three weeks, about the recent advancements of physics in the above mentioned areas.

Besides the scheduled courses and invited talks, poster and short communication sessions allowed a selected number of post-graduate students to present projects related to the work developpedat their home institutions. This provided a real interaction between latin americans which was one of our goals and the real essence of our school.

In addition three round tables were also organized under the titles "Future Trends of Desorption Methods of Mass Spectrometry", "Open Questions in Ion-Atom Collisions" and "Electron-Synchrotron Processes", *with the aim of presenting the state-of-the-art on these subjects.*

At the end of the third week a visit was organized to the National Laboratory for Synchrotron Light (LNLS) . This laboratory located in Campinas, Sao Paulo , is now under construction and is scheduled to be in operation in 1993.

We also express our gratitude to the important support received from our sponsors from Brazil and abroad for the realization of the School.

We are particularly indebted to our secretary Marcia Fernandes and to the Physical Chemistry Department staff (IQ/UFRJ) for their professionalism and dedication in dealing with the School's organization.

THE EDITORS

Ana Cecilia de A. e Souza
Enio Frota da Silveira
Jose Carlos Nogueira
Marco Antonio Chaer do Nascimento
Danilo de Paiva Almeida

SPONSORS

BRITISH COUNCIL	I. FISICA ROSARIO
CAPES	LNLS
CINVESTAV	OEA
CLAF	PETROBRAS
CNPq	PUC-RIO
CONACYT	SBF
DAAD	SBQ
FAPERJ	UFRJ
FAPESP	UFSCar
FRENCH EMBASSY	U.N. CORDOBA
IBM	VG INSTRUMENTS
ICTP	VG MICROTECH

CONTENTS

Preface..V

OPENING LECTURE

Collision Processes of Ions and Photons with Matter: an Overview of the History of Ion Emission Processes.......3
Ronald D. Macfarlane

COURSES

Ion Desorption Phenomena Induced by Various Types of Multiply Charged Projectiles and by Photons on Solid Surfaces..15
Y. L. Beyec

Electronic Sputtering of Large Organic Molecules and its Application in Biomolecular Mass Spectrometry.............59
Bo U. R. Sundqvist

Current Theoretical Methods for Electron-Molecule Scattering..94
R. K. Nesbet

Inner-Shell Studies of Molecules and Surfaces by Electron Impact..104
Adam P. Hitchcock

Ion Formation by Electron Impact..........................121
S. K. Srivastava

Synchrotron Radiation: Principles and Applications.......148
T. J. Morgan

INVITED LECTURES

Matrix Effects in Ion Desorption from Heavy Ion Tracks in Solids..161
J. Ai, R. Nord and R. D. Macfarlane

The Cluster-Solid Interaction: Fundamentals and Applications..171
M. A. Park, M. E. Inman, A. Y. Yau, B. D. Cox and E. A. Schweikert

Heavy-Particle Induced Electron Emission from Solids.....185
Raul A. Baragiola

Principles of Electron Stimulated Desorption.............199
Raul A. Baragiola

Dependence of Secondary Ion Emission from Organic Material on the Energy Loss of the Impacting Heavy Ion...214
Jerry E. Hunt and Karl Wien

Desorption Induced by Molecular and Atomic Ions..........230
C. V. Barros Leite, E. F. da Silveira, J. M. F. Jeronymo, R. R. Pinho, E. A. Schweikert and M. A. Park

Applications of the Generalized Multistructural (GMS) Wave Function to Photoelectron Spectra and Electron Scattering Processes....................................247
Marco Antonio Chaer do Nascimento

Main Features in the Atomic Ionization by Ion Impact.....268
C. R. Garibotti and R. Pregliasco

Heavy Ion Cooler Rings and Their Application in Studies of Ion-Electron-Photon Interactions......................283
R. Schuch

Electron-Multiply Charged Ion Collisions.................301
N. V. de Castro Faria

Phenomena in Low Energy Electron Scattering by Atoms and Molecules Studied by Means of a Multi-Angle Parallel Detection Electron Spectrometer................311
D. Roy, D. Tremblay and D. Dubé

The measurement of Ionization Cross Sections: Ar and CF_4 as Case Studies.......................................329
R. A. Bonham, M. R. Bruce and Ce Ma

Investigation of NO^+, SF^+ and SF^{2+} by Means of Translational Energy Spectroscopy.........................343
M. Hamdan

Experimental Workstation for the Brazilian National Synchrotron Light Laboratory..............................350
Aldo Craievich

New Materials Processing with Ion Beams..................358
I. J. R. Baumvol

CONTRIBUTED PAPERS

Absolute Elastic Differential Cross Sections for Electron Scattering from SF_6 in the 75-1000 eV Range..............371
J. C. Nogueira and M. J. Dallavalli

Study of Gas Phase Products Formed By Flash-Vacuum Pyrolysis of Halomethanes Using Electron Energy-Loss Spectroscopy..378
I. M. Xavier Jr., K. N. Walzl, M. R. Giorgi and A. Kuppermann

Laser Excited Fluorescence Spectrum of PbI................388
P. Patiño, A. Tovar and D. Iacocca

A Distorted Wave Model for Electron Capture in Large Angle Collisions with Correct Asymptotic Conditions.....393
O. A. Fojon and J. M. Maidagan

Saddle Electron Emission from Ionizing H^+ and $^3He^{2+}$ Collisions on Ne..401
S. Suárez, G. Bernardi, C. R. Garibotti and W. Meckbach

Neutralization of Slow Ions Before Collision with Surfaces..410
F. P. Santos, M. Matos and E. F. da Silveira

Improvements in the Description of the Background Spectrum in EPMA......................................420
Jorge Rubio, Gustavo Castellano and José Riveros

Critical Behavior of the Weakly Diluted Ising Antiferromagnets and Ferromagnets.........................428
Gloria M. Buendía and Julio F. Fernandez

Theoretical Double Photoionization Study of Helium.......437
V. D. Rodríguez, M. A. Kornberg and C. A. Falcón

Distorted Wave Calculations for Double Electron Transfer..447
Alejandra E. Martinez, Roberto D. Rivarola, Robert Gayet and Jocelyn Hanssen

Electron Capture from Molecular Targets by Ion Impact....457
Silvia Corchs, Roberto Rivarola and J. H. McGuire

Ion Beam Analysis of Aluminium Nitride Thin Films........466
F. C. Stedile, I. J. R. Baumvol, W. H. Schreiner and F. L. Freire Jr.

The Different Steps in the Ion Beam Modification of Poly(ethylene terephthalate) (PET)........................474
R. M. Papaléo, M. A de Araújo and R. P. Livi

Simple Model for Convoy Electrons Transport..............485
W. R. Cravero, M. L. Martiarena and C. R. Garibotti

APPENDIX

Suggestions for Next Desorption Studies..................497

Developing a Program Based on PDMS........................497
Ronald Macfarlane

Suggestions for New Experiments..........................499
Yvon Le Beyec

PDMS for M < 1000 u.......................................501
Raul Baragiola

OPENING LECTURE

COLLISION PROCESSES OF IONS AND PHOTONS WITH MATTER: AN OVERVIEW OF THE HISTORY OF ION EMISSION PROCESSES

RONALD D. MACFARLANE
Department of Chemistry, Texas A&M University
College Station, Texas, 77843, U. S. A.

ABSTRACT

This overview begins with a statement of the processes of ion emission from matter induced by ion or photon impact at the atomic level and under conditions where the interaction is so explosive and complex that the process can only be treated statistically . This will be followed by a review of some the milestones during the past 20 years that have brought us to our current level of understanding of the subject and to applications to analytical problems ranging from materials science to biotechnology.

1. AN ORIENTATION

This Summer School covers a wide range of topics under the heading of "Collision processes of ion, positron, electron, and photon beams with matter" that will be covered during the 3 weeks of its duration. These topics are not only diverse in character but also represent the full spectrum of the level of detail and control that is utilized in developing the science of particle/solid interactions. At one end of the spectrum, the researcher has carefully selected the quantum state of the impacting particle, a particular well-defined structure of the matrix being excited and follows the evolution of the interaction of the particle with matter as it proceeds through a succession of quantum states to the final state. Photons, ions, and particles that are emitted, (their multiplicities, energies, directions and correlations with the incident particle) are the clues that are given to the experimentalist from which mechanistic details are formed. These are topics that are examples of the kinds of research projects that will carried out at the new Labotarorio Nacional de Luz Sincrotron in Campinas and which are the subjects to be covered in the latter part of the Summer School session.

At the other end of the spectrum of studies carried out, the ion/surface interaction is characterized by a much more violent, explosive encounter generally with an ill-defined matrix. The first week's lectures are devoted to topics relating to these kinds of studies.

When a high energy complex particle or energetic photon pulse collides with the surface of a matrix, it produces a localized excitation region of very high energy density, equivalent to tens of thousands of degrees Kelvin. Thousands of neutral particles, ions, electrons, and photons are ejected leaving behind submicron-sized craters on the surface. The complexity of the interaction obscures the microscopic detail to one better described in terms of statistical processes. The motivation for understanding as much as we can of these complicated, somewhat uncontrolled interactions is in application to analytical problems. An analysis of the products desorbed during these more violent particle/solid interactions gives information on the chemical composition of matter as well as the distribution of components within the matrix. The fields of research influenced by our ability to analyze matter by particle bombardment range from materials science to biotechnology and our capability of improving the analytical power of these methods is dependent on the continued evolution of our knowledge of the statistical nature of these complex interactions and its connection with the fundamental physical processes at the molecular level.

2. PARTICLE-INDUCED ION DESORPTION

The lecturers for the first week's lectures will give examples of the wide range of incident particles that are used in initiating the desorption of ions from surfaces. They include atomic and molecular ions with keV to 100 MeV energies and pulsed laser-induced photon beams ranging from UV to IR energies with variable energy densities. Ion emission, induced by ion or photon beam impact on a surface, is generally only a small component of the spectrum of particles emitted accounting for as little as 0.1 % of the total number of particles emitted [1]. These ions are, however, the most convenient for chemical analysis because they can be directly analyzed by mass spectrometry. But, for the studying the mechanism of particle-induced desorption, the neutral species that are ejected can be playing an important role in influencing the dynamics of ion formation and desorption.

3. HISTORICAL REVIEW OF STUDIES OF PARTICLE-INDUCED ION DESORPTION

3.1 The Origin of SIMS and LD.

The first detailed studies of particle-induced ion emission were carried out in the 1930's by Sloane and Press [2] using keV atomic ions impacting on surfaces containing adsorbed organic molecules. They observed that atomic and molecular ions characteristic of fragments of the adsorbed species were ejected from the surface. During the next 3 decades, this phenomenon was developed into a powerful surface analytical method and was given the name Secondary Ion Mass Spectrometry (SIMS) by Benninghoven [3], one of the important contributors to its development. In the first applications of SIMS, the emphasis was on the emission of atomic ions for elemental analysis. Molecular ions were also observed in mass spectra as a part of an undesirable "background" of ions. In 1972, the first laser-induced ion emission results (laser desorption, LD) were reported by Mumma and Vastola [4] using a time-of-flight mass spectrometer. They reported spectra similar to those observed in SIMS experiments.

During that period, classical mass spectrometry was being used routinely with remarkable success for the analysis of species that could be volatilized by heating and then ionized in a gas phase ionization process. Exact mass measurements of the molecular ion of the intact molecule gave the elemental composition of the molecule and the fragmentation pattern gave information on the chemical structure.

3.2 Solving the Problem of Gas Phase Ion Formation for Involatile, Thermally-unstable Molecules; The Origin of ^{252}Cf-PDMS and FAB.

The power of classical mass spectrometry could not be applied to the large class of polar, thermally unstable molecules that could not be volatilized without decomposing. This limitation was considered to be a major obstacle seemingly unresolvable. The solution to the problem came by way of a chance observation in a nuclear physics experiment at Texas A&M in 1973. It was found that when energetic and highly charged ions such as those produced in the spontaneous fission of ^{252}Cf (e.g. $^{120}Sn^{20+}$, 105 MeV energy) impacts a solid matrix of these involatile, thermally unstable species, a detectable component of the desorbed ions includes intact molecular ions of the molecule. This phenomena was given the name ^{252}Cf-plasma desorption mass spectrometry (^{252}Cf-PDMS) [5]. The first spectrum that was recorded is shown in figure 1.

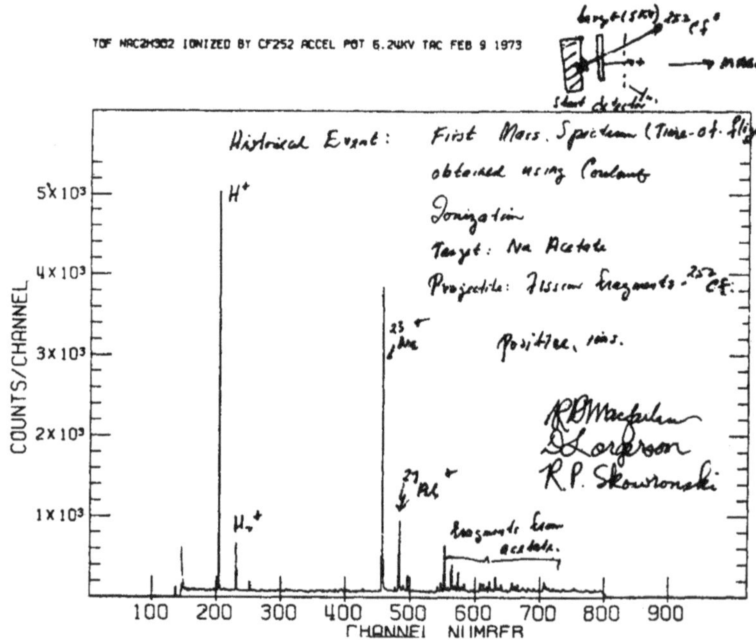

Figure 1: First ^{252}Cf-PDMS spectrum.

It was soon recognized by others that the essential ingredient of the success of ^{252}Cf-PDMS for the formation of gas phase molecular ions of these species lay with the high energy density created within the matrix and the short time scale for the excitation. These features are also a characteristic of SIMS and pulsed laser excitation and consequently, these techniques, which had previously been developed for surface analysis, were also found to be capable of desorbing intact molecular ions of biological molecules [6,7].

These new developments have taken place during the past 18 years. Since then, refinements in the instrumentation have led to the commercialization of ^{252}Cf-PDMS, SIMS and laser desorption for the analysis of biomolecules using the time-of-flight method. A schematic diagram of a ^{252}Cf-PDMS system is shown in figure 2.

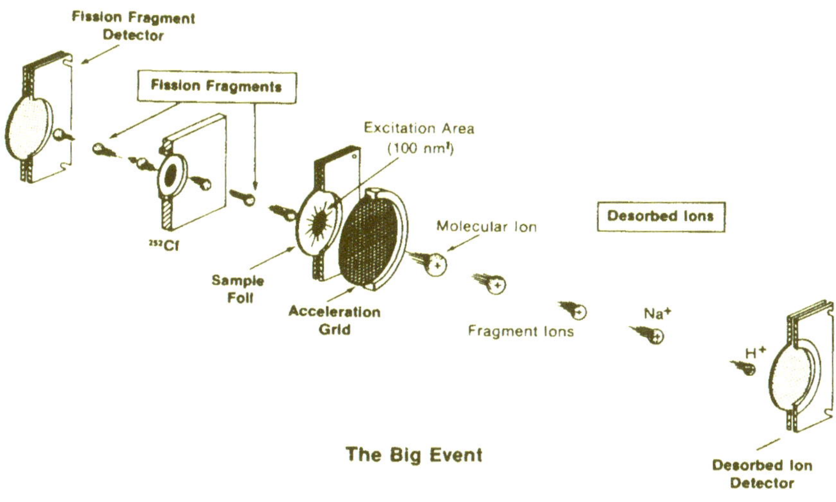

Figure 2: A schematic diagram of a ^{252}Cf-PDMS system showing the use of both fission fragments in the time-of-flight mass spectrometer [8].

Improved sensitivity and mass resolution, and the development of ion microprobes to obtain information on identification as well as location of surface molecules are part of an on-going effort to improving the application of the particle-induced ion desorption process to surface analysis.

A recent spectrum of a peptide showing a complex fragmentation pattern from which the amino acid sequence can be elucidated is shown in Figure 3. This is an example of an application of ^{252}Cf-PDMS to a problem in biotechnology where molecular mass and amino acid sequence information is needed to verify the authenticity of a synthetic product.

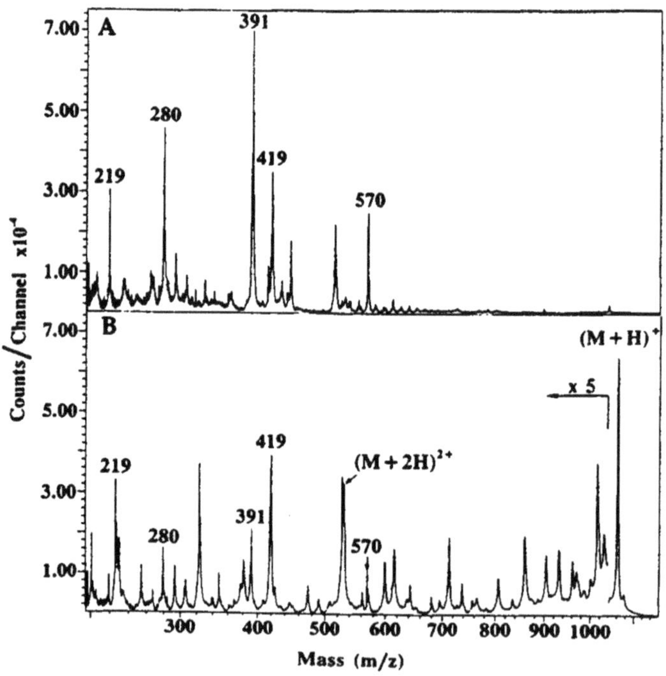

Figure 3: ^{252}Cf-PDMS spectrum of A.) a nitrocellulose matrix and B.) the same matrix with bradykinin adsorbed on it. In addition to the singly- and doubly protonated molecular ion, 37 sequence-related fragment ions were observed indicating that considerable excitation to vibrationally-unstable states is occuring in the heavy ion track [9].

3.3 First Mechanistic Studies for SIMS and ^{252}Cf-PDMS

The SIMS method utilizes atomic ions in the keV energy regime as the incident species. Recently, the technique has been extended to include incident molecular ions as well [10]. At these energies, the primary energy loss is by atomic collisions between the incident ion and the atoms of the matrix and the energy loss process is referred to as nuclear stopping. Sigmund has developed a linear cascade theory that gives a good account of the dynamics of the emission of atomic ions [11]. However, an explanation of the mechanism of molecular ion emission which is characterized by a much lower initial kinetic energy distribution has been somewhat problematic although some progress has been made by Garrison by studying the evolution of the cascade to later stages of development [12]. A key feature of SIMS theory is that a mechanism for transfer of kinetic energy to surface molecules is built into the description of the linear cascade process. This is not so for ^{252}Cf-PDMS.

The ^{252}Cf-PDMS method operates at much higher energies than SIMS; energies dominated by electronic excitations rather than direct collisonal processes. The first question confronting the ^{252}Cf-PDMS researcher was how electronic excitation could be converted to kinetic energy so that ion desorption could occur. Part of the answer to this question came from the observation by Wien [13] that unlike SIMS, ion emission from clean metal surfaces was strongly suppressed but the addition of an insulating layer on the surface restored the high ion emission yield. While ^{252}Cf-PDMS does poorly relative to SIMS for ion desorption from metals, it produces orders of magnitude higher molecular ion yields from insulating layers, particularly for organic species and the range of biomolecules desorbed as molecular ions extends to much higher masses (45,000 u) than what is observed for SIMS [14].

The first experiments carried out to learn something about the dynamics of ion emission surfaces excited by ^{252}Cf-fission fragments were performed in the laboratory of Wien [15]. They measured initial kinetic energy and angular distributions under ^{252}Cf-PDMS and verified the statistical nature of the dynamics of ion formation. But it wasn't until heavy ion particle accelerators were brought into the picture that a much more complete picture of the role of the incident ion emerged.

3.4 Accelerator Beams of Heavy Ions Reveal Much of the Ion Emission Dynamics.

It was clear from the beginning of the ^{252}Cf-PDMS studies that many of the features of the mechanism of ion desorption were connected with the properties of the incident ion, particularly velocity and charge state. These are the variables that control the magnitude of the electronic stopping. The problem with using ^{252}Cf-fission fragments for mechanistic studies is that there is a distribution of ion velocities and charge states and it is difficult to change these variables for fission fragments. Consequently, studies were initiated at several nuclear physics laboratories that had high energy heavy ion accelerators. The first of these studies was carried out by Voit et al at Erlangen [16], Sundqvist et al at Uppsala [17], LeBeyec et al at Orsay [18], and Wien et al at Darmstadt [19]. From these studies, the important factors relating to the incident beam that influence ion desorption yields were mapped out. Three of these distinguished scientists are participating in this Summer School. The research that remains to be done essentially involves the influence of the matrix on the ion desorption process. This is the topic I will comment on in the paper I will present later in the Summer School.

5. ACKNOWLEDGEMENTS

I am grateful to the Organizing Committee, specifically, Drs. A. C. A. Souza and E. F. Silveira, for the opportunity to participate in this Summer School. I would also like to extend my appreciation to the Brazilian CNPq, the U.S. National Institutes of Health, and the Welch

Foundation for their financial support of my research and my participation in this Summer School.

6. REFERENCES

[1] Salehpour, M., Hakansson, P., Sundqvist, B., and Widdiyasekera, S., Nucl. Instr. Meth. Phys. Res. B13, 278 (1986).

[2] Sloane, R. H., and Press, R., Proc. Royal Soc., A154, 538 (1938).

[3] Benninghoven, A., Z. Physik 230, 403 (1970).

[4] Mumma, R. O., and Vastola, F. J., Org. Mass Spectrom. 6, 1373 (1972).

[5] Torgerson, D. F., Skowronski, R. P., and Macfarlane, R. D., Biochem. Biophys. Res. Commun. 60, 616 (1974).

[6] Benninghoven, A., Jaspers, D., and Sichtermann, W., Appl. Phys. 11, 35 (1976).

[7] Posthumus, M. A., Kistemaker, P. C., Meuzelaar, H. L. C., and Ten Noever de Brau, M. C., Anal. Chem. 50, 985 (1978).

[8] Macfarlane, R. D., Anal. Chem. 55, 1247A (1983).

[9] Bunk, D. M., and Macfarlane, R. D., J. Am. Soc. Mass Spectrom. (in press).

[10] Blain, M. G., Della-Negra, S., Joret, H., LeBeyec, Y., and Schweikert, E. A., J. Vac. Sci. Technology, A8, 2265 (1990).

[11] Sigmund, P., Rev. Roum. Phys., 17, 1079 (1972).

[12] Garrison, B., Int. J. Mass Spectrom. Ion Phys., 53, 243 (1983).

[13] Wien, K., Review, 7th Int. Workshop Inelastic Ion Surface Collisions; Krakow, Sept. 1988, Report IKDA 88/21, Darmstadt Technische Hochschule, Germany.

[14] Jonsson, G., Hedin, A., Hakansson, P., Sundqvist, B., Bennich, H., and Roepstorff, P., Rapid Commun. Mass Spectrom. 3, 190 (1989).

[15] Furstenau, N., Knippelberg, W., Krueger, F. R., Weiss, G., and Wien, K., Z. Naturforsch. 32a, 711 (1977).

[16] Duck, P., Treu, W., Galster, W., Frohlich, H., and Voit, H., Nucl. Instr. Meth. 168, 601 (1980).

[17] Hedin, A., Hakansson, P., Salehpour, M., and Sundqvist, B., Phys. Rev. B35, 179 (1982).

[18] Della- Negra, S., Jacquet, D., Lorthiois, I., Le Beyec, Y., Becker, O., and Wien, K., Int. J. Mass Spectrom. Ion Phys., 53, 215 (1983).

[19] Becker, O., Knippelberg, W., and Wien, K., Physika Scripta T6, 117 (1983).

COURSES

ION DESORPTION PHENOMENA INDUCED BY VARIOUS TYPES OF MULTIPLY CHARGED PROJECTILES AND BY PHOTONS ON SOLID SURFACES

Y. LE BEYEC

Institut de Physique Nucléaire, 91406 - Orsay Cedex, France

Ion desorption phenomena are the general denomination for the emission of ions (atomic, molecular, clusters) from solid samples - generally insulating organic or inorganic materials - bombarded by accelerated ions or electrons, or by photons (laser).

The understanding of the desorption mechanisms is to day an attractive problem of fundamental physics linked to energy deposit and transfer in solids. The great potential of the desorption processes, to investigate interactions between fast ions and materials and to develop analytical tools, explains the increasing number of scientists and laboratories devoting their interest to ion desorption.

New experiments require the construction of specific equipments and the use of advanced technologies but, with the exception of beams from accelerator, relatively simple and light experimental facilities are necessary. The wide range of options to study ion desorption and the large number of possible applications - that has demonstrated the interdisciplinarity of the field - make the all domain, a good choice, for emerging research groups.

It is a great pleasure to participate into the "Escola Latino Americana de Fisica, ELAF91" and to have the opportunity to talk on this subject. I have summarized, in the following, some of the work done in our laboratory during the past recent years. I have tried to show, by various types of experiments and results that this field of research is a source of fruitful developments. I did not cover, neither the applications that would have required a rather long paragraph, nor the instrumental part which is, of course, also very important. The I.P.N. (Institut de Physique Nucléaire Orsay) is known for its contribution to many other laboratories and colleagues.

I believe in the rapid growing of the field - including companies involved in instrumentation - and, in the future of massive ion impacts. All domains of applications will gain from fundamental experiments.

The field of desorption-ionization of secondary species induced by fast and slow ions in a solid is extremely large (fast and slow will be defined later). A collision betweeen a two atom system or more generally a two-body system is relatively simple to describe even if the interacting potentials are not completely known. When a large number of atoms and complex ions are involved in multiple collisions, understanding the detailed processes that occur in a short time after the initial interaction is an ambitious goal. The complex sequence of dynamical events in the solid that finally lead for example, to the ejection into the gas phase of intact molecules is poorly understood. In recent years however a certain number of efforts in several laboratories have contributed to elucidate the basic phenomena. A large variety of applications has been developed and has demonstrated the interdisciplinarity of the field. Until now mainly single atoms or ions (sometimes multiply charged ions) were used to interact with solid materials. More recently, cluster ions have been used as projectiles. Small power density lasers have also been used recently to eject intact molecular ions from the surface.

In all cases, whatever the initial triggering process, the energy that is deposited in the solid must be coupled to atomic and molecular motion.

In these lectures, I will summarize some of the results obtained in experiments performed at our laboratory, Institut de Physique Nucléaire Orsay, France (Ion-Surface and Material Interactions Group). In these experiments we have used several types of projectiles at various energies.

I. GENERAL REMARKS

For simplification it is generally considered that there are two regimes of primary ion velocities that correspond to two distinct classes of mechanisms of interaction. The "critical" value of velocity that "separates" the two regimes is the Bohr velocity $V_0 = \frac{e^2}{\hbar}$. When the velocity of the primary ion (PI) V_p is small, target electrons move much faster than the projectile and the primary ions interact with the whole atoms. Atomic collisions take place. When the velocity of the PI is large compared to the electron velocity in the target atoms ($V_p \gg V_0$), the PI "sees the electrons at rest", and there are interactions with electrons. In this case, electronic collisions occur. There is of course an overlapping region of primary ion velocities where the two processes are both present.

In the following, experiments will be described in the two regimes of velocity. V_0 is about 0.2 cm/nsec and corresponds to an energy per mass unit of 25 keV/u ($E_{MeV/u} = \frac{1}{2} V^2$ with V in cm/nsec). The quantity E/u is a practical unit that will be employed in the next sections.

Energy loss mechanisms are therefore related to the two regimes of velocities and correspond to the so-called nuclear stopping power ($V_{PI} < V_0$) and electronic stopping power ($V_{PI} > V_0$). The nuclear stopping power is based on elastic two particle collisions between M_1 and M_2, with M_2 at rest. $E_{M_1 i}$ is the initial energy, $E_{M_1 f}$ is the final energy of M_1 after having transferred an energy E_2 to M_2. φ_1 and φ_2 are the scattering and recoil angle in the laboratory system. θ is the center of mass scattering angle and E_{CM} the energy in the center of mass system. Classical conservation laws lead to the following basic formulas :

$$E_{recoil} = E_2 = E_{M_1 i} - E_{M_1 f}$$

$$E_{CM} = \frac{4 M_1 M_2}{(M_1 + M_2)^2} \times E_{M_1 i} \quad \text{and } E_{recoil} = E_{CM}.\sin^2 \theta/2$$

with $$\theta = \pi - 2p \int_{R_{min}}^{\infty} \frac{d R/R^2}{\sqrt{1 - V_{(R)}/E_R - P^2/R^2}}$$

where R is the distance between two nuclei, R_{min} the minimum distance of approach, $V_{(R)}$ the interatomic potential and p the impact parameter. Different values can be assumed for $V_{(R)}$. The nuclear stropping power dE/dx can be approximated through most of the keV energy regions as [1] :

$$\frac{dE}{dx} \simeq 1.3 \pi \text{ Na } Z_1 Z_2 e^2 \frac{M_1}{M_1 + M_2}$$

where N is the target atomic density (atoms/cm^3), Z_1, Z_2 atomic numbers, a is a numerical constant $\simeq 0.5 (Z_1^{2/3} + Z_2^{2/3})^{-1/2}$.

Simple expressions have also been derived by W.Wilson et al [3] for the nuclear stopping :

$$\frac{dE}{dx} = S_n(\varepsilon) \,.\, \pi a^2 \gamma N \,.\, \frac{\varepsilon}{E}$$

with $$\varepsilon = \frac{aM_2E}{Z_1Z_2e^2(M_1 + M_2)} \qquad \gamma = \frac{4M_1M_2}{(M + M_2)^2}$$

and $$S_n = \frac{0.5 \ln (1 + \varepsilon)}{\varepsilon - A\varepsilon^B} \qquad A = 0.1412 \text{ and } B = 0.42059$$

Multiple scattering, path length distributions, displacement damages, and three dimensional evaluation make a complete calculation very complex. A very convenient P.C. program is now available : "TRIM" [3] that allows calculation of most of the needed simulations with a variety of projectiles and targets. Standard sputtering theory is used. An example of calculation is given in Fig. 1.

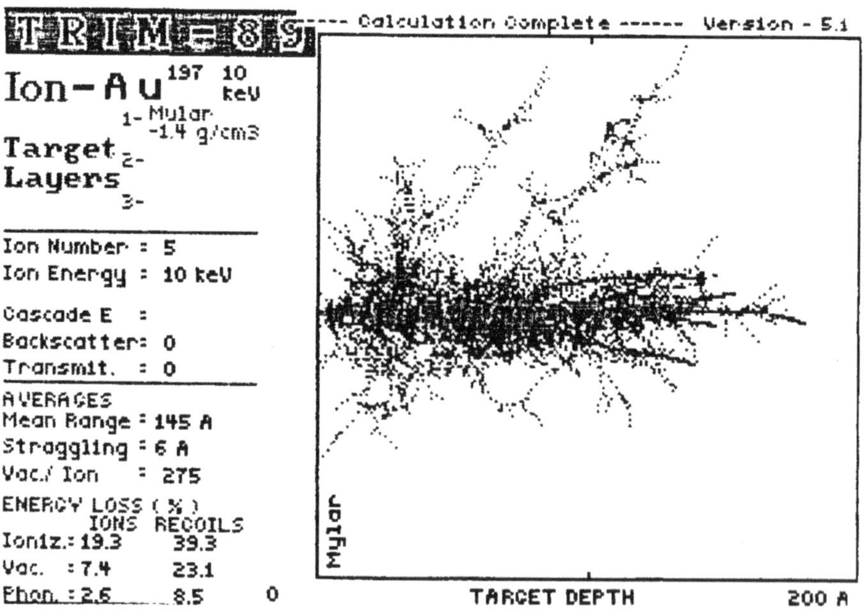

Figure 1 : Example of output with the program TRIM.

At large velocity the energy loss is based on collisions between a primary ion with charge Z_1 at velocity V and an electron "at rest" at a distance b from the PI trajectory. The maximum acting force is Z_1e^2/b^2 with an estimated time of collision of 2b/V. The transferred momentum is roughly equal to $\Delta p \simeq \frac{2Z_1e^2}{b\,V}$ and the energy absorbed in the collision $\delta E \simeq \frac{\Delta p^2}{2M} = \frac{2Z_1e^4}{MV^2b^2}$. If electrons in a solid have a volumetric density of n_0, then the energy loss in a thickness dx at a distance b from the PI ion path is dE = $\delta E \times 2\pi b db.dx.n_0$. Between minimum distances b_{min} and b_{max} we have :

$$\frac{dE}{dx} = \frac{4\pi Z_1^2e^4n_0}{MV^2}\int_{b_{min}}^{b_{max}}\frac{db}{b}$$

b_{max} and b_{min} can be approximated according to the PI velocities [4]. An approximated expression, for a projectile with charge Z_1 and an atom with Z_2 electrons, is given by the Bethe formula [5] :

$$\frac{dE}{dx} = \frac{4\pi Z_1^2e^4n_0Z_2}{MV^2}\left\{\log\frac{2MV^2}{I}\right\}$$

Empirically, I is the mean ionization energy and $I \simeq 10\ Z_2$ (eV) for $Z_2 \gg 1$. Z_2 is the target atomic number.

More complete formulas can be derived from a more detailed description of the energy loss. In any case, theoretical values have to be adapted to experimental values, and for practical purposes, adjustable parameters have been introduced into the formula's derivation.

The charge $q_1 = Z_1$ of the primary ion is a very critical parameter since this parameter changes rapidly when the primary ion penetrates in a solid. There is even a charge distribution probability according to the cross sections of gain or loss of electrons by a fast ion traveling through the material. Electronic stopping power values have been tabulated by several authors [6-8]. Frequently, there are large discrepancies between these values. In addition, comparisons between calculated and experimental values show differences that can reach 30 to 40 %. Experiments are clearly needed to obtain reliable values.

Our goal is not to develop here energy loss theories but to emphasize the distinction between the two modes of energy loss. We are interested in the final step, that follows the energy loss and transfer mechanisms, i.e. the emission of secondary species from the surface.

Atomic collision cascades and the sputtering theory can reproduce the emission of atoms relatively well, but as already mentioned, the simultaneous emission of an assembly of atoms (as a molecule) is more difficult to explain.

II. ENERGETIC PROJECTILES : MEV IONS AND MOLECULAR ION EMISSION

In the first part I will discuss results with fast ions above 0.2 MeV/u. When a solid material is bombarded by such energetic particles, molecular ions ion-clusters and neutrals are ejected from the surface[1]. The yield of emission, which can be very important, ranges from 0.1 to 10^5 per impact depending on the material, the amount of energy loss, the type of surface,etc.

We will distinguish the entrance channel and the exit channel in the interaction. In the entrance channel we have several parameters :

- the type of ions, the velocity, the charge state (and equilibrium charge state),
- the angle of incidence with the surface,
- the nature of the surface and material.

In the exit channel we may consider :

- the type of secondary ions, i.e. atomic, molecular ions, cluster ions...
- the type of neutrals,
- the secondary ion yield, i.e. the number of SI per primary ions,

~~~~~~~~~~~~~~~~

[1]Prof. R. Macfarlane from Texas A&M has initiated this field. A special issue of "Int. J. of Mass Spectrometry and Ions Processes" has been dedicated to R. Macfarlane, vol. 53 (1983).
~~~~~~~~~~~~~~~~

- the secondary electron yield,
- the energy and angular distribution of the emitted species,
- the stability of the molecular ions.

The phenomena observed in the exit channel correspond to the solid's response to the interaction. They depend on the parameters in the entrance channel. Therefore, a study of the variation of these parameters gives some insight into the mechanisms.

II.1. Experimental method

Heavy ion beams from accelerators were used with energies per mass unit between 0.3 and 5 MeV. Ions of ^{32}S to ^{235}U have been delivered by a cyclotron, a Tandem machine in Orsay and a linear accelerator in Darmstadt. Experiments have also been performed at Uppsala, Erlangen, Argonne, Catania. I suggest that the readers look at ref. [9-14] for more information. Briefly, a beam of heavy ions is selected in mass and energy by the accelerator before entering a reaction chamber where targets are positioned in the center. Variation of the energy is achieved either by changing the machine parameters or by inserting metallic absorber foils in the beam prior to the target. For experiments with a specific charge state, the beam passes first through a thin foil of carbon (10 μ) or gold (10 μ) to produce a distribution of charge states. The charge state desired is selected by a magnet, and the resulting low intensity ion beam (100 to 500 ions/sec) is collimated to 1 mm diameter before striking the target. Fig. 2 shows a schematic diagram of the experimental geometry. The energy of the primary ion is monitored and controlled by a Si detector after passing through the target. A vacuum of 10^{-7} torr is maintained in the system by a cryopump. These are counting experiments, and for each ion hitting the target, the secondary ions emitted from the target surface are identified and counted. Secondary ion yields are measured according to the entrance channel parameters.

Time of flight (TOF) mass spectrometry is used to identify secondary ions. This technique is a very simple and efficient technique. Several developments and pieces of semi-commercial equipment have been made at the Institut ; e.g. fast electronics, time digital converters and data acquisition systems. TOF mass spectrometers for analytical applications have even been built. More details on the techniques and the electronics can be obtained from ref. [15-17].

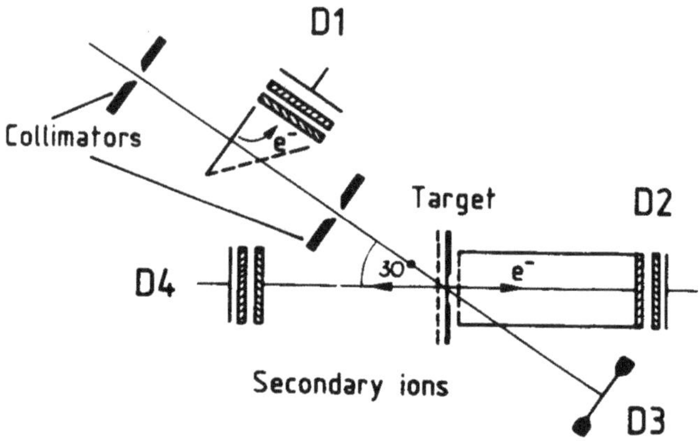

Figure 2
Experimental arrangement for on beam experiments
◦ *D1, D2, D4 are micro-channel plate detectors for time signals*
◦ *D3 is a silicon detector for energy signals and time signals*

In experiments with accelerator-beams, time of flight distances shorter than 20 cm were generally used for secondary ions and secondary electrons as well.

Targets consist of thin films of materials deposited on metallic backing by evaporation under vacuum or by electrospray. Different types of sample compounds have been used (inorganic and organic).

II.2. P.I. velocity dependence

The SI yield as well as the total sputtering yield depend on the PI velocity. The velocity dependence has been observed in many cases with targets of condensed gases, H_2O films [18-20], inorganic films, and organic films [21- 23,11,13]. An example, already shown in ref. [24], is presented below in Fig. 3. A rapid increase of the SI yield is observed above 0.4 cm/nsec followed by a maximum at about 1-1.2 cm/nsec. Similar trends have been observed for several compounds. The parameter "velocity" is not a simple parameter because the primary ion charge state varies with the PI velocity. As seen before, dE/dx is proportional to q^2/V^2, and therefore the variation of yield Y also reflects the variation of Y with the primary ion

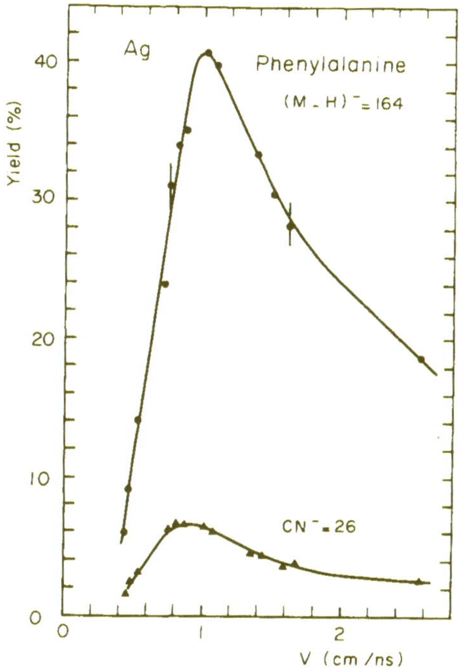

Figure 3
Variation of the desorption yield of molecular ion $(M-H)^-$ from a phenylalanine sample as function of velocity of Ag ions.

charge state. It has been found in many cases that $Y \propto \left(\frac{dE}{dX}\right)^2$ in the increasing part of the yield curve. When the velocity exceeds that corresponding to the maximum of the yield curve ($\simeq$ 1.1 cm/nsec) there is no simple relation between the yield and the energy loss. In addition, the maximum yield appears at a PI velocity that is lower that the velocity at which maximum dE/dx occurs.

More interesting information is obtained from experiments in which the parameter incident-charge-state q_1 is varied. Before going into more details, let us consider how the charge state of a fast ion varies when it enters a solid material.

II.3. P.I. charge state dependence

Fast ions traveling in a solid rapidly reach highly excited states. Several processes can change the ionic charge state of an ion moving through the solid. Theoretically, several approximations must be made to estimate charge state values of a moving ion in a solid and experiments able to "photograph" the ionic charge state value in matter do not exist. However, we will describe in the next chapter an experiment that gives information on the charge variation inside solids.

After an ion has penetrated a certain distance x below the surface of a solid, its charge state reaches an "equilibrium value" $\langle q_{eq} \rangle$. An example of variation of charge state in a solid is shown by Fig. 4 which presents Kr at 1 MeV/u traveling through carbon. It must be noted that for heavier PI, like ^{127}I, the equilibrium charge state is preceeded by a pre-equilibrium charge state (see following chapter).

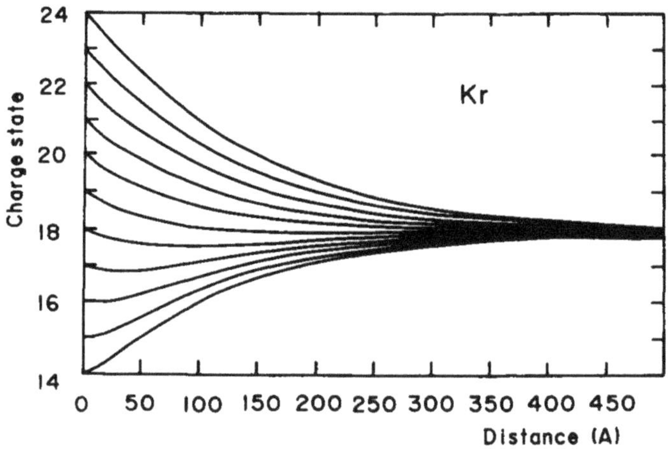

Figure 4
Calculated variation of the ionization state of Kr ions at 1 MeV/u passing through carbon material.

As seen in Fig. 4, $^{24+}$Kr becomes $^{21+}$Kr after 100 Å and $^{19+}$Kr at 200 Å. The equilibrium value 18^+ is reached beyond 450 Å. This curve has been calculated by Maynard [25]. Several semi-empirical formulas are generally employed to calculate the equilibrium charge state value of a given projectile in a given material [26,27]. Shima and collaborators [28] have established a semi-empirical relation that gives satisfactory values of equilibrium charge state :

$$q_{Shima} = Z_1\left[1 - \exp\left(- 1.25x+0.32x^2-0.11x^3\right)\right].\left[1 - 0.0019(Z_2-6)\sqrt{x}+10^{-5}(Z_2-6)^2x\right]$$

with $x = \frac{v}{v_0 Z_1^{0.45}}$, $v_0 = 3.6.10^8$ cm/s and Z_1 and Z_2 are atomic numbers of projectiles target.

It is obvious from the expression dE/dx that the energy loss in the solid is proportional to $<q_{eq}>^2$ once the $<q_{eq}>$ value is reached. But what takes place before this state is reached ? It happens that the charge state does vary along the path of the primary ion (according to Fig. 4 for example for Kr in carbon) and that the energy loss dE/dx is essentially proportional to $q_1^2(x)$. $q_1(x)$ is the value of the ionic charge state for an initial charge state q_1 after a penetration of a distance x in the solid. Since an equilibrium value is reached within a few hundreds Angströms, the ion velocity does not change but, as seen above, the charge state and therefore the dE/dx can change enormously.

Reality is not as simple because there exists, inside the solid and outside the solid, a charge distribution. For energy loss estimation, the mean charge state is generally considered.

Instead of using the results of Maynard's calculation, it is possible (for atomic number below ≃ 40) to calculate q(x) with an exponential dependence proposed by Bohr [29] :

$$q(x) - q_{eq} = (q_0 - q_{eq}) \exp - \frac{x}{d}$$

where q_0 is the initial charge state, q_{eq} the equilibrium charge state, x the distance below the surface in Angströms, and d the distance of relaxation to equilibrium.

A fast PI ion with a charge state 30^+ deposits in the first 100 or 200 Å of a solid more energy than the same PI with a charge state of only 10^+. As a consequence, it is obvious that since energy losses are different, yields of secondary ions emitted from the surface are also different. Fig. 5 illustrates this charge state dependence. The material was a thin organic film of the molecule phenylalanine ($C_9H_{11}O_2N_4$, mw = 165). The incident charge state of the primary ion Ar at 1 MeV/u was varied between 7 and 17. For Ar, the equilibrium charge state at 1 MeV/u is around 12.

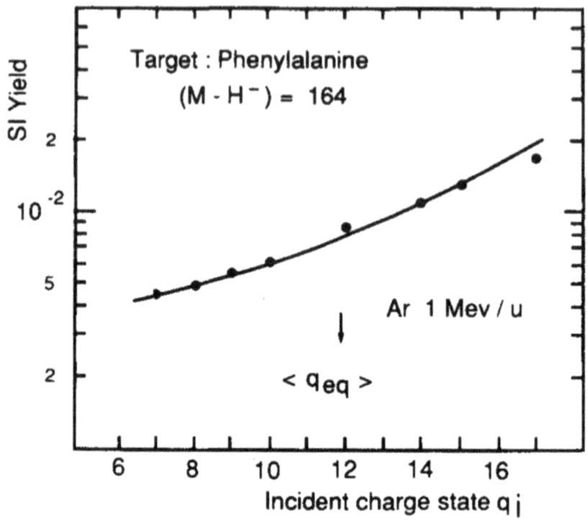

Figure 5
Variation of the emission yield as function of the incident charge state of Ar projectiles at 1 MeV/u.

The charge state of an Ar^{q_1} ion with $q_1 = 7$ increases as the ion penetrates in the solid. Therefore, the SI yield corresponding to $q_1 = 7$ includes also effects of energy dissipation inside the solid with larger charge state. If the incident charge state is 16^+, the charge state inside decreases and the SI yield includes effects of dE/dx with smaller charge state. Interactions below the surface contribute to a smaller rate of variation of $Y_{SI} = f(q_1)$. Projectiles with a smaller atomic number like ^{20}Ne or a larger atomic number like ^{84}Kr have different equilibrium charge state values. They generate respectively smaller and larger SI yield than Ar PI. Fig. 6 shows the results.

A large number of different primary ions has been used in Orsay and Darmstadt. Fig. 7 summarizes the results. It is observed that different PI with the same incident charge state and the same velocity provide different values of SI yield. The reason is that their charge states vary in a different way below the surface and contribute to a different amount of energy deposited in the solid.

Fig. 8 shows clearly how the charge state of 3 different ions Ne, Ar and Kr, with the same initial charge state 10^+, varies in a carbon foil.

An attempt has been made to reproduce the variation of the relative SI yields as a function of the incident charge state [30,31]. This is independant of any model of energy transfer but it is supposed that there is a contribution of the energy deposited at the depth x below the surface. Thus, at this distance x, the contribution to the total yield depends on the

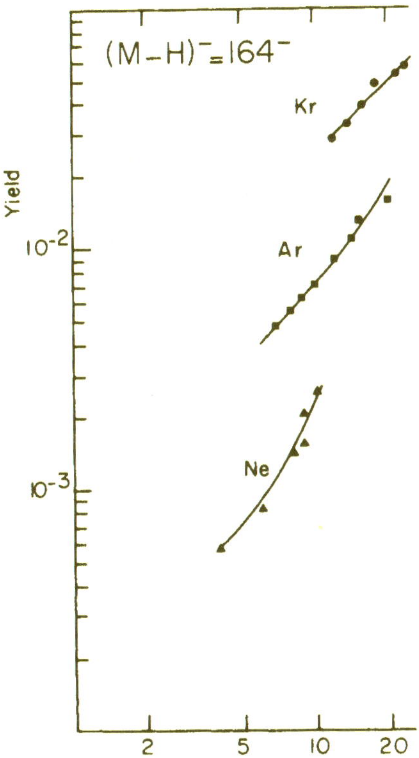

Figure 6
Relative yield of $(M-H)^-$ for the projectiles Kr, Ar, Ne at the same velocity (1 MeV/u) as function of their incident charge state.

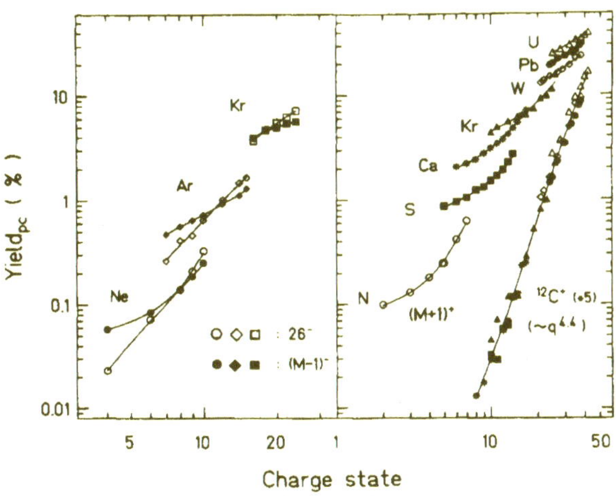

Figure 7
Similar results as in Fig. 6 for different projectiles. Results from the right part have been obtained in Darmstadt (K.Wien et al).

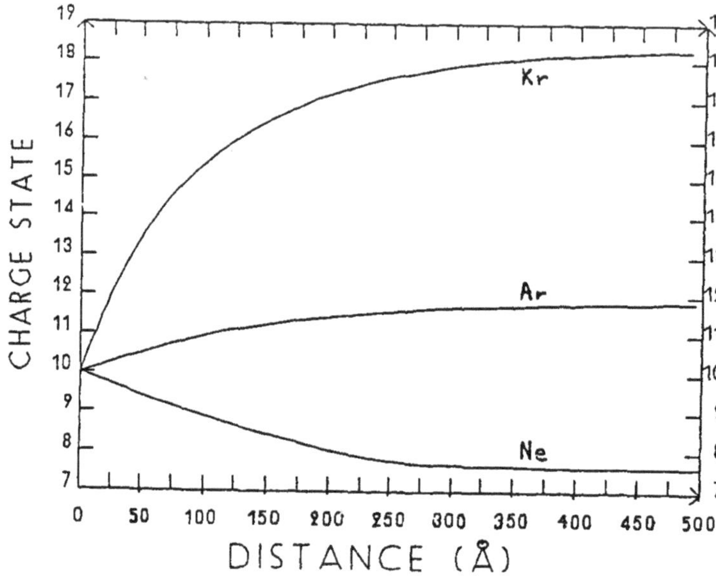

Figure 8
Calculated variation of the ionization for three ions (Ne, Ar, Kr) at the same velocity entering a carbon foil with a charge state 10^+.

charge state variation $q(x) = f(q_1,x)$. Let us call $\frac{dE}{dx}(x,q_1)$ the energy deposited at x, then the contribution dY(x) to the yield at x can be written as :

$$dY(x) = A\left[\frac{dE}{dx}(x,q_1)\right]^n \psi(x)\, dx$$

A step function $\psi(x) = \left[1 + \exp\left(\frac{x-d}{\Delta d}\right)^{-1}\right]$, where d is an interaction depth that depends on $\frac{dE}{dx}$ values, has been used to characterize the desorption probability at a distance x. We have assumed the relation $d = B\left[\frac{dE}{dx}(q_1,x)\right]^{n'}$ with $\Delta d = 0{,}1$ d. Therefore one gets :

$$Y_{q_1} = A\int\left[\frac{dE}{dx}(x,q_1)\right]^n \times \left[\frac{1 + \exp(x - B\left(\frac{dE}{dx}(q_1,x)\right)^{n'}}{0{,}1\ B\left(\frac{dE}{dx}(q_1,x)\right)^{n'}}\right] \qquad (1)$$

If $q_1 = q_{eq}$ = cte, we obtain $Y_{q_{eq}} = A.B.(q_{eq})^{n+n'}$.

In Fig. 9 the fit of the shape of one curve (Kr for example) allows us to obtain values of B and n'. From the experimental relationship

$$Y_{q_{(eq)}} \propto C\, q_{eq}^{\;n+n'}$$

deduced from Fig. 9, we obtained n+n' and C = AB. These parameters are the same for any primary ions on the same target n' values are generally around 1/2 and n values around 2.

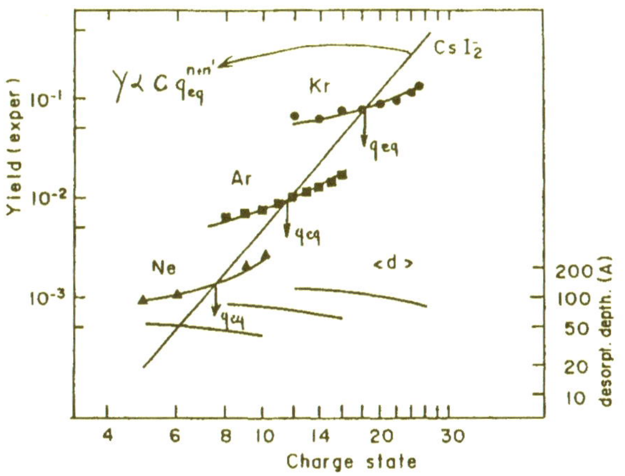

Figure 9 : Experimental and calculated curves for $q_{eq} \Rightarrow Y \propto C\, q(eq)^{n+n'}$.

This approach is a phenomenological approach and does not explain the basic processes. It however indicates that there is a certain interaction depth in the desorption processes and that the energy deposited inside plays a role in the desorption processes. All of the preceding is valid for molecular ions and atomic or cluster ions emitted from a film of material. Interaction depths of 90 ± 20 Å and 300 ± 80 Å for ^{32}S and ^{127}I have been estimated by this procedure. During these studies on SI by multiply charged ions, angles of impact have been varied as well as thicknesses of material.

II.4. Depth of emission

The increase of secondary yields with the thickness of samples was first been studied with fission fragments from ^{252}Cf [32]. These experiments have shown that intact molecular ions can be emitted from underneath the upper layers. Recently, Langmuir-Blodgett (LB) films have been used to obtain quantitative results as function of the sample thickness [33,34]. These results were consistent with a simple picture of emission from a "crater" with a depth of about 200 Å.

Experiments with beams of ^{32}S and ^{127}I at 1 MeV/u, in their equilibrium charge states were used to bombard L.B. layers of nM_1 on $n'M_2$. The number of top layers n has been varied from 2 to 12. Molecular ion yields from upper and underlying layers were measured simultaneously for each target. Fig. 10 shows the results for $^{127}I^{24+}$ projectiles for n' = 6 layers. The yield of the top layer saturates at n ≃ 10 and the yield from the "bottom" layer decreases.

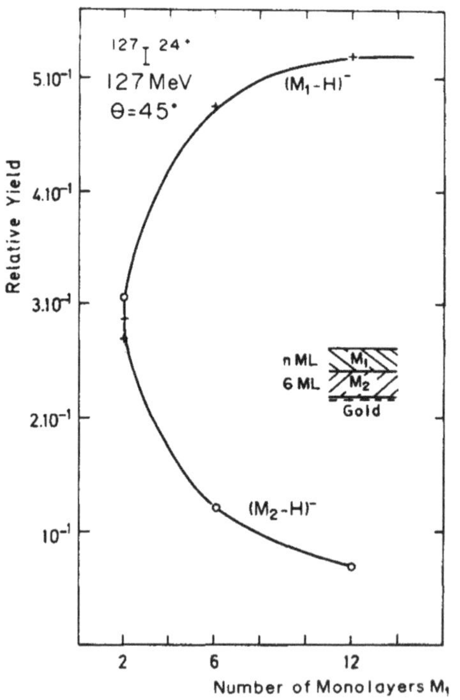

Figure 10
Yield variation from the top layer and bottom layer molecular ions as function of the number of top layers.

The saturation depth is rougly the same with ^{32}S and ^{127}I although the electronic stopping power of the projectiles are different

$$\left(\frac{dE}{dx}\,{}^{127}I \simeq 3\,\frac{dE}{dx}\,{}^{32}S\right).$$

The absolute secondary ion yield of emission is, however, 10 times larger with ^{127}I than ^{32}S. That is consistent with many other experiments where a $(dE/dx)^n$ dependence of secondary ion yields with n ≃ 2 - 3 is reported [35]. If we apply the simple picture of crater emission (conical volume for example), the cone would have the same depth for ^{32}S and ^{127}I but a larger surface cross section in the case of ^{127}I ions, which have a higher dE/dx value.

II.5. Incident angle dependence of the electronic desorption yield [36]

Experiments with different incident charge states and different angles of impact have been performed. The charge state effect has been described previously, but here we also varied the angles θ. Fig. 11 shows an example of results for a LB target made of two different kinds of molecules. The beam was ^{127}I at 127 MeV. For an impact angle of 20°, the charge state effect is clearly visible while at grazing angle, around 78°, there is no effect of charge state. At 20° the slope of yield variation with q_1 (incident charge state) is very steep. The deposited energy loss that contributes to the SI emission depends on the charge state variations along the track. Qualitatively, at this angle, the interaction distance (see above) is at least of the order of the distance traveled for equilibration (200-300 Å). As θ increases, the influence of the primary ion charge state vanishes and the yield tends to be flat. The PI travels a long distance (close to the surface) compared to the equilibrium charge state distance, and consequently, the secondary emission depends mainly on the energy loss with the charge state equilibrated. The memory of the incident charge state is lost. The geometrical shape of the volume of emission is expected to vary considerably with the angle of projectile impact and is probably very asymmetric for larger angles.

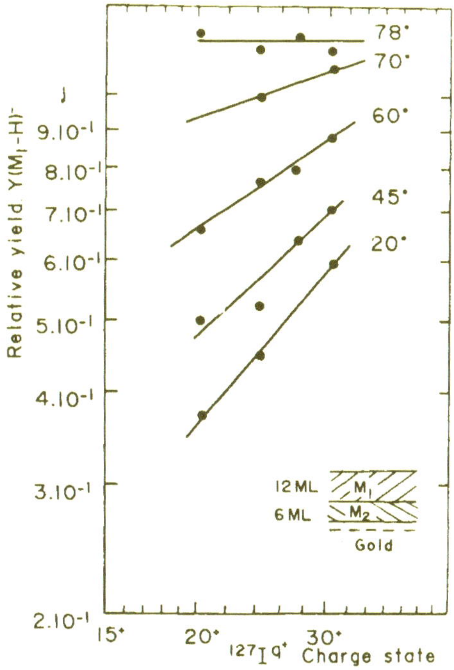

Figure 11

Experimental yield of the molecular ions M-H⁻ as function of the angle of incidence and the charge state of the primary ions of ^{127}I at 127 MeV.

Others results have been shown in reference [36]. A shock wave model or a pressure pulse model was used to tentatively reproduce the data. Reasonable qualitative agreement was obtained.

These results also show that, with MeV ions, a simple relation- ship between the emission yield and the angle of impact does not exist. $Y = f(\theta)$ depends on the incident charge state.

All the results presented above have been obtained with secondary ions. Will they be the same for neutral species emitted from the surface ? This is a very important question because in all theoretical models the ionization step is not treated. There is therefore a need for new experimental results on neutral emission. These experiments might be more difficult, but they are necessary to confirm that the observation of secondary ions does not give a false picture of the general mechanism of emission.

Some experiments with neutrals have already been performed by the Uppsala Group [37,38].

Desorption models have been developed by several groups, and it would be interesting to make a systematic comparison [see 37,38]. The general picture is that, during the passage of a fast ion through the solid, intense ionization and excitation occurs in and just outside the nuclear track. Electrons carry energy far into the material ; the electronic excitation density is assumed to decrease as 1/r with the distance r from the track. This fast process ($\simeq 10^{-15}$ sec) must generate a coherent transfer of momentum to the volume of molecules that escape from the bulk. The angular distribution of emitted molecules is asymmetric with respect to the primary ion trajectory [39]. This strongly favors a fluid dynamical model.

There is a long time delay between the electronic energy deposition in the solid (a fast ion travels 10 Å in 10^{-16} sec) and the molecular ion emission. Molecular vibration and expansion take 10^{-12}-10^{-13} sec ; therefore molecules are ejected into vacuum 3 or 4 orders of magnitude later. During all this time, a large number of complex processes take place in the material. Experimental access to these intermediate phenomena is very difficult.

However, at the very beginning of the interaction, shortly after the passage of the PI, light ions like H^+ or fragments like $^{12}C^+$ are emitted from the solid surface and the study of their emission has been applied to various problems (see IV).

III. EMISSION OF HYDROGEN IONS FROM SOLID SURFACES

In experiments using multiply charged MeV ions as projectiles, it has been shown that H^+ ion emission follows a very simple pattern. For a given projectile velocity the emission yield of H^+ is independent of the bombarding ion and therefore of the total energy lost in the foil. It depends only on the primary ion charge state when it crosses the surface, and not on the charge state variation below the surface. Fig. 12 shows the yield variation

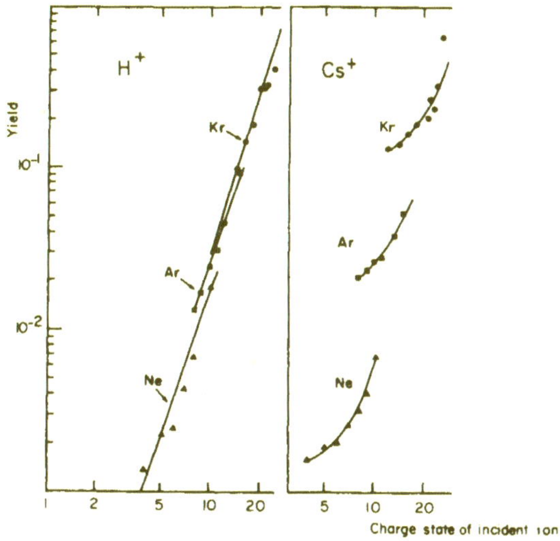

Figure 12
Left side : variation of H^+ yield with the incident charge state of the three projectiles Ne, Ar, Kr.
Right side : variation of Cs^+ from a CsI deposit. A very marked difference is seen.

of H^+ for projectiles of Ne, Ar, and Kr at 1 MeV/u. At the same incident charge state, the same yield values are obtained. For comparison the yield of Cs^+ from a CsI sample is also shown. Another example is given in Fig. 13 with Au and I projectiles at 0.5 MeV/u. There is an overlapp of charge states between 22^+ and 27^+. H^+ yields are also the same. An application of this experimental observation has been to use the value of the H^+ yield to measure the incident charge state of fast projectiles.

To investigate the role of the charge state on H^+ ion emission, primary ions with a much lower energy have been used [40,41], for example Tl^q ions at 110 keV ($\simeq$ 0.5 keV/u, i.e. 1000 times less of energy) and Ar^q at 18 keV. In order to be able to compare MeV ions and keV ions, the measurements were performed under exactly the same experimental conditions, i.e., with the same target and in the same reaction chamber continuously kept under vacuum. Only the projectiles were changed.

Fig. 14 shows the experimental set up. A thin target foil was positioned in the center of the reaction chamber. Ions of Au and I at 0.5 MeV/u were obtained from the Tandem machine. The charge selection was made by a magnet prior to the target. Multiply charged keV ions were obtained from a small alpha radioactive source. Let us describe the experimental arrangement. It is easy to obtain multicharged MeV ion from accelerators. It is not as common to generate 100 keV ions multiply charged. As an ion source we have used recoiling atoms resulting from radioactive alpha decays of ^{212}Po and ^{212}Bi. It is easy to prepare radioactive sources of thoron emanation on thin gold or aluminium foils. The radioactive decay serie leads to ^{212}Po and

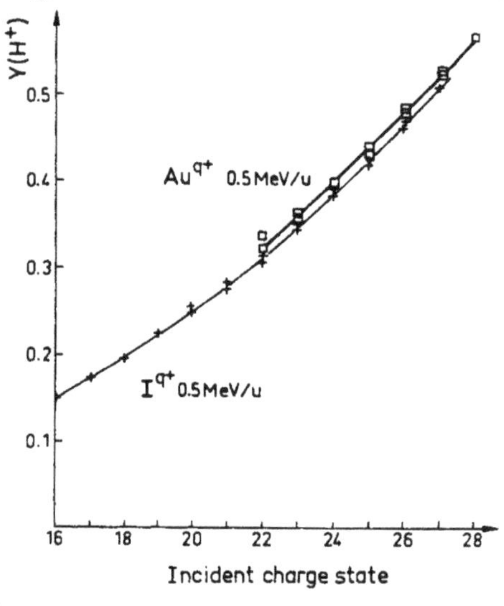

Figure 13
Emission of H^+ ions from the surface of the same carbon foil bombarded with ^{127}I and ^{197}Au at the same velocity

^{212}Bi, which are alpha emitters. ^{212}Po emits an 8.78 MeV α particle and ^{212}Bi a 6.05 MeV α particle. For every α emission there is a recoiling nucleus with an energy $E_{recoil} = 4\ E_{\alpha}/208$. Due to the internal conversion of the 39.8 keV gamma ray, a charge state distribution of Tl atoms is generated and can be measured by time of flight.

It is therefore rather easy and simple to generate multicharged keV ions. In Fig. 14, an α source is mounted at 20 cm from the target. A recoiling atom with a certain charge state is associated with each α decay and the α particle is detected by a silicon detector that provides a start signal. The recoiling nuclei is accelerated by the difference in potential applied between the source (3 to 8 keV) and a 98 % transmission grid at zero potential. A thin grid (at ground) is placed 3 mm from the target surface, which itself is at the same potential as the source. Because of the different charge states, the accelerated recoils have different energies, and the measurement of their TOF between the source and the target allows clearly separation of the different charge states that appear as peaks in Fig. 15. The charge state of each incident ion hitting the target is thus identified and selected. It is an event by event experiment. The TOF of the correlated secondary ions are accumulated in different computer memories according to the selected charge state.

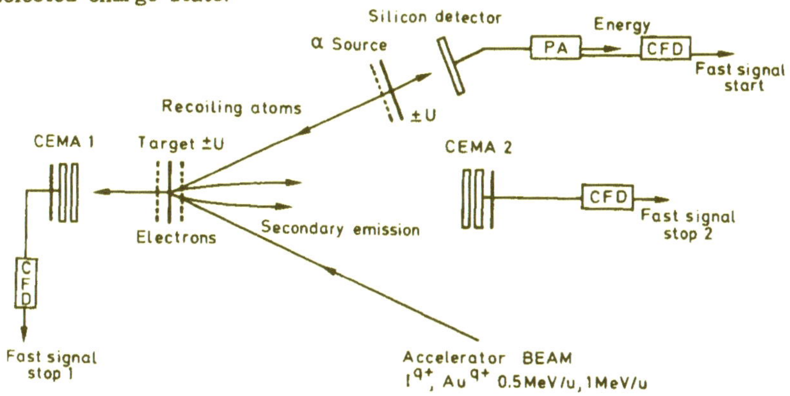

Figure 14

Experimental arrangement for the bombardment of the same target with projectiles at 0.5 MeV/u and 0.5 keV/u.

In Fig. 14, keV ions and MeV ions are shot at the target at an angle of 25°. Where do H^+ ions come from ? Hydrogen, water, organic contaminants at the surface ?

A mass spectrum recorded just after cleaning the foil by sputtering with Ar ions from and Ar gun did not exhibit any H^+ lines. However, after a few minutes, H^+ ions were present again and the yield saturated after 10 minutes. Negative ions at 16(O^-) and 17(OH^-) were also present in the negative spectra. This indicates the presence of water at the surface.

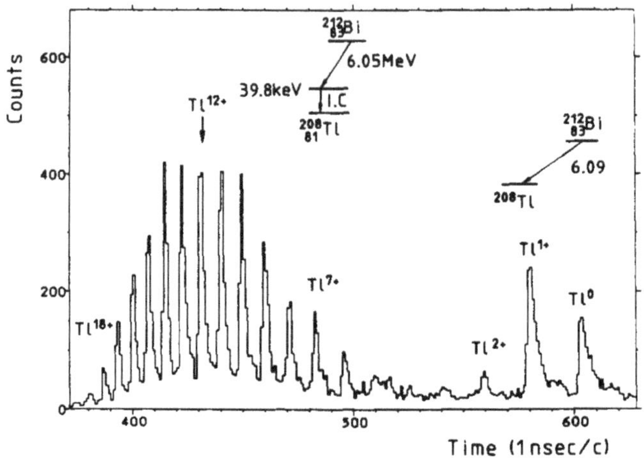

Figure 15

Charge state distributions of the primary ions of ^{208}Tl measured by time of flight.

After heating a UHV system (at Texas A&M) and cleaning the target, a very small amount of D_2O was introduced in the chamber. Almost immediately, ion peaks of D_1^+, D_2^+, D_3^+ were observed in the mass spectra [42].

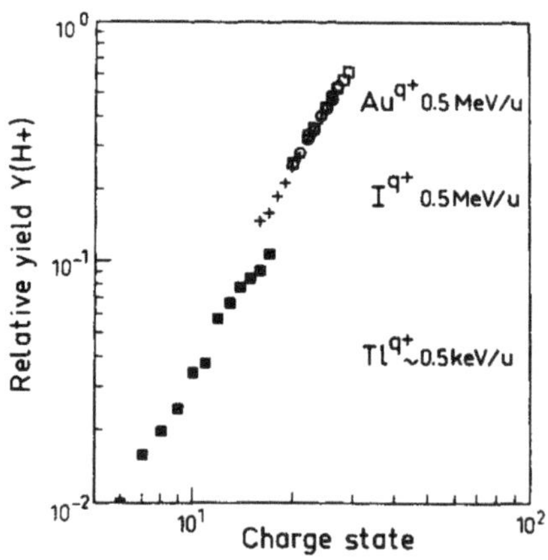

Figure 16
Relative yield of H^+ with three projectiles as function of their incident charge state.

Fig. 16 shows results obtained on H^+ yield as a function of the incident charge state for the three types of projectiles (Au and I as shown before) and ^{205}Tl. It is observed that for Tl ions having an energy 1000 times smaller ($v = 3.10^7$ cm/sec), the emission yield of H^+ remains about the same. The trend in the variation is the same, and in all cases a power law $Y(H^+) \propto q_i^3$ has been established.

There is no direct evaluation of the mechanism of emission. It has been emphasized that when a multiply charged ion approaches a surface at a velocity about 2.10^7 cm/sec, a large number of electrons are captured in outer shells at distances 25-50 Å from the surface [43]. Auger cascades occur in the decay processes. The time for the first step of the cascade is around 10^{-14} - 10^{-15} sec and X rays are emitted once the projectile has reached the surface. The triggering process of H^+ emission occurs before any charge state variation, and the question is wether or not the mechanisms are the same for keV and MeV ions. It is only within the first ten Angströms through the solid that electron losses and captures change the ionization state of fast ions, and these 10 Å are traveled in 10^{-16} sec. This gives a time scale of 10^{-16} sec for the triggering process.

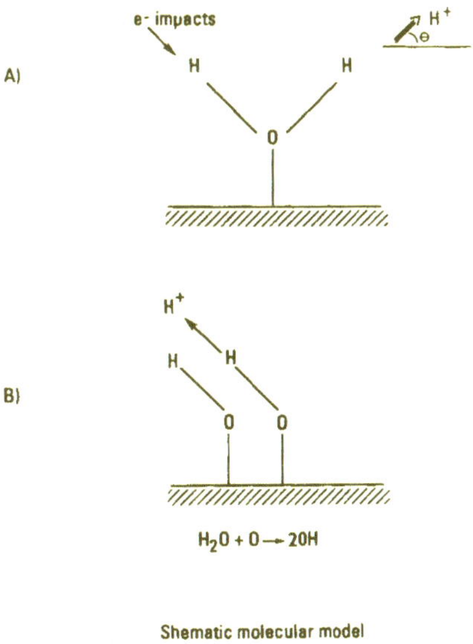

Figure 17 : Schematic molecular representation of attachment.

When a large charge state approaches a surface, electrons are ejected from the surface. In keV experiments, as said before, it seems that electrons escape from the surface before impact, and it could happen that the projectile is partially neutralized before hitting the surface. In any case, fast displacements of electrons take place at the surface where water molecules are attached. Electronic excitation of a surface molecular bond can occur in 10^{-16} sec. Excited states can be repulsive and the electronic energy (≃ coulomb energy) can be converted into nuclear motion along the direction of the bond that is ruptured. This picture is used to explain the preferential direction of emission of fragments from small molecules adsorbed at surfaces and "desorbed" by electron by electron impact (ESD = electron stimulated desorption). Fig. 17 shows one type of attachment of a water molecule to a surface. Angular distribution measurements of emitted H^+ can give information on bonding to the surface as well as on molecular structure [44].

In our case the primary charged particles arrive at a certain angle θ. Electron displacement at the surface could excite surface bonding states and interatomic bonding states. With well known surface structures (in ultra high vacuum) it would be interesting to perform experiments similar to ESD experiments and verify if angular distributions of H^+ emission would yield information on molecular bonding.

It is likely that a local positive electric charge results from the electronic emission (with keV and MeV ions). A rather strong coulomb respulsion can therefore accelerate the H^+ ions. This would explained the H^+ energy distribution that has an average value around 7-8 eV and that extends to large value ($\approx$ 20 eV). Several groups have measured H^+ energy distribution with fission fragments or ion beams [45-49]. A coulomb repulsion model is proposed by the Uppsala group [49]. The q^3 dependence of the H^+ emission yield can also be explained by a coulomb repulsion [50].

No energy measurement has been made for low energy multiply charged projectiles. This is obviously one of the experiments to be made. In addition, the initial energy measurement of H^+ as a function of the primary ion charge state is also another (or simultaneous) experiment to be performed. Light secondary ions like D^+ should also be used.

A first test experiment was tried at Orsay with a ^{127}I beam at 1 MeV/u but the results were not satisfactory. However, it seems that H^+ ions of higher energy are ejected by more highly charged projectiles. This should be repeated with keV and MeV ions because, if the mechanisms of emission are similar with keV and MeV projectiles, the energy spectra should, of course, be the same.

IV. APPLICATIONS OF THE PHYSICS OF EMISSION OF H^+ TO THE DETERMINATION OF EQUILIBRIUM (AND PREEQUILIBRIUM) CHARGE STATE OF A FAST IONS IN A SOLID

The behavior of the charge states of energetic ions passing through matter is a difficult problem of considerable theoretical interest. Energy losses in solid and gas depend on charge state and on its variation in the first layers in the case of desorption. Energy loss measurements in plasma are being made in several laboratories in connection with basic aspects of fusion induced by fast heavy ions [51] leading to a renewal of activity for these questions.

One of the questions raised a long time ago by Betz was : Is there a difference of ionic projectile charge states inside and outside stripper targets ? This question deals with electron loss and gain in the solid, excitation time, and differences of equilibrium charge states between gas and solids. According to Betz and Grodzins [52], Auger transitions take place once the ion leaves the exit surface of a target. This process of de-excitation from multiexcited states that are populated inside the solid would account for the enhancement of the projectile charge state after passage through a solid target. Until now, the equilibrium charge state of a fast ion was determined by means of a magnetic spectrometer after the beam had passed the target. This means that the ions have traveled for a long time in vacuum after exiting from the surface.

Since the yield of H^+ is strongly dependent on the charge state of the projectile, this yield has been used, after calibration, as a sensitive probe of projectile charge state at the surface.

Fig. 18 illustrates the experimental setup. Several targets can be successively put in the irradiation position without opening the vacuum chamber. The target and the MCP detectors are mounted on a rotating platform. Therefore the beam can bombard the same target surface either directly or from the back after passage through the target. In both cases the secondary ions originate from the same area. About 70 cm upstream from the target, a foil (equilibrium charge state foil) of the same nature and thickness as the target can be inserted into the beam. At the highest beam velocity used (corresponding to 1.5 MeV/u) the time of flight of an ion projectile is about 40 nsec.

In the "Secondary Ion TOF chamber" carbon foils of 20-40 $\mu g/cm^2$ and gold foils of 170 $\mu g/cm^2$ are also used. For charge state measurements inside a solid, very thin foils of nitrocellulose have been prepared as targets (100 Å to 500 Å).

The experimental procedure is as follows :

i) Measurement of the calibration yield curve $Y(H^+) = f(q_1)$. For each incident charge state the H^+ yield per incident ion is measured, for example curve 1 in Fig. 19.

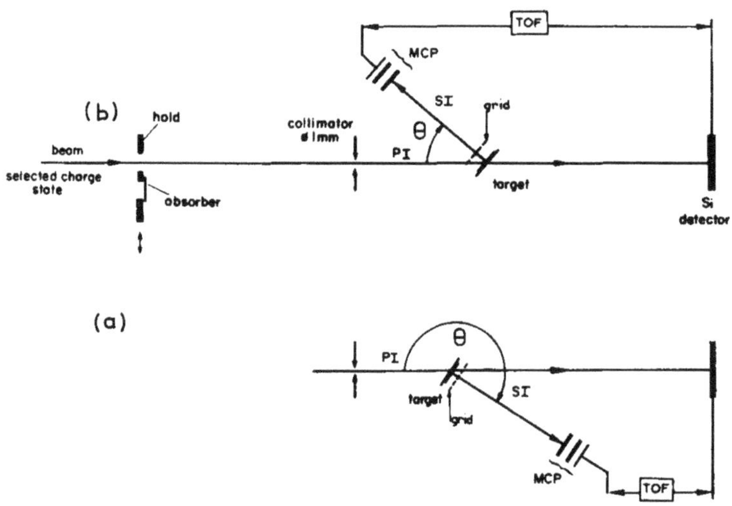

Figure 18
Experimental arrangement for charge state measurements of fast ions in solids (Tandem Orsay).

ii) In a second series of measurements, the "equilibrium charge state foil" is inserted into the beam and the H^+ yield (from the same point of the target) is again measured as a function of the incident charge state hitting the equilibrium charge state foil. If the equilibrium charge state is reached in this foil the H^+ yield curve exhibits a flat dependence versus

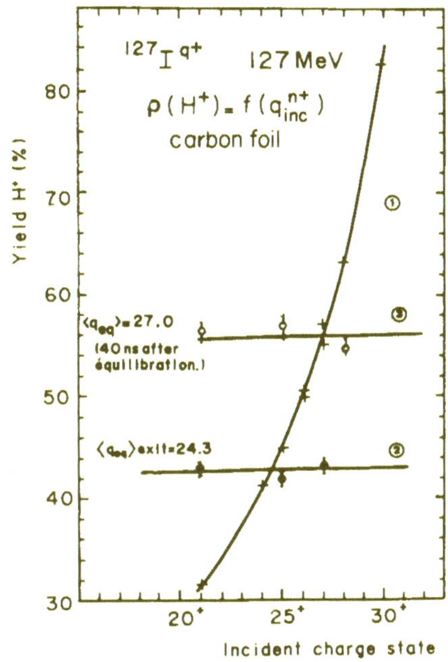

Figure 19
Experimental results on equilibrium charge states (see text).

the charge state. This is shown with curve 3 in Fig. 19. The intersection between curve 1 and 3 gives the equilibrium charge state of an ion more than 40 nsec after passage through a solid foil ("$\langle q_{eq} \rangle$" in the figure).

iii) The third step is to turn the target by 180°, remove the "equilibrium charge state foil", and measure once more the H^+ yield curve as a function of incident charge states. Results are shown in curve 2 in Fig. 19.

The crossing between curve 1 and 2 gives the equilibrium charge state measured at the exit surface of the target when the projectile leaves the surface ("$\langle q_{eq} \rangle$ exit" in the figures). The results from Fig. 19 indicate that the charge state at the exit surface is smaller than the charge state measured more than 40 nsec later. When the target is thick enough to achieve charge state equilibrium, the H^+ yield measured at the exit surface does not depend on the projectile charge state at the entrance surface and $Y(H^+)_{exit} = f(q_1)$ is flat as shown by curve 2 in Fig. 19. This means that the memory of the entrance charge state is lost. If, however, the target thickness is smaller than the equilibrium distance, the equilibrium charge state is not reached, and the H^+ yield increases with q_1. This case will be shown in the following section where very thin targets are used. The experimental yields were reproducible, their values being checked several times. During the experiments, the target chamber was not opened to air, all the targets were then kept under vacuum.

As demonstrated in Fig. 19, outside a carbon foil the equilibrium charges states $<q_{eq}>$ of ^{127}I was found to be 27 whereas inside the foil $<q_{eq}>$ exit is only 24.3. These values are average values which have not been corrected for the charge state distribution. However, based on the calibration curve $Y(H^+)$ which varies as q^3, and the shape of the charge state distributions [53], we have calculated that the mean equilibrium charge state values extracted from the present results are overestimated by only a few percent.

Several different ion beams and targets have been investigated by using the H^+ method. The increase of charge state of a fast ion traveling through a foil is observed for projectiles with masses above 40. Let us recall that equilibrium charge states in gases are smaller than those in solids [54]. For example, ^{127}I at 110 MeV the difference between a gas and a carbon target is about 7 charge units. The post foil increase is only about 3 charge units, and in general the increase $\Delta\bar{q}$ is only 10-15 % of the equilibrium charge at the exit surface of a solid. Therefore only part of the difference between gases and solids can be due to post-deexcitation. Fig. 20 is a schematic view of charge state variation inside and outside solid targets according to Bohr and Lindhard [55], Betz and Grodzins [52], and experimental results [53].

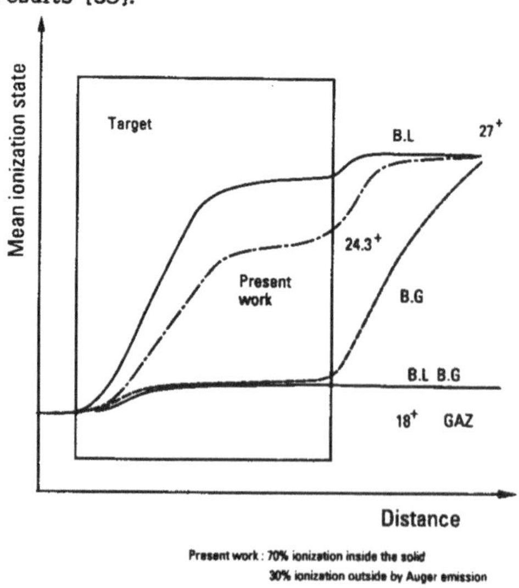

Figure 20
Shematic representation of charge state variation inside and outside a target.
- *Bohr and Linhard theory (BL)*
- *Present experimental result*
- *Betz and Grodzins (BG).*

The variation in a gas target is also given.

Experimental and theoretical mean equilibrium charge states are presented in ref. [53].

Theoretical calculations of charge states inside matter have been pursued by G.Maynard and C.Deutsch [25] who have considered an average atom model with 10 levels of number quantum n populated by Pn electrons. The mean ionization states <q> can be written as :

$$\langle q\rangle = Z - \sum_{n1}^{n=10} Pn \qquad \text{and} \qquad \frac{d\langle q\rangle}{dt} = - \sum_{n1}^{n=1} Pn$$

$$\text{with } P_n = - P_n \left(l(n+n) + \sum_{k\neq n} (2k^2-P_k)\ l\ (n,k)\right) + (2n^2 - P_n) \left(g(n,n)\right)$$

$$+ \sum_{k\neq n} P_k\ g(k,n)$$

where, if $k\neq n$, $l(n,k) = g(n,k)$ represents the rate coefficient for one electron transfer from level n to k ; g(n,n) and l(n,n) are rate coefficient for recombination and ionization [25]. The rate of shell occupancy or vacancy varies differently for the different shells, as a function of time (and of distance). In other words the hole population is taken into account. Therefore, according to the total shell population probability, the total ionization state can be represented by the calculated curves of Fig. 21 for ^{127}I projectiles at 1 MeV/u passing through a carbon foil. To verify the prediction, experiments based on H^+ yield measurements at the exit surface of thin foils have been performed. This procedure allows the charge state of a projectile to be determined after a certain thickness of material.

By changing the thickness, it is therefore possible to determine mean ionization charge states as a function of the distance traversed. Self supported targets of carbon foils from 250 to 4000 Angströms have been bombarded by primary ions with charge states selected by a magnet (between 19 and 28). Fig. 22 shows the experimental variation of the ionization state of ^{127}I. It is shown that whatever the initial ionization state, a common value is reached after 400-500 Angströms (pre-equilibrium charge state). The memory of the entrance channel is lost, but the complete charge state equilibration is not yet reached. This is clearly observed in Fig. 22 where the charge state increases beyong 500 Å. Full equilibration is attained at about 2000 Angströms.

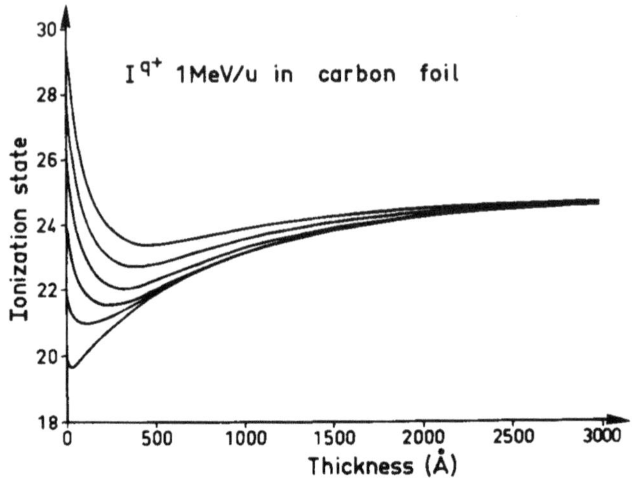

Figure 21
Theoretical predictions for the charge state variation inside a carbon foil of a ^{127}I projectile.

These experiments are the first ones to give information on variation of charge state in matter. Others measurements with different projectiles and targets would facilitate a comparison with theoretical models and refine our understanding of complex collisions and excitation in solids. This example shows that ion desorption phenomena have been unexpectedly applied to another field in physics.

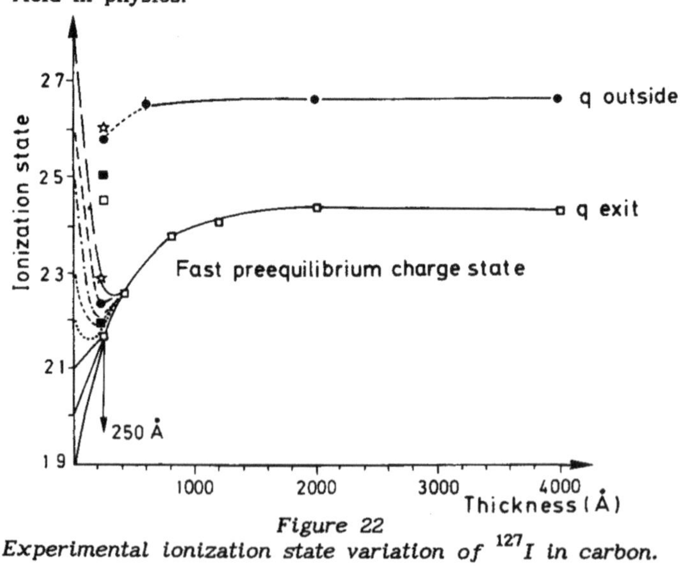

Figure 22
Experimental ionization state variation of ^{127}I in carbon.

V. FROM SINGLE ATOMS TO CLUSTER IONS AS PROJECTILES

KeV ions have been used for many years as projectiles to produce secondary ions from inorganic surfaces. It is only since the past decade that this method of producing ions from non volatile samples has been used for organic solids [56]. With polyatomic ions as projectiles, an enhancement of the secondary ion yield was observed when compared to single atom ions [57-62].

A systematic study of secondary ion yields from different targets bombarded by various projectiles has been undertaken at our laboratory, following the first experiments performed by the Uppsala and Texas Groups [63,64]. The experimental method has been much improved. Organic molecular ions, polyatomic ions like CsI clusters and metallic clusters of gold have been used as projectiles. Table 1 shows the different projectiles and targets. Bombarding energies were between a few keV and 56 keV.

Fig. 23 shows the experimental set-up for organic and CsI cluster projectiles. These are produced from a compound deposited on a thin foil and desorbed by impact of one of the fission fragments. The other fission fragment is detected and activates the timing electronics. These "primary ions" are accelerated by the potential U_1 and strike a target after a free field distance. Upon impact, secondary ions and electrons are emitted from the target. Electrons are very fast and therefore, the time between the spontaneous fission events and the impact of electrons at MCP2 (see Fig. 23) gives the TOF of primary ions.

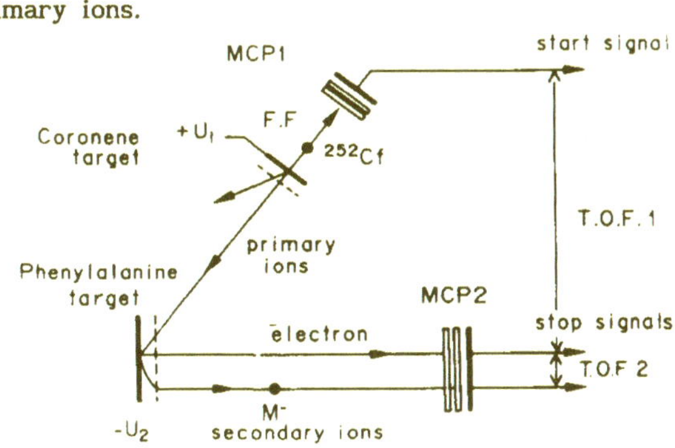

Figure 23
Experimental arrangement for impacts of large clusters produced by fission fragments.

PRIMARY IONS

ORGANIC PROJECTILES

MASS	COMPOUND
73^+	$(CH_3)Si^+$
147^+	$(CH_3)_3SiOSi(CH_3)_2^+$
120^+	M - $COOH^+$ fragment of phenylalanine
166^+	$[C_6H_5CH_2CH(NH_2)COOH+H]^+$ phenylalanine
331^+	$(2M+H)^+$ phenylalanine dimer
300^+	$C_{24}H_{12}^+$ coronene
598^+	$2(M-H)^+$ coronene dimer

ATOMIC AND CLUSTER PROJECTILES

Cs^+, Cs_2I^+, $Cs_3I_2^+$ } with CsI target and Cs ion gun

Au_n^+ n = 1 to 5 with Gold LMIS pulsed
Au_n^{++} n = 1 and 3

TARGETS

Gold

$R_4SiW_{12}O_{40}$

CsI

Organic films : Phenylalanine
Dinitrostilbene
Lipid EG
Erythromicin

ENERGY

5 keV < E_{impact} < 30 keV
60 keV for Au_n^{++}

Table 1

Electronic coincidence window can be set on TOF 1 events (for example on a molecular ion TOF peak) and this time window serves as a t_0 marker for the time signals that are in the same event. An example is given in Fig. 24 a-b. Fig. 24 is the sum of primary ions and secondary ions. Fig. 24.b shows the SI spectrum of ions emitted by the impact of a molecule of mass 300 (coronene).

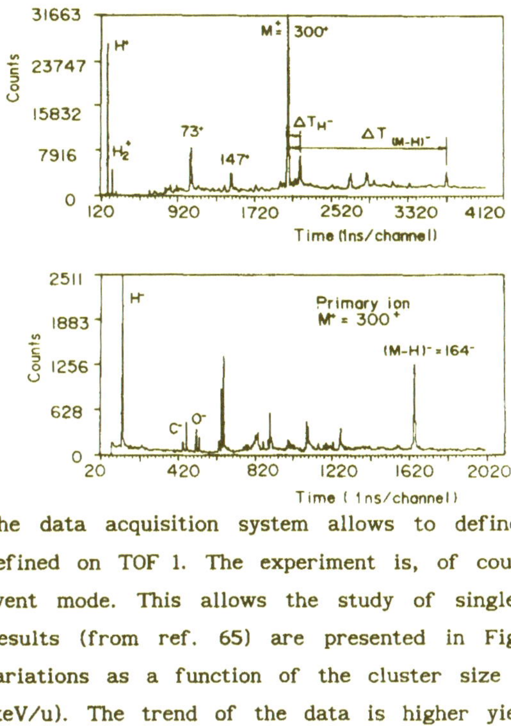

Figure 24
Cumulative time of flight spectrum (top).
Correlated time of flight spectrum obtained with a time window on M = 300.

The data acquisition system allows to define several time windows to be defined on TOF 1. The experiment is, of course, performed in an event by event mode. This allows the study of single projectile-target interactions. Results (from ref. 65) are presented in Fig. 25.a, which shows SI yield variations as a function of the cluster size and the energy per mass unit (keV/u). The trend of the data is higher yields for more complex primary ions. The yield of the molecular ion $M\text{-}H^-$ = 164 from the phenylalanine target increases by a factor of 20 upon going from Cs^+ to Cs_2I^+. That is a yield enhancement of about 7. We define the yield enhancement factor ε for a homonuclear projectile A_n having n constituents by :

$$\varepsilon = \frac{Y\ A_n(E)}{nY\ A(E/n)}$$

where $Y\ A_n(E)$ is the secondary ion yield for the cluster projectile at energy E and $Y\ A(E/n)$ the yield for the constituent atom of the homonuclear projectile at equal velocity.

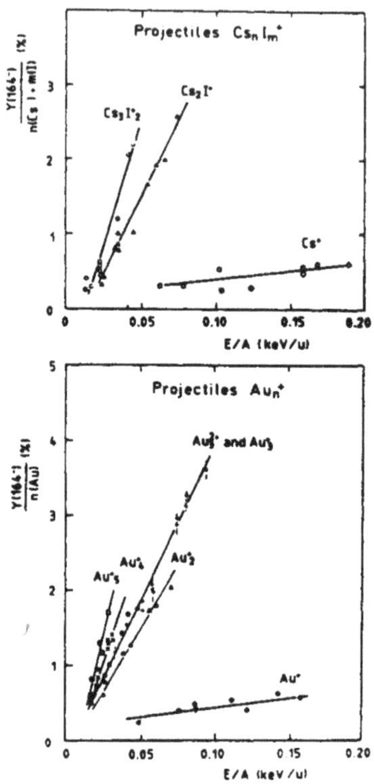

Figure 25
- *Relative yield per constituent of the cluster projectile as function of keV/u units ($\propto v^2$).*
- *The results with gold clusters have been obtained with a liquid metal ion source.*

Similar effects have been observed with clusters of gold atoms bombarding targets. In this case, a liquid metal ion source (LMIS) built by P. Sudraud and G. Ben Assayag [66] has been used[1]. A pulsation of the extracted focused beam allows easy identification, by time of flight measurements, of the different cluster ions produced in the source [67-68]. With a double

~~~~~~~~~~~~~~~~

[1]The simple design for the ion source can be obtained from P.Sudraud, Orsay Physics, Orsay
~~~~~~~~~~~~~~~~

pulsation it is possible to select a Au cluster of a given mass with which to bombard a target. The intensity of the ion source is adjusted to a small value in order to have only one or none primary ion per pulse. The exact number of cluster ions hitting the target surface can be accurately known. Fig. 25.b shows the variation of the molecular ion yield of $(M-H)^-$ from phenylalanine for the different projectiles Au_1, Au_2^+, Au_3^+, Au_4^+, Au_5^+ as function of keV/u.

Yield variation

The general trend of variation is very similar to that observed with CsI clusters. The variation of yield with the energy of impact is very steep and it would certainly be very interesting to increase the kinetic energy of those high mass clusters.

The three types of projectiles ; organic molecules (composed of light elements), ionic clusters (mass of constituent about 130), and gold clusters (mass of constituent equal to 197) all produce the same yield enhancements.

The SI yields are proportional to the square of the projectile velocity V and follow a power law of the number of cluster constituents n. This can be expressed with the following equation :

$$Y(SI) = K(SI).n^{\alpha}.V^2 \qquad (6)$$

- $\alpha = 2$ for SI atomic ions I^-, Au^- and organic fragments and molecules,
- $\alpha = 3$ for SI cluster as $(CsI)_nI^-$, $Au_kI_l^-$.

Let us consider the mass B of the constituent for a simple homonuclear cluster projectile B_n. We can write :

$$Y_{B_n}(SI) = k(B).n^2.B^2.V^2 = k(B).M^2.V^2 = k(B).P^2 \qquad (7)$$

Here, the constant k depends on the SI and the mass of the constituent of the cluster. In this case, the yield is proportional to the square of the projectile momentum P. This simple expression suggests that momentum plays a major role in the desorption process.

Although the processes of energy deposition by keV cluster ions and MeV ions are different, it is of interest to compare the desorption yields of both types of projectiles. The secondary ion emission reflects the amount of energy that is brought into the medium over a certain distance below the surface in an undefined volume after energy damping. A rough estimate indicates that for a fast ion (M = 100, E = 60 MeV) the amount of energy available for desorption is about 100 keV if one considers a desorption depth of 200 Å. On the other hand, a gold cluster Au_6 with an energy of 60 keV deposits its energy over a distance between 150 and 200 Å. The linear energy loss values are thus not so different. Table 2 give a comparison of secondary ion yield between different cluster projectiles and fission fragments (F.F.) from ^{252}Cf. It is known that atomic ions in the keV energy regime are much less efficient than MeV ions for desorption-ionization (a factor of 20 to 100 less per projectile for organic compounds). It is interesting to note that gold clusters of Au_5 at 27 keV and Au_3^{2+} at 60 keV give secondary yield values close to those obtained with fission fragments for polyatomic samples. For a monoatomic target (Gold foil), cluster ions generate a much larger yield of Au^- than F.F.

P.I. / ENERGY / TARGET IONS	Au^+	Au_3^+	Au_5^+	Au_3^{2+}	FF ^{252}Cf
	27 KeV	27 keV	27 keV	60 KeV	0.5 MeV/u
LIPID EG MW = 528^-	7.10^{-4}	$0.87.10^{-2}$	$1.60.10^{-2}$	$\simeq 2.10^{-2}$	$\simeq 6.10^{-2}$
$R_4SiW_{12}O_{40}$ MW = 2821^- MW = 3086^-	$0.2.10^{-2}$ $0.15.10^{-2}$	$0.8.10^{-2}$ $0.4.10^{-2}$	$1.2.10^{-2}$ $0.6.10^{-2}$	$\simeq 1.6.10^{-2}$ $\simeq 1.10^{-2}$	$2.7.10^{-2}$ $2.2.10^{-2}$
GOLD MW^- = 197^-	6.10^{-3}	$1.3.10^{-2}$	2.10^{-2}	$2.6.10^{-2}$	$< 10^{-4}$

Table 2

A new project [69] for high energetic cluster ions is being prepared at the Orsay Institut. A large amount of energy could be deposited in a short time on a small area. As in the case of a laser, it is possible to calculate the "power density" for a fast molecular ion striking a surface. For example, a molecular ion of mass 20000 daltons (with a mean radius of 50 Å) at an energy of 4 MeV will deposit its energy in a time around 5.10^{-14} sec. That gives a power density of 10^{14} W/cm^2. If the energy reaches 400 MeV, the power density could be around 10^{16}-10^{17} W/cm^2. The most powerful picosecond lasers are in this power range. Energy loss processes, simulation of high current density beams, fluid dynamical pressures, surface target modifications, and interstellar collisions simulation will all be within the scope of studies with energetic clusters.

VI. LASER DESORPTION

It has already been demonstrated that very large biomolecules can be introduced in the gas phase by shooting with UV laser light on solid samples [70-72]. Molecules with masses larger than 100,000 were observed [72-78]. Very peculiar conditions have been established in the preparation of the samples, which, of course, are linked to emission processes.

For example, a dilute solution of the analyte (10^{-4}-10^{-5} mole/l) is mixed with a more concentrated solution of the matrix. A small amount of analyte/matrix is dried on a metallic sample holder (molar ratio $\simeq 10^{-4}$). Matrices as nicotinic acid or cinnamic acid have proved to be very useful and efficient. A very small amount of material is necessary (femtomole range). The method will certainly be very useful in the future and several laboratories are working on its development. Small and inexpensive UV lasers are now available, and analytical instruments can be easily built. Because of their ability to measure large masses, time of flight instruments are certainly very well adapted to matrix assisted laser desorption. An example of a TOF spectrum with masses up to 150,000 is shown in Fig. 26 from results obtained by the Rockefeller group [79].

At Orsay, various TOF spectra have been obtained with the multidesorption probe TOF system TOF.23. A pulse N2 laser is used. An interesting spectrum (Fig. 27) recorded with TOF23 (IPN-Orsay) shows that molecular cluster ions of large molecules can be emitted by laser desorption. The number of molecules in the clusters depends on the mode of preparation of the sample and on the laser power. In Fig. 27 cluster ions of insulin molecules with m/z larger than 42000 are observed.

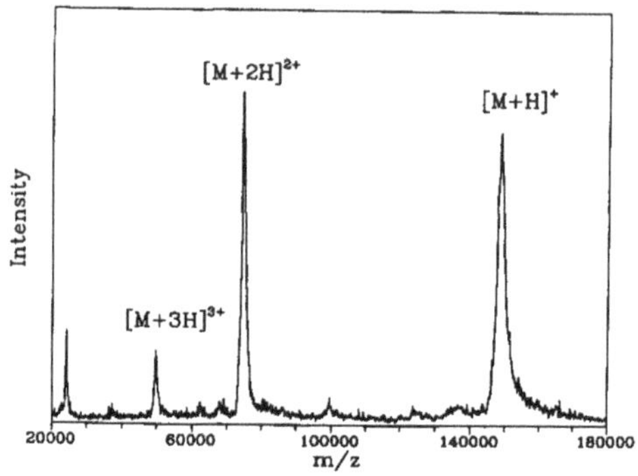

Figure 26
Mass spectrum of intact molecules ejected from a sample. Results from B.Chait and R.Beavis (Rockfeller University).

In TOF spectra, a large number of low mass ions and neutrals are visible. They are formed at the surface of the sample (narrow peaks) and in the acceleration space (broad peaks). The rather broad mass region at apparent masses below 1000 is probably due to in-flight decompositions of matrix clusters in the acceleration region. The molecular ion peak shape of large molecules indicates that these ions are released from the surface of the solid sample at an early stage of the gas phase transition.

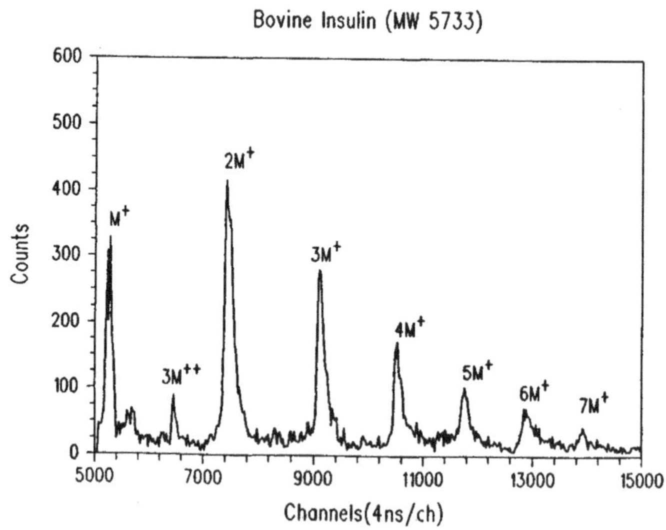

Figure 27

TOF mass spectrum of insulin cluster ions produced by matrix assisted laser desorption (Orsay)

Intact molecular ions are emitted from the sample if the power density at the surface is small. This is defined by $P = W/S\tau$ where W is the energy per pulse, τ the duration of pulse, S the area of impact. It has been shown that a power density of around 10^6 watt/cm^2 can be considered as a desorption threshold value. The role of the matrix is very important since the laser wavelength must be strongly absorbed by the matrix molecules. Almost all of the energy is supposed to be absorbed in internal vibrations of the host molecules (matrix) and transfered to lattice vibrations on a time scale of 10^{-12} sec. The energy density in the lattice vibrations must increase very quickly to cause a "sublimation spike" producing an ejection of material containing the embedded large molecules into the gas phase. To be ejected, a substantial amount of momentum has to be transfered to the molecules.

Theoretical models [80,81] of matrix assisted laser desorptions try to explain the spectacular experimental results. As with keV and MeV ions these models are certainly useful and can serve as a guide line to experimentalists who are still working to improve the method.

This recent discovery has opened new directions of research. Besides the theoretical aspects, the field of biology should benefit from the progress made to identify very large molecules. The detection of large molecular ions with small velocity is a crucial problem and efforts are being developed in several laboratories. Laser desorption experiments are rather simple to perform, and the equipment is now rather inexpensive. Time of flight techniques are mainly used.

The experimental results presented here, as well as new ideas to pursue this work in fundamental and applied physics and chemistry, are due to collective efforts from several people.

I would like to thank deeply my close collaborators S. DELLA-NEGRA, A. BRUNELLE, J. DEPAUW, C. DEPRUN. H. JORET has also been very much involved in this work. Collaborations with groups in France and other countries have been extremely fruitful and still contribute to the general dynamical processes of the development of the field.

I would like to acknowledge especially K. WIEN (Darmstadt), E. HILF (Oldenburg University), E.A. SCHWEIKERT (Texas A&M University), F. ROLLGEN (Bonn University), K.G. STANDING (Winnipeg University), I. BITENSKY (Elect. Inst. Tashkent), E. DA SILVEIRA (PUC Rio), G.B. BATISTA (PUC Rio) for their active participation in experiments or in discussions and interpretations.

I also thank G. BOLBACH (Inst. Curie, Paris), Y. HOPPILLIARD (Ecole Polytechnique, Palaiseau), P. SUDRAUD (Orsay Physics, Orsay), G. BEN ASSAYAG (CNET, Bagneux), G. MAYNARD and C. DEUTSCH (Orsay University).

This work would have not been done without the support of several good people at the laboratory, from the Electronic Workshop and Mechanical Workshop.

As usual, with the same efficiency and shrewdness, Evelyne D. has taken care of the manuscript and figures.

I do not forget my friend Douglas Barofsky (Oregon State University) who has kindly polished the english after, sometimes, long and interesting discussions on the physics and on the role of experimentalists !

The general support of IN2P3-CNRS is also greatly aknowledged.

Y. LE BEYEC

References

[1] P. Sigmund, Report Fysik Institut, Odense Univ. 1972, *"Lectures on Collision Theory..."*, p. 3.

[2] W.D. Wilson, L.G. Haggmark and J.P. Biersack, *Phys. Rev.* **B15** (1977) 2458.

[3] J.F. Ziegler, J.P. Biersack and H. Littmark, *"The stopping and ranges of ions in solids"*, vol. 1, Pergamon Press, New York (1985).

[4] R.H. Ritchie, *Nucl. Instr. and Meth.* **198** 5(1982) 1.

[5] H.A. Bethe, *Ann. Phys.* **5** (1930) 325.
H.A. Bethe, *Z. Phys.* **76** (1932) 293.

[6] L.C. Northcliffe, R.F. Schilling, *Nucl. Data Table* **A7** (1970) 233.

[7] J.F. Ziegler, *"The stopping and ranges of ions in matter"*, vol. 4, Pergamon Press, NY (1977).

[8] F. Hubert, R. Bimbot and H. Gauvin, *Nucl. Data Tables,* vol. **46** (1990) 1

[9] P. Håkansson, I. Kamensky, B. Sundqvist, *Nucl. Instr. and Meth.* **198** (1982) 43.

[10] P. Dück, W. Tren, H. Fröhlich and H. Voit, *Nucl. Instr. and Meth.* **191** (1981) 245.

[11] F. Riggi, *Nucl. Instr. and Meth.* **B43** (1989) 520.

[12] S. Della-Negra, D. Jacquet, I. Lorthiois, Y. Le Beyec, O. Becker, K. Wien, *Int. J. of Mass Spectrometry and Ions Physics* **53** (1983) 215.

[13] M. Salehpour and J.E. Hunt, *J. Phys.* **C2**, 50 (1989) 55.

[14] O. Becker, S. Della-Negra, Y. Le Beyec and K. Wien, *Nucl. Instr. and Meth.* **B16** (1986) 321.

[15] S. Della-Negra, Y. Le Beyec, *SIMS VI 1987*, Edited by A.Benninghoven, H.Huber, H.W.Werner, J.Willey Sons (1988) p. 251

[16] S. Della-Negra, J. Depauw, H. Joret, Y. Le Beyec, *J. Phys.* **C2**, T50 (1989) 63..

[17] Y. Le Beyec, *"Advances in Mass Spectrometry"*, vol. 11A (1989) 126, Ed. by P.Longevialle, Heyder and Son, London.

[18] W.L. Brown, W.M. Augustyniak, L.J. Lanzerotti, R.E. Johnson, R. Ewatt, *Phys. Rev. Lett.* **45**, 1632 (1980).

[19] F. Bensenbacher, F. Bottiger, O. Graversen, J. Hansen, H. Sorensen, *Nucl. Instr. and Meth.* **49** (1980) 203.

[20] W.L. Brown and R.E. Johnson, *Nucl. Instr. and Meth.* **B13** (1986) 295.

[21] L.E. Seiberling, J.E. Griffith and R.A. Tombrello, *Rad. Effects* **52** (1980) 201.

[22] H. Voit, B. Nees, E. Niechler and H. Frölich, *Int. J. of Mass Spectrometry and Ions Physics* **53** (1983) 201.

[23] P. Håkansson and B. Sundqvist, *Rad. Effects* 69 (1982) 179.

[24] S. Della-Negra, Y. Le Beyec, B. Monard, K. Standing, K. Wien, *Nucl. Instr. and Meth.* **B32** (1988) 360.

[25] G. Maynard and C. Deutsch, *Proc. of the First Workshop on Physics of Small Systems*, Ed. by E.Hilf and K.Wien, Springer Series **269** (1986) 94.

[26] E. Baron and B. Delaunay, *Phys. Rev.* **A12** (1975) 40.

[27] V.S. Nikolaev and I.S. Dmitriev, *Phys. Lett.* **28A** (1968) 277.

[28] K. Shima, T. Ishihara and T. Mikumo, *Nucl. Instr. and Meth.* **200** (1982) 605.

[29] N. Bohr, *Phys. Rev.* 59 (1941) 270.

[30] B. Monart, Thesis, Université Paris Sud (1988).

[31] K. Wien, O. Becker, W. Guthier, S. Della-Negra, Y. Le Beyec, B. Monart, K. Standing, G. Maynard and C. Deutsch, *Int. J. of Mass Spectrometry and Ions Processes* **78** (1987) 273.

[32] I. Danigel, H. Jungclas and L. Schmidt, *Int. J. of Mass Spectrometry and Ions Processes* **52** (1983) 223.

[33] G. Bolbach, S. Della-Negra, C. Deprun, Y. Le Beyec, *Rapid Communication in Mass Spectrom.* **1** (1987) 22.

[34] G. Save, P. Håkansson, B. Sundqvist, U. Johnson, E. Soderstöms, E. Lindquist and J. Berg, *Appl. Phys. Lett.* **51**, 1379 (1987).

[35] K. Wien, *Radiation Effects and Defects in Solids,* **109** (1987) 137.

[36] S. Della-Negra, J. Depauw, H. Joret, Y. Le Beyec, I. Bitensky, G. Bolbach, M. Galera and K. Wien, *Nucl. Instr. and Meth.* **B52** (1990) 121.

[37] B. Sundqvist, this conference.

[38] Proc. of the Orsay Workshop, *J. Phys.* **C2** (1989).

[39] D. Fenyo, B.U.R. Sundqvist and B.R. Karlsson, *Phys. Rev.* **B42** (1990) 1895.

[40] S. Della-Negra, J. Depauw, H. Joret, Y. Le Beyec, E.A. Schweikert, *Phys. Rev. Lett.* **6010** (1988) 948.

[41] M. Benguerba, S. Della-Negra, H. Joret, Y. Le Beyec, *Int. J. of Mass Spectrometry and Ions Processes* (in press).

[42] M. Blain, S. Della-Negra, M. Park, E.A. Schweikert, unpublished results.

[43] E.D. Donets, *Nucl. Instr. and Meth.* **B9** (1985) 532.

[44] T.E. Madey, *Science* **234** (1986) 316.

[45] R. Moshammer, R. Matthäus, K. Wien and G. Bolbach, IFOS V, Edited by A.Hedin, B.Sundqvist and A.Benninghoven, J.Willey and Sons (1990), in press.

[46] O. Becker, *Nucl. Instr. and Meth.* **198** (1982) 53.

[47] S. Widdiyasekera, P. Hakansson, B. Sundqvist, *Nucl. Instr. and Meth.* **B33** (1988) 836.

[48] R. Macfarlane, private communication and Jacobs, D.L., Ph.D. Thesis, Texas A&M University (1988).

[49] D. Fenyo, Thesis Uppsalla (1991) and to be published.

[50] I. Bitensky, E. Parilis, S. Della-Negra, Y. Le Beyec, to be published.

[51] Proc. of the workshop "Atomic Physics for Ion Driven Fusion IV", Edited by C.Deutsch, *J. Phys.* **C7** (1988).

[52] H. Betz and L. Grodzins, *Phys. Rev. Lett.* **25** (1970) 211.
H. Betz, *Nucl. Instr. and Meth.* **132** (1976) 19.

[53] A. Brunelle, S. Della-Negra, J. Depauw, H. Joret, Y. Le Beyec, K. Wien, *Nucl. Instr. and Meth.* **B43** (1989) 484.

[54] H.D. Betz, *Rev. Mod. Phys.* **44** (1972) 465.

[55] N. Bohr and J. Lindhard, *Mat. Fys. Medd. Danske Vid. Selsk* **26** (1954) 7.

[56] A. Benninghoven, *Org. Mass Spectrom.* **12** (1977) 595.

[57] H.H. Andersen and H.L. Bay, *J. Appl. Phys.* **46** (1975) 246.

[58] D.A. Thompson, S.S. Hohar, *Appl. Phys. Lett.* **34** (1979) 342.

[59] W. Reuter, *Anal. Chem.* **59** (1988) 2081.

[60] P.C. Zalm and L.J. Becker, *J. Appl. Phys.* **56** (1984) 220.

[61] J. Beuhler and L. Friedman, *Chem. Rev.* **86** (1986) 521.

[62] J.P. Thomas, D.E. Filpus-Luyckx, M. Fallavier, E.A. Schweikert, *Phys. Rev. Lett.* **55**, 1 (1985) 103.

[63] I. Kamensky, P. Håkansson, B. Sundqvist, C. McNeal and R. Macfarlane, *Nucl. Instr. and Meth.* **198** (1982) 65.

[64] M.G. Blain, E.A. Schweikert and E.F. da Silveira, *J. Phys.* **50**, C2 (1989) 85.

[65] M.C. Blain, S. Della-Negra, H. Joret, Y. Le Beyec and E.A.Schweikert, *Phys. Rev. Lett.* **63**, 15 (1989) 1625.

[66] P. Sudraud and G. Ben Assayag, Orsay Physics Company, Orsay.

[67] A. Brunelle, Thesis Orsay Paris Sud (1990).

[68] M. Benguerba, A. Brunelle, S. Della-Negra, J. Depauw, H. Joret, Y. Le Beyec, M.G. Blain, E.A. Schweikert, G. Ben Assayag, P. Sudraud, IPNO-DRE-91-16.

[69] S. Della-Negra, D. Gardès, Y. Le Beyec, Projet ORION, IPN Orsay (1990).

[70] K. Tanaka, H. Waki, Y. Ido, S. Akita, Y. Yoshida and Y. Yoshida, *Rapid Communication in Mass Spectrom.* **2** (1988) 151.

[71] M. Karas and F. Hillenkamp, *Anal. Chem.* **60** (1988) 2299.

[72] M. Karas, U. Bohr and F. Hillenkamp, *Int. J. of Mass Spectrometry and Ions Processes* **92** (1989) 231.

[73] M. Salehpour, I. Perera, J. Kjellberg, A. Hedin, M.A. Islamian, P. Hakansson and B.U.R. Sundqvist, *Rapid Communication in Mass Spectrom.* **3** (1989) 259.

[74] R.W. Nelson, M.J. Rainbow, D.E. Lohr and P. Williams, *Science* **246** (1989) 1585.

[75] M. Karas, U. Bahr, A. Ingendoh and F. Hillenkamp, *Angew. Chem.* (Int. Ed., Engl) **28** (1989) 760.

[76] R.C. Beavis and T. Chait, *Rapid Communication in Mass Spectrom.* **3** (1989) 233.

[77] R.C. Beavis and T. Chait, *Rapid Communication in Mass Spectrom.* **3** (1989) 432.

[78] R.C. Beavis and T. Chait, *Rapid Communication in Mass Spectrom.* **3** (1989) 432.

[79] R. Beavis and B. Chait, *Proc. of the 38th ASMS Conference* (1990) p. 26.

[80] A. Vertès, R. Gigbels, R.D. Levine, *Rapid Communication in Mass Spectrom.* **4** (1990) 6.

[81] R.E. Johnson, S. Banergee, A. Hedin, D. Fenyo and B.U.R. Sundqvist, *Proc. of the "Minaki Meeting on Large Molecules"*, Canada (1990), in press.

ELECTRONIC SPUTTERING OF LARGE ORGANIC MOLECULES AND ITS APPLICATION IN BIOMOLECULAR MASS SPECTROMETRY

BO U. R. SUNDQVIST
Division of Ion Physics, Department of Radiation Sciences
University of Uppsala, Box 535, 751 21 Uppsala, Sweden

ABSTRACT

This is a review of research which has its origin in the discovery of Plasma Desorption Mass Spectrometry (PDMS). Two main fields of research have developed, namely fundamental studies of the ejection process at fast ion impact and studies of applications of the new mass spectrometric technique. In this review the emphasis will be on the process of electronic sputtering of organic solids but also applications of this process in biomolecular mass spectrometry will be discussed.

1. INTRODUCTION

This is a review of electronic sputtering from bioorganic films. Examples will also be given to illustrate how this process has been exploited in mass spectrometric studies of biomolecules, such as proteins. The original discovery, which made the way for the development of the field covered in this review, was made by Macfarlane and coworkers in 1974 [1]. The important observation was that fissions fragments can desorb and ionize polar molecules, like small peptides, without fragmentation. The authors, who used the time-of-flight method for mass analysis of the secondary ions, called the method Plasma Desorption Mass Spectrometry (PDMS). As a result of this discovery two related fields of research developed. One field of studies deals with mechanisms for ejection of material at the impact of a fast heavy ion, i. e. electronic sputtering and the other field deals with mass spectrometric studies of ionized ejecta produced at impact of fast heavy ions, like fission fragments, on solid samples of organic molecules, such as biomolecules.

The first area of research is related to the general problem of understanding erosion of samples under bombardment of materials with particles faster than the Bohr velocity. Such energetic particles initially interact mainly with the electrons in a solid. The process of electronic sputtering is also of interest in astro physics in order to understand the erosion of icy objects in space by particle bombardment.

In the field of mass spectroscopy (MS) of large organic molecules, such as biomolecules, there are other competing methods for producing ionized gas phase biomolecules and for mass analysis of ionized ejecta. Until a couple of years ago electronic sputtering, i.e. PDMS, was the most successful MS method for biomolecules because it was used to produce the largest molecules intact in gas phase. Recently there have appeared other methods which seem to be more efficient in this particular sense, namely Matrix assisted laser desorption mass spectrometry [2] and Electrospray mass spectrometry [3]. However still PDMS (and to some extent SIMS (Secondary Ion Mass Spectroscopy)) is the only surface analytical method which probes macromolecular-surface interactions.

In this review the basic sputtering problem will be dealt with and in particular a number of experiments will be discussed which have lead to a certain level of understanding of the ejection mechanisms involved. Experiments made in our group at Uppsala will be mainly covered and used as illustrations. The basis for the theoretical understanding will be described and also in this case the work in which our group has been involved will be described in some detail. However as for the experimental part references to other groups active in the field will also be made.

2. ELECTRONIC SPUTTERING

Energetic particles erode a solid surface and this phenomenon is called sputtering [4]. The process was first observed more than hundred years ago in discharge tubes as erosion of cathodes. The process has been extensively studied at keV energies both theoretically and experimentally [4]. When the interaction of a charged particle and a solid is discussed it is convenient to use the Bohr velocity ($v_0 = 2.2 \times 10^8$ cm/ns) as a border line between particles which mainly interact with the atoms of a solid and particles which interact mainly with the electrons of the medium. The latter particles are often labelled fast ions ($v > v_0$) and the former slow ($v < v_0$). The main interaction of slow particles with solids is described in terms of two body collisions between screened nuclei and sputtering due to slow particles is therefore called nuclear sputtering. Slow particles will erode any solid surface.

In the electronic stopping region there is no erosion for solids with a fast electronic relaxation, like metals. However if the life times of excited electronic states are long, as is the case in insulators, there may be time for the electronic energy to be transferred to atomic motion and there will eventually be erosion of a surface. Sputtering in this velocity regime is called electronic sputtering.

This review is mainly concerned with organic solids eroded by fast ions. In Fig. 1 the energy loss of a heavy particle (^{127}I) in a model organic solid is shown as an illustration. The solid has the stoichiometric composition C_3H_5NO and a density of 1 g/cm^3. The nuclear energy loss in the organic material was calculated using ref. 4, Eq. 2.2.1 and a Lenz-Jenzen potential. The electronic stopping powers were taken from data tables in ref. 5.

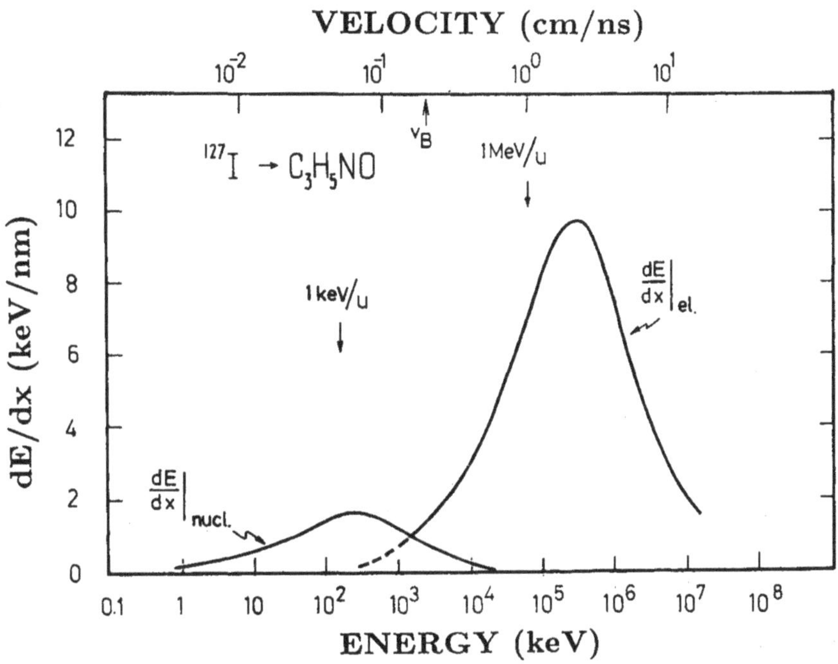

Fig. 1 Nuclear and electronic energy loss per unit length for ^{127}I in a solid of a model protein, C_3H_5NO, with a density, $\rho = 1$ g/cm^3. The nuclear energy loss in the organic material was calculated using Ref. 4, Eq. (2.2.1) and a Lenz-Jensen potential. The electronic stopping powers were taken from data tables [5].

To be able to explain sputtering due to electronic excitation we have to understand how the electronic energy is converted into atomic and molecular motion. When the fast ion travels through a solid it produces a cylindrical region around the track with intense ionization and excitation due to the direct interaction of the charge of the passing ion. This region is called the infra-track [6]. The radius of the infra-track is the Bohr adiabatic radius, which is proportional to the velocity of the particle. In the velocity region discussed here, i.e. particles of mass 100 u and energies of the order of 100 MeV the infra-track radius is of the order of a few tenth of a nm. In some of the infra-track interactions energetic (keV) secondary electrons are produced, so called δ-electrons. When they interact with the medium they give rise to a so called ultra-track with lower ionization density. The ultra-track radius is determined by the projected range of the δ–electrons and is roughly proportional to the square of the primary ion velocity.

For a long time it has been known that fast particles like fission fragments cause radiation damage in certain insulators, so called track forming materials [7]. One explanation for this which has been put forward is that the positive ion-core, formed initially after the passage of a fast ion, "explodes" because of the Coulomb repulsion. In agreement with experimental observations [7] this track formation was supposed to be effective only in materials with slow electronic relaxation such as insulators. In 1976 Haff [8] predicted sputtering effects in the electronic stopping regime on the basis of the Coulomb explosion model. The question of track formation was resurrected with the discovery of electronic sputtering a few years later. Around 1980 several studies were published on sputtering yields in the electronic sputtering regime for materials like alkali halides [9], ices [10-12], condensed noble gases [13,14], metal halides [15] and layers of biomolecules [16,17,18].

3. ELECTRONIC SPUTTERING FROM INORGANIC SOLIDS

Models for fast ion erosion of inorganics have been reviewed by Johnson and Brown [19, 20] and Schou [21]. Brown et al. [10] also considered Coulomb explosion mechanism for sputtering [7] in connection with work on fast ion sputtering of ices. The model was later questioned by Ritchie and Claussen [22] who argued that in molecular solids there is probably not enough time for the positive ions to become accelerated appreciably before the electrons return to the core and neutralize it. Therefore they proposed that the returning electrons create a plasma fed with energy from the excited states produced in the region around the core. In the plasma there would be time enough for the ions to be heated by the electrons and for the core to expand , causing momentum and energy transfer to the surrounding region. In the Watson-Tombrello model [23] the concept of an expanding lattice driven by residual Coulomb forces is also used. In this high-excitation-density regime Itoh and Nakayama [24] have discussed the local lattice distortion caused by the vicinity of two excited states.

More recently a model based on the repulsive decay of localized excited states was used to explain sample thickness dependence of ion and electronic excitation sputtering of condensed layers of Ar [25-27] and the sputtering of molecular solids in the low ionization regime [28]. Most excitations of fast particles produce Ar^+, which diffuses rapidly (it may diffuse tens of nano-meters before coming localized), but can gain energy by pairing with neutral Ar to form Ar_2^+. As electrons are available, this system can repulsively decay to Ar• + Ar with a release of 1 eV in kinetic energy to the two Ar atoms. If this happens at a surface it can lead to sputtering.

Also Ar• has an attractive state formed with a ground state of Ar, i.e. the excimer, Ar_2•. The only way out of this potential minimum is a vertical transition, which gives rise to the dominant eximer luminescent band. As Ar atom separation in the eximer state is much less than the equilibrium lattice spacing in solid Ar, potential energy is again released into kinetic energy of Ar atoms. This is a second chance for sputtering if the decay takes place close to the surface. In such a decay , energy is released via recoils, and a mini cascade is produced over a small volume, i.e. a few atomic distances [29]. For ionized molecules in a solid such a process will always occur with electronic recombination [19,20,26]. For lower lying states the process is always in competition with photon emission.

Exiton models for ion and electron sputtering of alkali-halides were discussed earlier by Townsend [30].

The basic microscopic properties of organic solids is often not so well known and detailed models for energy conversion schemes are lacking. In the case of electronic sputtering of bioorganic solids the main interest and focus have been to try to understand the ejection of "intact" fairly large biomolecular ions at the impact of fast heavy ions like fission fragments. A fast heavy ion like a fission fragment deposits almost 1 keV per Å when interacting with an organic solid. There is consequently enough energy involved to fragment a labile biomolecule. In spite of this large energy deposition, fairly large labile molecules are softly ionized and ejected. In the following the present knowledge about the mechanisms involved will be summarized. Mechanisms for desorption of large organic molecules has been reviewed by Johnson [31]

4. EARLY EXPERIMENTS ON ELECTRONIC SPUTTERING OF BIOMOLECULES

The interest in trying to understand electronic sputtering emanated originally from two rather disconnected areas, namely astro physics (erosion of condensed gases by fast ions) and biomedical mass spectrometry (PDMS as discovered by Macfarlane and coworkers). The direct connection between

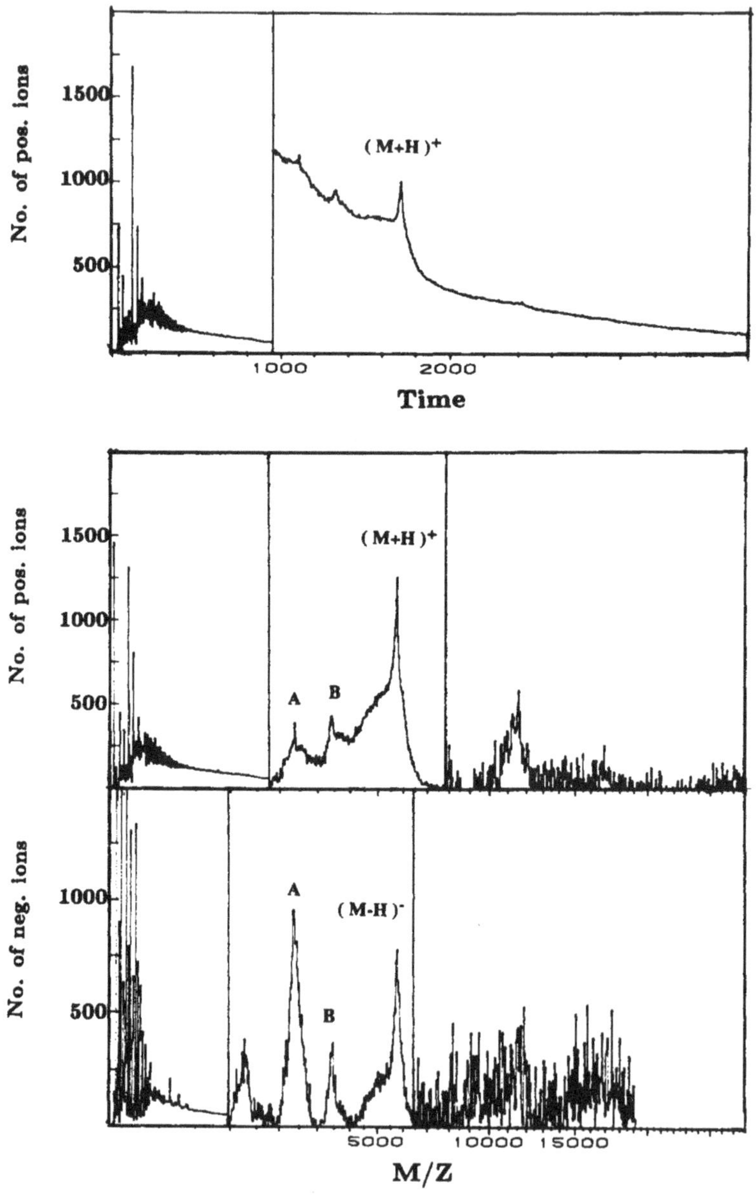

Fig. 2 PDMS (TOF) spectra of of bovine insulin (MW 5733). A. Raw spectrum of positive ions. B. Positive ions, background subtracted. C. Negative ions, background sub-tracted.

PDMS ejection and electronic energy deposition by fast ion impact was demonstrated independently and about the same time, around 1980, by experiments at Erlangen [16] and Uppsala [17]. The first international meeting discussing electronic sputtering of biomolecules and its connection to erosion of condensed gases was held at Uppsala in 1981 [32]. Since then there are two series of symposia where electronic sputtering has been one of the main topics, namely the IFOS meetings [33] and a series of symposia on impact phenomena at Wangerooge [34], Orsay [35] and Minaki [36]. In addition a biannual series of symposia on PDMS and related topics have been organized by the Texas group [37].

Most experimental data on secondary ion yields were originally taken with time of flight analysis of the mass spectrum of secondary ions ejected. The yield (number of secondary ions of one kind per incident primary particle) of secondary ions and mass spectra of ejecta were studied. Today there is a fairly large amount of experimental data on secondary ion yields of biomolecules as a function of various primary ion parameters such as velocity [16,18,38], angle of incidence [39,40] and charge state [41,42,43]. Most data concern yield of molecular ions.

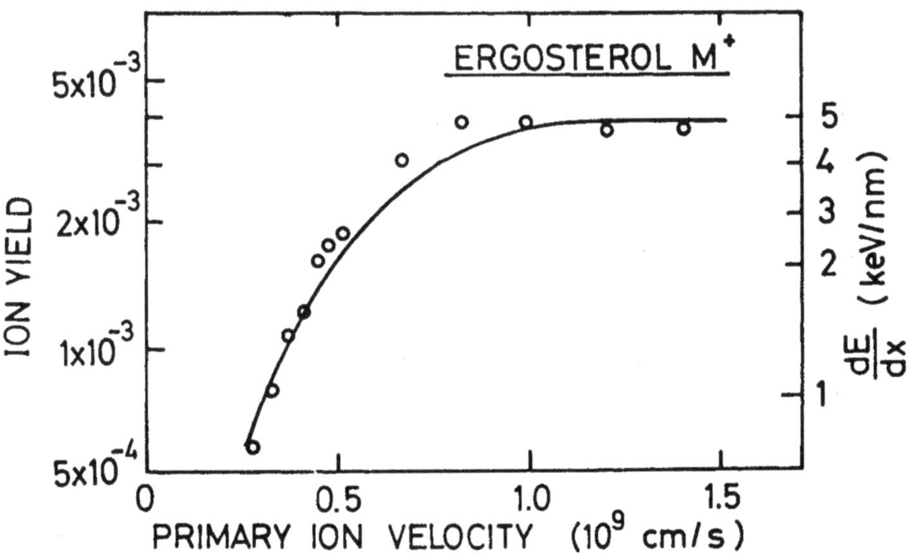

Fig. 3 Yields of molecular ions of ergosterol ($C_{28}H_{44}O$) as a function of the velocity of ^{63}Cu ions in the electronic stopping regime [18]. The electronic stopping powers (full line) were taken from data tables [5].

In Fig 2 mass spectra from a sample of bovine insulin (MW 5733) is shown as an example. In the electronic stopping regime there are in fact more experimental data of this kind available than in the nuclear stopping regime. A general early experience was that the primary beam parameters do not influence the relative peak hights in the resulting mass spectra, only the absolute yields of the various secondary ions are influenced. As illustrated for one particular example the ion yields are in some case very closely related to the electronic stopping power (See Fig. 3). Illustrations of the charge state influence on secondary ion yields can be found in the paper by LeBeyec in these proceedings [44]. Although various models were suggested to describe the secondary ion yields by electronic sputtering the lack of sputtering yields involving the total amount of ejecta (neutrals) were lacking for a long time.

5. SAMPLE PREPARATION METHODS USED IN THE FIELD OF ELECTRONIC SPUTTERING OF BIOMOLECULES

All the sample preparation methods discussed here have the first stage in common namely the preparation of a solution of the sample molecules. The simplest method used early on was to deposit a droplet with the solution with a microsyringe onto a backing and to allow the solvent to evaporate. In the spincasting technique [45,46] the backing is rotating at a high speed and then a droplet is deposited in the centre of the backing surface. This results in a rather homogeneous layer of molecules. In the electrospray method [47] the solvent should be easy to evaporate. The solution is electrosprayed onto a backing (usually in air). A voltage of a few kV is applied to an electrode in a capillary tube which is filled with the molecule solution. The tube is placed at a distance of about 1-2 cm from the backing foil which is held at ground potential. The small droplets formed in the spray process are converted to a dry layer on the backing. The layer consists of μm sized particles. This spray technique is less sample consuming than spin casting but the uniformity of the films is not as good as in spin casting. All these methods will result in multilayer samples dominated by molecule-molecule interactions. A method which is very useful for mass spectrometric studies and which allows for more controlled experimental conditions is to use a suitable substrate backing and to adsorb the sample molecules to the substrate. Once adsorbed the molecules can be made to stick so well that small organic contaminents and salts can be rinsed away. The first demonstration of this principle was made by Jordan et al [48] and Nafion was used as backing. Nitrocellulose was introduced by Jonsson et al [49] and has turned out to very useful in this sense. In Fig 4 a comparison of mass spectra from a multilayer sample and nitrocellulose adsorbed sample of porcine trypsin (MW 23 463) are shown. The main

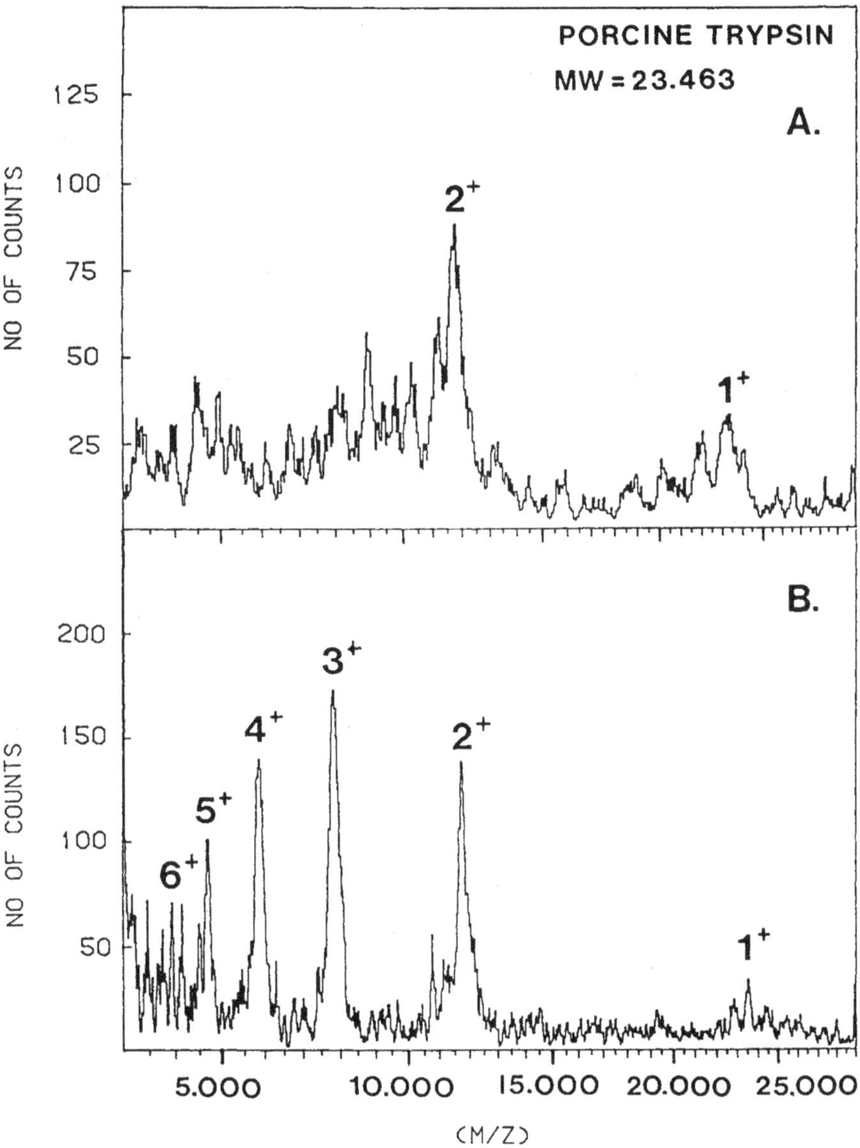

Fig. 4 PDMS (TOF) mass spectra of positive ions of porcine trypsin (MW 23 463), A. For an electrosprayed sample (multilayer) . B. For molecules adsorbed to a nitrocellulose substrate. In both cases a smooth background is subtracted.

experimental observations are that the internal energies of ejected molecular ions is lower in the case of nitrocellulose and that molecular ions of higher degree of protonization are observed. This method has turned out to be a useful way of preparing a monolayer and even sub-monolayer samples of peptides and proteins [50].

In the field of electronic sputtering the Langmuir-Blodgett technique has also been used to prepare ordered structures of organic molecules and samples of well defined thickness. The classical procedure of the Langmuir-Blodgett technique [51,52] is that a small amount amphipatic molecules, like fatty acids, is spread on the surface of pure water contained in a trough. The surface is subsequently compressed by a barrier giving a stacked monolayer of deposited molecules on the surface. In order to obtain the required sample film thickness, i.e. a certain number of monolayers, a substrate (backing) is repeatedly dipped and raised through the subphase, catching one monolayer at each penetration of the molecular/water interface.

6. DAMAGE CROSS SECTIONS FOR EJECTION OF SECONDARY IONS IN ELECTRONIC SPUTTERING OF BIOMOLECULES

When an ion hits a sample of organic molecules a certain area around the point of a single particle impact may be modified or destroyed so that the original molecules in that area will not appear as molecular ions when this area is hit by another primary particle (see Fig 5) The area σ is called disappearance or damage cross section.

Strictly speaking a certain volume of the sample will be destroyed. For low velocity impact that volume will have the approximate shape of a half sphere and for a high velocity ion with a long range the volume is more like a cylinder. In the following only damage on the surface is taken into account. If Y is the yield of molecular ions from a fresh sample, the yield $Y(\Phi)$ can be written

$$Y(\Phi)) = Y(0)\,[\,1-\delta A/A] \qquad (1)$$

where $\delta A = \delta A(\Phi,\sigma)$ is the damaged part of A caused by ΦA incident ions. The damage cross section for each ion is σ. In the expression for δA one must include a correction for overlapping areas. This can be done in an approximate way as shown by Salehpour [53] to give

$$\delta A/A = 1\text{-}\exp(-\sigma\Phi) \qquad (2)$$

i.e.

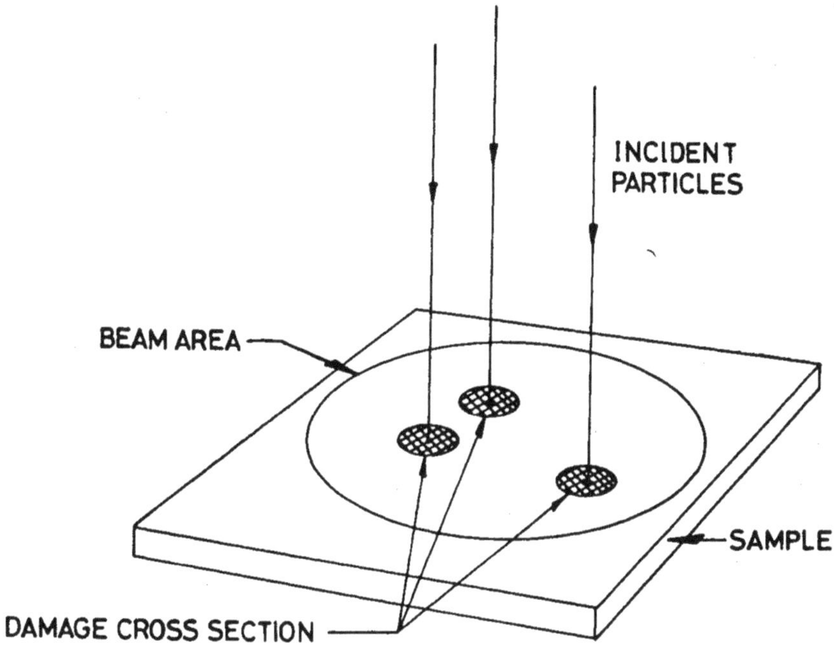

Fig. 5 An illustration to the damage cross section concept

$$Y(\Phi) = Y(0) \exp(-\sigma\Phi) \quad (3)$$

In Fig 6 from ref. [53] the experimental data from a damage cross section determination of 90 MeV $^{127}I^{14+}$ on valine are shown. Equation (3) was used to fit the dashed line to the data and a value of $\sigma=6.8\pm1.8 \times 10^{-13}$ cm^2 was extracted. Similar results were obtained for other amino acids and peptides [53]. It is important to note that the damage cross section will also reflect the damage to the sample molecules caused by "late" effects of radiation. The term late is meant to denote times long after ejection and ionization. In fact damage cross sections are not directly correlated with the area of removed material around the point of impact (crater formation). In general damage cross sections are larger than the areas of sample removal.

Sichtermann et al [54] have measured damage cross sections for low energy primary ions (2 keV Ar) and molecules like amino acids. The cross sections were found to be of the order of 10^{-14} cm^2. Consequently for solid organic surfaces, SIMS type analysis with fast and slow primaries can only

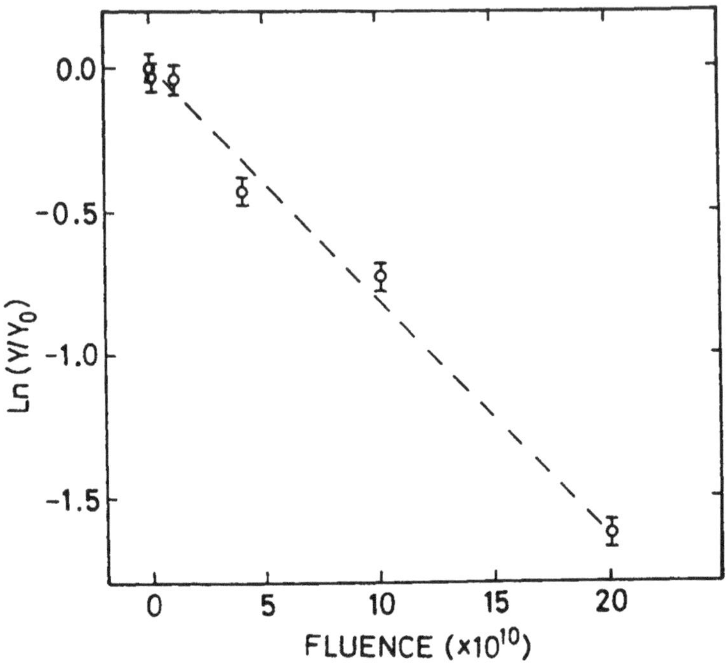

Fig. 6 Normalized intensities of molecular ions of valine (MW 117) as a function of the primary ion fluence for 90 MeV $^{127}I^{14+}$. The dashed line is a least square fit of eq. (3) to the data. The damage cross section extracted from the graph is (6.8 ± 1.8) × 10^{-13} cm^2 [53].

be run with low primary fluences. This mode of operation is often called static SIMS.

The Uppsala group has studied damage cross sections for bovine insulin (5734 u) [55] for the same primary ions as for valine mentioned above for multi-layer targets and for insulin adsorbed to a nitrocellulose backing and damage cross section radii of 130±20 and 210±30 Å were extracted, respectively. This can also be compared to the case of valine mentioned above, where the cross section radius is 47±7 Å. Fenyö et al [56] have studied damage cross sections from Langmuir-Blodgett films of arachidic acid. Two layers of arachidic acid (312.5 u) were positioned at three different depths (at the surface, with four and eight layers of stearic acid (284.5 u) on top) in a film consisting of 17 layers in total. Two different primary particle beams , namely 72.9 MeV $^{127}I^{14+}$ and $^{32}S^{4+}$ were used. The cross sections radii were 83, 82 and 82 and 33, 22 and 24 Å respectively. It is interesting to note that for $^{32}S^{4+}$ primaries the damage radius is smaller for deeper layers.

7. MOLECULAR SIZE EFFECTS IN ELECTRONIC SPUTTERING OF PEPTIDES AND PROTEINS

In a number of early experiments on secondary ion yields in electronic sputtering the scaling of yields with energy deposited (dE/dx) was studied. As pointed out above the ion track geometry changes with velocity of the primary particle. The radius of the infra track scales linearly with the velocity of the primary particle and the radius of the ultra track with the square of the same velocity. Therefore it was suggested by Håkansson et al [57] to study scaling with energy density rather than by dE/dx. Different primary ions with different dE/dx but the same velocity (same track geometry) were used. Håkansson et al [57] found an interesting difference in scaling with energy density for molecules of different size. These results were interpreted in the so called ion-track model by Hedin et al [58] and later by Lucchese [59] in a thermal model. In both models the ejection probabability is determined from the energy density in the surface region sufficient to overcome molecular binding. Furthermore in both models the inner track area can be so high that the molecule is fragmented. Both models describe qualitatively the "size" effect.

8. MEASUREMENTS OF ELECTRONIC SPUTTERING YIELDS OF NEUTRALS OF ORGANIC MOLECULES

So far there is very little experimental information available on neutral yields in electronic sputtering of organics. The only measurements published so far are those by the Uppsala group [60, 61]. These studies are based on the collector principle. In Fig. 7 this principle is schematically shown. A carefully cleaned Si slice is used as a collector in front of the biomolecular sample which is bombarded with a fast ion beam. In these studies thick electrosprayed layers of the amino acids valine (u) and leucine (u) were used. Beams from the Uppsala tandem accelerator were used to induce sputtering and part of the sputtered material was collected on the collector. The collected material was analyzed in two ways. In order to estimate the absolute yields the collector was washed with HCl and the solution amino- acid analyzed [62]. This procedure was used to determine the absolute of amount collected. This method gives the amount of whole amino acids of a particular type. The second type of analysis was to use PDMS of the collector surface. The latter method was used for the determination of relative yields in an experiment where the scaling of sputtering yield was studied as a function of energy density in ion tracks for

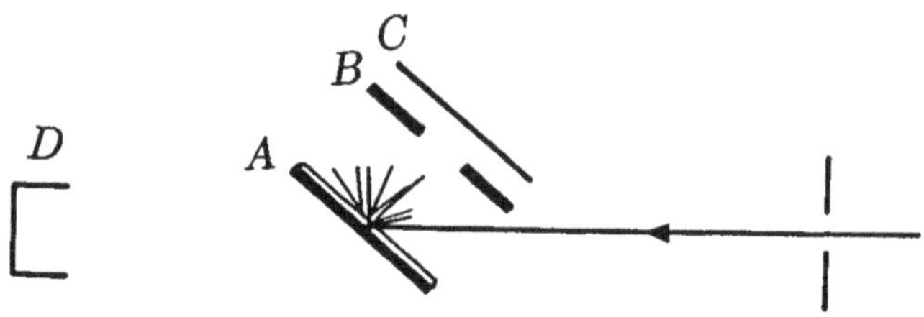

Fig. 7 The principle of the collector method used in Ref. 60 to measure the yield of neutral ejecta in eletronic sputtering of organic molecules. A- target, B-collimator, C-movable collector, D-Faraday cup.

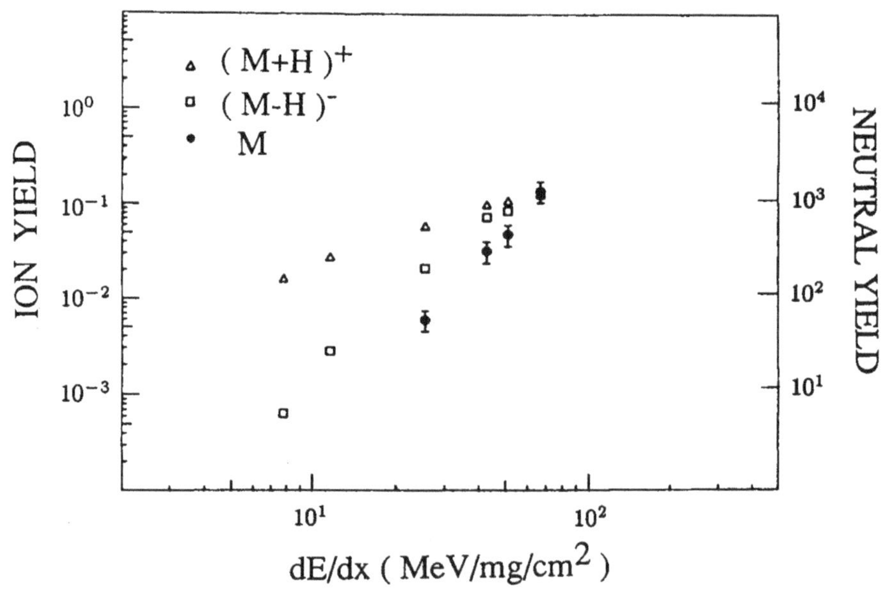

Fig. 8 Yield of intact neutral molecules (right scale), positive (left scale) and negative (multiplied by a factor 10) molecular ions of the molecule leucine as a function of electronic stopping power (for details see ref. 61).

the ions used to bombard the sample film (see sect. 7 above). The basic assumptions used was that the sticking probability of the collector was 1. In order to get enough amount of collected material to analyze for, it was important to irradiate the sample as much as possible without damaging the surface. The maximum dose is related to the damage cross section discussed above. Such a cross section measurement was made for the case of incident 78.2 MeV ^{127}I and intact neutral ejection of leucine and the cross section was found to be about the same as for intact ejection of molecular ions. The results of these studies are summarized in Fig 8 where the yields of neutral molecules, positive and negative molecular ions of leucine are shown as a function of stopping power of four different primary ions i.e . 78.2 MeV I, 48.7 MeV Br, 33.7 MeV Ni and 19.7 MeV S. The velocity (1.11 cm/ns) of the ions were selected to be the same for the reasons described above. That means that the horizontal scale in Fig. 8 can be looked upon as an energy density scale. The main finding was that the neutral yield scales with the third power of the energy density.

9. STUDIES OF ELECTRONIC SPUTTERING FROM LANGMUIR-BLODGETT FILMS OF FATTY ACIDS

In the section on sample preparation above, the principle how to prepare a Langmuir-Blodgett film was described. In studies of electronic

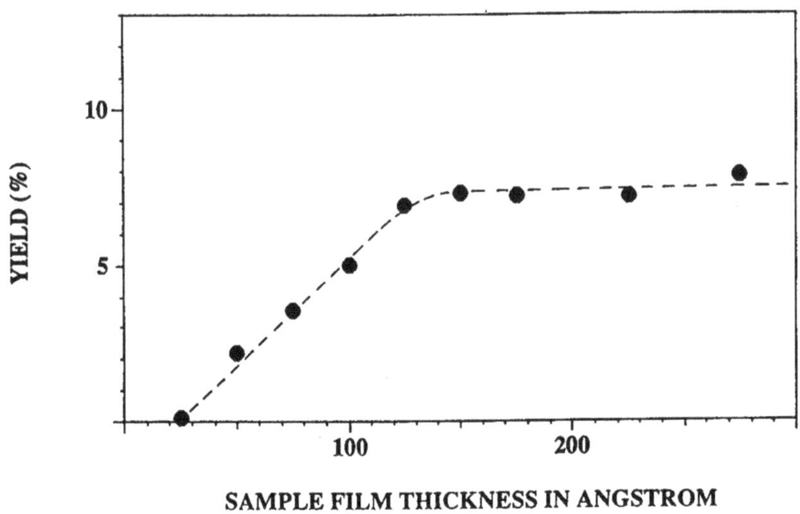

Fig. 9 The yield of negative molecular ions of arachidic acid as a function of film thickness [52].

sputtering this technique has been exploited in studies of sample film-thickness dependence of secondary ion yields as a function of various primary ion parameters [46,52,63]. Another application has been to study samples of mixed films, i.e. films made up of films of different fatty acids [64, 65]. The fatty acids used differ in mass, i.e. the number of CH_2 groups. Here these two different uses of LB films will be illustrated. In the first case film thickness dependence in for 70 MeV ^{127}I on films of arachidic acid is discussed [61]. Fig. 9 shows the yield of negative molecular ions of arachidic acid as a function of number of monolayers in the film. The yield is saturating at about 4 layers. This dependence is similar to that fund by Säve et al [46] for molecular ion yields and spin-coated films of valine and insulin.

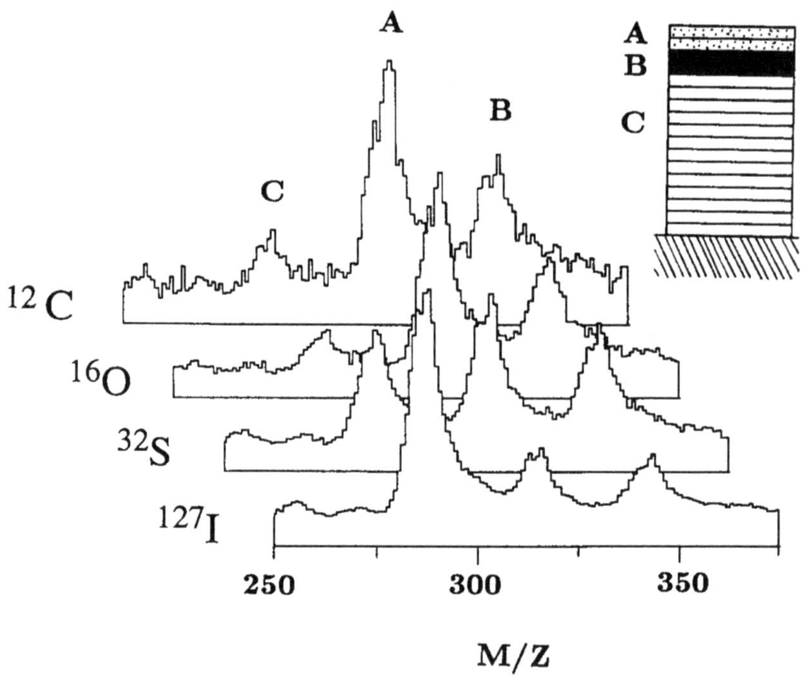

Fig. 10 Molecular ion region of negative secondary ion mass spectra of a Langmuir-Blodgett film, containing three different fatty acids for four different primary ions, namely 78.2 MeV ^{127}I, 19.7 MeV ^{32}S, 9.9 MeV ^{16}O and 7.4 MeV ^{12}C. The corresponding stopping powers are 66.7, 25.4, 11.4 and 7.8 MeV/mg/cm^2. The sample molecules are arachidic acid (MW 312.5 , A, 2 layers), behenic acid (MW 340.6 , B, 2 layers) and stearic acid (MW 284.5, C, 13 layers). The mass spectra are normalized with respect to the total yield of A, B and C [64].

In the case of mixed films Säve et al [64] have used the sample structure shown in Fig. 10. The three different fatty acids (described in the figure caption) were ordered in the way shown. The yield of secondary ions was measured for four different primary ions, all of the same velocity. The idea was again to vary the energy density by using different primary ions of the same velocity. If no crater was formed one would expect only molecular ions from the first layer to contribute to the mass spectrum of secondary ions. In Fig. 10 the part of the mass spectra corresponding to the region of the molecular ions for the four cases are shown. In the upper case (i.e. for the primary ions with the lowest energy deposition), secondary ions from the lower layers show up in addition to the top layer. Here one may argue that maybe the films are not perfect in terms of coverage. For the other cases of higher energy density, however, the relative molecular ion intensities change strongly. The integrated intensities in the four spectra have been normalized to each other. The results strongly suggest that molecular ions ejected partly originate from below the surface layer. The conclusion is that craters are formed. Similar conclusions were reached by the Orsay group from the same type of experiment [64]. An open question at present is whether such craters can be observed microscopically. Considering the results described, it is also likely that the film-thickness dependence mentioned above can be explained by the formation of craters.

10. AXIAL AND RADIAL VELOCITY DISTRIBUTIONS OF SECONDARY IONS IN ELECTRONIC SPUTTERING OF ORGANIC MOLECULES

Initial velocity distributions of ejecta often give information about the sputtering mechanism. It is convenient to distinguish between the radial and axial velocity components of the initial velocity. In the case of electronic sputtering of organics no such data are yet available for neutral ejecta. For light molecular and fragment ions there is a limited sets of data on axial velocity distributions [66-68]. For larger organic molecular ions the same kind of data are even more scarce. Widdiyasekera et al [69] and Jacobs and Macfarlane [70] have made measurements for molecular ions of bovine insulin (MW 5733).

At Uppsala quite extensive measurements have been made in recent years on radial velocity distributions of a number of secondary ions for different samples [71,72]. In straight time of flight spectrometers, pairs of deflection plates were installed in the field free region immediately after the acceleration region. The intensity of secondary ions is studied as a function of deflection voltage. The method to convert such data to radial velocity distributions is given in ref. 73. As an example of the typical findings Fig 11 shows the results for the case of molecular ions from renin substrate (MW 1801) bombarded with 72 MeV $^{127}I^{14+}$ at different angles of incidence. The intensity of secondary ions is shown as a function of radial velocity. In any

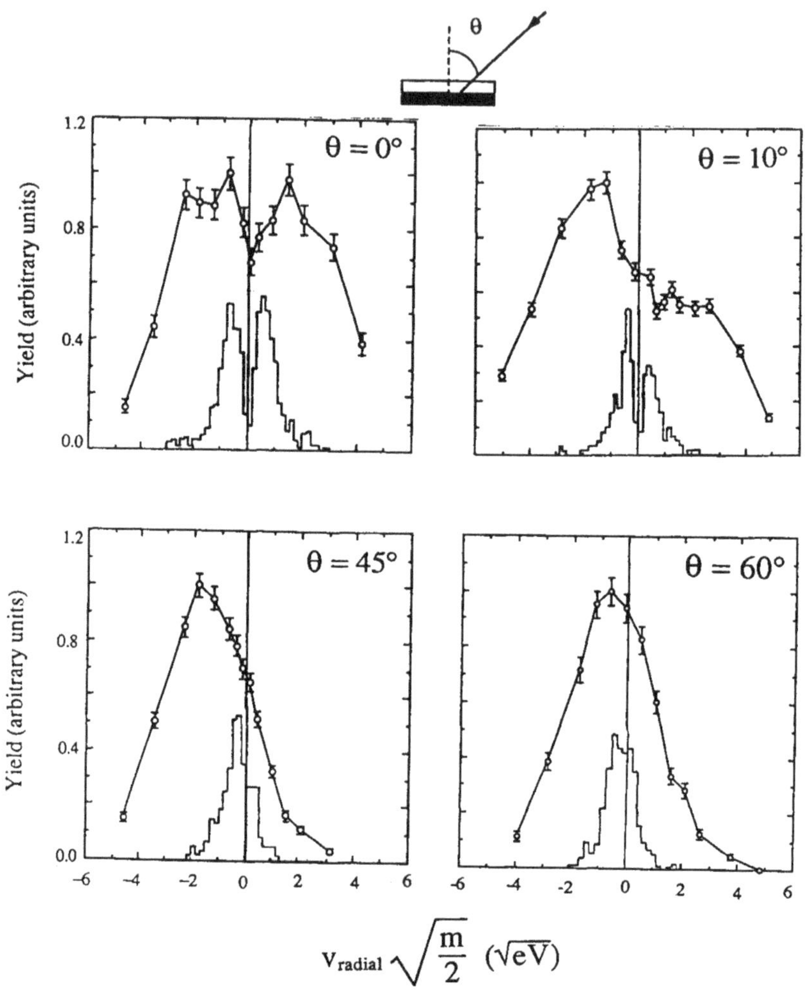

Fig. 11 Experimental and simulated (ejected non-expanded molecules) radial velocity distributions of renin substrate (MW 1801). In the experiment 72.3 MeV $^{127}I^{14+}$ ions at 0°, 10°, 45° and 60°angle of incidence were used [73].

thermal process (evaporation of particles from the surface one by one) one would expect a distribution of ejection angles peaked at the normal to the sample surface. The striking finding for the molecular ions of large organic molecules is that there is a correlation between the average ejection angle and the direction of the incoming primary ion. By using the axial velocity distributions studied by Widdiyasekera et al [69] and radial velocity distributions of the described kind one can estimate the ejection angle at the peak of the distribution for bovine insulin to be 50±10 degrees for a primary particle angle of impact of 45 degrees.

Most of the fragment ions have been shown by Fenyö et al [73] to have a symmetric distribution of ejection angles around the normal to the sample surface. However recently Brinkmalm et al [74] have found that carbon clusters, like C_{60}, produced by electronic sputtering from polyvinylidene ($[CH_2 - CF_2]_n$), are ejected backwards along the direction of the incoming particle as in a jet. In this case both fragmentation and rearrangement is needed. The surprising finding is that forming such cage like carbon structures can happen on a ns time scale.

The fact that large molecules ejected "remember" the direction of the incoming beam indicates that there is a direct momentum transfer from the expanding track core on the time scale of the ejection time of the molecule and the velocity of sound in the molecular solid. This seems to happen in a direct energy transfer process much like a mechanical shock wave [75] or a pressure pulse [76].

In a recent and elegant experiment the Darmstadt group [77] have used a position sensitive stop detector in the time of flight spectrometer. The non normal ejection of dimer ions of valine has been demonstrated, confirming the Uppsala results.

The direct ejection of secondary ions of large organic molecules have been exploited by Demirev et al [78,79] to study the ionization mechanism in electronic sputtering (PDMS). Earlier suggestions that post ejection collisions in the gas phase close to the surface (selvedge) cause ionization in a chemical ionization sense [80] can be ruled out on the basis of these studies. It rather seems that the ions are created during the ejection.

In the case of the labile molecular ions these are softly pushed out by the radial expansion of the track while the carbon clusters seem to form in the more energetic inner part of the track and pushed out by the axial expansion of the track.

11. MOLECULAR DYNAMICS SIMULATIONS OF ELECTRONIC SPUTTERING OF LARGE ORGANIC MOLECULES

To test the ideas born by the observation that molecular ions of large organics are ejected as if pushed out by the expanding track core Molecular dynamics simulations based on such a model were made. Similar simulations have been made earlier by Garrison and Srinivasan [81] in the case of nuclear sputtering of organics and laser ablation of polymers. For the particular case of electronic sputtering Hilf et al [82] have attempted a simulation of sputtering from an LB film. The Uppsala model [83,84] is based on the assumption that fast ions cause a track cylinder of molecules to expand very rapidly. In fact the idea of fission fragments giving rise to an expanding track core has been discussed earlier by Goldanskii et al [85]. In the Uppsala model the expansion is thought to be due to either a (i) Coulomb explosion of the infra-track [7,8], (ii) excited molecules, produced initially or as a result of electron-hole recombination that can decay or the excitation energy may relax to the internal vibrational modes of the molecules [86] and /or (iii) excitation of low lying vibrational modes by the ultra-track low energy electrons [87], or to some combination of the above factors. The expansion creates a compression wave that will propagate through the solid and, upon reaching the surface, may cause ejection of molecules. The molecules are assumed to interact with each other via two-body interactions, which are described by a Lennard-Jones potential. At time zero, the interaction is shifted by the expansion so that the interaction becomes repulsive. Newton's equation is solved for an ensemble of molecules according to the procedure of Verlet [88]. The parameters in the potential have been chosen so that the velocity of sound and the cohesive

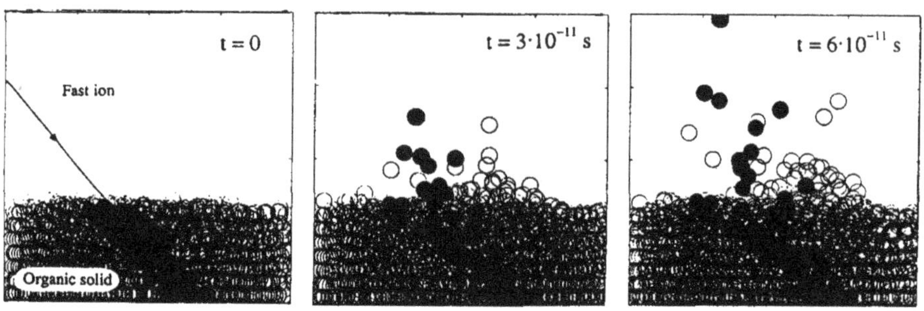

Fig. 12 Molecular dynamics simulation showing the sample at 0, 30 and 60 ps after the expansion of a cylindrical track. Expanded and non expanded molecules are shown as filled and unfilled circles, respectively. A side projection of the cylinder is shown. The angle of incidence of the fast ion is 45°.

energy of the molecular solid fit experimentally known quantities. To achieve this, a hard core had to be introduced in the potential. The ensemble of about 2000 molecules was arranged in a cylindrical matrix. In the calculations made so far, neutral and spherical molecules of mass 10 000 u were used. The reader is referred to papers by Fenyö et al [83,84,89] for details on the simulations.

To date, mainly axial and radial velocity distributions of ejecta, total sputtering yields, and craters have been studied for different energy deposited in the repulsive energy. In Fig. 12 the development of a sputtering event for the case of a 45 degree angle of incidence is shown for three different times. As can be seen from this figure, there is an asymmetry in the ejection angle of non-expanded molecules. In Fig. 11 also the simulated distributions are shown. The experimental distributions are qualitatively reproduced and as shown by Fenyö [89] the discrepancy between simulation and experiment is mainly due to the smooth surface used in the simulations as compared to the experimental samples. The scaling of the total sputtering yield with the third power of the deposited energy is nicely reproduced, at least for the highest energy densities. The simulations have also stimulated the derivation of an analytical sputtering formula for pressure pulse sputtering, described below.

12. PRESSURE PULSE SPUTTERING OF LARGE ORGANIC MOLECULES

Johnson et al [76] have described how a sum of energized small volumes (spherical impulses) stimulated by a number of means can describe sputtering of molecules over a broad range of dE/dx. At low excitation energies, individual impulses can act to eject species if these impulses are sufficiently energetic over an area comparable to the species size. This `linear´ regime is well established both in nuclear and electronic sputtering. When these energized volumes come close enough to each other in time and space, they can act cooperatively. This is the "diffusive" spike regime of sputtering for which the sputtering yield scaling is quadratic in dE/dx, and this is also well-established experimentally. The new finding is that for large energy densities, the volume ejection mechanism dominates. The energy gradient created will cause a pressure pulse which will give an outward and radial momentum to a volume of material at the surface. The energy density in the solid was calculated as a superposition of contributions from each point along the track. Although the formulas were derived for a process for which the energy transfer is described by the diffusion equation, the results were shown to be the same for impulses that do not spread (delta functions).

In the case of a particle that has a high penetration depth (the track geometry is cylindrical) and normal incidence, analytical approximations were derived for the sputtering yield and the angular ejection effect. Further assumptions are that molecules having an impulse larger than a critical impulse, determined by the material cohesive energy, are ejected. The sputtering yield for an infinitely thick sample can be written:

$$Y \propto \frac{1}{n_M}\left[\frac{\beta}{U}\left(\frac{dE}{dx}\right)_{eff}\right]^3 \qquad (4)$$

where n_M is the molecular density, $(dE/dx)_{eff}$ the effective stopping power, b is (C_p/C_v-1) for a gas, U is the cohesive energy of the solid and M is the molecular mass. This approximation gives the same dependence on the energy deposition and the cohesive energy as the simulations. The scaling of the yield with dE/dx is an effect of the geometry of the track whereas the amount of material ejected depends of the nature the energizing process. For the assumptions above, one can derive expressions for the momentum direction of the ejecta for a molecule on the surface which qualitatively is in agreement with the experiment.

In summary, the conclusions from the Johnson et al [76] study are that large molecules cannot be ejected by the linear process (individual impulses), the large energy density for a spike is detrimental to the molecule and that only the pressure pulse region is efficient and the yield will show the (dE/dx) dependence. The development of this picture was very much inspired by the early radial velocity experiments [71], the molecular dynamics simulations described [83] and the shock wave model of Parilis and Bitensky [75].

13. COMPARISON OF FAST AND SLOW PARTICLE SPUTTERING OF LARGE ORGANIC MOLECULES

A fission fragment deposits about ten times as much energy per unit length than a typical keV particle as used in Fast Atom Bombardment (FAB) or SIMS. From the results in the earlier section, however, one may expect mass spectra in the different techniques to be similar. The absolute yields may be different and, therefore, also the largest intact molecule one could study. Although the track geometry can be different in the two velocity regimes, the nature of the pressure pulse pushing the molecule out of the surface should be the same.

Unfortunately, there are very few data available for neutral ejection in the keV regime. In a recent review by Sundqvist [90], the total sputtering yields from liquid glycerol (used as a liquid matrix in FAB) were summarized, but those data do not give sufficient information in this context. In an experiment Ens et al [91] have compared yields of molecular ions of peptides and proteins for keV and MeV bombardment. This experiment gives some hint as to whether the speculations above are correct.

As shown in the inset in Fig. 13, fission fragments from a Cf-source bombarded samples from the back, and at 66 degrees angle of incidence from the front a pulsed beam of 18 keV Cs^+ ions hit the sample. Two sets of samples (prepared by the electro-spray method, multilayer, and adsorption to nitrocellulose [49], monolayer) of peptides of different mass were used. The results were that the mass spectra looked very much the same, but the main differences observed were in the absolute yields. I Fig 13 the yields are compared. The yields are normalized for the lightest molecule. The absolute yield for that molecule, leucine enkephalin, is ten times larger for fission fragments than for 18 keV Cs^+ ions. The conclusion is that fission fragments in general give larger yields. It is also evident that for nitrocellulose the keV ions give more comparable yields for monolayer

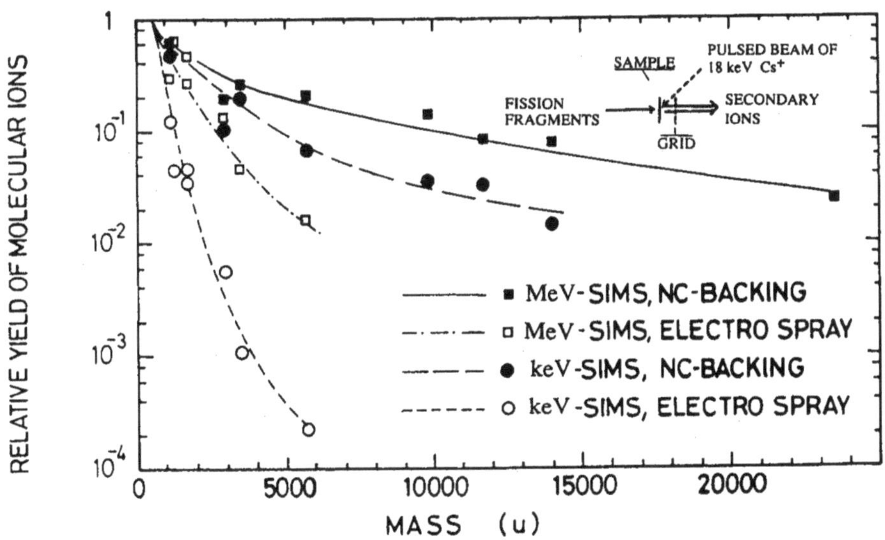

Fig. 13 Positive molecular ion yields (normalized to that of leucine-enkephalin) for a series of peptides and proteins for fission fragments and 18 keV Cs^+ ions [91]. The fission fragment yield for leucine enkephalin is ten times larger than that for 18 keV Cs^+ ions.

samples, and this may be a geometry effect. In the MeV case, the crater is deeper, whereas the surface area is not so different in the two cases. Experiments with keV ions on LB films by Bolbach et al [92] support the idea that in the keV regime, the surface layer is mainly affected.

The earlier mentioned experiments on bovine insulin showing the asymmetrical radial velocity distribution have been repeated by Ens et al [93] for 18 keV Cs^+ ions. In this case, the distributions measured were symmetric with respect to the normal of the surface. To test if this is an effect of the smaller dE/dx of the Cs^+ ions, as compared to the fission fragments, the experiment was repeated but with a fast ion (with a cylindrical track geometry) with the same total dE/dx as that of the Cs^+ ions, namely 12 MeV ^{16}O. It was found that the shift of the ejection angle was the same as with the 72 MeV ^{127}I ions. The main difference between the two probes (fast and slow) for ejection of large molecules seems to be the geometry of the track. It is therefore likely that the pressure pulse mechanism is also efficient in the keV regime.

14. EJECTION OF HYDROGEN IONS AT EARLY TIMES IN ELECTRONIC SPUTTERING FROM ORGANIC SOLIDS

So far the discussion has almost exclusively been devoted to the ejection of fairly large secondary ions. From the Molecular dynamics simulations, one can see that the ejection time of these large objects is fairly long, namely 10-100 ps. Another aspect of the discussion above is that those large "intact" ions are ejected from outside the infra-track, otherwise the molecules would fragment. In addition to large slow molecular ion ejection there are large numbers of small ions, probably emanating from the infra-track, ejected at very early times, possibly even influenced by the Coulombic repulsion of the non-neutralized track core . In fact Widdiyasekera et al [69] have observed an influence of the non conducting sample film thickness on the average axial velocity of ejected hydrogen ions in electronic sputtering. Velocity distributions for Li^+ and heavier secondary ions were not influenced by the thickness of the sample film. Recently Fenyö et al [94] have studied the radial velocity distributions for H^+ and H_2^+. Also in this study the distributions were influenced by the sample film thickness. A simple model based on the idea of neutralization of the charged infra-track explains some of the effects described. The infra-track is charged for a short time before the track is neutralized via the conducting sample substrate. Neutralization times of the order of 0.1 ps have to be assumed to fit the experimental data with the model.

The effect of sample thickness on velocity distributions of H^+ ions has important implications in PDMS applications. The mass calibration of PDMS spectra are often partly based on the H^+ peak. The proper way to deal with this problem is to use a TOF system with an electrostatic mirror which compensates for initial velocity differences of secondary ions [95].

15. PLASMA DESORPTION MASS SPECTROMETRY (PDMS) OR SECONDARY ION MASS SPECTROMETRY (SIMS) BASED ON ELECTRONIC SPUTTERING

The mass spectrometric method based on electronic sputtering is called Plasma Desorption Mass Spectrometry (PDMS). The mass analysis is based on the time-of-flight (TOF) method. In Fig. 14 the principle is shown. Since

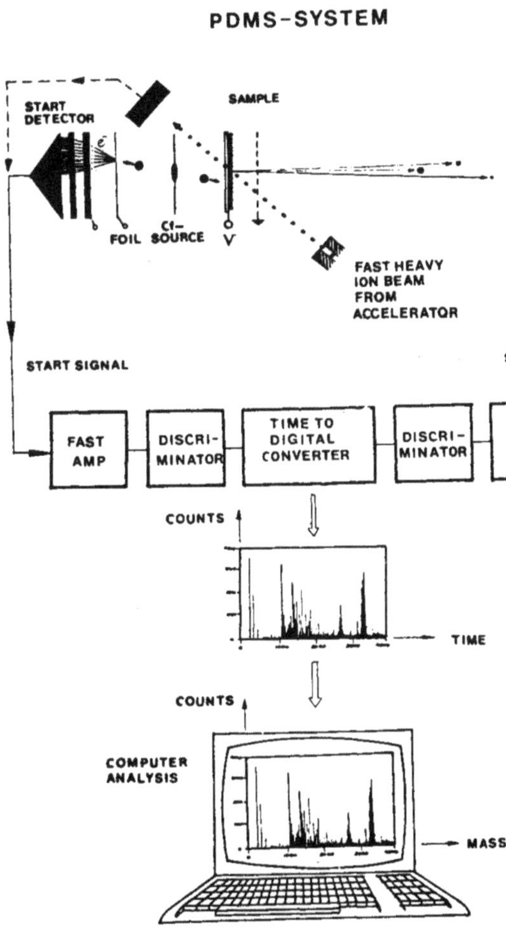

Fig. 14 Principle of a PDMS spectrometer

a few years there are commercial spectrometers available. PDMS was reviewed a few years ago by Macfarlane and Sundqvist [96].

The practical use of PDMS is at present to a large extent based on the use of nitrocellulose (NC) as a substrate for adsorption of peptides and proteins [49].The adsorption to a surface and the nice feature of PDMS that a sample is not consumed by an analysis allows for multiple use of a sample and a sequence of analyses can be performed. After the initial molecular weight determination a spectrum of proteolytic enzymes can be applied and the protein can be taken to pieces in a controlled way and its structure extracted [97]. At present the 'magic' of nitrocellulose is far from understood and no other substrates with features close to those of nitrocellulose have been found. In most applications nitrocellulose is electrosprayed onto a backing and the adsorption of proteins is made on the rough electrosprayed nitrocellulose surface.

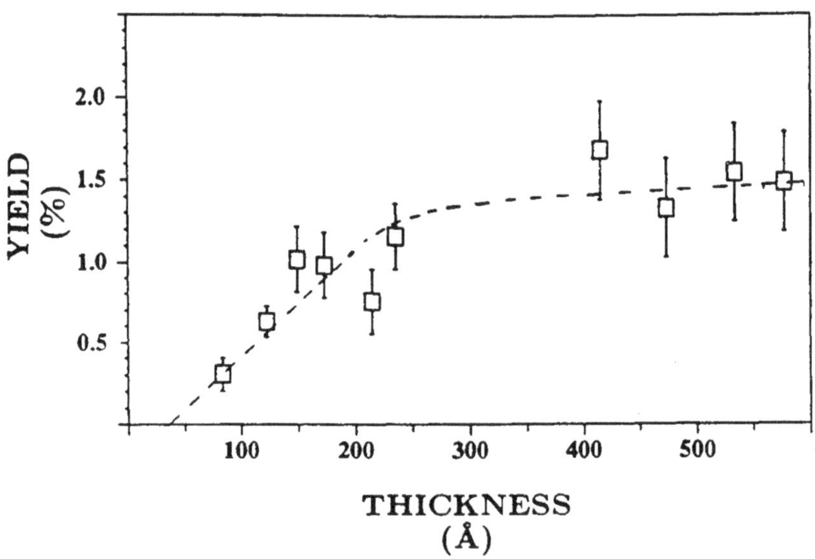

Fig. 15 The molecular ion yield of bovine insulin $(M+H)^+$ as a function of the film-thickness of spin-coated nitrocellulose. The dotted line is a guide to the eye.

In order to make controlled experiments on mechanisms for ejection and ionization it a natural physics approach to try to define the initial state of the protein as well as possible. Therefore the following experiment was initiated at Uppsala. Nitrocellulose was spin coated onto very smooth Si-

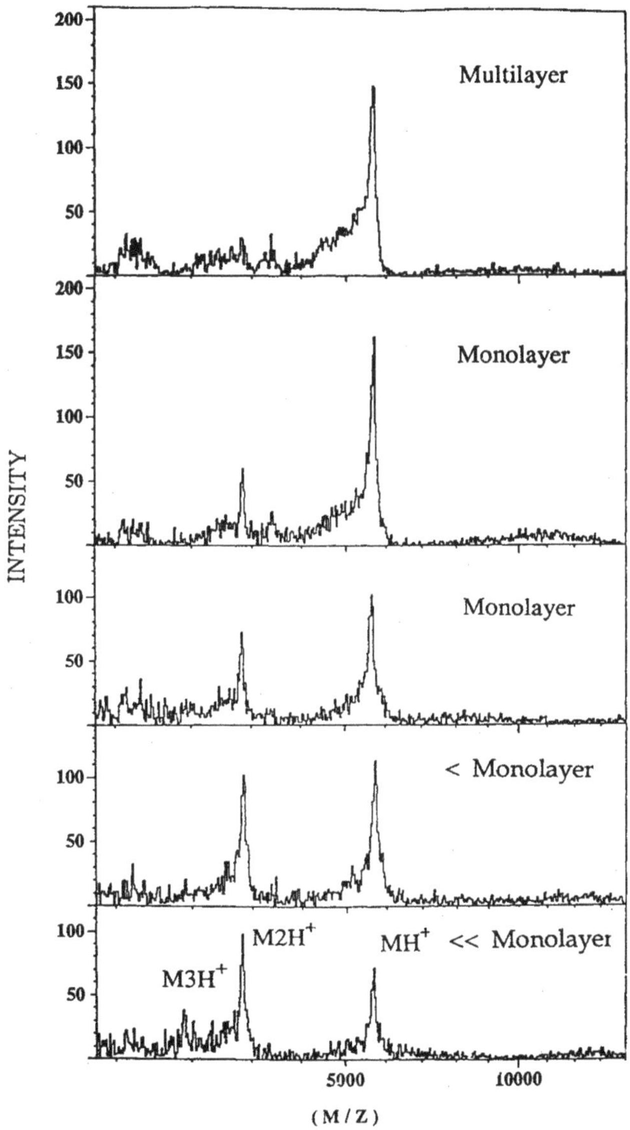

Fig. 16 Positive ion spectra of bovine insulin for different amounts of insulin applied followed by a rinsing step. The amount of insulin left (adsorbed) was measured by ellipsometry.

slices [98]. Using ellipsometry [99] the thickness of the spin-coated layers can be measured with fairly good accuracy. The homogeneity of the films was checked with Scanning Electron Microscopy and Scanning Tunneling Microscopy and little structure down to a resolution of about 50 Å was resolved. The first experiment with these NC-backings was to apply a monolayer amount of bovine insulin and study the yield of molecular ions of insulin as a function of the NC-film thickness. The results are shown in Fig. 15. The yield saturates at about 200-300 Å [98]. This is in agreement with the results mentioned earlier on LB films. In the mechanism picture given above it means that the expansion needs a certain NC-film thickness to be effective on surface adsorbed insulins. In the second experiment a certain film-thickness of NC was chosen, namely 450 Å for all backings used. Then PDMS spectra (positive and negative ions) were studied as a function of applied amount of insulin [98]. The sample was rinsed after the application of a certain amount of insulin. Also ellipsometry was used to estimate the amount of insulin adsorbed, i.e. insulin left on the surface after rinsing. It was found that the film thickness saturates at about monolayer coverage.

In the spectra in Fig. 16 it can be seen that there is a trend that the smaller amounts applied on the surface, the higher is the relative intensity of the multiply charged molecular ions. The interpretation is that if isolated insulin molecules are adsorbed to nitrocellulose the multiply charged molecular ions are enhanced, while for larger amounts applied when the insulin molecules can interact with each other (multilayer case), the lower charged states are favoured. In fact the corresponding results for the negative molecular ions indicate that in order to see molecular ions at all one needs multi-layers. These results have been checked for other molecules like, lysozyme, phospholipase A2 and trypsin. The trends are the same but the self association properties of insulin are the strongest of all cases studied. One possible interpretation of these results is that if the proteins are adsorbed isolated from each other they are multiply protonized species with small counter ions nearby. In the case of multi-layers the interchange of protons between proteins becomes important and less multiply charged ions are observed and negatively (deprotonized) charged proteins appear. Further experiments are needed to fully understand these mechanisms but these studies illustrate the potential of using a SIMS method like PDMS to study the adsorption properties of macromolecules on surfaces. It also shows that the desorption/ionization probability is very sensitive to bonding to the surface and may be possible to exploit in further studies of this kind. This study also illustrates that if proper substrates are found one can tune and improve the desorption results much further than what has been demonstrated up till now.

PDMS has mainly been used in protein chemistry so far although there are other applications [97]. Three areas where PDMS have proved to be particularly useful are for characterization of proteins by molecular weight determination, structural studies and chemical synthesis of peptides and proteins. Up to the present mass limit of PDMS, namely about 40 000 u [100], it has replaced the standard methods like SDS-gel electrophoresis and gel permeation chromatography for molecular weight determination in the laboratories where PDMS is used [97]. Here only one example of such an early study will be mentioned. During sequence determination of a structural protein from cuticle of the migratory locust, locusta migratoria, a discrepancy was found between the molecular weight calculated from the sequence data (15 323) and that estimated by SDS-gel electrophoresis (21 600). The protein contains several repeated sequences and the discrepancy might indicate that part of the sequence was missing or that the protein contained a large prosthetic group. The molecular weight of the protein was then determined by PDMS to be 15 329±50 which confirmed the sequence data and showed that the molecular weight determination based on SDS-electrophoresis was too high [101]. The PDMS spectrum is shown in Fig. 17.

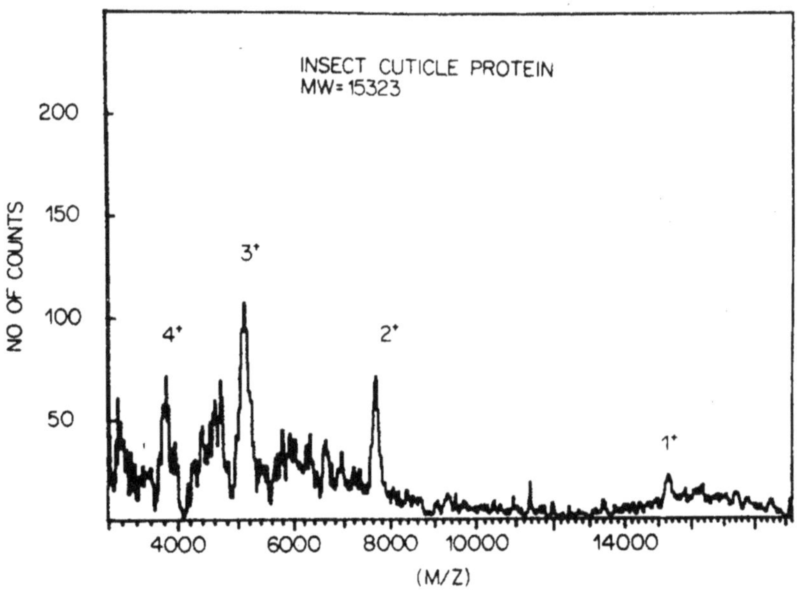

Fig. 17 PDMS spectrum of positive ions of the insect cuticle protein showing a set of multiply charged molecular ion peaks. The spectrum can be used to determine the molecular weight of this protein to 15 329 ± 50 u [101].

Modern techniques for sequence analysis are developed to meet an increasing demand to minimize the amount of sample required. Thus today approximately 100 pmole of a protein having 95% purity or better is regarded as a 'standard amount' for sequence analysis, which will include chemical and/or enzymatic fragmentation and chromatographic separation prior to analysis. By application of PDMS analysis on the various fractions obtained, usually micro bore HPLC, the analyst can easily decide which fractions are containing single species and thus be selected for further analysis. Furthermore, information on the molecular size of a peptide selected for sequence analysis is of great help in determining the number of steps to be extracted for a particular sample. In addition conventional sequence analysis can often be combined with PDMS analysis before and after in situ chemical and/or enzymatic fragmentation [97]. Analysis of the progress of synthesis is conventionally performed by analytical HPLC, amino acid analysis and sequence analysis. Though HPLC is fast and a versatile procedure, which for analytical purposes requires minute amounts of sample, the results must in most cases be supplemented by additional analysis of each peak obtained. In many cases PDMS is superior to analytical HPLC to elucidate the status of a crude peptide/protein fraction obtained after chemical synthesis. The procedure is simple and analysis may be performed immediately after deblocking the product from the resin. Contrary to the results obtained by HPLC the PDMS spectrum observed after a few minutes run will provide information about not only the desired product but also the eventual presence of truncated peptides and of still blocked material [102].

16. ACKNOWLEDGEMENTS

The author greatly acknowledges valuable comments and criticism by P. Håkansson, D. Fenyö, P. Roepstorff and R.E. Johnson.

17. REFERENCES

1. D.F. Torgerson, R.P. Skowronski and R.D. Macfarlane, Biophys. Res. Commun. **60** (1974)616
2. M. Karas and F. Hillenkamp, Anal. Chem. **60** (1988) 2299
3. M. Meng, M. Mann, and J. B. Fenn, Z. Phys. **D10** (1988) 361
4. P. Sigmund in *Sputtering by Particle Bombardment I,* ed. by R. Behrisch, Topics Appl. Phys., Vol. 47 (Springer, Berlin, Heidelberg 1981) p. 9
5. J. F. Ziegler (ed.) : *Handbook of Stopping Cross-Sections for Energetic Ions in All Elements* (Pergamon Press, New York 1980)

6. W. Brandt and R. H. Ritchie in *Physical Mechanisms in Radiation Biology*, ed. by R. D. Cooper and R. D. Wood (USEC Technical Inf. Center, Oak Ridge, Tennesse 1974) p. 20
7. R. L. Fleischer, P. Price and R. M. Walker, *Nuclear Tracks in Solids* (University Press, Berkeley, 1975)
8. P. K. Haff, Appl. Phys. Lett. **29** (1976) 473
9. J. P. Biersack and E. Satner, Nucl. Instrum. Methods **132** (1976) 229
10. W. L. Brown, L. J. Lanzerotti, J. M. Poate, W. M. Augustyniak, Phys. Rev. Lett. **40** (1978) 1027
11. W. L. Brown, W. M. Augustyniak, E. Brady, B. Cooper, L. J. Lanzarotti, A. Ramirez, B. Evatt and R. E. Johnson, Nucl. Instrum. Methods **170** (1980) 321
12. W. L. Brown, W. M. Agustyniak, L. J. Lanzerotti and R. E. Johnson and R. Evatt, Phys. Rev. Lett. **45** (1980) 1632
13. R. W. Ollerhead, J. Böttiger, J. A. Davies, J. Lècuyer, H. K. Haugen and N. Matsumani, Radiat Eff. **49** (1980) 203
14. F. Besenbacher, J. Böttiger, O. Graversen, J. L. Hansen and H. Sörensen, Nucl. Instrum. Methods **191** (1981) 221
15. J. E. Griffith, R. A. Weller, L.E. Seiberling, T. A. Tombrello, Radiat. Eff. **51** (1980) 223
16. P. Duck, W. Treu, H. Fröhlich, W. Galster, H. Voit, Surf. Sci. **95** (1980) 603
17. P. Håkansson, A. Johansson, I. Kamensky, B. Sundqvist, J. Fohlman and P. A. Peterson, IEEE Trans. Nucl. Sci. **NS-28** (1981) 1776
18. P. Håkansson and B. Sundqvist, Radiat. Eff. **61** (1982) 179
19. R. E. Johnson and W.L. Brown, Nucl. Instrum. Methods, **198** (1982) 103
20. R. E. Johnson and W. L. Brown, Nucl. Instrum. Methods **209/210** (1983) 469
21. J. Schou, Nucl. Instrum. Methods **B27** (1987) 188
22. R. H. Ritchie and C. Claussen, Nucl. Instrum. Methods **198** (1982) 133
23. C. C. Watson and T. A. Tombrello, Radiat. Eff. **89** (1985) 263
24. N. Itoh and T. Nakayama, Phys. Lett. **A92** (1982) 471
25. C. T. Reimann, R. E. Johnson and W. L. Brown, Phys. Rev. Lett. **53** (1984) 600
26. W. L. Brown, C. T. Reimann and R. E. Johnson, Nucl. Instrum. Methods **B19/20** (1987) 9
27. P. Borgersen, J. Schou, H. Sörensen and C. Claussen, Appl. Phys. **29** (1982) 222
28. F. L. Rook, R. E. Johnson and W. L. Brown, Surf. Science **164** (1985) 625
29. B. J. Garrison and B. E. Johnson, Surf. Science **148** (1984) 388

30. P. Townsend in *Sputtering by Particle Bombardment II*, ed. by R. Behrisch, Topics Appl. Phys., Vol. 52 (Springer, Berlin, Heidelberg 1981) p. 147
31. R. E. Johnson, Int. J. Mass Spectrom. Ion Proc. **78** (1987) 357
32. B. Sundqvist (ed.): *Proc. of the Nordic Symp. on Ion Induced Desorption of Molecules from Bioorganic Solids*, Nucl. Instrum. Methods **198** (1982)
33. A. Benninghoven (ed.): *Ion Formation from Organic Solids (IFOS II)*, Springer Ser. Chem. Phys. Vol. 25 (Springer, Berlin, Heidelberg, 1983)
A. Benninghoven (ed.): *Ion Formation from Organic Solids (IFOS III)*, Springer Ser. Phys. Vol. 9 (Springer, Berlin, Heidelberg, 1986)
A. Benninghoven (ed.): *Ion Formation from Organic Solids (IFOS IV)*, (Wiley & Sons, Chichester, New York, Brisbane, Toronto, Singapore, 1989)
A. Hedin, B. U. R. Sundqvist and A. Benninghoven (eds.): *Ion Formation from Organic Solids (IFOS V)*, (Wiley &Sons, Chichester, New York, Brisbane, Toronto, Singapore, 1990)
34. Y. Le Beyec, S. Della Negra and J. P. Thomas (eds.): *Proc. from Sec. Int. Workshop on MeV and keV Ions and Cluster Interactions with Surfaces and Materials*, J. de Phys. **C2** (1989)
35. E. R. Hilf, F. Kammer and K. Wien (eds.): *PDMS and Clusters*, Springer Ser. Phys. 269, (Springer, Berlin, Heidelberg, New York, London, Paris, Tokyo, 1987)
E. R. Hilf and W. Tuszynski (eds.): *Proc. from Workshop on Physics on Small Systems*, (World Scientific, Singapore 1990)
36. K. Standing and W. Ens (eds.): *Proc. from NATO workshop on Methods and Mechanisms for Producing Ions from Large Molecules*, (To be published by Plenum Press, 1991)
37. C. McNeal (ed.): *Proc. from Texas Symp. on Particle Induced Desorption Mass Spectrometry*, Int. J. Mass Spectrom. Ion Phys. 53(1983)
C. McNeal (ed.): *Mass Spectrometry in the Analysis of Large Molecules*, (Wiley & Sons, Chichester, New York, Brisbane, Toronto, Singapore, 1986)
C. McNeal (ed.): *The Analysis of Peptides and Proteins by Mass Spectrometry*, (Wiley & Sons, Chichester, New York, Brisbane, Toronto, Singapore, 1988)
38. S. DellaNegra, D. Jacquet, I. Lorthiois, Y. LeBeyec, O. Becker and K. Wien, Int. J. Mass Spectrom. Ion Phys. **53** (1983) 215
39. P. Håkansson, I. Kamensky and B. Sundqvist, Surf. Science **116** (1982) 302
40. E. Nieschler, B. Nees, N. Bischof, H. Fröhlich, W. Tiereth and H. Voit, Surf. Science **145** (1984) 294

41. P. Håkansson, E. Jayasinghe, A. Johansson, I. Kamensky and B. Sundqvist, Phys. Rev. Lett. **47** (1981) 1227
42. W. Guthier, O. Becker, S. DellaNegra, W. Knippelberg, Y. LeBeyec. U. Weikert, K. Wien, P. Wieser and R. Wurster, Int. J. Mass Spectrom. Ion Phys. **53** (1983) 185
43. E. Nischler, B. Nees, N. Bischof, H. Fröhlich, W. Tierath and H. Voit, Radiat. Eff. **83** (1984) 121
44. Y. LeBeyec (these proceedings)
45. L. F. Thompson, M. J. Bowden (eds.): *Introduction to Microlitography* (Am. Chem. Soc. Washington DC, 1983) p. 187
46. G. Säve, P. Håkansson, B. U. R. Sundqvist, U. Jönsson, G. Olofson, M. Malmquist, Anal. Chem. **59** (1987) 2059
47. C. J. McNeal, R. D. Macfarlane and E. L. Thurston, Anal. Chem. **51** (1979) 2036
48. E. A. Jordan, R. D. Macfarlane, C. R. Martin and C. J. McNeal, Int. J. Mass Spectrom. Ion Proc. **53** (1983) 345
49. G. Jonsson, P. Håkansson, B. Sundqvist, P. Roepstorff, P. Carlsen, K. E. Johansson, I. Kamensky and M. Lindberg, Anal. Chem. **58** (1986) 1084
50. G. Jonsson, G. Brinkmalm, B.U.R. Sundqvist, A. Hedin and P. Håkansson, Internal Report, **TSL-ISV 21**, Uppsala 1989
51. K. B. Blodgett, J. Am. Chem. Soc. **57** (1935) 1007
52. G. Säve, P. Håkansson, B. U. R. Sundqvist, E. Söderström, S. E. Lindqvist and J. Berg, Int. J. Mass Spectrom. and Ion Proc. **78** (1987) 259
53. M. Salehpour, P. Håkansson, B. Sundqvist, Nucl. Instrum. Methods **B2** (1984) 752
54. W. Sichtermann and A. Benninghoven, Int. J. Mass Spectrom. Ion Phys. **40** (1981) 177
55. B. U. R. Sundqvist, A. Ariaratne, W. Ens, D. Fenyö, A. Hedin, P. Håkansson, G. Jonsson and S. Widdiyasekera, *Ion Formation from Organic Solids (IFOS IV)*, ed. by A. Benninghoven (J. Wiley & Sons, Chichester, July 1989) p. 51
56. B. U. R. Sundqvist, D. Fenyö, P. Håkansson, G. Säve, S. E. Lindqvist, E. Söderström and J. Berg, *Proc. from SIMS VI*, Versailles 1987, ed. by A.M. Huber, A. Benninghoven, H. W. Werner and G. Slodzian (J. Wiley & Sons, Chichester, 1988) p. 615
57. P. Håkansson, I. Kamensky, M. Salehpour, B. Sundqvist and S. Widdiyasekera, Radiat. Eff. **80** (1984) 141
58. A. Hedin, P. Håkansson, B. Sundqvist and R. E. Johnson, Phys. Rev. **B31** (1985) 1780
59. R. R. Lucchese, J. Chem. Phys. **86** (1987) 443
60. M. Salehpour, P. Håkansson, B. Sundqvist and S. Widdiyasekera, Nucl. Instrum. Meth. **B13** (1986) 278
61. A. Hedin, P. Håkansson, M. Salehpour and B.U. R. Sundqvist, Phys. Rev. **B35** (1987) 7377

62. H. Spackman, W. H. Stein and S. Moore, Anal. Chem. **30** (1958) 1190
63. G.Säve, P. Håkansson, B. U. R. Sundqvist and U. Jönsson, Nucl. Instrum. Meth. **B26** (1987) 571
64. G. Säve, P. Håkansson, B. U. R. Sundqvist, E. Söderström, S. E. Lindqvist and J. Berg, Appl. Phys. Lett. **51** (1987) 1379
65. G. Bolbach, S. Della-Negra, C. Deprun, Y. LeBeyec and K. G. Standing, Rap. Commun. Mass Spectrom. **1** (1987) 22
66. R. D. Macfarlane and D. F. Torgerson, Phys. Rev. Lett. **36** (1976) 486
67. N. Fürstenau, W. Knippelberg, F. R. Krüger, G. Weiss and K. Wien, Z. Naturforsch. **32a** (1977) 711
68. P. Dück, W. Treu, H. Fröhlich, W. Galster and H. Voit, Surf. Science, **95** (1980) 603
69. S. Widdiyasekera, P. Håkansson and B. U. R. Sundqvist, Nucl. Instrum. Meth. **B33** (1988) 836
70. D. L. Jacobs and R. D. Macfarlane, *36th Ann Conf. Mass Spectrom. and Allied Topics*, San Francisco, 1988, Book of abstracts
71. W. Ens, B. U. R. Sundqvist, A. Hedin, P. Håkansson and G. Jonsson, Phys. Rev. **B39** (1989) 763
72. W. Ens, B. U. R. Sundqvist, P. Håkansson, D. Fenyö, A. Hedin and G. Jonsson, J. de Physique, **C2** (1989) 9
73. D. Fenyö, A. Hedin, P. Håkansson and B. U. R. Sundqvist, Int. J. Mass Spectrom. Ion Proc. **100** (1990) 63
74. G. Brinkmalm, D. Barofsky, P. Demirev, D. Fenyö, P. Håkansson, R. E. Johnson, C. T. Reimann and B. U. R. Sundqvist (submitted to Nature, July 1991)
75. I. S. Bitensky and E. S. Parilis, Nucl. Instrum. Meth. **B21** (1987) 26
76. R. E. Johnson, B. U. R. Sundqvist, A. Hedin and D. Fenyö, Phys. Rev. **B40** (1989) 49
77. R. Mosshammer, R. Matthäus, K. Wien, Y. LeBeyec and G. Bolbach, *Ion Formation from Organic Solids (IFOS V)*, ed. by A. Hedin, B.U. R. Sundqvist and A. Benninghoven (J. Wiley & Sons, Chichester, 1990) p. 17
78. P. Demirev, D. Fenyö, P. Håkansson and B. U. R. Sundqvist, Org. Mass Spectrom. **26** (1991) 471
79. P. Demirev, D. Fenyö, P. Håkansson and B. U. R. Sundqvist (Accepted for publication in Int. J. Mass Spectrom. Ion Proc. 1991)
80. K. L. Busch and R. G. Cooks, Science **218** (1982) 247
81. G. J. Garrison and R. J. Srinivasan, Appl. Phys. **57** (1985) 2909
82. E. R. Hilf, H. F. Kammer and B. Nitzschmann, *Ion Formation from Organic Solids (IFOS IV)* , ed. by A. Benninghoven (J.Wiley & Sons, Chichester, July 1989) p. 97
83. D. Fenyö, B. U. R. Sundqvist, B. Karlsson and R. E. Johnson, Phys. Rev. **B42** (1990) 1895

84. D. Fenyö, B. U. R. Sundqvist, B. Karlsson and R. E. Johnson, J. de Physique **C2** (1989) 33
85. V. I. Goldanskii, E. Ya. Lantsburg and P. A. Yampolskii, JETP Lett. **21** (1975) 166
86. R. E. Johnson and B. Sundqvist, Int. J. Mass Spectrom. Ion Phys. 53 (1983) 337
87. P. Williams and B. U. R. Sundqvist, Phys. Rev. Lett. **58** (1987) 1031
88. L. Verlet, Phys. Rev. **159** (1967) 98
89. D. Fenyö (preprint 1991, Dept. of Rad. Sci., Uppsala University, Uppsala, Sweden)
90. B.U.R. Sundqvist, Adv. in Mass Spectrom. **11A** (1989) 363
91. W. Ens, P. Håkansson and B. U. R. Sundqvist, *Proc. of SIMS VI*, Versailles 1987, ed. by A.M. Huber, A. Benninghoven, H. W. Werner and G. Slodzian (J. Wiley & Sons, Chichester, 1988) p. 623
92. G. Bolbach, R. Beavis, W. Ens, D. E. Main, E. Schueler and K. G. Standing, *33rd Ann. Conf. Mass Spectrometry and Allied Topics*, San Diego 1985, Book of abstracts.
93. W. Ens, B. U. R. Sundqvist, P. Håkansson, D. Fenyö, A. Hedin and G. Jonsson, J. de Physique **C2** (1989) 9
94. D. Fenyö, , P. Håkansson, B. U. R. Sundqvist and R. E. Johnson (preprint 1991, Dept. of Rad. Sci., Uppsala University, Uppsala, Sweden)
95. B. A. Mamyrin, V. I. Karataev, D. V. Shmikk and V. A. Zagulin, Sov. Phys. JETP, **37** (9173) 4594.
96. B.U.R. Sundqvist and R.D. Macfarlane, Mass Spectrom. Rev. **4** (1985) 21
97. P. Roepstorff, K. Klarskov and P. Højrup, *Methods in Protein Sequence Analysis*, ed. by B. Wittmann-Liebold (Springer Verlag, Berlin, 1989) p. 191
98. G. Jonsson, G. Brinkmalm, B.U.R. Sundqvist, A. Hedin and P. Håkansson, Internal Report, TSL-ISV 21, Uppsala 1989
98. M.A. Azzam and N.M. Bashara, *Ellipsometry and Polarized Light*, (North-Holland Publ. Co, Amsterdam, 1977)
100. G. Jonsson, A. Hedin, P. Håkansson, B. U. R. Sundqvist, H. Bennich and P. Roepstorff, Rap. Commun. Mass Spectrom. **6** (1989) 190
101. P. Roepstorff, P. Højrup, B.U.R. Sundqvist, G. Jonsson, P. Håkansson, S.O. Andersen and K.-E. Johansson, Biom. Environ. Mass Spectrom. **13** (1986) 689
102. G. Lindeberg, Å. Engström, A.G. Craig and H. Bennich, *Proceedings of Peptides 1988*, ed. by G. Jung and E. Bayer (W. de Gruyter, Berlin, New York, 1989) p. 121

CURRENT THEORETICAL METHODS FOR ELECTRON-MOLECULE SCATTERING

R. K. NESBET
IBM Almaden Research Center
San Jose, CA 95120, USA

ABSTRACT

The present status of theoretical methods for accurate calculation of electron-molecule scattering cross sections is reviewed. Alternative methods for fixed-nuclei scattering are described and compared on the basis of their treatment of the physical processes involved in scattering. Vibrational and rotational effects are important in low-energy scattering, requiring theoretical treatment of rovibrational coupling. Adiabatic methods are reviewed, and several nonadiabatic methods are discussed.

1. INTRODUCTION: FORMALISM OF SCATTERING THEORY

This field was reviewed as of 1980 by Lane.[1] Variational methods for atoms, generally extendable to fixed-nuclei electron-molecule scattering, were reviewed by Nesbet.[2] In 1986 Gianturco and Jain[3] reviewed theory for polyatomic target molecules. Near-threshold scattering, for which rovibrational coupling is important, was reviewed more recently by Morrison.[4]

1.1 Many-electron wave functions

A particular solution, denoted by index q, of the N+1-electron Schrödinger equation at specified total energy E can be expressed in the form

$$\Psi_q = \sum_p \mathcal{A}\Theta_p\psi_{pq} + \sum_\mu \Phi_\mu c_{\mu q}, \tag{1}$$

where Φ_μ is an N+1-electron correlation function, Θ_p is an N-electron target state, and ψ_{pq} is a channel orbital wave function. The radial part of ψ has a canonical form with coefficients α_{ipq}, where $i=(0,1)$ denotes asymptotic form (sin,cos), respectively, with argument $kr - \ell\pi/2$ for neutral open channels.

1.2 Scattering matrices and cross sections

Standard scattering matrices are defined in terms of the matrix α of these asymptotic coefficients. The open-channel reactance matrix,

$$K = \alpha_1\alpha_0^{-1}, \tag{2}$$

is real and symmetric. The scattering matrix,

$$S = (I + iK)(I - iK)^{-1}, \tag{3}$$

which is unitary, is used to define the transition matrix,

$$T = (S - I)/2i. \qquad (4)$$

A typical state-to-state integral cross section is given by

$$\sigma_{q\to p} = \frac{4\pi}{k_q^2} |T_{pq}|^2. \qquad (5)$$

Total cross sections are obtained by summing over final states and by averaging over degenerate initial states. Differential cross sections are sums of squared absolute values of state-to-state transition amplitudes, each expressed as a sum of T-matrix elements multiplied by angular coupling coefficients.

1.3 Close-coupling equations

Close-coupling equations are coupled integrodifferential equations obtained by projecting the N + 1-electron Schrödinger equation onto specified target states, which may correspond either to open or closed scattering channels. If these target states are represented explicitly in Eq.(1), the Hamiltonian matrix can be reduced by partitioning to produce an effective Hamiltonian $\tilde{H}_{PP}$ acting only in the subspace of target state functions. This procedure is equivalent to the variational condition that the projection of the Schrödinger equation onto the space of correlation functions should vanish. This effective Hamiltonian contains a matrix optical potential.[2] The resulting close-coupling equations at total energy E are

$$\sum_q (\Theta_p | \tilde{H}_{PP} - E | \mathscr{A} \Theta_q) \psi_{qs}(\varepsilon_q; \underset{\sim}{r}) = 0, \qquad (6)$$

for all values of index p. Here ε_q is a channel energy. For large r, these equations reduce to a system of coupled ordinary differential equations containing static multipole potentials and asymptotic forms of multipole polarization potentials, expressed by constants times spherical harmonics divided by powers of r.

2. METHODS FOR FIXED-NUCLEI SCATTERING

2.1 Subdivision of regions

The interaction of a scattered electron with a molecule is characterized by three distinct ranges of distance r from the center of target mass or charge. In region (i), inside the molecular radius r_0, the scattered electron is part of an N + 1-electron system. Exchange and correlation must be treated as for a bound state. In region (ii), between r_0 and an R-matrix radius r_1, asymptotic multipole potentials become valid, but exchange integrals cannot be neglected. In region (iii), outside r_1, asymptotic potentials are valid and exchange can be neglected, reducing the computational problem to solution of a set of coupled ordinary differential equations. Vibronic coupling is strong in region (i), but rotation can be neglected. In region (iii), potentials are not large compared to rotational level spacings. This requires a transformation from the body frame (BF) to the laboratory frame (LF) somewhere in region (ii). Dynamical effects of target polarization response may be important in region (ii), but can be modeled by a static polarization potential in region (iii).

2.2 Analytic methods for the asymptotic region

If the wave function computed in regions (i) and (ii) is used to construct the R-matrix at r_1, defined as the multichannel generalization of the reciprocal of a radial logarithmic derivative, analytic methods can be used to carry the solution out to infinite r. The coupled ordinary differential equations in region (iii) can be solved in principle by an asymptotic expansion, in the form of inverse powers of r multiplying standard sin/cos functions,[5] or multiplying Coulomb/Bessel functions.[6] These expansions define divergent series, but can be evaluated analytically by converting the series to continued fractions.[7] If the Gailitis radius r_G, inside which this purely analytic procedure loses accuracy, is greater than the R-matrix radius r_1, then it is accurate and efficient to extend the R-matrix out to r_G by the R-matrix propagator method.[8,9]

2.3 ADW and Born approximations for high-order partial waves

It is inefficient and unnecessary to use basis orbitals with high angular quantum numbers in regions (i) and (ii). If $\ell + 1/2$ is large compared with kr_0, an orbital wave function is inside the classical turning radius in region (i) and can be neglected. In this case, interaction with the target molecule is described by asymptotic multipole potentials. For high ℓ-values, scattering matrix elements can be computed accurately by the first Born approximation (FBA). For lower values, or if off-diagonal multipole potentials are strong, analytic methods can be applied to the multichannel problem, using these multipole potentials in the asymptotic distorted-wave method (ADW),[10] to complement low-ℓ terms computed directly.

2.4 Static and dynamic polarization

A target molecule is polarized by the electric field of a scattered electron. The response due to dipole polarizability α lowers total system energy by $-\alpha/2r^4$ for large r. Polarization response is a correlation effect, mutual virtual multipole excitation of the target and the incident electron. A compact formulation of this effect uses polarization pseudostates of the target. A pseudostate is the normalized state function proportional to the first-order multipole perturbation of a given target state. This dynamical treatment of polarization response is required only for small r. In region (iii) only static polarization potentials are relevant.

2.5 Static models

Many early calculations of electron-molecule scattering used static models of both polarization and exchange.[1] Until recently, this was the only feasible method for polyatomic targets.[3] For low-energy scattering, it has been very difficult to define a satisfactory local exchange potential. As computational power becomes more easily available, the direct solution of integrodifferential close-coupling equations with nonlocal exchange has become possible. Recent calculations by this method, still using model potentials, give low-energy results for polyatomic target molecules that are in good agreement with recent experimental data. The improvement over prior theoretical results is especially striking for Ramsauer minima, for methane and other small polyatomics.[11]

2.6 Hulthén-Kohn variational method

The Hulthén-Kohn variational method, for the real symmetric K-matrix, was applied effectively to electron-atom scattering. This formalism has been described in detail elsewhere.[2] The great practical advantage of this methodology is that it makes use of matrix elements of the N- or N+1-electron Hamiltonian in a local orbital basis exactly in the form implemented for accurate calculations in atomic physics and quantum chemistry. The major difficulty is that two-electron integrals must be evaluated in a mixed basis combining the usual short-range Slater or Gaussian orbitals with continuum functions, asymptotically Coulomb/Bessel functions. There is no efficient analytic method to evaluate these integrals for molecules.

A secondary problem, solved or bypassed in various ways in electron-atom scattering calculations, is that the Kohn variational principle has anomalies that occur at energy values which depend on the choice of basis representation. This problem has been bypassed recently by computing the complex-valued S-matrix or T-matrix, rather than the K-matrix, moving the energy-anomaly into the complex plane. A recent implementation of this 'complex Kohn' method has been combined with powerful direct quadrature procedures, exploiting the efficiency of modern computers, to produce a viable method for scattering by polyatomic target molecules.[12]

2.7 Schwinger variational method

The one-electron model Schrödinger equation is equivalent to an integral equation, the Lippmann-Schwinger equation, which builds in asymptotic boundary conditions by incorporating an appropriate Green function integral operator. In the Schwinger variational principle, derived elsewhere and extended to a multichannel formalism,[2] variational trial functions are used only in the nonzero range of the potential function. If the Green function corresponds to the asymptotic close-coupling equations, the residual potential function is short-range, and only localized orbital basis functions are required. A fundamental difficulty is that integrals of the operator product VGV must be evaluated, where G is the Green function and V is the residual potential, which may itself be nonlocal. It has not been found feasible to evaluate these integrals if long-range potentials are included in the Green function. Because of this, the inherent power of this formalism has been difficult to exploit. The Schwinger Multichannel (SMC) method, implemented for polyatomic molecules by McKoy and collaborators,[13] is an alternative version that uses the free-electron Green function.

2.8 R-matrix theory

The Wigner-Eisenbud R-matrix was adapted to the long-range potentials characteristic of electron scattering by Burke and collaborators, with many applications to electron-atom scattering,[14] and more recently to electron-molecule scattering. Because this formalism separates short-range and long-range scattering, it takes full advantage of the efficient analytic methods available in the asymptotic region, while exploiting standard atomic physics and quantum chemistry methodology for the correlated N + 1-electron problem in the inner region. For efficient convergence of the variational version of this method, mixed orbital basis sets of standard localized functions and of oscillatory continuum functions must be used. Efficient analytic methods are not available to evaluate the required two-electron integrals, truncated at the R-matrix radius r_1, for Gaussian-continuum mixed basis sets. Applications to polyatomic molecules have been limited by this technical problem. It has recently been found that continuum functions, inside r_1, can be expanded efficiently as linear combinations of Gaussian basis functions. This may remove the computational bottleneck.

Following earlier variational derivations,[2,15] the standard formalism in spherical geometry has been generalized to a variational principle valid for an enclosed space of arbitrary shape.[16] A linear integral operator $\mathcal{R}$ is defined over the surface σ of an enclosed volume τ such that, for regular solutions ψ of the Schrödinger equation in τ at specified energy ε,

$$\psi(\sigma) = \int \mathcal{R}(\sigma, \sigma')\nabla_n \psi(\sigma')d\sigma'. \tag{7}$$

The stationary value of the kernel of $\mathcal{R}$ in a basis of linearly independent functions ϕ_α, regular in τ, is

$$\mathcal{R}(\sigma, \sigma') = \frac{1}{2} \sum_{\alpha\beta} \phi_\alpha(\sigma) A^{-1}_{\alpha\beta} \phi^*_\beta(\sigma'). \tag{8}$$

Matrix $A_{\alpha\beta}$ here is a Bloch-modified Hamiltonian matrix in which kinetic energy terms have been integrated by parts to give the Hermitian form $\frac{1}{2}\nabla\phi^*_\alpha \bullet \nabla\phi_\beta$. The Wigner-Eisenbud R-matrix is derived from the $\mathcal{R}$ kernel on a sphere by projecting onto spherical harmonics and dividing by the sphere radius. For polyatomic molecules of arbitrary shape, the variational $\mathcal{R}$ kernel can be fitted to asymptotic continuum functions on a bounding ellipsoid or polyhedron to determine the standard fixed-nuclei scattering matrices.

3. ADIABATIC METHODS FOR ROVIBRATIONAL COUPLING

3.1 Rovibrational close-coupling

The rovibrational close-coupling equations take the same form as Eq.(6), if rotational and vibrational kinetic energy operators are included in the effective Hamiltonian, and if the target states are generalized to rovibronic functions. The resulting coupled equations contain rovibronic matrix elements of the multipole potentials of the target molecule.[17] The rovibronic sublevels are so numerous that the number of coupled scattering channels prohibits direct solution of these equations in most applications.

3.2 Adiabatic theory, EMA

Vibronic close-coupling equations and the corresponding fixed-nuclei equations are equivalent if the electronic scattering energy parameter ε in the fixed-nuclei equations is replaced by an operator in the nuclear vibration coordinates (q),[18]

$$\varepsilon_{op} = E - H_q. \tag{9}$$

The fixed-nuclei electronic wave function, parametrized by (ε, q), becomes an operator in the nuclear coordinates. Since the asymptotic form of the electronic wave function is expressed by standard continuum functions multiplied by a coefficient scattering matrix, this matrix becomes an operator in the nuclear coordinates. Matrix elements of this operator, in a basis of rovibrational wave functions, define the rovibrational scattering matrix. If ε_{op} is treated as a c-number parameter, rovibrational matrix elements of the fixed-nuclei scattering matrix define the standard adiabatic approximation to the corresponding scattering matrix.[19] This approximation is valid when the scattering energy is large compared to rovibrational level spacings.

In the energy-modified adiabatic approximation (EMA),[18] ε_{op} is treated as an operator, introducing simplifying assumptions for particular applications. Considering the vibronic K-matrix, in the absence of a scattering resonance the fixed-nuclei matrix $K(\varepsilon,q)$ is converted to a vibronic matrix $(\mu|K(\varepsilon_{\mu\nu}, q)|\nu)$, assuming slow variation with energy, replacing ε_{op} by $\varepsilon_{\mu\nu}$, the geometric mean of $(E - E_\mu)$ and $(E - E_\nu)$. This is justified for diagonal elements, since ε_{op} has the quantized value $E - E_\mu$ in vibrational state μ. The geometric mean formula is qualitatively correct at threshold for short-range potentials.

3.3 Phase matrix methods

Since powerful analytic methods are available to solve the asymptotic rovibrational close-coupling equations, it is efficient to convert the fixed-nuclei R-matrix directly to a vibronic or rovibronic R-matrix. With a suitable choice of basis functions, the R-matrix is real and symmetric, but it necessarily has poles as a function of the fixed-nuclei scattering energy parameter ε. Energy dependence cannot be considered to be weak near such a pole. For this reason, it is preferable to convert the R-matrix to the corresponding phase matrix, defined by $\tan \Phi = R$, to adjust eigenphases by multiples of π to assure smooth interpolation, and then to use the EMA method to compute a vibronic Φ-matrix. This process defines the EMAP method. It is complemented by direct application of the ADW method to the vibronic close-coupling equations in region (ii), out to the R-matrix radius. EMAP and vibronic ADW (VADW) combine to produce the vibronic R-matrix.

3.4 MEAN and Born closure

Given the vibronic R-matrix, the body-frame(BF) K-matrix is computed by fitting to asymptotic vibronic close-coupling solutions at the R-matrix boundary. The 'multipole-extracted adiabatic-nuclei' method (MEAN) of Norcross and Padial computes total and differential cross sections in the laboratory frame(LF), starting from the BF vibronic K-matrix.[20] Angular-momentum transfer analysis is used to modify the FBA LF differential cross section, indexed by rovibrational channels, by removing its lowest order angular momentum terms, replaced by the corresponding partial-wave terms computed from the BF vibronic K-matrix. This Born-closure procedure makes use of the rotational adiabatic approximation to convert angular momentum transfer matrix elements from vibronic to rovibronic representation.

4. NONADIABATIC ROVIBRATIONAL METHODS

Qualitative failure of the standard adiabatic approximation in the threshold energy range is corrected by the EMA method, but important problems remain. In particular, because the EMA formula for off-diagonal elements is not quantitative, the magnitudes of excitation cross sections are not well defined. Moreover, if off-diagonal multipole potentials are large, close-coupling equations lead to indirect or higher-order channel coupling that is not correctly represented by FBA or EMA matrix elements. Nonadiabatic theoretical methods are needed for these reasons.

4.1 Rovibrational close-coupling with EMAP

The EMAP vibronic R-matrix can be transformed to the LF(vjl) representation at the R-matrix boundary and matched to solutions of the rovibrational close-coupling equations. This rotational frame transformation is described in detail by Chandra.[21] LF Born closure is used.

4.2 Vibronic R-matrix theory

A general theory of vibrational excitation and dissociative attachment was derived from R-matrix theory for the vibronic Hamiltonian by Schneider et al.[22] The variational principle for the $\mathscr{R}$ operator, in the 4-dimensional hyperspace of coordinates (r,θ,ϕ,q), provides an alternative derivation. Here q is a mass-weighted vibrational coordinate of the form $M^{1/2}(R-R_e)$ for reduced mass M. The enclosed volume τ is the hypercylinder defined by $q_0 \leq q \leq q_1$ and $0 \leq r \leq r_1$, for all values of electronic coordinates (θ,ϕ). Trial wave functions are assumed to vanish at $q = q_0$ and to be regular at $r = 0$. The outer hypersurface consists of two sectors: σ_q, on which $q = q_1$ and $\nabla_n = \partial/\partial q$; and σ_r, on which $r = r_1$ and $\nabla_n = \partial/\partial r$. The $\mathscr{R}$ kernel is given by Eq.(8), if the vibronic A-matrix includes Bloch-modified vibrational kinetic energy terms $1/2(\partial\phi_\alpha^*/\partial q)(\partial\phi_\beta/\partial q)$. The generalized $\mathscr{R}$-matrix is obtained by projecting $\mathscr{R}$ onto $\chi_v(q)Y_{lm}(\theta,\phi)$ on σ_r and onto $\phi_d(r,\theta,\phi)$ on σ_q, where $\phi_d(\underset{\sim}{r})$ is an N+1-electron wave function for a dissociating state or stable negative ion. The resulting submatrices are $\mathscr{R}_{vlm,v'l'm'}$, vibronic excitation; $\mathscr{R}_{vlm,d}$, associative detachment; $\mathscr{R}_{d,vlm}$, dissociative attachment; and $\mathscr{R}_{d,d}$, ion-atom collision.

4.3 Method of Schneider et al

The proposal is to use Born-Oppenheimer basis states,[22]

$$\phi_{kp}(\underset{\sim}{r},q) = \psi_k(\underset{\sim}{r},q)\theta_{kp}(q), \tag{10}$$

where

$$\{T_e + V_{qe}\}\psi_k = \psi_k E_k(q), \tag{11}$$

$$\{T_q + V_q + E_k(q)\}\theta_{kp} = \theta_{kp}\varepsilon_{kp}. \tag{12}$$

Schneider et al select eigensolutions of these equations that have zero normal gradients on the boundary surfaces. This boundary condition characterizes poles of the R-matrix. Calculations of vibrational excitation in N_2 showed the well-known 'boomerang' structure of overlapping resonances.[23]

This procedure requires, in principle, a complete set of vibrational functions θ_{kp} for each electronic eigenstate ψ_k, including a vibrational continuum. This requirement is difficult to implement, and convergence of a truncated expansion is difficult to assess. The choice of R-matrix pole states is unnatural from the point of view of the variational theory. As a variational basis, these states are complete only for functions that satisfy the same imposed boundary condition, which is correct for the R-matrix only at the pole energies. In practice, the basis expansion is completed at each output energy by a Buttle correction.[24] The variational derivation shows that vibronic basis functions must be linearly independent, but not constrained on the boundary hypersurface.

4.4 Vibrational frame transformation

'Vibrational frame transformation' theory provides an alternative method of converting a fixed-nuclei K-matrix into the vibronic K-matrix.[25] For different values of the vibrational coordinate q electronic states are associated by requiring them to have a common K-matrix eigenphase. This determines a set of energy functions $\varepsilon_{\alpha\gamma}(q)$ for each eigenchannel α. Vibrational eigenstates and total eigenenergies are computed from the vibrational Schrödinger equation with Hamiltonian $H_q + \varepsilon_{\alpha\gamma}(q)$. The selected eigenphase must be varied so that some energy level v′ of this equation coincides with the specified total energy E. Calculations of vibrational excitation in N_2 produced the well-known overlapping resonance structure.

This formalism is inherently computationally inefficient. A distinct set of vibrational eigenfunctions is required for each index pair $\alpha\gamma$, and these calculations must be iterated in order to satisfy the energy-matching condition. Use of the fixed-nuclei K-matrix in this procedure fails to exploit the efficient analytic methods currently available to integrate vibronic or rovibronic close-coupling equations in the asymptotic region. Moreover, because molecular rotation suppresses nonisotropic long-range potentials, in particular a static dipole potential, it is in principle incorrect to compute the fixed-nuclei K-matrix without taking molecular rotation into account.

4.5 Nonadiabatic vibronic phase matrix method

For reasons stated above, it is desirable to develop a nonadiabatic method to convert a given fixed-nuclei R-matrix into the vibronic R-matrix. A nonadiabatic method (NADP), using the phase matrix Φ, has recently been implemented.[26] A precursor resonance state is defined within the R-matrix radius when for fixed vibrational coordinate q the energy-derivative of the trace of the Φ-matrix has a relatively sharp maximum as a function of scattering energy ε. Such a maximum, traced over the range of q, defines an energy function $\varepsilon_{res}(q)$. In order to separate Φ into a resonant part Φ_1 and a background part Φ_0, a matrix eigenvalue problem is solved in the form[27]

$$\frac{d\Phi}{d\varepsilon}\Big|_{res} y = y\left(\frac{2}{\Gamma}\right), \tag{13}$$

where the maximum eigenvalue is to be selected. This determines the width function $\Gamma(q)$ and the unit projection vector y(q). The background matrix Φ_0 is defined by subtracting the Breit-Wigner matrix determined by the resonance values of y,ε,and Γ from the original Φ-matrix. This procedure is repeated until all overlapping resonances of the same global symmetry are removed. The residual Φ_0 is converted to a vibronic matrix by the EMAP method. The vibronic matrix corresponding to Φ_1 is computed from the matrix expression

$$(v|\Phi_1|v') = \tan^{-1}\sum_s (v|y\Gamma^{1/2}|s)\frac{1}{2(E_s - E)}(s|\Gamma^{1/2}y^{\dagger}|v'), \tag{14}$$

where index s refers to an eigenfunction of the effective vibrational Hamiltonian $H_q + \varepsilon_{res}(q)$. In principle, a complete set is required for each precursor resonance in the energy range of interest. The vibronic phase matrices are added and converted to a vibronic R-matrix as in the EMAP method.

REFERENCES

1. Lane, N.F., *Rev.Mod.Phys.***52**,29 (1980).
2. Nesbet, R.K., *Variational Methods in Electron-Atom Scattering Theory* (Plenum, New York,1980).
3. Gianturco, F. and Jain, A., *Phys.Rept.***143**,347 (1986).
4. Morrison, M.A., *Adv.At.Mol.Phys.***24**,51 (1988).
5. Burke, P.G. and Schey, H.M., *Phys.Rev.***126**,147 (1962).
6. Gailitis, M., *J.Phys.B.***9**,843 (1976).
7. Noble, C.J. and Nesbet, R.K., *Comput.Phys.Commun.***33**,399 (1984);
   Nesbet, R.K., *Comput.Phys.Commun.***32**,341 (1984).
8. Light, J.C. and Walker, R.B., *J.Chem.Phys.***65**,4272 (1976).
9. Baluja, K.L., Burke, P.G. and Morgan, L.A., *Comput.Phys.Commun.***27**,299 (1982).
10. Nesbet, R.K., *J.Phys.B.***17**,L897 (1984).
11. McNaughton, P., Thompson, D.G. and Jain, A., *J.Phys.B.***23**,2405 (1990);
    Jain, A., Invited lecture, XVII ICPEAC (1991).
12. Rescigno, T.N., McCurdy, C.W. and Schneider, B., *Phys.Rev.Lett.***63**,248 (1989);
    Rescigno, T.N., Invited lecture, XVII ICPEAC (1991).
13. Takatsuka, K. and McKoy, V., *Phys.Rev.A***24**,2473 (1981);
    Takatsuka, K. and McKoy, V., *Phys.Rev.A***30**,1734 (1984);
    Lima, M.A.P. and McKoy, V., *Phys.Rev.A***38**,501 (1988).
14. Burke, P.G. and Robb, W.D., *Adv.At.Mol.Phys.***11**,143 (1975).
15. Jackson, J.L., *Phys.Rev.***83**,301 (1951);
    Oberoi, R.S. and Nesbet, R.K., *Phys.Rev.A***8**,215 (1973); *Phys.Rev.A***9**,2804 (1974).
16. Nesbet, R.K., *Phys.Rev.B***30**,4230 (1984); *Phys.Rev.A***38**,4955 (1988).
17. Chandra, N. and Temkin, A., *Phys.Rev.A***13**,188 (1976).
18. Nesbet, R.K., *Phys.Rev.A***19**,551 (1979).
19. Chase, D.M., *Phys.Rev.***104**,838 (1956).
20. Norcross, D.W. and Padial, N.T., *Phys.Rev.A***25**,226 (1982);
    Fano, U. and Dill, D., *Phys.Rev.A***6**,185 (1972).
21. Chandra, N., *J.Phys.B.***8**,1338 (1975); *Phys.Rev.A***16**,80 (1977);
    Chang, E.S. and Fano, U., *Phys.Rev.A***6**,173 (1972).
22. Schneider, B., LeDourneuf, M. and Burke, P.G., *J.Phys.B.***12**,L365 (1979).
23. Schneider, B., LeDourneuf, M. and Vo Ky Lan, *Phys.Rev.Lett.***43**,1926 (1979).
24. Buttle, P.J.A., *Phys.Rev.***160**,719 (1967).
25. Greene, C.H. and Jungen, Ch., *Phys.Rev.Lett.***55**,1066 (1985);
    Greene, C.H., *Comments At.Mol.Phys.***23**,209 (1989).
26. Nesbet, R.K., Program RVRMAT, NRES option, unpublished (1991).
27. Smith, F.T., *Phys.Rev.***118**,349 (1960).

INNER-SHELL STUDIES OF MOLECULES AND SURFACES BY ELECTRON IMPACT

ADAM P. HITCHCOCK
Dept. of Chemistry and Institute for Materials Research,
McMaster University, Hamilton, Ontario, Canada L8S 4M1

ABSTRACT

A progress report is given on recent experimental studies of core excitation of free and surface adsorbed molecules by inelastic electron scattering. The use of inner shell electron energy loss spectroscopy (ISEELS) to investigate dipole and non-dipole excitations and to measure absolute electric dipole oscillator strengths are reviewed. Comparisons are made with valence shell EELS. A model of inner-shell spectra based on local structure is outlined. The spectra of organometallic compounds are discussed as intermediates between those of free molecules and those of surface absorbed molecules. Recent advances in the inner shell spectroscopy of chemisorbed molecules by reflection near-edge energy-loss fine structure (NEELFS) spectroscopy are described.

1. Introduction

Inner-shell excitation spectroscopy is being used increasingly for fundamental and applied studies of a wide range of materials, because of the very large growth in synchrotron radiation based experiments[1-3] as well as the development of efficient energy loss spectrometers coupled to transmission electron microscopes.[4] Inner shell electron energy loss (ISEEL) spectroscopy of free molecules has been carried out in several laboratories over the past 15 years.[5-9] This work is providing very useful information on the fundamentals of electronic excitations involving core electrons as well as model studies for more applied projects, carried out by X-ray absorption (XAS) or electron energy loss techniques. Recently it has been shown that reflection EELS can be used to study core excitation of sub-monolayer concentrations of chemisorbed atoms[10,11] and molecules.[12,13] The present capabilities and future expectations for this technique are discussed. While a number of groups - Brion (Vancouver), Stefani (Rome), King (Manchester), Moore (Maryland), de Souza (Rio de Janeiro) - are actively measuring ISEELS of gases, most of the examples presented in this article are taken from work done at McMaster simply for convenience. I apologize for not giving adequate coverage to the important contributions of other groups.

2. Experimental Aspects

Fig. 1 presents a schematic of the inner-shell electron energy loss spectrometer (ISEELS) at McMaster that is used for survey studies of core excitation in gas phase molecules.[8] This spectrometer uses fast incident electrons (>2500 eV impact energy), small scattering angles (2°), and an unmonochromated electron beam which limits the ultimate resolution to 0.4 eV (typically it is operated at 0.6 eV fwhm in order to increase signal). Even so, the resolution of this instrument competes favourably with that of many current synchrotron radiation soft X-ray monochromators. A distinct advantage of electron impact over optical spectroscopy in the area of core excitation is that the resolution is constant over all energy losses when deceleration is used. The high resolution ISEEL spectrometer at the University of British Columbia (UBC)[14] employs a monochromated incident electron beam and routinely achieves better than 65 meV fwhm instrumental resolution for any core excitation below 1000 eV. The resolution

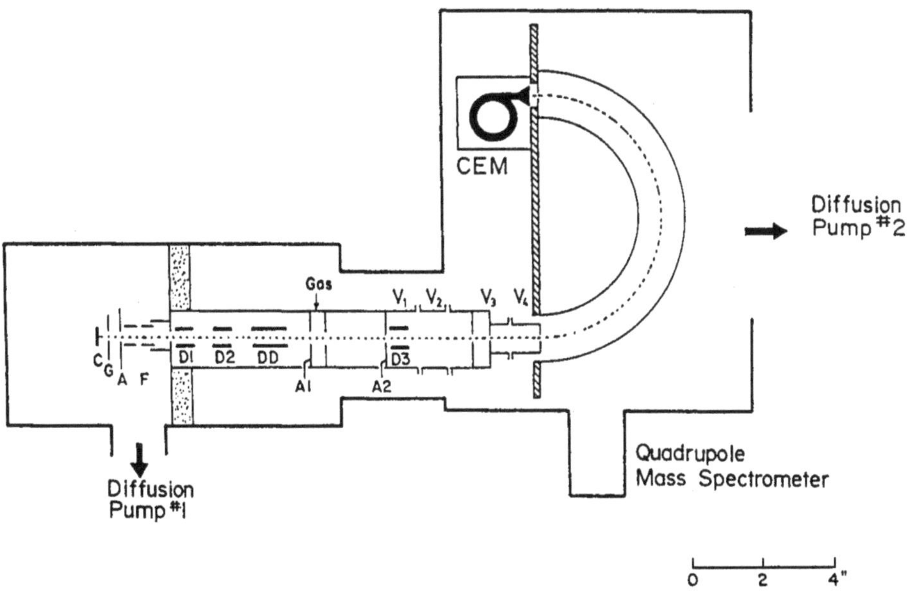

Fig. 1 Schematic of the electron spectrometer at McMaster University used to record inner shell electron energy loss spectra (ISEELS) of gases. Typical operating conditions are 2.5 keV final electron energy, 2° scattering angle (achieved by electrostatic deflection with D2 and DD, prior to the gas cell); 1-10 μA incident beam current, 40 eV analyzer pass energy and 0.6 eV fwhm overall energy resolution. The residual gas analyzer (RGA) is used to check sample purity.

advantage, that until recently has been enjoyed by ISEELS, is currently in favour of photoabsorption since extremely high resolution in the soft X-ray (< 50 meV) with high signal rates has become available at several soft X-ray beam lines in the past 3 years.[15,16] High resolution is always of interest in all spectroscopies in order to distinguish closely spaced spectral features. However the intrinsically short lifetimes of core hole states[17] limit the intrinsic linewidths to 50 meV or greater for core states above 100 eV. Thus state-of-the-art ISEELS and XAS instruments are now capable of recording all potentially-resolvable core excitation features.

3. Fundamentals of Inelastic Electron Scattering for Inner-shell Excitation

While the majority of ISEELS studies of gases have been carried out under electron scattering conditions which favour electric dipole transitions, the technique is also capable of exciting states not accessible by single photon absorption. For fast incident electrons the relationship between scattering conditions and effective selection rules can be understood through the Bethe-Born theory of inelastic scattering.[18,19] This expresses the cross-section measured at a given momentum transfer (K) in terms of a generalised oscillator strength, (GOS, df(K)/dE for continuum states) which is related to the measured differential cross-section ($d\sigma/d\Omega$) for inelastic scattering at energy loss, E into solid angle, Ω by:

$$df(K)/dE = (E/2)(k_o/k_1)K^2(d\sigma/d\Omega)$$

Here k_o, k_1 and K are the incident, scattered and transferred momenta ($K^2 = k_o^2 + k_1^2 - 2k_ok_1\cos\theta$). Within the first Born approximation the GOS can be expanded in terms of K^n-weighted electric multipole matrix elements:

$$df(K)/dE = df_0/dE + (\epsilon_2 - \epsilon_1.\epsilon_3).K^2 + B.K^4 + \ldots$$

where df_0/dE is the (continuum) optical oscillator strength (OOS) containing the electric dipole matrix element (ϵ_1) and B contains further higher order multipole matrix elements. This expression indicates that the GOS converges to the OOS in the limit of zero momentum transfer (achieved at $\theta=0°$ and infinite impact energy). The contributions of non-dipole excitations increase with increasing momentum transfer which occurs with larger scattering angle or lower impact energy. The momentum transfer under typical dipole-dominated ISEELS conditions (1-3 keV, $0<\theta<2°$) is actually quite significant, amounting to 1.8 Å^{-1} at 300 eV (C 1s), 2.2 Å^{-1} at 410 eV (N 1s) and 2.8 Å^{-1} at 540 eV (O 1s) for 2.5 keV final electron energy and 2° scattering angle. At these K-values the

K^2 term containing the quadrupole matrix element is given a greater weight than the dipole term (df_o/dE) in the Bethe-Born expansion of the GOS. In principle, one needs to measure energy loss spectra at a range of momentum transfers and extrapolate to zero K in order to obtain accurate OOS from GOS. This has been done in a few cases for inner shell excitation.[20] However in many cases it suffices to work at sufficiently small-K that the GOS becomes a good approximation to the OOS. This turns out to be true for inner-shell excitation even under scattering conditions where the momentum transfer is quite large. This fortunate situation arises because core orbitals are spatially small. Due to the influence of the r^n-weighting on the overlap of the initial and final state wavefunctions contained in the transition matrix element, the r^1-weighted dipole transition moment ($<1s|er|V^*>$) is generally much larger than the r^2-weighted quadrupole transition moment ($<1s|er^2|V^*>$) (V^* is the virtual valence orbital to which the core electron is excited).[5] df_0/dE is expected to dominate the GOS up to $K \approx 1/R_c$, where R_c is the average radius of the core wavefunction.[21] Thus for example, non-dipole transitions in O 1s excitation are expected to be negligible until $K > 10$ Å^{-1}. Certainly direct comparisons of molecular core spectra recorded by ISEELS and true photoabsorption are generally in good agreement over a wide range.[22,23] A recent example of such a comparison for quite a complex molecule[23] is presented in Fig. 2. It should be noted that, while there is a good match between photoabsorption and ISEELS in the majority of spectra recorded under "dipole scattering conditions" (E_o > 1 keV; θ <5°), there are a few, very interesting cases where inner-shell electric quadrupole transitions have been identified through comparisons between low-K ISEELS and optical studies.[24,25]

When the incident electron energy is no longer large relative to the transition energy, there are strong interactions between the target and the scattered electron so that the first Born approximation no longer holds and the above treatment can no longer be applied. At large scattering angles (>5°) and intermediate energies (2-5 times the transition energy) quadrupole transitions become increasingly important. As the impact energy approaches the threshold for the electronic transition the probability of an "exchange" between the incident and a target electron increases. This leads to the detection of spin-exchange transitions. Such studies can be particularly useful for distinguishing valence and Rydberg core-excited states[26] since an appreciable singlet-triplet splitting only occurs for excitations to compact virtual valence orbitals which have good spatial overlap with the core. An example of the detection of triplet states in the C 1s spectra of CO and $Fe(CO)_5$ is presented in Fig. 3.

Based on the Bethe-Born theory[19], dipole-regime ISEEL spectra can be converted to approximate optical oscillator strengths. A systematic procedure for this is now used routinely at McMaster.[27] This involves background subtraction to isolate signal from a single core edge, kinematic and transmission function correction, followed by normalization to calculated atomic oscillator strengths[28] in the far continuum (>25 eV above the IP), where molecular resonance effects are negligible. This procedure allows systematic quantitative comparison of spectra from different core edges in the same or different molecules. Although the same normalisation procedure can be applied to soft X-ray photoabsorption spectra, in many cases the derivation of accurate absolute intensities is more difficult with photon spectra than with electron impact spectra. A number of factors can be involved. In certain spectral regions problems arise with accurate measurement of the incident photon flux. This frequently occurs in the C 1s region (280-300 eV) where absorption of the X-rays by carbon contamination on optical surfaces is often so strong that stray light (10-100 eV) and higher order components (600, 900 eV ..) dominate the photon flux delivered by the monochromator, resulting in greatly distorted spectra. When recording in a true absorption mode, it is often very difficult to use a pressure which is sufficiently low that the most intense features are not affected by

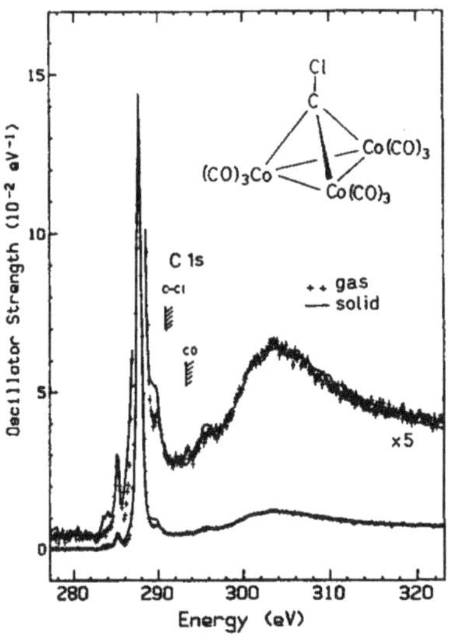

Fig 2 Comparison of C 1s optical oscillator strength spectra of chloro-alkylidyne tricobalt nonacarbonyl derived from gas phase ISEELS and X-ray absorption (XAS) of the solid, recorded by total electron yield.[23] The excellent agreement indicates that electric dipole transitions dominate ISEELS under the scattering conditions described in Fig. 1.

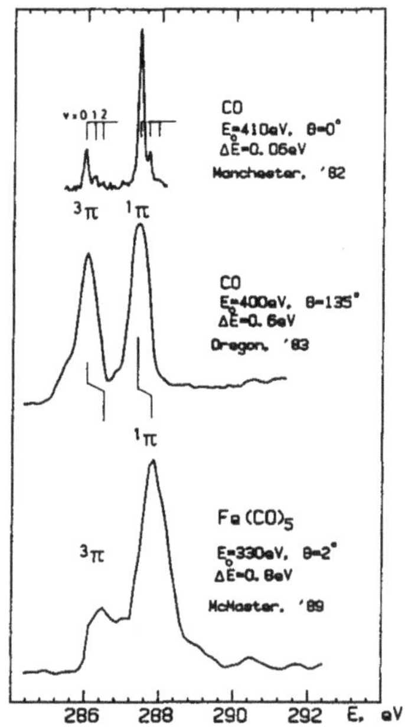

Fig. 3 ISEELS near threshold detects spin exchange transitions. In the top curve vibrational structure is resolved with a monochromatised incident beam[72] even under low cross-section near threshold conditions. The middle spectrum illustrates that spin-exchange is greatly enhanced at large scattering angles.[73] The lower curve illustrates that $^1\pi$-$^3\pi$ separations are chemically sensitive - the value decreases by 0.15 eV when CO is bound to Fe.

absorption saturation and yet still have enough pressure for adequate statistical precision to detect weak features. Ion or electron yield detection is frequently used to minimize the absorption saturation problem. However there are fundamental differences between yield and true absorption spectra although these are often ignored. A major factor is an energy and/or state dependence of the degree of multiple ionisation. Since higher degrees of ionisation produce more ions and electrons, the absorption spectra are distorted in favour of multiply ionized states. These effects have been placed on a quantitative basis in several recent studies.[23,29]

4 Comparison of Valence- and Inner-shell Excitation by Electron Impact

Fig. 4 presents a comparison of the valence and inner-shell electron energy loss spectra of norbornene, methylnorbornene and trifluoromethylnorbornene.[30] It is clear that the C 1s edge spectra show significantly larger variation among these three molecules than the valence shell spectra recorded at the same resolution. Of course the availability

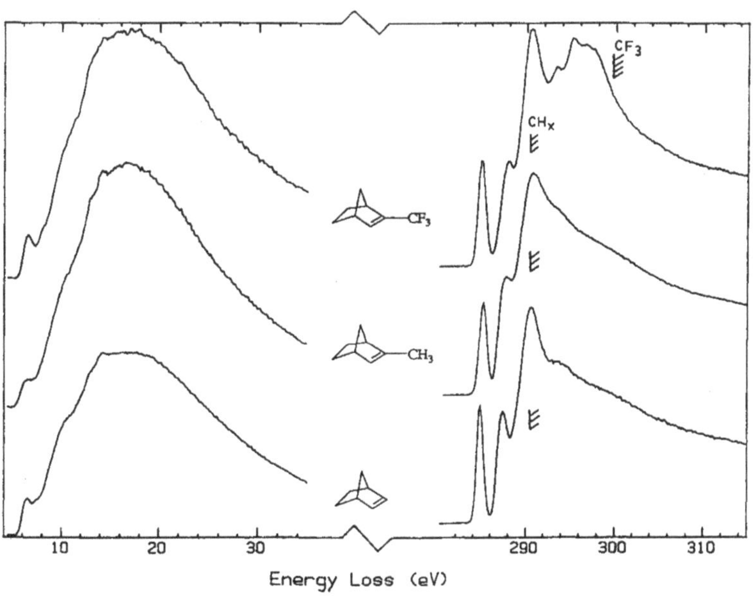

Fig. 4 Comparison of the energy loss spectra of three norbornenes[30] to demonstrate that inner-shell (ISEELS) has significant advantages over valence shell (VSEELS) spectroscopy for characterising large molecules. The valence region shows only subtle changes in the $\pi\text{-->}\pi^*$ band at 6.5 eV while the C 1s region is much more structured and chemically sensitive.

of extremely high resolution optical techniques below 10 eV means that, for those systems with stable valence excited states, rotational and even nuclear spin energy levels can be resolved, providing a wealth of information that core spectroscopy will never be able to achieve. However there are many systems where the high resolution structure is very complex[31] or where there are only band spectra because of (pre)dissociation. In these cases core spectra are probably superior to valence spectra for molecular fingerprinting.

5. Absolute Oscillator Strengths for Core Excitation

A valuable aspect of the systematic derivation of absolute oscillator strengths (OS) from core excitation spectra is the ability to compare intensities of transitions to a common final orbital from different core levels in a single molecule and to compare corresponding transitions in related molecules. For example, 1s-->π^* oscillator strengths have been used to map the spatial distributions of π^* orbitals in the substituted carbonyls, $HCONH_2$, HCOOH and HCOF.[32] The results of this study are presented in Fig. 5, in comparison to orbital distributions calculated by HAM/3 semiempirical quantum calculations.[33] Both the calculations and experimental OS indicate that the π^*(C=O) orbital in all three species has a larger density on carbon than oxygen and about 6% contribution from X $2p_z$, (X=N,O,F).[32]

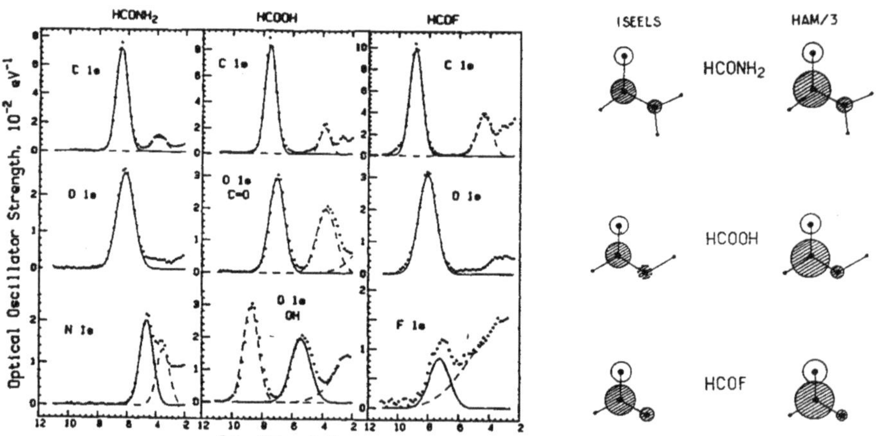

Fig. 5 Example of determinationm of an orbital spatial distribution from experimental oscillator strengths.[32] **left**: the 1s-->π^* region of the C 1s, O 1s and X 1s (X = N,O,F) spectra of formamide, formic acid and formyl fluoride. **right**: experimental and calculated (HAM/3) estimates of the shape of the π^*(CO) orbital. The circle diameters are proportional to $f_{\pi^*}/(0.185E\langle X2p|\mu|X1s\rangle)$. Experiment and calculation predict similar trends in π^*(CO) delocalisation onto the X atom.

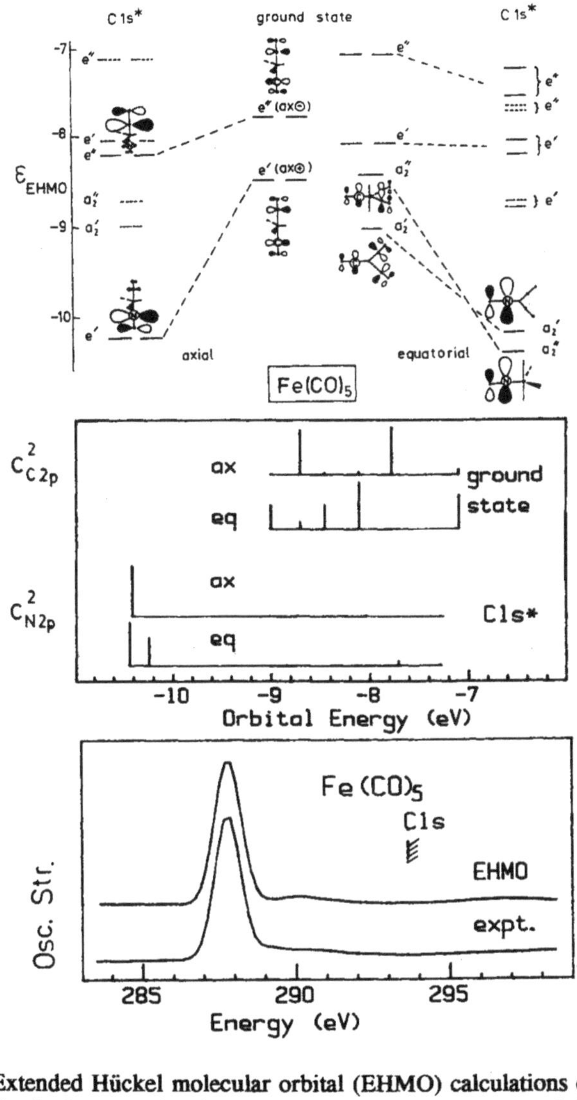

Fig. 6 Extended Hückel molecular orbital (EHMO) calculations of the π^*(CO) orbitals of $Fe(CO)_5$ in the ground and C 1s excited state. The upper panel sketches the MO spatial distributions for the axial and equatorial components. The centre panel plots the π^*(CO) energies against the 2p LCAO coefficients at the C (g.s.) or N (C1s*). Note the stabilisation and spatial localisation of the π^*(CO) orbital upon C 1s excitation. The lower panel compares the EHMO simulated spectrum (aligned at the π^* resonance) to the experiment spectrum.

Recently we have found that extended Hückel molecular orbital (EHMO) calculations[34] can be a particularly versatile tool to assist spectral interpretation, particularly for compounds containing metal atoms since EHMO is one of the few applicable packages. Comparisons with experimental spectra using as realistic as possible spectral simulations based on EHMO, are an important part in gaining confidence in the accuracy and thus relevance of the EHMO results. In carrying out these calculations the often strong influence of the core hole (assumed to be localised[35,36]) is taken into account by a combination of the equivalent core (Z+1) approximation and an additional valence charge. This equivalent ionic core virtual orbital approximation (EICVOM) was first proposed by Schwarz.[37] The positive charge accounts for additional relaxation of the valence electron distributions in the excited states. The energies and LCAO coefficients of the virtual orbitals of the cation approximate the term values and intensities of core -> valence transitions. An illustration of the importance of core hole relaxation and how well EHT simulates this effect is given for $Fe(CO)_5$ in Fig. 6. Several other applications of EHT to the interpretation of ISEEL spectra of organometallic compounds have been published recently.[38-40]

6. **Relationships between Core Spectra and Molecular Structure**

During the 1980's empirical rules relating the positions of core-- > σ^* excitations and bond lengths to adjacent atoms have been proposed, tested[41,42], explained from first principles[43,44], criticised[45] and defended.[46] While the correlation has notable problems in molecules with delocalised electronic structure (e.g. CO_2 and aromatic compounds),[46] it does provide a useful guide to understanding the inner-shell spectra of many compounds and has been found useful for predicting intramolecular bond lengths in surface adsorbates with an estimated accuracy of 0.05 Å.

In many cases a rather good treatment of the core excitation spectrum of a complex molecule with several different functional groups can be obtained by considering it as the weighted sum of the core spectra of the constituent parts. This assumes that core excitation is dominated by the local structure surrounding the core excited atom, as predicted by the bond length correlation as well as the occurrence of strong, low-lying π^* resonances at unsaturated centres. This is reasonable except in cases with extensive delocalisation and mixing of the electronic structure of the individual functional groups. This "building block" approach has been outlined and tested by Outka and Stöhr,[47] who have also characterised some of the exceptions associated with delocalisation.[48]

An area where the "building block" approach is expected to be useful is in modelling core spectra of polymers through small molecule analogs. An example where ISEEL spectra of small molecules are combined to help understand the C 1s and O 1s spectra of polyethylene terephthalate (recorded by transmission electron microscopy (TEM-EELS)[49] is given in Fig. 7.

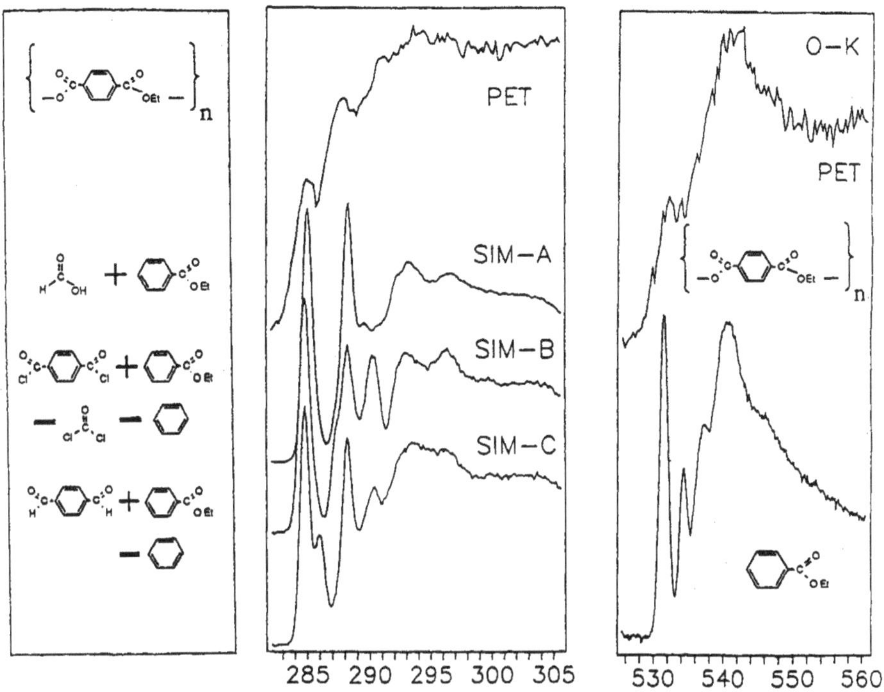

Fig. 7 Application of molecular ISEELS to aid polymer analysis. The upper curves are the C 1s and O 1s spectra of polyethlyeneterephthalate (PET) recorded with parallel-EELS in a transmission electron microscope (TEM). The lower curves are simulations based on the indicated linear combinations of small molecule spectra which approximate the monomer structure of PET.[49]

7. Core Excitation of Organometallic Compounds: Models for Chemisorption

In almost all cases studied to date, the positions of core-excitation features of condensed or surface adsorbed molecules are essentially identical to those in the ISEEL or X-ray absorption spectrum of the gas phase molecule, provided the molecular structure is unchanged in the condensed state.[50-53] The independence of the excitation energy on phase (gas, solid or monolayer) can be explained by the self-shielding character of core-excited states[54,55] - i.e. the bulk of the shielding of the core hole is accomplished by the core excited electron so that the valence electron distribution is largely undisturbed. This occurs even for very short lived (1s,σ^*) continuum resonances, whose lifetime is typically $< 10^{-16}$ s, based on the resonance width.

The paucity of visible changes to core excitation spectra associated with the chemisorption bond has been a surprising aspect of comparisons between the NEXAFS spectra of chemisorbates and gas phase molecular core excitation spectra.[50-52] In some cases additional weak shoulders and pre-peaks have been noted in the NEXAFS spectra which may be excitations to unoccupied metal-adsorbate antibonding orbitals. We are currently carrying out systematic studies of the core excitation spectroscopy of organometallics in order to investigate the effects of metal-ligand bonding on unoccupied MOs as probed by core excitation. So far results on carbonyls,[55-58] metallocenes[38] and $CpCo(CO)_2$ [59] have been published. While this work was initiated in order to better understand the spectroscopy of organic chemisorbates on metal surfaces, it is turning out to be of considerable interest as a probe of the electronic structure of the complexes themselves.

An interesting aspect is studies of the influence different ligands have on each other when combined in the same complex. Deviations between simulations based on the experimental spectra of pure ligand complexes, and the experimental spectrum of mixed ligand complexes can be used to investigate charge shifts among ligands. An example for $CpCo(CO)_2$ is shown in Fig. 8.[59] The deviations between the simulation (based on the C 1s spectra of $CoCp_2$ and $Co_2(CO)_8$) and the experimental C 1s spectrum of $CpCo(CO)_2$ indicate that there is a net charge transfer from Cp to CO. This is consistent with the reduction in the CO stretching frequency from 2025 in $Co_2(CO)_8$ to 1980 cm^{-1} in $CpCo(CO)_2$,[60] due to increased Co-->CO backbonding.

8. NEELFS: NEXAFS in the Home Laboratory

Inner shell excitation by reflection EELS has been studied for over 10 years. The status of this field has been reviewed recently.[61,62] The apparatus used at McMaster to record reflection electron energy loss spectra of clean and adsorbate covered surfaces is presented in Fig. 9.[63] Impact energies of 1 to 2.5 keV and scattering angles of 6-20° and 90° are available. In addition the UHV apparatus has a number of other surface analysis techniques to complement the reflection EELS.

A central theme in ISEELS research is its use to help interpret the inner-shell spectra of condensed phases (surface adsorbates, polymers etc). While these latter are most frequently obtained by X-ray excited NEXAFS,[64] reflection electron energy loss spectroscopy offers a lab-based alternative to study core excitation spectra of the condensed state. If reflection EELS of chemisorbed molecules is practical this will allow NEXAFS-type studies of surface adsorbates in the more comfortable, much more widely accessible and frequently better-equipped environment of the home surface science laboratory. We have recently developed an apparatus for such studies and have reported the first near-edge energy-loss fine-structure NEELFS spectra of chemisorbates.[12,13]

The momentum transfer vector in inelastic electron scattering plays a similar role to the electric vector in optical polarization studies. Thus the dependence of the intensity of NEELFS spectral features on the scattering angle should contain the same information

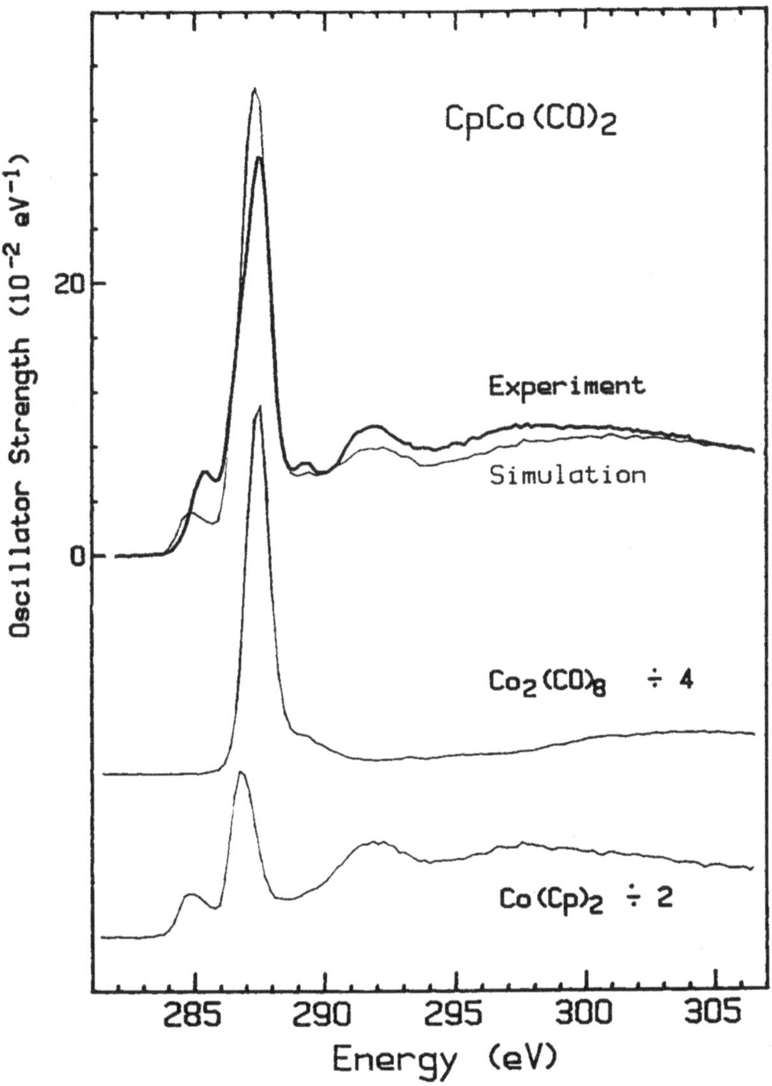

Fig. 8 Comparison of the C 1s spectra of $CpCo(CO)_2$ ($Cp = \eta^5$-C_5H_5) and a simulation based on the spectra of $Co_2(CO)_8$ and $Co(Cp)_2$. The deviation between experiment and simulation in the intensities of the first two peaks is consistent with charge transfer from Cp to CO increasing the C 1s-->π*(Co-Cp) and decreasing the intensity of the C 1s-->π*(CO) transitions.[59]

as the polarization dependence of NEXAFS. However these is still an important question as to whether or not the momentum transfer direction in the inelastic collision which produces the NEELFS signal is sufficiently well-defined to obtain a directional sensitivity to the electronic transition. Plural scattering events dominate the inelastic reflection of high energy electrons from a solid surface.[65,66] At large scattering angles it is widely believed that the inner-shell component of the spectra mainly arises from loss-elastic (LE) or elastic-loss (EL) plural scattering processes, in which a small angle inelastic scattering is combined with a large angle elastic scattering to produce the observed signal. Such processes would lead to a range of q directions for the inelastic step and thus a reduced directional sensitivity. However studies of the C 1s spectrum of graphite suggest that the LE/EL model does not apply at scattering angles below 20°.[67] Our very recent results on the coverage dependence of the N 1s spectrum of pyridine chemisorbed on Ag(100) and Ag(111)[13] clearly indicate that direct, single inelastic scattering dominates the small angle reflection EELS of adsorbates and that the q-direction is well defined. The N 1s NEELFS spectra of pyridine adsorbed on Ag(100) at 1, 2 and 5 Langmuir exposure are presented in Fig. 10. A small beam current (0.2 μA) and short scans (10 minutes maximum on the

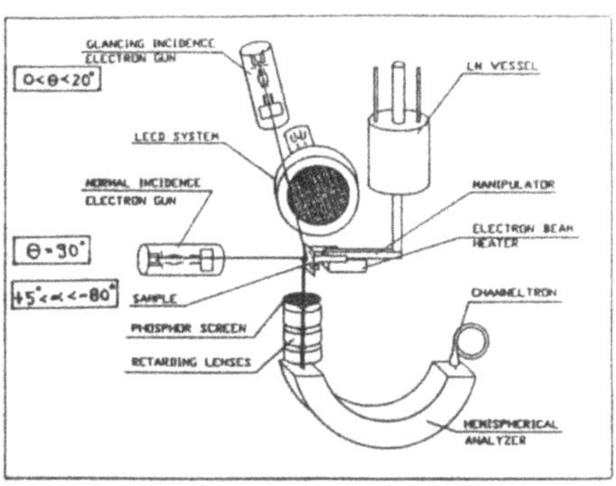

Fig. 9 Ultrahigh vacuum (UHV) apparatus for reflection EELS of surfaces at McMaster.[63] Sample temperatures can be adjusted from 50 to 800 K; beam current of 0.2 μA, impact energy of 2000 eV and typical recording times of 10-30 minutes are used in obtaining spectra of submonolayer chemisorbates.

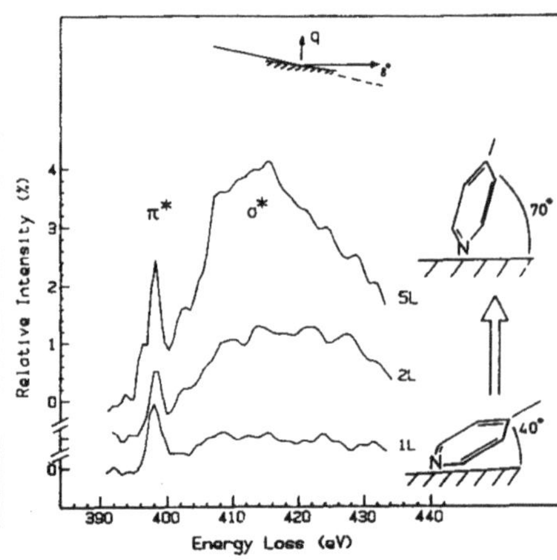

Fig. 10 N 1s NEELFS of pyridine adsorbed on Ag(100) at 80 K at 3 different exposures (coverages).[13] A Langmuir (1 L = 10^{-6}torr.sec) exposure produces about 0.2 of the saturated monolayer. The indicated orientation change with increasing coverage causes the N 1s NEELFS to change from a π^*-dominated to a σ^*-dominated spectrum.

same spot of the surface) were used to ensure that beam damage effects were minimal. The change from a π^* to a σ^* dominated spectrum tracks a surface phase transition in which the pyridine molecule changes from an orientation predominantly parallel to the surface to one with a high tilt angle between the molecular and the surface planes. NEXAFS has been used to follow a corresponding orientational phase transition of pyridine on the Ag(111) surface.[68] With this development there now exists in non-synchrotron radiation laboratories, the capability to study the core excitation spectroscopy of a molecule both free and in the chemisorbed state.

9. Future Directions

In coming years dipole-regime gas phase ISEEL spectroscopy will be applied to more challenging systems such as molecular beams, transients, low-volatility species etc. Since the vapour pressure required for ISEELS is about an order-of-magnitude less than that needed for photoelectron spectroscopy (PES), the technique should be applicable to the many transient species already studied by PES. At several institutions variable angle ISEEL spectrometers are being used for systematic GOS measurements. The recent work of the de Souza group at IQ-UFRJ[69] is particularly notable in this area. Such measurements will provide an improved understanding of non-dipole core-excitations and the momentum transfer dependence of inner-shell electronic excitations, a relatively unexplored area in which electron impact has unique capabilities. Dipole regime investigations will be increasingly shadowed by high resolution gas phase X-ray studies with synchrotron radiation. This implies that increased activity in non-dipole core excitation studies is a sensible way to proceed.

In the area of core excitation studies of surface adsorbed molecules, although NEXAFS is already a very popular spectroscopy with an increasingly large number of devotees, reflection electron energy loss spectroscopy will likely experience rapid development and will play an important role in surface chemistry and physics. Parallel detection schemes will be essential to allow spectral acquisition in a time short compared to electron beam damage. With improvements currently being implemented reflection EELS with variable incident angle could provide a lab-based technique that has a capability similar to that of NEXAFS for determining the orientation of surface adsorbed species. There is also recent work in using reflection-ISEELS for spatially resolved surface studies in a reflection electron microscope[70] and its application for in-situ layer-by-layer characterisation of thin films grown by molecular beam epitaxy (MBE).[71] Such practical developments must be complemented by further fundamental studies in order to fully exploit the information contained in core excitation spectra obtained by reflection-EELS. The future for both gas phase and surface inner shell energy loss spectroscopy appears particularly promising.

Acknowledgements: Financial support for this research was supplied by the Natural Sciences and Engineering Research Council (NSERC, Canada), the Ontario Centre for Materials Research, Dow Chemical USA. The donors of the Petroleum Research foundation of the American Chemical Society are also acknowledged. Discussions and collaborative research projects with E. Rightor, W.J. Leigh, M.J. McGlinchey, E. Rühl and J. Stöhr have made large contributions to the work described above. The experimental program of core excitation at McMaster has been performed with the capable assistance of Dr. T. Tyliszczak, Dr. P. Aebi, Dr. I. Ishii, D.C. Newbury, R. McLaren, R.S. Dewitte, S.G. Urquhart, J.T. Francis and A.T. Wen.

References

1. Koch, E.E. and Sonntag, B.F. *Synchrotron Radiation*, Topics in Current Physics 10, 269 (Springer, Heidelberg, 1979).
2. Brown, F.C., 'Inner-shell Threshold Spectra', Ch. 4, *Synchrotron Radiation Research*, (ed. S. Winnick and S. Doniach). Plenum, NY, 1980.
3. Nenner, I. and Beswick, J.A., *Photodissociation and Photoionisation*, Chapter 6 in *Handbook on Synchrotron Radiation*, Vol. 2 (G. Marr, ed) Elsevier, Holland, 1987; Morin, P. and Nenner, I., Physica Scripta T17, 171 (1987).
4. R.F. Egerton, *Electron Energy Loss Spectroscopy* (Plenum, 1988); Spence, J.C.H., Ultramicroscopy 18, 165 (1985); Proceedings of the 1990 Tahoe EELS Workshop, Microscopy, Microanalysis and Microstructures (1991).
5. Read, F.H., J. Phys. 39, C1-82 (1978); King, G.C., Lecture Notes in Chemistry 35, 162 (1984).
6. Hitchcock, A.P., J. Electron Spectrosc. 25, 245 (1982). (updates available from the author).
7. Brion, C.E., Daviel, S.D., Sodhi, R.N.S. and Hitchcock, A.P., Third Int. Conf. on X-ray and Atomic Inner-Shell Physics, AIP Conference Proceedings 94, 429 (1982); Brion, C.E., Com. At. Mol. Phys. 16, 249 (1985).
8. Hitchcock, A.P., Ultramicroscopy, 28, 165 (1989); Hitchcock, A.P. Physica Scripta T31, 159 (1990).
9. Benitez, A., Moore, J.H. and Tossell, J.A., J. Chem. Phys. 88, 6691 (1988).
10. De Crescenzi, M., Antonangeli, F., Bellini, C. and Rosei, R., Phys. Rev. Lett. 50, 1949 (1983).
11. Luo, B. and Urban, J., J. Phys. Conden. Matter 3, 2873 (1991).
12. Hitchcock, A.P., Esposto, F. and Tyliszczak, T., Phys. Rev. Lett. 62, 2551 (1989).
13. Tyliszczak, T. and Hitchcock, A.P., J. Vac. Sci. Tech. A 8, 2552 (1990).
14. Daviel, S., Brion, C.E. and Hitchcock, A.P., Rev. Sci. Inst. 55, 182 (1984).
15. Chen, C.T. and Sette F., Physica Scripta T 31, 119 (1990).
16. Domke, M., Xue, C., Puschmann, A., Mandel, T., Hudson, E., Shirley, D.A. and Kaindl, G., Chem. Phys. Lett. 173, 122 (1990).
17. Keski-Rahkonen, O. and Krause, M.O., At. D. Nucl. D. Tables 14 139 (1974).
18. Bethe, H., Ann. Physik. 5, 325 (1930).

19. Inokuti, M., Rev. Mod. Phys. 43, 297 (1971).
20. Camilloni, R., Fainelli, E., Petracelli, G. and Stefani, G., J. Phys. B 20, 1839 (1987).
21. Saldin, D.K. and Yao, J.M., Phys. Rev. B 41, 52 (1990).
22. McLaren, R., Clark, S.A., Ishii, I. and Hitchcock, A.P., Phys. Rev. A 36, 1683 (1987).
23. Johnson, A.L., Walter, W.K., Perez Jigato, M., King, D.A., Norman, D., Rühl, E., Schmale, C., Baumgärtel, H., McGlinchey, M.J. and Hitchcock, A.P. J. Chem. Soc. Faraday Transactions (to be submitted).
24. Hitchcock, A.P. and Brion, C.E., Chem. Phys. 33, 55 (1978).
25. Sodhi, R.N.S. and Brion, C.E., J. Electron Spect. 37, 125 (1985).
26. Harrison, I. and King, G.C., J. Electron Spectr. 43, 155 (1987).
27. McLaren, R., Ishii, I., Hitchcock, A.P. and Robin, M.B., J. Chem. Phys. 87, 4344 (1987).
28. Henke, B.L., Lee, P., Tanaka, T.J., Shimabukuro, R.L. and Fujikawa, B.K., At. Data and Nucl. Data Tables 27, 1 (1982).
29. Hitchcock, A.P., Lablanquie, P. Morin, P., Simon, M., Lizon a Lugrin, E., Thiry, P. and Nenner, I., Phys. Rev. A 36, 2448 (1988).
30. Wen, A.T., Hitchcock, A.P., Werstiuk, N.H., Nguyen N. and Leigh, W.J., Can. J. Chem. 68 1967 (1990).
31. Douglas, A.E. and Greening, F.R., Can. J. Phys. 57, 1650 (1979); Hitchcock, A.P., Williams, G.R.J., Brion, C.E. and Langhoff, P.W., Chem. Phys. 88, 65 (1984).
32. Ishii, I. and Hitchcock, A.P., J. Chem. Phys. 87, 830 (1987).
33. Lindholm, E., and Asbrink, L., Lecture notes in Chemistry 38: *Molecular Orbitals and Their Energies Studied by the Semiempirical HAM Method* (Springer, Berlin, 1985). The HAM/3 program for IBM-PC and compatibles can be obtained as program QCMP005 from Quantum Chemical Program Exchange (QCPE), Creative Arts Bldg 181, Indiana University, Bloomington, IN 47405 USA.
34. The program Forticon8 based on the Extended Hückel MO method is available in IBM-PC version as program QCMP011 from QCPE.
35. Cederbaum, L.S. and Domcke, W., J. Chem. Phys. 66, 5084 (1977).
36. Bradshaw, A.M. and Somers, J., Physica Scripta T 31, 188 (1990).
37. Schwarz, W.H.E., Chem. Phys. 11 217 (1975).
38. Rühl, E. and Hitchcock, A.P., J. Am. Chem. Soc. 111, 5069 (1989).
39. Hitchcock, A.P., Wen, A.T. and Ruhl, E., Chem. Phys. 147, 51 (1990).
40. Wen, A.T., Rühl, E. and Hitchcock, A.P., J. Electron. Spectrosc. (1991) in press.
41. Sette, F., Stöhr, J. and Hitchcock, A.P., J. Chem. Phys. 81, 4906 (1984).
42. Kreile, J., Schweig, A. and Thiel, V., Chem. Phys. Lett. 108, 259 (1984).
43. Arvanitis, D., Rabus, H., Wenzel, L. and Baberschke, K., Z. Phys. D 11, 219 (1989).
44. Sheehey, J.A., Gil, T.J., Winstead, C.L., Farren, R.E. and Langhoff, P.W., J. Chem. Phys. 91, 1796 (1989); P.W. Langhoff, Physica Scripta T31, 179 (1990).

45. Piancastelli, M.N., Lindle, D.W., Ferrett, T.A. and Shirley, D.A., J. Chem. Phys. 86, 2765 (1987); ibid, J. Chem. Phys. 87, 3255 (1987).
46. Hitchcock, A.P. and Stöhr, J., J. Chem. Phys. 87, 3523 (1987).
47. Stöhr, J. and Outka, D.A. J. Chem. Phys. 88, 3539 (1988).
48. Outka, D., Stöhr, J., Madix, R.J., Rotermund, H.H., Hermsmeier, B. and Solomon, J., Surf. Sci. 185, 53 (1987).
49. Rightor, E., Hitchcock, A., Urquhart, S.G. and Wen, A., Proc. EMSA conference (San Francisco, 1991); Urquhart, S.G., Rightor, E., Hitchcock, A. to be published.
50. Horsley, J.A., Stöhr, J., Hitchcock, A.P., Newbury, D.C., Johnson, A.L. and Sette, F., J. Chem. Phys. 83, 6099 (1985).
51. Hitchcock, A.P., Newbury, D.C., Ishii, I., Stöhr, J., Horsley, J.A., Redwing, R.D., Johnson, A.L. and Sette, F., J. Chem. Phys. 85, 4849 (1986).
52. Hitchcock, A.P., Horsley, J.A. and Stöhr, J., J. Chem. Phys. 85, 4835 (1986).
53. Liu, A.C., Stöhr, J., Friend, C.M. and Madix, R.J., Surf. Sci. 235, 107 (1990).
54. Wurth, W., Coulman, D., Puschmann, A., Menzel, D. and Umbach, E., Phys. Rev. B 41, 12933 (1990).
55. Rühl, E. and Hitchcock, A.P., J. Am. Chem. Soc. 111, 2614 (1989).
56. Sondericker, D., Fu, Z., Bradley, J. and Eberhardt, W., J. Chem. Phys. 92, 2203 (1990).
57. Cooper, G., Sze, K.H. and Brion, C.E., J. Am. Chem. Soc. 111, 5051 (1989); ibid, J. Am. Chem. Soc. 112, 4121 (1990).
58. Hitchcock, A.P., Wen, A.T. and Rühl, E., J. Electron Spectrosc. 51, 653 (1990).
59. Rühl, E., Wen, A.T. and Hitchcock, A.P., J. Electron. Spec. (1991) in press.
60. Chen, H.W., Jolly, W.L., Kopf, J. and Lee. T.H., J. Am. Chem. Soc. 101, 2607 (1979).
61. De Crescenzi, M. J. Vac. Sci. Tech. A5, 869 (1987).
62. De Crescenzi, M. CRC Crit. Rev. Sol. St. & Materials Science 15, 279 (1989).
63. Tyliszczak T., De Crescenzi, M., Hitchcock, A.P. Phys. Rev. B 38 10,664 (1988).
64. Stöhr, J., Z. Phys. B 61, 439 (1985); Stöhr, J. and Outka, D.A., J. Vac. Sci. Tech. A 5 919 (1987); Stöhr, J., *NEXAFS Spectroscopy* (Springer-Verlag, Heidelberg, 1991).
65. Saldin, D.K., Phys. Rev. Lett. 60, 1197 (1988).
66. Fujikawa, T. and Hedin, L., Phys. Rev. B 40, 11507 (1989).
67. Hitchcock, A.P. and Tyliszczak, T., Physica B 158, 631 (1989).
68. Bader, M., Haase, J., Franck, K.-H., Ocal, C. and Puschmann, A., J. Phys. (Paris) 47, C-8, 491 (1986); Phys. Rev. Lett. 56, 1921 (1986).
69. Bielschowsky, C.E., Roberty, H.M.B., and de Souza, G.G.B., J. Chem. Phys. (1991) in press.
70. Wang, Z.L. and Cowley, J.M., Surf. Sci. 193, 501 (1988).
71. Atwater, H.A. and Ahn, C.C., Appl. Phys. Lett. 58, 269 (1991).
72. Shaw, D.A., King, G.C., Cvejanovic, D. and Read, F.H., J. Phys. B 17, 2091 (1984).
73. Ungier, L. and Thomas, T.D., Chem. Phys. Lett. 96, 247 (1983).

ION FORMATION BY ELECTRON IMPACT

S. K. Srivastava
Jet Propulsion Laboratory, California Institute of Technology,
4800 Oak Grove Drive, Pasadena, California 91109, USA.

ABSTRACT

Various experimental methods used in the measurement of cross sections for the production of positive and negative ions by electron impact on atomic and molecular species are presented.

1. Introduction

When an energetic electron collides with a neutral atomic or molecular species a wide variety of processes can take place. Among them the following are of present interest:

$$e^- + A \dashrightarrow A^{n+} + (n+1)e^-, \quad (1)$$
$$e^- + MN \dashrightarrow MN^{n+} + (n+1)e^-, \quad (2)$$
$$\dashrightarrow M^{n+} + N + (n+1)e^-, \quad (3)$$
$$\dashrightarrow M + N^{n+} + (n+1)e^-, \quad (4)$$
$$\dashrightarrow M^{m+} + N^{n+} + (m+n+1)e^-, \quad (5)$$
$$\dashrightarrow M + N^-, \quad (6)$$
$$\dashrightarrow M^+ + N^-, \quad (7)$$

where e^- is the colliding electron, A the atomic species, MN the molecule with components M and N, m and n the degree of ionization, and + and - sign indicate the type of ion formed. The probability that any one of the above processes will take place upon collision of an electron with the target atom , A, or molecule MN, is related to a quantity called cross section, σ, which has dimensions of area. It, basically, determines the number of ions which are formed as a result of collisions. Equations (1) and (2) represent the situation where the colliding electron removes one or more electrons from the target and the cross sections, σ_p, for them are called either "direct ionization cross sections" or "partial ionization cross sections". Processes represented by equations (3), (4), and (5) pertain to the situation where the molecule, MN, dissociates into singly or multiply ionized components. The cross sections, σ_d, for them are generally called "dissociative ionization cross sections". If the cross sections are measured by counting individual ions then the sum of all partial ionization cross sections and dissociative ionization crosss sections for the production of positive ions are called "Total counting ionization cross section", σ_T, and are represented by :

$$\sigma_T = \Sigma_p \sigma_p + \Sigma_{i,n} \sigma_{i,n}, \tag{8}$$

where p represents the degree of ionization of species resulting from the direct ionization process, n the degree of ionization of the dissociated ion and i the species resulting from the dissociation of the molecule. On the other hand , if cross section is measured by recording the toatal ion current then the sum is called " Gross ionization cross section" and is represented by the following equation:

$$\sigma_T = \Sigma_p p \sigma_p + \Sigma_{i,n} n . \sigma_{i,n}. \tag{9}$$

Equation (6) pertains to a situation where the incident electron, instead of removing an electron from the target, attaches itself to it. It remains attached for a very short time and then causes dissociation of the molecule and attaches itself to one of the fragments. The cross section for this process, σ_a, is called "dissociative attachment cross section". Equation (7) presents a situation where the collision of an energetic electron results in an ion pair formation, which is generally known as "polar dissociation". In this review these cross sections will be represented by σ_i.

Figure 1 shows details of the collision process where the target is singly ionized and where states of polarization of the colliding electron and target are not considered. The incoming electron (primary electron) collides with the target, ionizes it, and is scattered at an angle, θ, with respect to the direction of the incident electron. The electron resulting from the ionization (secondary electron) leaves the target at an angle ϕ. The experiments which detect the two electrons in coincidence are known as e^-, $2e^-$ measurements. They throw light on the dynamics of collision process. They will not be discussed here. There have been several review papers in this field where details can be found[1]. Similarly, ionization phenomenon occurring near the threshold of ionization will also not be covered in the present review and the reader is referred to a paper by Read[2] for discussion on this subject. Processes represented by equations (1) through (5) are related to the situation where only the electrons from the outer shells of the target species are removed. Ionization related to the inner shell processes will also not be discussed here.

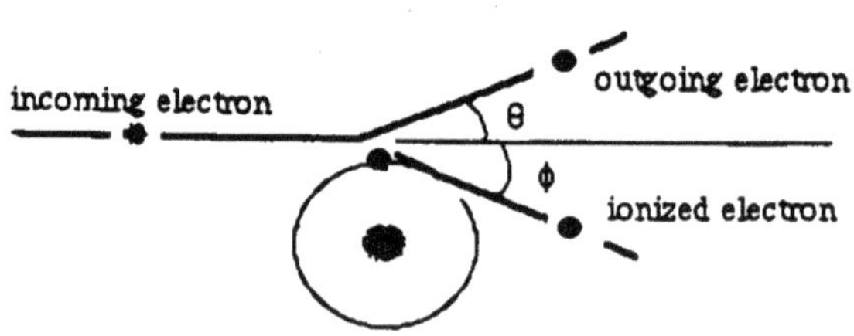

Fig. 1. Ionization process of a target species by a colliding electron.

Electron impact ionization and attachment processes are important not only from the point of view of basic physics but also from the point of view of their applications. The magnitudes of these cross sections are generally larger than others and play important roles in determining the conductivity of either a man made plasma or a naturally occurring one. They are important in calculating stopping powers of various materials and in analyzing spectra obtained by mass spectrometers. In order to understand the properties of low temperature plasmas, which are used for deposition or etching of semi-conductors, a knowledge of these cross sections is valuable. Their properties are important in designing efficient electric propulsion systems. These are just a few examples where ionization and attachment cross sections have applications.

In the following pages various experimental techniques which have been employed in the past and those which are currently being used, will be described. This review is written for graduate students of the Latin American universities in order to familiarise them with the experimental apparatus and techniques used in the measurement of electron impact ionization and attachment cross sections. Therefore, a comprehensive review of the literature on these subjects is not given here and the material presented is educational in nature.

2. Experimental Methods:

In the past, generally, the same instrumentation and techniques have been used for the measurement of cross sections for both the positive ions and negative ions. In the case of negative ions all electrical polarities on the various parts of the instruments are reversed from the arrangement for positive ions. Therefore, the following descriptions of the various parts of an apparatus, employed for cross section measurements, apply to both positive and negative ions. Negative ion cross sections can also be indirectly derived by swarm methods. These indirect methods for the measurement of either positive or negative ions will not be covered in this paper.

2.1 Apparatus: A typical arrangement of the various components of an experimental apparatus is shown in fig. 2. It consists of : 1. an electron gun which can produce a collimated beam of electrons (preferably energy selected), 2. a Faraday cup to trap the incident electron beam, 3. an ion source, 4. a mass selector when partial ionization cross sections are measured, 5. a charged particle detector, 7. an amplifier, and 8. a data recording system. In the following, general characteristics and design principles of each of these components are given.

2.1.1. Electron gun: A simple electron gun consists of two parts: an electron source or cathode region and a collimator or beam forming region. In general, the source of electrons consists of either a flat disc or a hairpin filament made of a suitable material which is heated to high temperatures to generate electrons by thermionic emission. There are two important considerations which are used in designing the electron source. The first one determines the amount of electron current, I_{max}, that can be generated by the source and supplied to the beam forming region without encountering the difficulties related to the space charge effects. It is well known[3] that the maximum current which can be passed through a tube of length L and diameter D against the forces of repulsion is given by:

$$I_{max} = 38.5\, E^{3/2}\, (D/L)^2, \tag{10}$$

where current is in micro amperes and energies E are in eV. The second one is concerned with the material of the filament. It has been shown[4], recently, that out of the three commonly used filament materials of tungsten, thoriated tungsten, and thoriated irridium, the thoriated tungsten generates electrons with the least amount of energy spread, i.e. they are more monochromatic than the ones emitted by other materials.

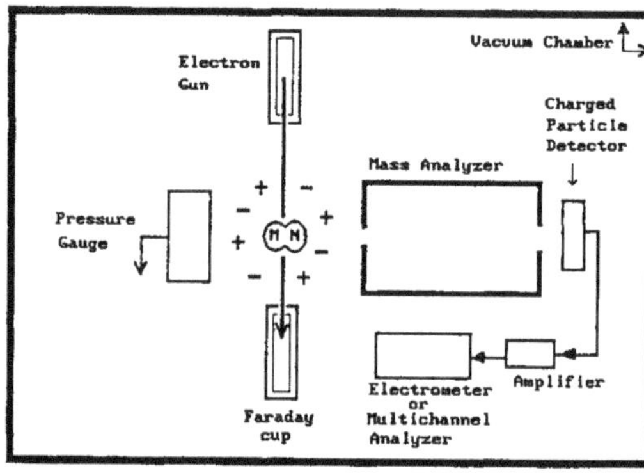

Fig. 2. A typical arrangement employed for the measurement of ionization cross sections.

The thermal energy spread, ΔE(eV), of electrons leaving the filament surface is given by:

$$\Delta E = 218.9\ 7\ T, \tag{11}$$

where T is the absolute temperature (not necessarily that of the filament surface) of the filament[5]. Therefore, the temperature of the filament plays an important role in determining the energy spread of the electron source.

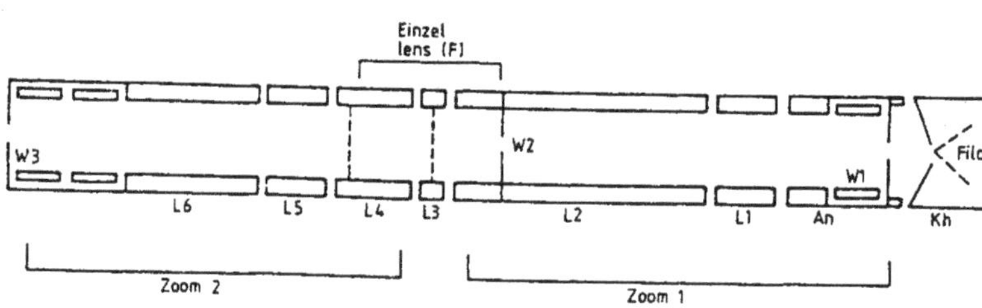

Fig. 3. A schemmatic diagram of a typical electrostatic electron gun.

Beam forming regions are of two types: those which utilize electrostatic lenses and those which employ magnetic fields for the collimation of the electron beam. A combination of the electron source and the beam forming region results in an electron gun. A simple type of electron gun does not use any energy dispersive device to generate an electron beam. A typical electron gun[4] of this type, based on electrostatic lenses, is shown in fig.3 which utilizes a series of cylindrical lenses. Figure 4 shows a typical magnetically collimated electron gun[6,7]. Both guns can generate a beam of electrons which have energy

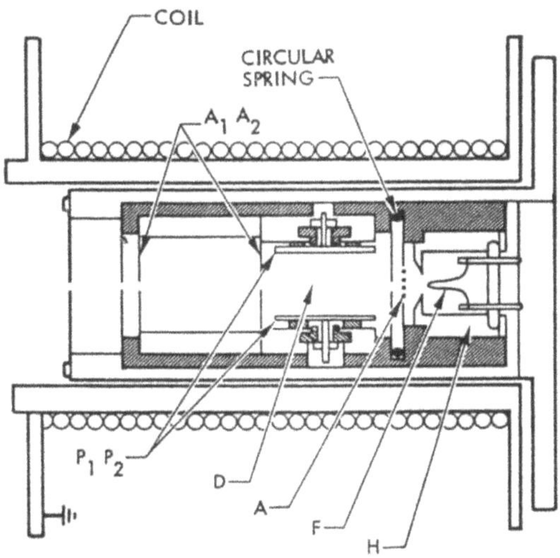

Fig. 4. A shematic diagram of a magnetically collimated electron gun.

spreads ranging from 0.25 eV to 0.5 eV. Lower energy spreads can be achieved by passing the electron beam through an energy dispersive device. Devices which depend on electrostatic methods of dispersion generally employ a hemispherical energy analyser[8]. There are many different designs of them but the basic one is that of Kuyatt and Simpson and the reader is suggested to consult their papers[9,10] for further details. Energy dispersion can also be achieved by the use of crossed electric and magnetic fields. Trajectories of electrons then depend on the energies of electrons and, thus, the dispersion is achieved. In a trochoidal monochromator the magnetic field is directed along the axis of the electron gun. A large number of papers have been published, in the past, on the design of this monochromator.However, references 11, 12, and 13 will give a good understanding of the principles of its design and operation. The configuration of electric and magnetic fields in the trochoidal monochromator differs from that of Wien's filter where the electron moves in a direction which is perpendicular to both the electric and the magnetic fields.

2.1.2. The Faraday cup: The beam of electrons, after leaving the electron gun, passes through the target species and then is collected by a device called "Faraday cup, cylinders, or cages" (fig. 1). Inside the Faraday cup these electrons can generate ions and secondary electrons by colliding to its walls. It is important that the primary electrons, secondary electrons, and ions do not escape the Faraday cup once they enter it. Therefore, this device should be carefully designed. First description of such a device dates back to 1895[14], 1896[15], and 1897[16], published by Perrin, and is shown in fig. 5. It consisted of two coaxial cylinders of which the outer cylinder shielded the inner one which collected all electrons. The basic design of this device has essentially remained the same since then with some minor modifications. Kuyatt has described 17 different types of Farady cups which have been used by various researchers in the past. However, none of them are suitable for the collection of intense beams of fast pulsed electrons. A Faraday cup for this purpose has been developed by Matsuzawa and Akitsu[19]. The design of the Faraday cup used in the author's laboratory is shown in fig. 6. The cover is grounded and provides a shield for the inner cylinder which is biased positively with respect to the outer one. The electrons are, therefore, accelerated to reach the inner parts of the second cylinder where a metal plate, made of molebdenum, is placed at an angle with respect to the axis of the cylinder. This metal plate is biased positively with respect to the inner cylinder. The circuit diagram for biasing the Faraday cup is shown in Fig. 6. V1 is always greater than V2 with typical voltages of 90 V and 10 V, respectively. Typical values of apertures in the two cylinders are 0.3 cm and 0.2 cm, respectively.

2.1.3. Ion source: With in this source ions are produced by electron impact on the target species of interest. These target species can be generated by two methods: i) either by filling the entire vacuum chamber (fig.1) inside which the instrument is placed, or ii) by forming a beam. The method of forming the beam of target species depends on the type of species itself. If the species is in the gaseous state at the room temperature and pressure then it can be formed by flowing the gas through either a single capillary or a capillary array. However, if the species is in the form of a liquid at NTP then it is filled in a small glass bulb which is gently heated to form vapor. This vapor is then connected to the input end of the capillary to generate a beam. For solids an electrically heated metal crucible is employed to form a beam. The crucible can either be heated by a heater wire or by the method of electron bombardment. A schematic diagram of an electron bombarded source is shown in fig. 7. For generating a target beam of radicals, fast neutral beams have beam employed by Freund and his co-workers[20]. In this method ions of neutral species of interest are formed in an electrical discharge which are then accelerated to high energies. They are subsequently passed through an alkali vapor cell where a charge exchange takes place. The emergent beam consists of both the neutral species and ions. The ions are swept out of the path of the neutral beam by the help of a deflector plate. Figure 8 shows a schematic diagram of this source.

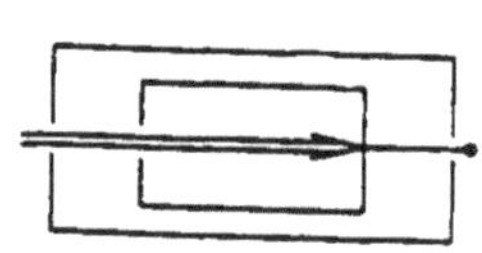

Fig. 5. Perrin's Faraday cup.

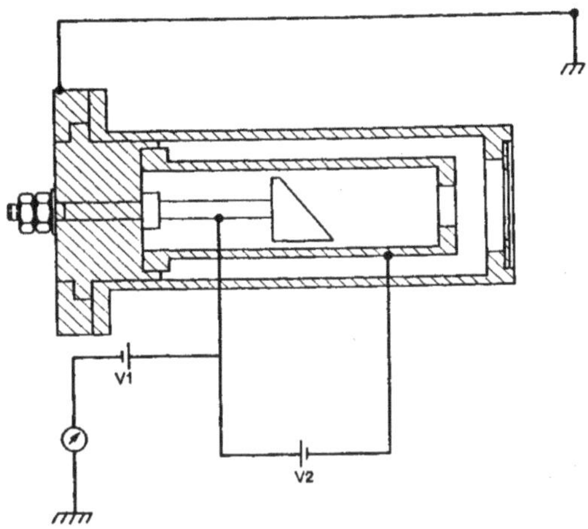

Fig. 6. Schematic diagram of the Faraday cup used by the author.

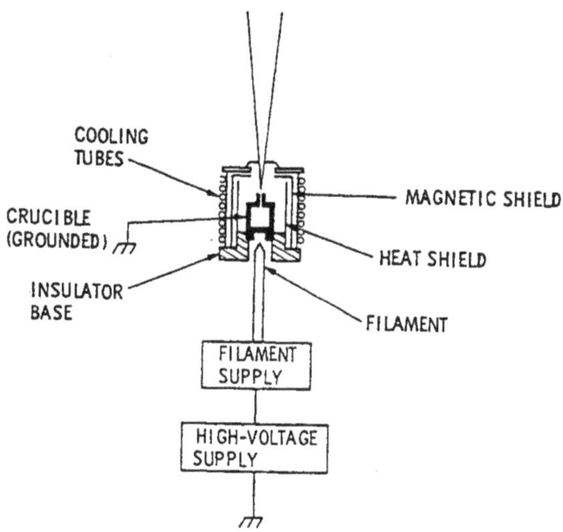

Fig. 7. A high temperature beam source used in author's laboratory

In the beam-beam geometry the beam of target molecules is, generally, crossed by a beam of electrons at an angle of 90°. As a result of collisions between the electrons and the target species ions are produced. For the measurement of total ionization cross sections (

eq. 8) all of these ions are collected without mass analysis. For the measurement of partial cross sections these ions are extracted out of the collision region and are focussed at the entrance aperture of mass analyzer (Fig. 1). The method of extraction of these ions for mass analysis differentiates the various ion sources. In the following these sources will be described.

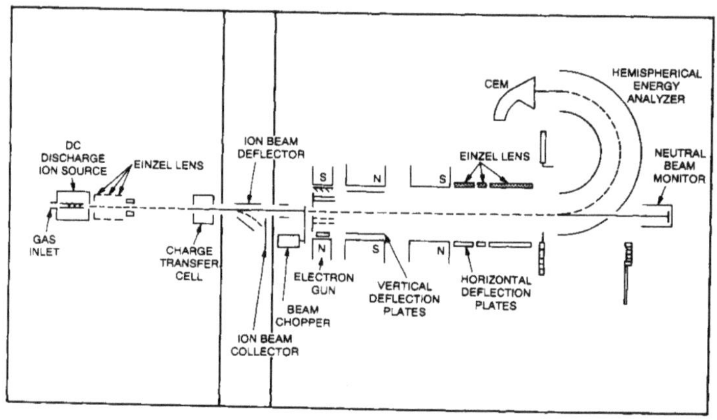

Fig. 8. Apparatus for the measurement of cross sections and for the production of a beam of radical species used by Freund and his co-workers.

2.1.3.1. Nier type of ion source: A schematic diagram of this source, as designed by Märk[21], is shown in fig. 9. The electron beam passes through the

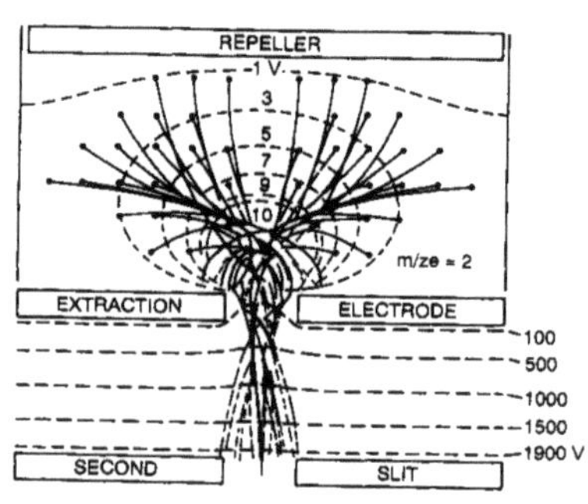

Fig. 9. Nier type of ion source design by Märk. Ion trajectories are shown.

source in a direction perpendicular to the page. An extraction voltage is applied between a repellar plate and an extraction electrode. The ions then pass through a second slit and are focussed at the entrance of the mass analyzer after passing through a set of electrostatic lenses. In fig. 9 the various ion trajectories are indicated. This type of ion source strongly depends on various parameters such as mass/charge ratio, kinetic energies of ions with which they are born, electron beam current and the space charge created by it, and extraction voltages. Although it has been employed in the past in several laboratories, Märk and Castleman[22] have shown that this type of source is not suitable either for the measurement of partial cross sections or cross section ratios. However, Märk and his colleagues have greatly improved its performance by employing penetrating extraction fields.

2.1.3.2. Pulsed extraction type of ion source: It was first shown by Rapp et al.[23] that very large extraction fields are required to extract ions from an ion source and to focus them onto the aperture of a mass analyzer for those ions which are born with kinetic energies as low as 8 eV. This motivated the design of a pulsed extraction scheme[24,25] where the electron gun is pulsed in such a way that the electron beam passes through the collision region during the period when the extraction voltage is zero. As soon as the electron pulse leaves the ion source the extraction voltage is applied. A schematic diagram of this type of ion source is shown in fig. 10. It consists of a pair of parallel wire

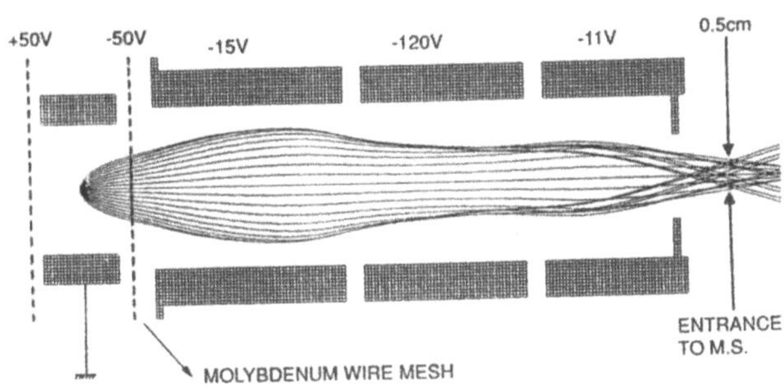

Fig. 10. Schematic diagram of a pulsed ion source employed by the author.

meshes.The region between the wire meshes is either filled with the target gas or by a target beam. The pulsing scheme for extracting ions is shown in fig. 11. Calculations for this ion source show that all ions born with energies up to 5 eV can be collected and focussed onto the entrance aperture of a mass analyzer by a pulse of about 50 V/cm.

2.1.4. Mass selector: For measuring partial cross sections, σ_p, dissociative ionization cross sections,σ_d, and dissociative attachment cross sections, σ_a, mass selection is necessary. Generally the following five types of instruments have been used in the past: i) magnetic sector mass spectrometer, ii) quadrupole mass spectrometer

(QMS), iii) time of flight mass spectrometer (TOFMS), iv) cycloidal mass spectrometer (CMS), and v) mass selector employing hemispherical electrostatic analyser. The principles of their designs will be briefly described in the following.

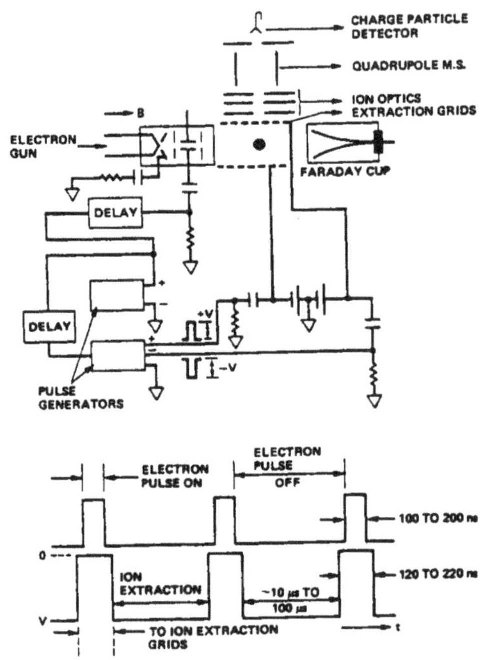

Fig. 11. Schematic diagram showing the pulsing scheme.

2.1.4.1 Magnetic sector mass spectrometer: When an ion of charge q and mass m enters a magnetic field, B, with a velocity, v, and in a direction perpendicular to the field lines it performs a circular motion. The radius, r, of this motion is given by the following relation:

$$r = (m/q).(v/B). \qquad (12)$$

It is clear from eq.(12) that the radius, r, depends on m/q. Therefore, each mass in a magnetic field performs a different trajectory and thus dispersion takes place. Alternatively, if all masses follow the same trajectory, then they must travel with different velocities. These two properties of mass selection are employed in the design of a magnetic sector mass spectrometer. The principle of this type of mass selector is illustrated in fig. 12 which has been employed by several researchers in the past for the measurement of ionization cross sections. One important characteristics of these spectrometers is that they provide high resolution for low masses and lower for the higher ones. They are bulky and need a carefully designed magnetic shield.

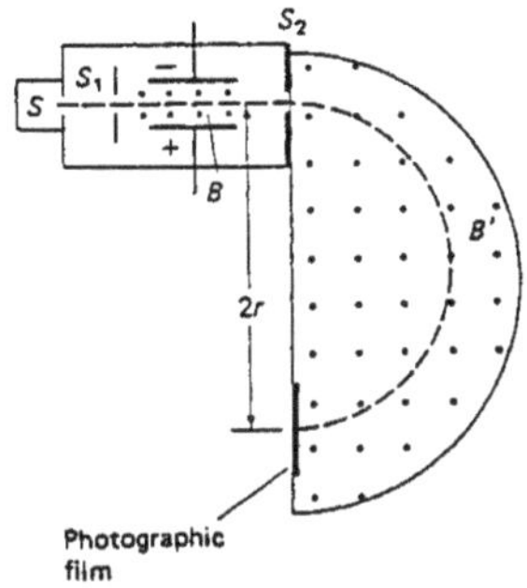

Fig. 12. Bainbridge mass spectrometer. The magnetic fields B and B' point out of the paper (indicated by the dots).

2.1.4.2. Quadrupole mass spectrometer: This spectrometer consists of four metal rods on which a combination of radio frequency (r.f.) and d.c. voltages are applied. A diagram illustrating the principle of its operation is shown in fig. 13. Ions enter

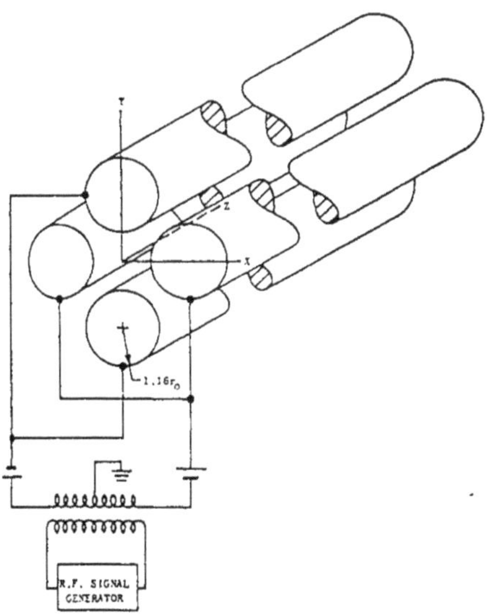

Fig. 13. Geometry and excitation schematic of a quadrupole mass spectrometer. The distance between the opposite rods is $2r_0$.

the space between the four rods with pre-determined velocities and follow trajectories given by following equations:

$$a = (8\ V_{dc}/r_0^2\omega^2)(q/m), \tag{13}$$

$$b = (4\ V_{rf}/r_0^2\omega^2)\ (q/m), \tag{14}$$

where a and b are dimensionless variables, q and m are the charge and mass of the ion, V_{dc} the voltage on the rods, ω the angular frequency of the rf voltage, and $2r_0$ is the distance between opposite rods. The mass resolution is determined by the ratio V_{dc}/V_{rf}. If V_{dc} is made equal to zero then the QMS loses its ability to resolve different masses and it becomes a good focussing lens to guide all masses to the other end of the rods. The mass selection is achieved by increasing or decreasing the values of dc and rf voltages simulaneously so that their ratio remains constant. The theory of the mass filter was first given by Paul et al.[26,27,28]. Many researchers have used this spectrometer for the measurement of cross sections[29]. As opposed to the magnetic sector mass spectrometer, the QMS has poor resolution for low masses and higher resolution for higher masses. It is lighter in weight than the magnetic sector mass spectrometers.

2.1.4.3. Time of Flight Mass Spectrometer (TOFMS): The principle of this spectrometer is illustrated in fig. 14. In this device ions of differing masses are

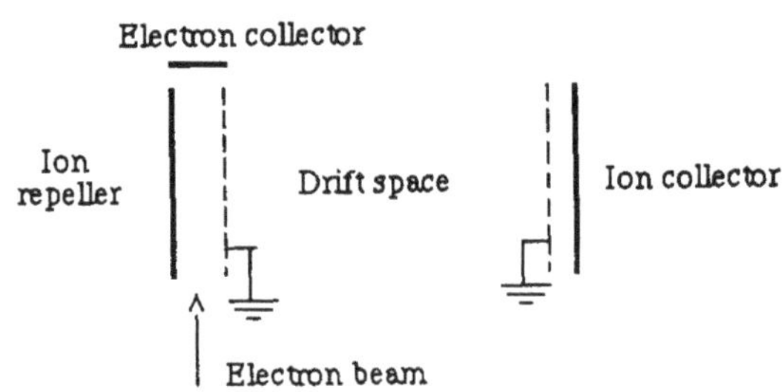

Fig. 14. Schematic diagram of time of flight mass spectrometer (TOFMS).

given the same energy and are allowed to traverse a field free space. Because of their differing velocities, the lightest ions arrive first at the end of the drift space where there is a collector. For cross section measurements TOF has certain advantages over QMS. First, the TOFMS eliminates, to a great extent, uncertainties in the relative values of ion intensities caused by high kinetic energies of ions formed in the dissociative ionization process. Second, if properly designed and operated, the transmission efficiency (i.e. the ratio of the number of ions entering the MS to the number detected at the exit end) of TOFMS can be made completely independent of mass/charge ratios of ions. In practice this condition is very difficult to obtain for QMS. Third, the TOFMS can record all ion intensities simultaneously. Therefore, rapid changes in the ion source conditions do not affect the relative intensities of ions with different mass/charge ratios.The TOFMS has better resolution for low mass/charge ratios. However, the main disadvantage of a TOFMS lies in the fact that it has, over all, a very poor mass resolution for higher masses and the dispersion is not linear with mass/charge ratio. Spectral analysis is difficult when ions possess initial kinetic energies. It has been used by several researchers[25,30] in the past for the measurement of cross sections. More details on the design of TOFMS can be found in a recent review paper by Price and Milnes[31].

2.1.4.4. Cycloidal mass spectrometer (CMS): This mass spectrometer is based on the principle that when an ion travels through a crossed electric and magnetic field it performs cycloidal motion. The first mass spectrometer, based on this principle, was built by Bleakney and Hipple[32,33]. Figure 15 presents trajectories of ions in

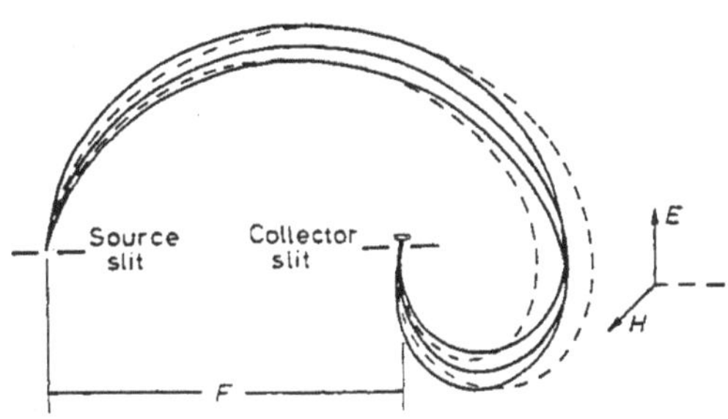

Fig. 15. The principle of cycloidal mass spectrometer. The ion trajectories are for one particular m/e.

the CMS. The basic equation relating the various parameters of the spectrometer is:

$$m/q = (F/2\pi).(B^2/E) \tag{15}$$

where B is the magnetic field strength, E the strength of the electric field, and F the focal distance. This type of mass spectrometer has been employed by Schram et al.[34], among others, for the measurement of ionization cross sections. The advantages of CMS are many. First, it has a perfect focussing condition which requires only two slits to define the resolution and, thus, has a very high throughput. Second, the ions are completely contained within the homogeneous fields and do not pass through any fringe field. Third, The shape and size of the ion trajectory allows the design of a compact instrument. However, this type of instrument is difficult to construct because one has to make sure that there are no fringe fields in the region where ions are present.

2.1.4.5. Hemispherical electrostatic analyzer employed as a mass spectrometer: Although this analyzer has been basically used for the energy analysis of charged particles, in the past, it has also been operated as a mass analyser[20]. The main advantage of this type of mass spectrometer lies in the fact that both mass and the kinetic energy of charged particles can be analysed simultaneously. However, for a high resolution work the spectrometer has very poor transmission properties.

2.1.5. Charged particle detectors: In general, two types of detectors have been used at the exit end of a mass selector. They are: i) a Faraday cup, and ii) an electron multiplier. The Faraday cup collects all charges and measures the total current without any amplification. An electron multiplier, on the other hand, amplifies the current. Usually this amplification factor is of the order of 10^7. There are two types of electron multipliers. One is known as channeltron or spiraltron and the other one is called channel plate. Both are based on the basic principle that electrons are produced when an energetic charged particle strikes their front surfaces. These electrons are accelerated through a narrow capillary like channel where a steep voltage gradient exists. The primary electrons collide with the walls of this channel and produce secondary electrons. Thus, the multiplication takes place and an amplified signal of the incident charged particles is obtained. In the analog mode of their operation the electron current is measured by an electrometer. In the pulse mode of operation, pulses of electrons are generated corresponding to each incident charged particle. These pulses are counted as numbers per second. Figure 16 shows these two types of devices along with electrical circuits generally used for counting positive or negative ions.

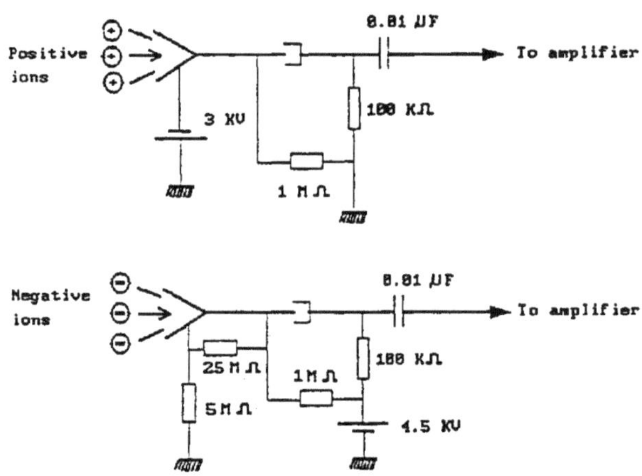

Fig. 16(a). Spiraltron electron multiplier.

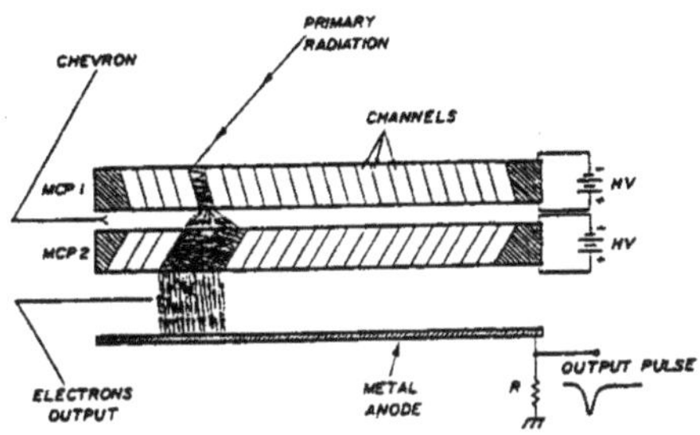

Fig. 16(b). Channel plate electron multiplier.

The conversion efficiency (i.e. ratio of incident current to the output current) of a channeltron and a channel plate strongly depends on the incident energy of the charged particle, its charge state, polarity of the charge, and the angle of incidence. Therefore, one has to carefully choose the biasing conditions to optimise the performance of the detector. Bruce and Bonham[35] have investigated in detail the problems associated with conversion efficiencies of channeltrons and channel plates and their effects on the accuracy of measured ionization data.

2.1.6. Amplifiers: The signal output from the charged particle detectors is small in magnitude. Therefore, it needs to be amplified before it is applied to a data recording system. In general amplifiers, used for this purpose, consist of two parts: preamplifier which is mounted very close to the detector and amplifier to which the output of the preamplifier is fed and can be located near the data recording system. There are two types of amplifiers: analog and pulse counting. Figure 17 shows a simple circuit diagram of an amplifier used by the author in the pulse counting mode. However, it can also be employed in the analog mode.

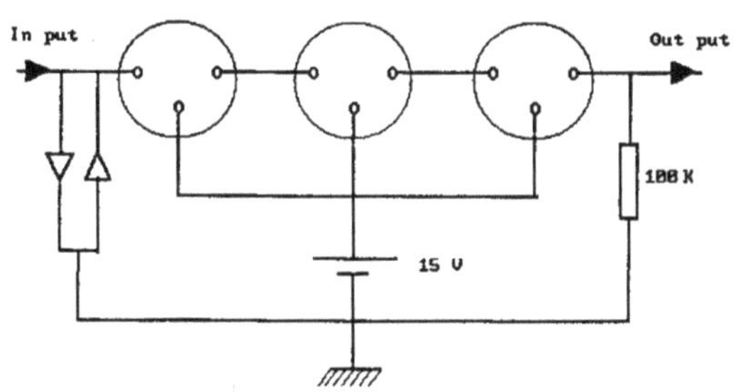

Fig. 17. Circuit diagram of a simple pulse amplifier using Avantek GPD 402.

2.1.7. Data recording system: The output of the amplifier is fed to a system which stores experimental data for future analysis. For analog outputs an electrometer in conjunction with an X-Y recorder is generally used for data storage. When data are in the form of pulses a multichannel scalar is employed. This device stores data as numbers of counts in an array of channels. Each channel is scanned for a certain amount of dwell time. The total number of channels employed for storing data depends on the requirements of the experiment. They are usually grouped in 256, 1024. 4056 etc.

2.2. Measurement of cross sections. There are several variations of the general apparatus, shown in fig. 1. They depend on the type of cross sections to be measured, i.e. σ_T, σ_p, σ_d, or σ_a. Each apparatus employs different sets of components described above. For the measurement of total ionization cross section, σ_T, mass analysis is not necessary. Therefore, the apparatus for the measurement of σ_T does not have a mass analyzer and uses analog techniques to record the ion signals. For the measurement of σ_p, σ_d, and σ_a mass analysis is necessary and requires particle counting techniques for data acquisition. In the following section the various instruments employed for these measurements will be described.

2.2.1. Total ionization cross sections. Although the measurement of cross sections began in 1920 the first reliable data were reported by Tate and Smith[36]. Apparatus used by them is shown in fig. 18. An electron beam was generated by heating a

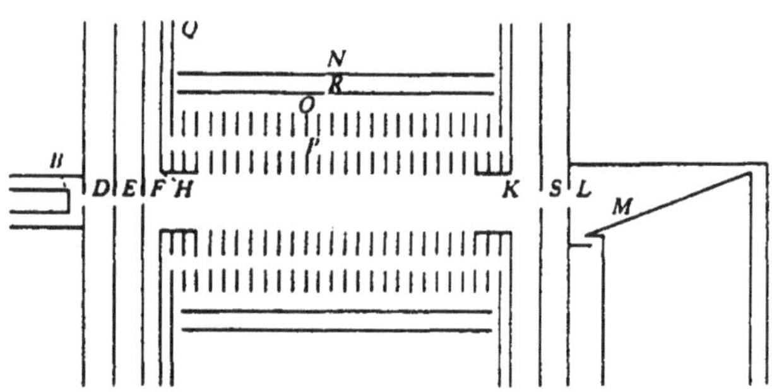

Fig. 18. Tate and Lozier's apparatus.

flat cathode of tungsten.The apparatus was filled with the gaseous species of interest and its pressure, p, was measured. The incident electron current was measured by a Faraday cup, M, and resulting ion current was obtained by collecting all ions by a set of annular rings, as shown in fig. 18. The value of cross section was then obtained from the following relation:

$$\sigma_T = I^+/ I^-. L. n, \quad (16)$$

where I^+ is the ion current, I^- electron beam current, L the path length travelled by the electron beam in the collision region, and n the number density of target species. Since all quantities in eq. (16) are experimentally determined the values of total ionization cross sections are absolute.

In the above apparatus the determination of L was not precise because no precaution was taken to correct the fringing fields at the ends of the ion source. This situation was improved by Rapp and his co-workers[37] and also by Schram et al.[38]. Their experimental arrangements are shown in figures 19 and 20, respectively. They placed guard rings at each end of the ion source to correct for fringing fields. Rapp and his co-workers[39] utilised their apparatus to measure cross sections for both positive and negative ions.

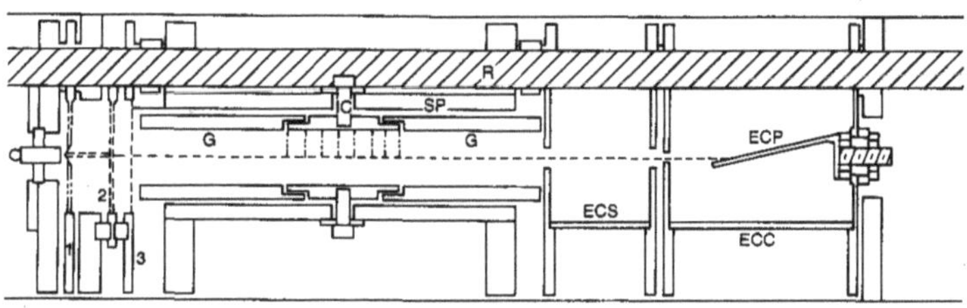

Fig. 19. Total ionization apparatus by Rapp and his co-workers.

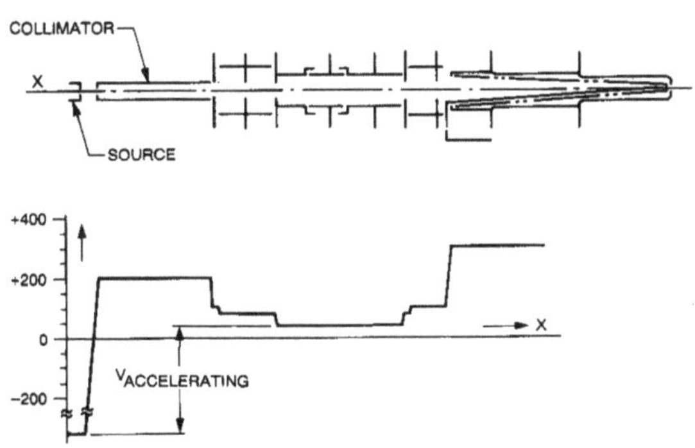

Fig. 20. Total ionization apparatus of Schram, DeHeer, and Kistemaker.

2.2.2 Partial, dissociative and attachment cross sections: For the measurement of these cross sections mass analysis is necessary. Therefore, they employ some type of mass analyser. Figure 8, 21 (a), 21 (b), 22 (a),a nd 22 (b) present different versions of the general apparatus shown in fig.1. The spectrometer employed by McConkey and his co-workers[40] uses a field free ion source and the ions drift into the

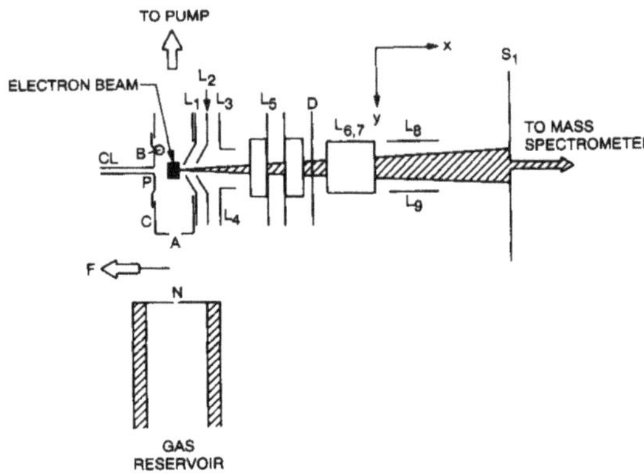

(a) Märk's apparatus

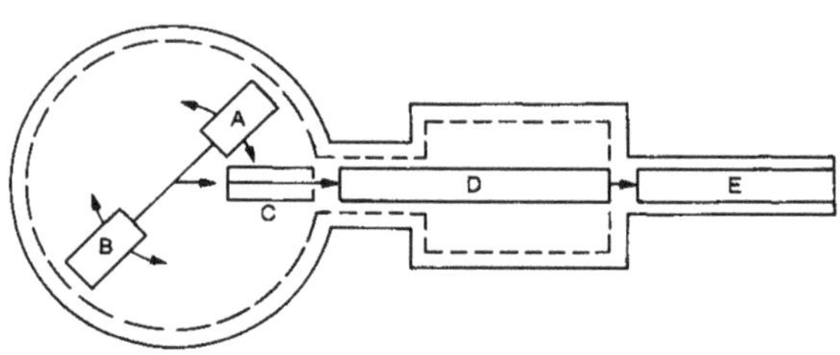

(b) McConkey's apparatus

Fig. 21.

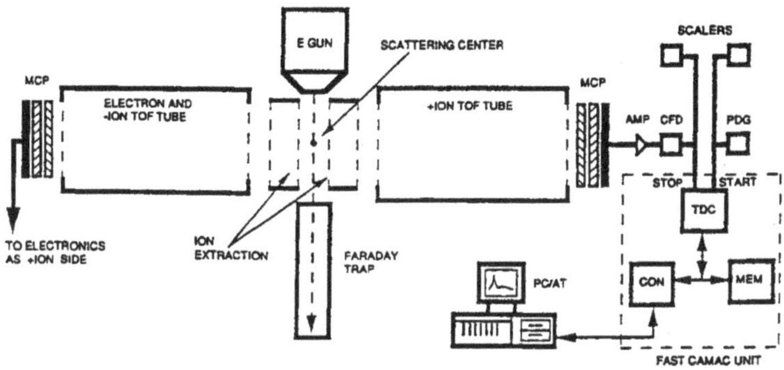

(a)

Bonham's apparatus

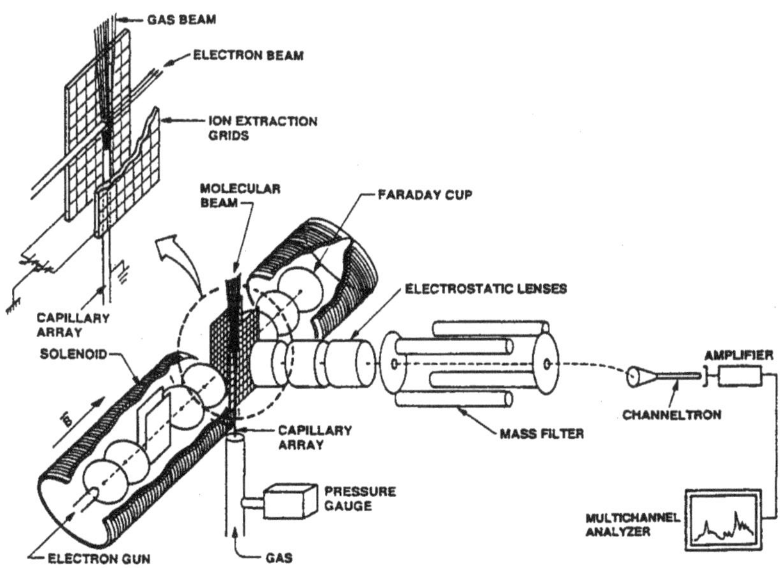

(b)
Srivastava's apparatus

Fig. 22

mass analyser. They used this spectrometer to measure partial and dissociative ionization cross sections for a number of atomic and molecular species. Märk's apparatus[21] has a Neir type of ion source and employes a magnetic sector mass spectrometer for mass analysis. The ion source of this apparatus has been discussed in section 2.1.3. Both apparatus, i.e. of Crowe and McConkey and of Märk, fill the ion source with the gaseous species of interest, pass the energy selected electron beam through it and then measure number of ions that are produced per second as a result of collisions. The apparatus of Freund and his co-workers[20], on the other hand, employes an ion source which produces a high energy beam of neutral species. This beam is crossed by an energy selected electron beam. The resulting ions are mass and energy selected by a hemispherical mass analyser and are counted as a function of electron beam energy. The apparatus of Bonham and his associates[25] and of Srivastava and his co-workers[24] are very similar to each other except that the latter group employes a quadrupole mass spectrometer as well as a time of flight mass spectrometer for mass analysis. Both use a static gas (ion chamber filled with the gaseous species) and crossed beams collision geometry. An energy selected beam of electrons is passed through the ion source. The ions are then collected by the pulsed extraction technique described in 2.1.3.2 and stored as a function of electron impact energy.

All of the above methods of data acquisition determine the ion count rate of a particular species as a function of electron impact energy. A plot of ion count rate as a function of electron impact energy is called ionization function curve. A typical ionization function curve is shown in fig. 25. These curves provide relative values of cross sections. In order to obtain the normalized values of cross sections various researchers have utilised different techniques. Since it is difficult to measure the absolute values of cross sections each one has used some type of normalization procedure to obtain cross sections. In the following, the procedure for obtaining normalised values of cross sections in the author's laboratory will be described in detail.

As shown in fig. 22 (b), a capillary array is used to generate a beam of target species. An electron beam is made to cross this target beam at 90°. As a result of collisions ions are produced. The ions are collected from the ion source, mass analysed and detected by the particle detector according to the methods which have already been described. The number of ions, $I(E_0)$, detected per second for an electron energy E_0 is given by:

$$I(E_0) = K(m).\ \sigma(E_0) \int_r \rho(r).\ f(r,E_0).\ \Delta\Omega(r).dr, \qquad (17)$$

where K(m) is a factor which depends on the combined efficiency of transmission of ions through the extraction grids, electrostatic lenses, quadrupole mass spectrometer, and the detection efficiency of charged particle detector, $\rho(r)$ and $f(r,E_0)$ are the target density and spatial electron flux distribution, respectively, and $\Delta\Omega(r)$ is the solid angle of detection for a collision point located at r within the interaction region of two beams.

As is clear from the above equation, in order to obtain $\sigma(E_0)$ from the experimentally measured parameter $I(E_0)$ we need to determine many other unknown factors. The exact determination of these parameters is almost an impossible task in commonly used experimental conditions. Therefore, for determining the unknown factors, some type of calibration procedure has to be followed. A calibration procedure using a relative flow technique was developed in the past for this purpose and has been described in detail by Srivastava et al.[41]. It is based on the fact that if the pressure behind the capillary array producing the molecular beam is low enough, the properties of the target beam such as spatial and velocity distribution of the molecules can be accurately predicted. Hence if we flow a gas of known cross section, $\sigma_s(E_o)$, for producing ions of mass, m_s, and determine the intensity of the ions, $I_s(E_o)$, at energy E_o of incident electron beam and the, flow rate, F_s, of the gas through the capillary array, we can relate them to the cross section $\sigma_u(E_o)$, for producing ions of mass, m_u, from the gas under study, $I_u(E_o)$, the measured intensity of ions, and the flow rate, F_u, of the gas by the following relation:

$$\sigma_u(E_o) = \sigma_s(E_o).\ [I_u(E_o)/I_s(E_o)].[\ M_s/M_u]^{1/2}.[F_s/F_u][K(m_s)/K(m_u)], \qquad (18)$$

where M_s and M_u are the molecular weights of the standard gas and the gas being studied, respectively. $K(m_s)$ and $K(m_u)$ are mass to charge dependent transmission and detection efficiencies of the system for ions of masses m_s and m_u, respectively.

For the determination of $\sigma_u(E_o)$ by using eq. (18) one has to know the accurate value of cross section, $\sigma_s(E_o)$, at least for one electron impact energy, E_o. For positive ions, cross sections[43] for electron impact ionization of He atoms are generally chosen. For the normalisation of negative ion cross sections, well established cross sections[44] for the dissociative attachment of electrons with O_2 are used. Besides the value of $\sigma_s(E_o)$ one also has to accurately measure flow rates of the two gases. These flow rates are usually very small and conventional flow meters do not provide accurate measurements. Therefore, in our laboratory a different method of measurement of the ratio of the two flow rates, F_s and F_u, was developed. These measurements are performed in the following way. First the flow of the standard gas is established by adjusting the leak valve which controls the flow rate. The intensity, $I_s(E_o)$, for this flow rate is then measured at an electron impact energy of E_o. Subsequently, without disturbing the setting on the leak valve, the inlet valve to the vacuum chamber is closed and the standard gas is quickly pumped out of the vacuum side of this valve. Pumping is then stopped and the rate of rise of pressure behind the capillary array is measured by a capacitance manometer. The rate of rise of the pressure behind the capillary array is directly proportional to the flow rate. This procedure is repeated for the gas under study. The ratio of two rates of rise in pressure behind the capillary array is equal to the ratio of flow rates in eq.(18). For further details on the flow rate measurements the reader is referred to a paper by Trajmar and Register[45].

It is evident from eq. (17) that absolute values of combined transmission and detection efficiency, K(m), of the instrument is a very difficult quantity to measure because then one has to measure absolute values of all other quantities in this equation. However, in eq. (18), fortunately, only relative values (i.e. with respect to one gas such as He) are needed. Since the ionization cross sections of rare gases are accurately known, we utilise their values and measure ion intensities and flow rates for each one of them. These quantities are substituted in eq. (18) and relative values of K(m) for rare gases are obtained. Then, the values of K(m) are plotted as a function of mass/charge ratios of rare gases. The values of K(m) for other masses are obtained by interpolation. One such curve for K(m) is shown in fig.23. It is clear from this figure that the value of K(m) varies appreciably over the mass range and passes through a maximum value of K(m).

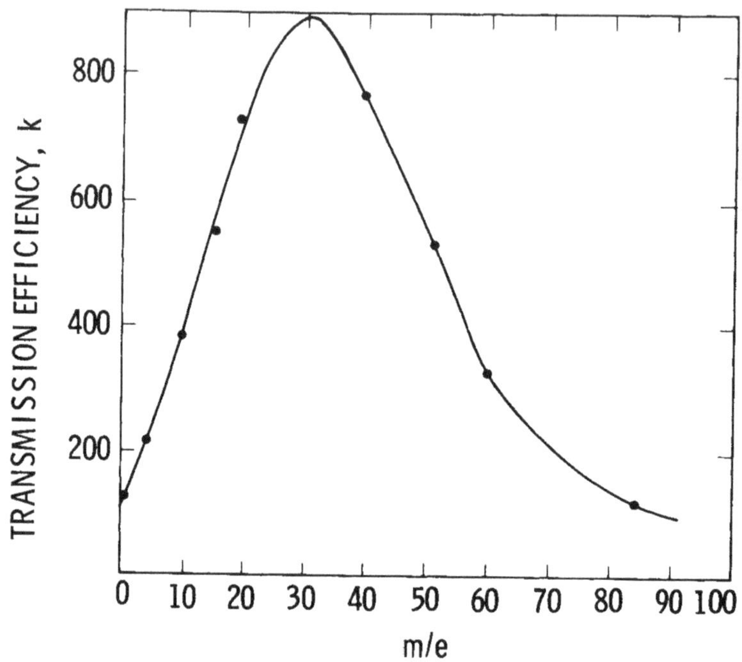

Fig. 23. Relative transmission efficiency curve of the ionization apparatus used by the author.

For absolute measurement of cross sections every quantity in eq. (17) must be determined. Which means that the absolute value of K(m) should also be measured. The author has devised a technique which can be utilised for this purpose. A schematic diagram illustrating the principle of this measurement is shown in fig. 24 and is described in greater detail in ref.46. In this method a gas of desired mass is formed into a beam by flowing it either through a capillary array or a hypodermic needle. This beam is directed into the entrance aperture of the instrument of which the value of K(m) is to be measured. A beam of electrons, with incident energy E_0 larger than the ionization potential of the gas beam is made to cross this gas beam. As a result of collisions of electrons with the gas beam ions are produced. For each new ion an electron loses an amount of energy which is equal to the ionization potential of the target species. The number, n_s, of scattered electrons which lose this amount of energy are counted by an electron energy loss analyser. This number tells us that at least n_s number of ions are generated in the ion source. At the same time these electrons are counted in coincidence with the ions which pass through the mass spectrometer and detected by the particle detector. Suppose n_c is the number of coincidence counts. Then, the ratio n_c/n_s is equal to the absolute value of transmission and detection efficiency of the apparatus.

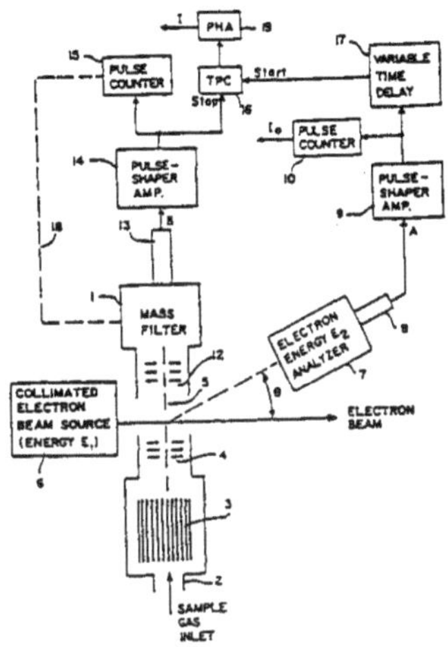

Fig. 24. Electron-ion coincidence technique for determining the absolute value of the transmission efficiency of a mass spectrometer.

3. Experimental Results

For the sake of illustration, N_2 (for positive ions) and N_2O (for negative ions) have been chosen for the discussion here. The main aim is to show the type of data that are obtained by the techniques described in the previous sections. In the following, each molecule is presented separately.

N_2: Results on this molecule have been published previously[24]. Figure 25 shows a typical ionization efficiency curve for single ionization. In this figure data from previous measurements are also presented for the sake of comparison. Results of our measurements are accurate to within ±8% and have been carried out for electron impact energies upto 1 KeV. It is clear from this figure that there are appreciable differences between the various data. These differences can be attributed to methods of normalization used by different experiments. In general, various measurements on cross sections for direct ionization for most molecules agree with each other within the experimental errors of each experiment.

However,for dissociative ionization, where ionized fragments are produced, the agreement between the various measurements for most molecules is, in general, very poor. This is illustrated by data presented in fig. 26 for the fragment ions of N_2. The reason for the disagreements is probably due to the fact that fragment ions of most molecules are born with appreciable amount of kinetic energies and, therefore, as discussed in section 2.1.3, it is difficult to collect all of them from the ionization region. Except for the two recent experimental arrangements[24,25] previous ion sources did not employ high extraction fields for the collection of all ions and, therefore, their data for the dissociative ionization are not reliable.

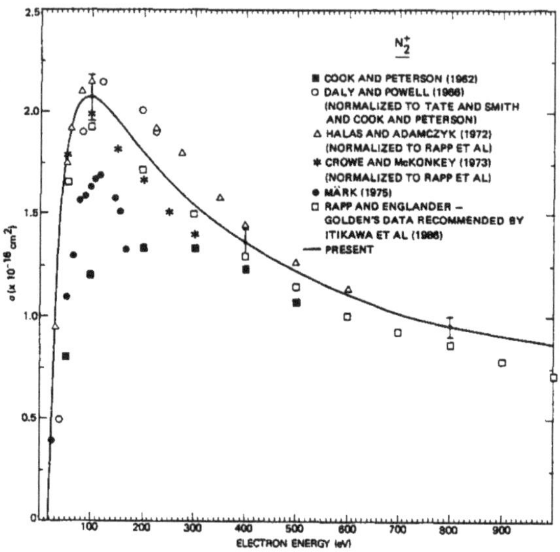

Fig. 25

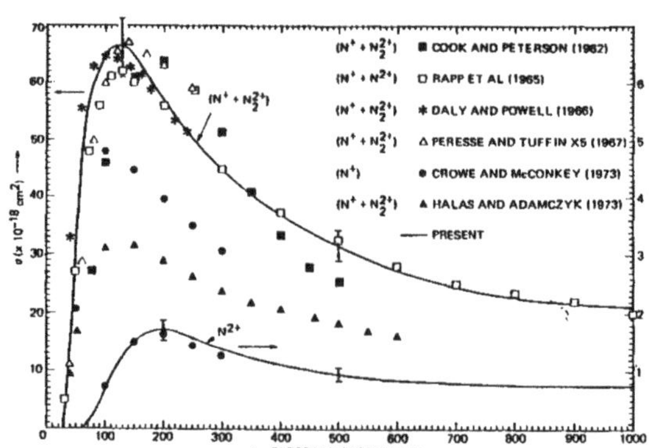

Fig. 26.

N_2O: A typical ionization efficiency curve for the production of negative ions from molecules is shown in fig. 27. It is illustrated by taking N_2O as an example. Spectral features 1,2,3,4 are due to the process of dissociative attachment represented by eq. (6).

They are sharp peaks and result from the resonance process. As the energy of the impacting electron increases the negative ion signal begins to monotonically increase once again. The negative ions produced at this energy result from polar dissociation of the molecule represented by eq. (7).

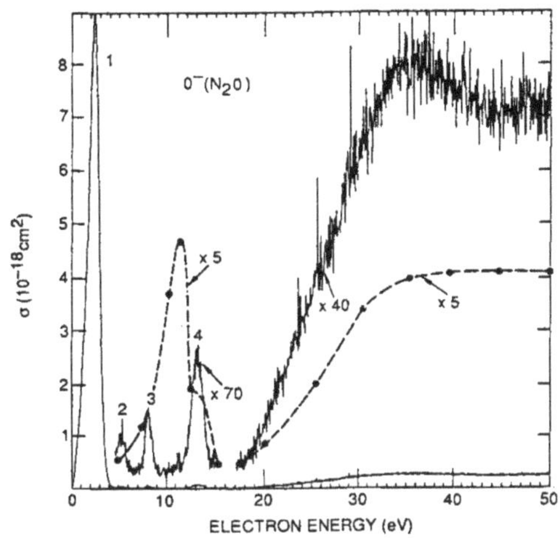

Fig. 27. O^- production by electron attachment to N_2O. Data generated in the author's laboratory along with those of Rapp and Briglia shown by dashed line and filled circles.

4. Future Needs

Cross sections for ionization and attachment of molecules from their excited states are expected to be large. However, there are very few measurements in this area. With the advent of tunable high power lasers it is hoped that the activity in this area of measurements will increase. In addition, most measurements have been performed with a very poor electron energy resolution. It is expected that higher energy resolution will reveal many interesting properties of atoms and molecules, specially near the threshold region of ionization.

5. Acknowledgement

This work was made possible by financial grants from the National Aeronautics and Space Administration, Air Force Office of Scientific Research, and NSF-Brazil collaborative research program.

6. References

1. P. J. O. Teubner, in "Electron Impact Ionization" ed. T. D. Märk and G. H. Dunn, Springer-Verlag Wien, New york., 1985; pp.89.

2. F. H. Read, in "Electron Impact Ionization" ed. T. D. Mark and G. H. Dunn, Springer-Verlag Wien, NewYork,1985; pp.42.
3. J. R. Pierce, J. Appl. Phys. 11 (1940) 518.
4. N. J. Mason and W. R. Newell, Meas. Sci. Technol. 1 (1990) 983.
5. K. Turvey, Eur. J. Phys. 11 (1990) 51.
6. M. A. Khakoo and S. K. Srivastava, J. Phys. E: Sci. Instrum. 17 (1984) 1008.
7. S. K. Srivastava, "Apparatus and Method for characterizing the transmission efficiency of a mass spectrometer", US patent #4,973,840.
8. E. M. Purcell, Phys. Rev., 54 (1938) 818.
9. C. E. Kuyatt and J. A Simpson, Rev. Sci. Instrum., 38 (1967) 103.
10. J. A. Simpson and C. E. Kuyatt, Rev. Sci. Instrum., 34 (1963) 265.
11. A. Stamatovic and G. J. Schulz, Rev. Sci. Instrum., 41 (1970) 423.
12. D. Roy, Rev. Sci. Instrum., 43 (1972) 535.
13. M. R. McMillan and J. H. Moore, Rev. Sci. Instrum., 51 (1980) 944.
14. J. Perrin, Compt. Rend., 121 (1895) 1130.
15. J. Perrin, Nature, 53 (1896) 298.
16. J. Perrin, Ann. Chem. Phys., 11 (1897) 503.
17. J. T. Tate, Phys. Rev., 17 (1921) 394.
18. C. E. Kuyatt, in "Methods of Experimental Physics, vol. 7, part A", ed. B. Bederson and E. Fite, (Academic Press, NewYork, 1968),pp.1.
19. H. Matsuzawa and T. Akitsu, Rev. Sci. Instrum., 58 (1987) 140.
20. T. R. Hayes, R. C. Wetzel, F. A. Baiocchi, and R. S. Freund, J. Chem. Phys., 88 (1988) 823.
21. T. D. Märk, in "Electron Impact Ionization", ed. T. D. Mark and G. H. Dunn (Springer-Verlag Wein, NewYork, 1985), pp.137.
22. T. D. Märk and A. W. Castleman, Jr., J. Phys. E: Sci. Instrum.., 13 (1980) 1121.
23. D. Rapp, P. Englander-Golden, and D. D. Briglia, J. Chem. Phys., 42 (1965) 4081.
24. E. Krishnakumar and S. K. Srivastava, J. Phys. B: At. Mol. Opt. Phys., 23 (1990) 1893.
25. M. R. Bruce and R. A. Bonham, private communication.
26. W. Paul and H. Steinwedel, Z. Naturforsch, 8a (1953) 448.
27. W. Paul and M. Raether, Z. Physik, 140 (1955) 262.
28. W. Paul, H. P. Reinhard, U. vonZahn, Z. Physik, 152 (1958) 143.
29. E. Krishnakumar and S. K. Srivastava, Phys. Rev. A41 (1990) 2245.
30. E. Krishnakumar and S. K. Srivastava, accepted for publication in Int. Jour. Mass Spectrom. and Ion Processes., 1991.
31. D. Price and G. J. Milnes, Int. Jour. Mass Spect. and Ion Processes, 99 (1990) 1.
32. J. A. Hipple and w. Bleakney, Phys. Rev., 49 (1976) 884.
33. W. Bleakney and J. A. Hipple, Phys. Rev., 53 (1938) 521.
34. B. L. Schram, B. Adamczyk, and A.J. H. Boerboom, J. Sci. Inst., 43 (1966) 638.
35. Ce Ma, M. R. Bruce, and R. A. Bonham, private communication.
36. J. T. Tate and P. T. Smith, Phys. rev., 39 (1932) 270.
37. D. Rapp and P. Englander-Golden, J. Chem. Phys., 43 (1965) 1464.
38. B. L. Schram, B. Adamczyk, and A. J. H. Boerboom, J. Sci. Instrum., 43 (1966) 638.
39. D. Rapp and D. D. Briglia, J. Chem. Phys., 43 (1965) 1480.
40. J. W. McConkey, A. Crowe, andM. A. Hender, Phys. Rev. Lett., 29 (1972) 1.

41. S. K. Srivastava, A. Chutjian and S. Trajmar, J. Chem. Phys., 63 (1975)2659.
42. R. T. Brinkmann and S. Trajmar, J. Phys. E, 14 (1981) 245.
43. E. Krishnakumar and S. K. Srivastava, J. Phys. B: At. Mol. Opt. Phys., 21 (1988) 1055.
44. D. Rapp and D. D. Briglia, J. Chem. Phys., 43 (1965) 1480.
45. S. Trajmar and D. F. Register, in "Electron Molecule Collisions", ed. I. Shimamura and K. Takayanagi, Plenum Publishing Corporation, 1984, pp. 427.
46. S. K. Srivastava, US patent #4,973,840.

SYNCHROTRON RADIATION: PRINCIPLES AND APPLICATIONS

T.J. Morgan
Physics Department
Wesleyan University
Middletown, CT 06459–0155

ABSTRACT

The present contribution reviews the basic physical principles of synchrotron radiation and outlines the historical development of the field. Emphasis is placed on third generation sources, and insertion devices in particular. Applications in atomic physics research are reviewed with emphasis on experiments on photoionization of excited atoms and photoionization of ionic targets. Recent results in these fields are discussed. Experimental developments in inverse photoionization and laser stimulated radiative recombination are summarized.

1. Introduction

The topic of synchrotron radiation is germain to this school for several reasons. Firstly, the level of activity of atomic and molecular physics research groups using synchrotron radiation has increased dramatically in recent years. Second, important new discoveries are occurring in the areas of the structure and dynamics of atoms and molecules probed by synchrotron radiation.[1] Third, the third generation of synchrotron radiation machines comprised of dedicated facilities using insertion devices is imminent in several locations throughout the world. Consequently, an explosion in research activity and productivity is impending and trained scientists are needed to enter the field. Fourth and finally, the Brazilian scientific community is hard at work on the design and construction of the first synchrotron radiation machine in Latin America at Campinas, Brazil with an expected turn–on–date of summer 1994. For all the above reasons, synchrotron radiation is a natural topic for a school whose audience is graduate students in physics and chemistry from all over Latin America.

As the title implies the lectures have been divided into two parts: the first half is devoted to an introduction to synchrotron radiation both from a historical and physical point of view while the second half is devoted to applications of synchrotron radiation in atomic and molecular physics research. Two particular experimental applications are emphasized: photoionization of laser prepared atoms and photoionization of ions. Both of these experiments constitute a third generation evolution in targets that can be explored feasibly using synchrotron radiation. First and second generation targets are rare gas atoms and metal vapors, respectively. The point here is that atomic and molecular gas phase research has been photon–flux starved for many years. As synchrotron radiation facilities have evolved so has the nature and complexity of the targets used.

2. Historical Development and Physical Principles

Synchrotron radiation is a unique and powerful tool for a wide variety of disciplines. One of the most important properties of synchrotron radiation is its broad tunable spectral range, which provides a versatile probe of the electronic structure and dynamics of matter. One particular book, Introduction to Synchrotron Radiation by G. Margaritondo (Oxford Press, 1988), served as the basis for much of the introductory material on synchrotron radiation in these lectures. The complexity of a synchrotron radiation facility was highlighted by pointing out in detail the components required to produce a storage ring. This complexity was contrasted by noting the small science nature of the facility. Most actual experiments are of the "table top" variety requiring two or three active experimenters at any given time. The efficiency and effectiveness of synchrotron radiation facilities lies in the fact that they can accommodate a large number of experiments simultaneously.

The basic physical principles associated with synchrotron radiation were discussed following closely the presentation given in Classical Electrodynamics by J.D. Jackson (John Wiley and Sons, Inc., 1962). Lecture one concluded with a historical survey of the evolution of synchrotron radiation facilities and an overview of several modern machines such as the operating SUPER–ACO at Orsay, France and the planned Advanced Light Source at Berkeley, California.

3. Insertion Devices and Spectroscopy

The future of synchrotron radiation looks very bright because of the incorporation of insertion devices into the design of storage rings.[2] Until recently useable light was produced by bending magnets in circular storage rings. Now, long straight sections are employed where light can be produced using insertion devices. Insertion devices are periodic–magnetic structures (usually an array of alternating permanent magnets) which deflect the electron trajectory resulting in the emission of radiation at each bend into an angle γ^{-1} where $\gamma^{-1} = (1 - v^2/c^2)^{1/2}$. There are two types of insertion devices, undulators and wigglers. In a wiggler, electrons are deflected through large angles producing a high flux of broad band radiation (a factor of 2N increase where N is the number of dipole magnets in the structure) due to the incoherent superposition of the emission from each bend. One can think of a wiggler as a collection of small bending magnets. In an undulator, electrons are deflected through very small angles. For 1.5 GeV electrons, $\gamma = 3000$ and $\gamma^{-1} \approx 0.3$ milliradians. The amplitude of deflection is of order 1 micron. Undulators produce a laser like coherent superposition of radiation which leads to interference and harmonic content of the radiation. The result is a small bright spot (increased by a factor of N^2) of nearly monochromatic radiation or put differently, a tunable soft x–ray picosecond strobe (since the electrons in the storage ring are bunched) with laser–like properties.

Lecture two concluded with a discussion and survey of absorption spectroscopy and photoelectron spectroscopy,[3] in preparation for two lectures dealing with the application of photoelectron spectroscopy and synchrotron radiation to atomic physics.

4. Photoionization of Excited Atom Targets

The purpose of this lecture is to introduce students to recent developments in the photoionization of atom targets prepared in well–defined excited states by laser radiation. The research discussed covers the period beginning in 1981 through the last experimental run to date in the summer of 1991. The work is carried out by a group in Orsay, France under the direction of Dr. Francois Wuilleumier. Until 1988 the experiments were done at the ACO facility and more recently on the new third generation machine SUPER–ACO. The author joined this group in 1986 while on sabbatical and has been involved in a National Science Foundation supported collaboration with this group since then.

The photoionization of atoms in selected excited states is of interest for several reasons.[4] It produces a well–defined initial state and serves as a good test of theory. It provides access to states forbidden from the ground state and allows study of the fundamental properties of otherwise not accessible autoionizing states, which provide insight into electron correlation effects in highly excited atoms. Applications exist in plasma physics and astrophysics and as a potential scheme for storage states for XUV lasers. Much work has been done on sodium[5] with the initial state in the one laser prepared electronic configuration $2p^6\ 3p\ ^2P_{3/2}$. Studies have involved both resonant and nonresonant one electron transition processes. Two electron–one synchrotron photon transitions have also been studied, where one observes both resonant and non–resonant correlation satellites. Incorporation of a second laser into the target state preparation scheme[6] provides access to many additional states, by two photon preparation; such as $2p^6\ 4p$, $2p^6\ 4d$, and $2p^6\ 5s$. Work has also been carried out on potassium atomic targets. The apparatus used in these studies is shown in Fig. 1.

The electron spectra resulting from the decay of highly excited autoionizing states such as $2p^5\ 3s\ 4p$, $2p^5\ 3s\ 4d$, and $2p^5\ 3s\ 5s$ have been observed.[7] A typical set of electron spectra is shown in Fig. 2. The excitation energies and the relative oscillator strengths of inner–shell transitions in the excited atoms have been measured. The relative intensity of satellite state production has also been measured and comparisons between the behavior in Na and K obtained. One interesting result[8] is that the ratio of Na(3p) to Na(3s) 2p main line photoionization is about 0.75 over the energy range 55 eV to 110 eV. Theory predicts unity. The variation of the binding energy of inner electrons with outer electron excitation has also been studied.[7] For example, the creation of a 2p core hole ($2p^6\ 3s \rightarrow 3p^5\ 3s\ 3s$) increases the 3s binding energy by 40%, and the binding energy continues to rise with increased excitation. The binding energy of the 3s electron in the electronic configuration $2p^6 3s$ is 5.14 eV while in the configuration $2p^5\ 3s\ 4d$ it is 11.5 eV. Another example is the binding energy of the 2p electron in the configurations $2p^6 3s$ and $2p^6\ 5s$ which changes from 38.0 eV to 44.9 eV, respectively. Clearly the binding energy of inner electrons is very sensitive to shielding and correlations between excited and core electrons.

Recently a new detection technique has been employed in these experiments. Instead of obtaining photoelectron spectra versus electron energy, one may look at a particular final ionic state by tracking the energy of the electrons associated with that state versus synchrotron photon energy. This allows observation of the interference between direct and resonance photoionization and measurement of accompanying Fano profiles.

5. Photoionization of Ionic Targets

The aim of this lecture was to present the very recent advances that have been made in the photoelectron spectroscopy of ionic targets.[9] Actually, the first observation of a photoelectron spectra emitted in the photoionization of an ion beam has just recently occurred[10] and was reported by the Orsay group in the 29th July 1991 issue of Physical Review Letters, less than one month before these lectures.

Very little is known experimentally about the photoexcitation and photoionization of ions. A few photoabsorption cross sections have been measured for light elements[11] and alkali and alkaline–earth metal atoms,[12-15] using resonantly laser driven and dual plasma laser techniques. In the mid 1980's the first experiments combining synchrotron radiation and ion beams took place at Daresbury.[16-20] These experiments used the technique of ion spectroscopy to measure resonant photoionization cross sections in several targets including Ca^+ and Ba^+.

The initial goal of the Orsay work which I am focussing on here is to prove the feasibility of electron spectrometry experiments on positive ions using synchrotron radiation. The experimental set–up, called PISA (Photoionization of Ions at SUPER–ACO), is shown in Fig. 3 and can be segmented into three parts: the ion source, the photon source, and the electron spectrometer. A plasma discharge ion source developed at the Institute de Physique Nucleaire at Orsay was used to provide up to 100μA of singly–charged ions between 15 keV and 40 keV. The ions were mass and charged selected and focussed to the source volume of the electron spectrometer by a triplet lens system. The photons were provided by the SU6 undulator on SUPER–ACO at Orsay. Using a toroidal grating monochrometer, up to 2×10^{13} photons/sec in 0.3% bandwidth between 15 eV and 150 eV was obtained. The electron spectrometer consists of a cylindrical mirror analyzer (CMA) detecting electrons emitted at the magic angle. The length of the interaction region is 1 cm and the transmission of the spectrometer is about 0.5% for a resolution of 1%. The two beams counter propagate through the source volume of the electron spectrometer. The apparatus is shown in Fig. 3 called PISA (Photoionization of Ions at SUPER–ACO).

To estimate the signal expected in this experiment one needs to first calculate the target ion density d given by $d = n/Av$ where n is the number of ions/sec., A is the cross sectional area of the beam and v the velocity. For 20 keV Ca^+, $v \sim 3 \times 10^7$ cm/sec. With an ion current of 100μA and a beam radius of 1mm we obtain $d = 6.4 \times 10^8$ cm^{-3}. We may now estimate the number of events per second τ produced in the source volume of the electron spectrometer. $\tau = Nd\sigma\ell$ where N is the number of photons/sec, σ the photoionization cross section and ℓ the length of the interaction volume. At this point one needs to distinguish between direct continuum photoionization and resonant photoionization. Since, in the latter case, the process is a bound–bound transition, not all the photons from the undulator are useable compared to direct photoionization. Using for the continuum case $N=2 \times 10^{13}$ sec^{-1} and $\sigma = 1$ Mb we obtain $\tau = 10^4$ sec^{-1}. In the resonant case $N = 2 \times 10^{12}$ sec^{-1}, $\sigma = 100$ Mb and so $\tau = 10^5$ sec^{-1}. However only 0.5% are transmitted through the cylindrical mirror analyzer and detected. Therefore, we expect counting rates of about 50 sec^{-1} and 500 sec^{-1}. It is clear the experiment is feasible provided the background noise level is tolerable. Since the Daresbury work had identified a very intense resonant photoexcitation (σ = 2000Mb) at $h\nu$ = 33.2 eV we decided to begin the photoelectron experiments at this energy. The process under study is $Ca^{+}(3p^{6}4s) + h\nu(33.2eV) \rightarrow Ca^{+*}(3p^{5}3d4s) \rightarrow Ca^{2+}(3p^{6}) + e^{-}(21.3eV)$ via autoionization. Several corrections to the energies are needed. Due to the motion of the ions there is a Doppler shift of the synchrotron radiation by +0.03 eV in the ion frame. Therefore monochromatized photons of 33.17 eV are used. There is also an electron energy shift of +3 eV due to the velocity of the ions. Finally, the charge of the ion beam provides a potential which shifts the energy of the electrons by –6 eV.

The photoelectron spectra emitted in the photoionization of Ca^{+} at $h\nu$ = 33.2 eV has been measured[10] and the values observed for the kinetic energy and for the intensity of the photoelectron line are in good agreement with the predicted values. Also, the excitation function of the autoionizing line versus photon energy has been measured. Examples of data obtained are found in Fig. 4 and Fig. 5. The success of this feasibility experiment opens up wide opportunities for similar photoionization and Auger studies on multiply charged ions.[21]

6. Inverse Photoionization and Stimulated Radiative Recombination

The lecture series concluded with a discussion of the inverse process of photoionization, radiative recombination. This process has been of recent interest because of the possibility of stimulating it via laser radiation and thus demonstrating the feasibility of producing antihydrogen by using this scheme with antimatter. The normal matter process is, $h\nu + e^{-} + X^{+} \rightarrow X^{0} + 2h\nu$. It has been recently demonstrated by two groups that this is indeed feasible. One experiment[22] measured the spontaneous radiative recombination of protons and electrons to produce hydrogen atoms with principle quantum numbers $8 \lesssim n \lesssim 19$. This experiment used a merged beam apparatus in combination with electric field ionization detection. Also, infra–red radiation from a CO_2 laser was used to induce stimulated radiative recombination to the n=11 and n=12 states of hydrogen. An example of the data is shown in Fig. 6. In the second experiment,[23] stimulated radiative recombination to the n=2 level of atomic hydrogen was observed in an ion storage ring.

7. Acknowledgments

The support of the Brazilian Science Council and the United States National Science Foundation International Program Division is gratefully acknowledged.

8. References

1. B. Crasemann and F. Wuilleumier, Physics Today 37, 1 (1984)
2. G. Brown, K. Halbach, and J. Harris, Physics Today 34, 50 (1981)
3. J. Wynne, Physics Today 36, 52 (1983)
4. F.J. Wuilleumier, D.L. Ederer, and J.L. Picque, Advances in Atomic and Molecular Physics 23, 197 (1988)
5. J.M. Bizau, F. Wuilleumier, D.L. Ederer, J.C. Keller, J.L. LeGouet, J.L. Picque, B. Carre, and P.M. Koch, Physical Review Letters 55, 1281 (1985)
6. D. Cubaynes, J.M. Bizau, T.J. Morgan, F.J. Wuilleumier, M. Ferray, F. Gounand, P. D'Oliveira, and P.R. Fournier, Journal de Physique 48, 513 (1987)
7. M. Ferray, F. Gounand, P. D'Oliveira, P.R. Fournier, D. Cubaynes, J.M. Bizau, T.J. Morgan, and F.J. Wuilleumier, Physical Review Letters 59, 2040 (1987)
8. D. Cubaynes, J.M. Bizau, F.J. Wuilleumier, B. Carre, and F. Gounand, Physical Review 63, 2460 (1989). See also: B. Carre, P.D'Oliviera, M. Ferray, P. Fournier, F. Gounand, D. Cubaynes, J.M. Bizau, and F.J. Wuilleumier, Zeit. Phys. D, Atoms, Molecules and Clusters 15, 117 (1990) and M. Richter, J.M. Bizau, D.Cubaynes, T. Menzel, F.J. Wuilleumier, and B. Carre, Eusophysics Letters 12, 35 (1990)
9. F.J. Wuilleumier, Proceedings of the ESRF Workshop: "Photon–Ion Interactions," Grenoble 8–9 December 1988, p. 231 (Ed. by F.J. Wuilleumier and E. Kallne)
10. J.M. Bizau, D. Cubaynes, M. Richter, F.J. Wuilleumier, J. Obert, J.C. Putany, T.J. Morgan, E. Kallne, S. Sorensen, and A. Damany, Physical Review Letters 67, 576 (1991)
11. P.K. Caroll and E.T. Kennedy, Physical Review Letters 38, 1068 (1977)
12. T.B. Lucatorto and T.J. McIlrath, Physical Review Letters 37, 428 (1976)
13. T.J. McIlrath and T.B. Lucatorto, Physical Review Letters 38, 1390 (1977)
14. T.B. Lucatorto, T.J. McIlrath, J. Sugar, and S.M. Younger, Physical Review Letters 47, 1124 (1981)
15. J.W. Cooper, C.W. Clark, C.L. Cromer, T.B. Lucatorto, B. Sonntag, and F.S. Tomkins, Physical Review A 35, 3970 (1987)
16. I.C. Lyon, B. Peart, J.B. West, and K. Dolder, Journal of Physics B 19, 4137 (1986)
17. I.C. Lyon, B. Peart, and K. Dolder, Journal of Physics B 20, 1925 (1987)
18. I.C. Lyon, B. Peart, K. Dolder, and J.B. West, Journal of Physics B 20, 1471 (1987)
19. B. Peart and I.C. Lyon, Journal of Physics B 20, L673 (1987)
20. B. Peart, I.C. Lyon, and K. Dolder, Journal of Physics B 20, 5403 (1987)
21. F.J. Wuilleumier and E. Kallne, Proceedings of the ESRF Workshop: "Photon–Ion Interactions, Grenoble 8–9 December 1988, p. 251 (Ed. by F.J. Wuilleumier and E. Kallne)
22. F.B. Yousif, P. Vander Donk, Z. Kucherovsky, J. Reis, E. Brannen, J.B.A. Mitchell, and T.J. Morgan, Physical Review Letters 67, 26 (1991)
23. U. Schramm, J. Berger, M. Grieser, D. Habs, E. Jaeschke, G. Kilgus, D. Schwalm, A. Wolf, R. Neumann, and R. Schuch, Physical Review Letters 67, 22 (1991)

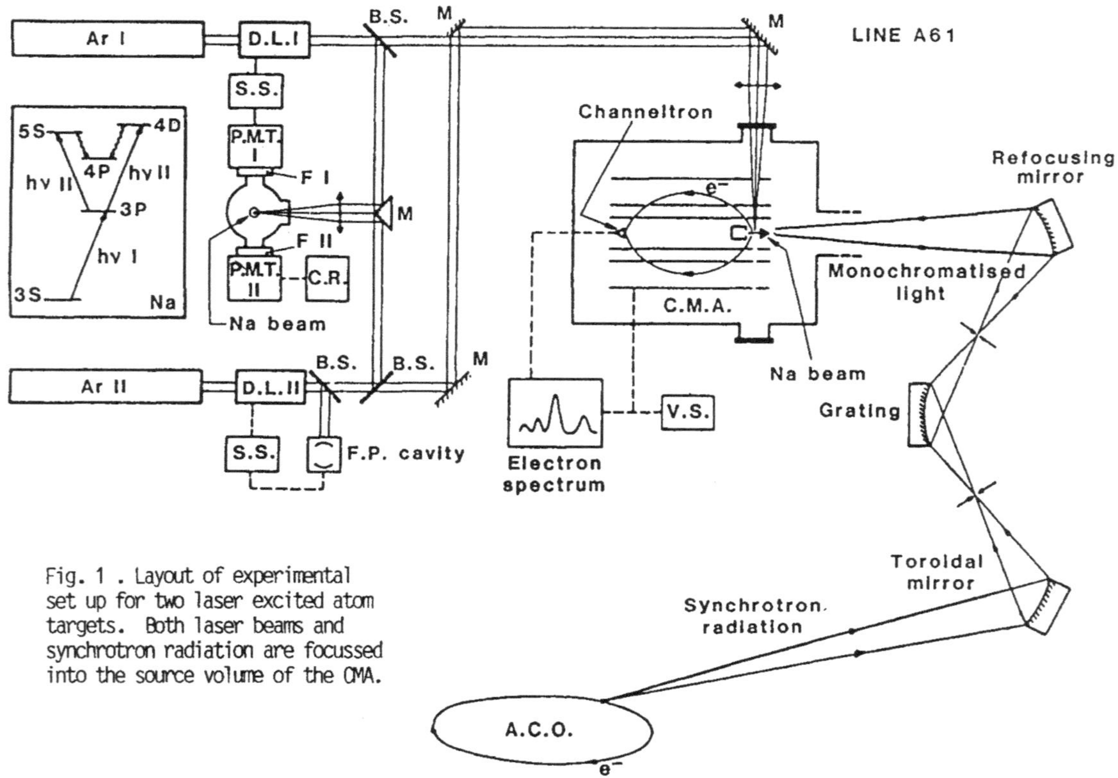

Fig. 1 . Layout of experimental set up for two laser excited atom targets. Both laser beams and synchrotron radiation are focussed into the source volume of the CMA.

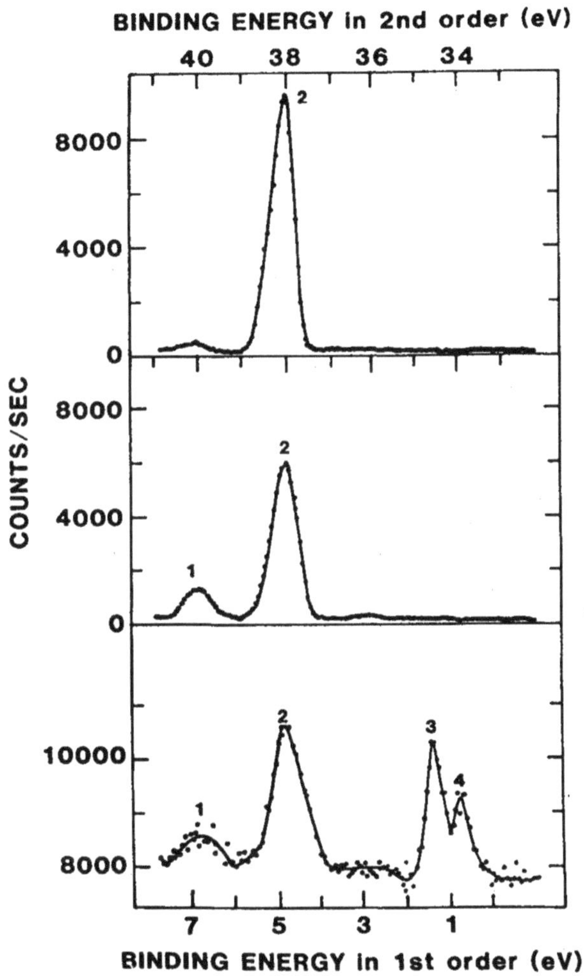

FIG. 2. Top: Photoelectron spectrum of Na taken at $h\nu = 33.06$ eV with both lasers turned off. Peak 2 is due to photoionization of the $2p$-shell electrons by photons of energy equal to twice $h\nu$ or 66.12 eV. Middle: Spectrum taken at the same $h\nu$ with laser I turned on; peak 1 is due to photoionization of the $2p$-shell electrons in $2p^6 3p$ atoms. Bottom: Spectrum taken at $h\nu$ with both lasers turned on; peaks 3 and 4 are due to autoionization of $2p^5 3s 4p\ ^2D$ and $2p^5 3s 4d\ ^2F$ states, respectively.

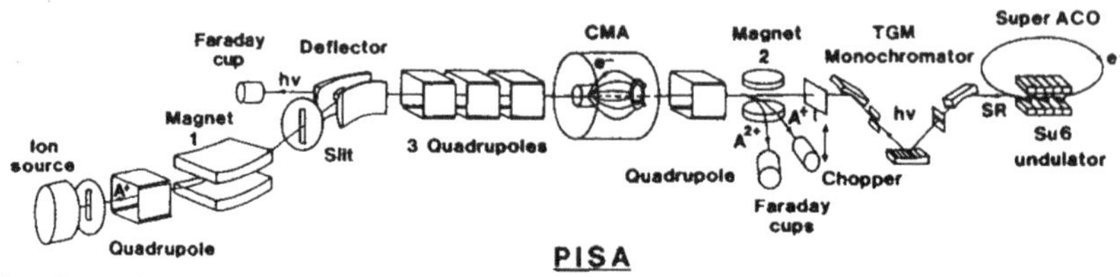

FIG. 3. Layout of the experimental setup. See text for description. Ions propagate from left to right, photons travel the other way. Both beams are focused into the source volume of the CMA.

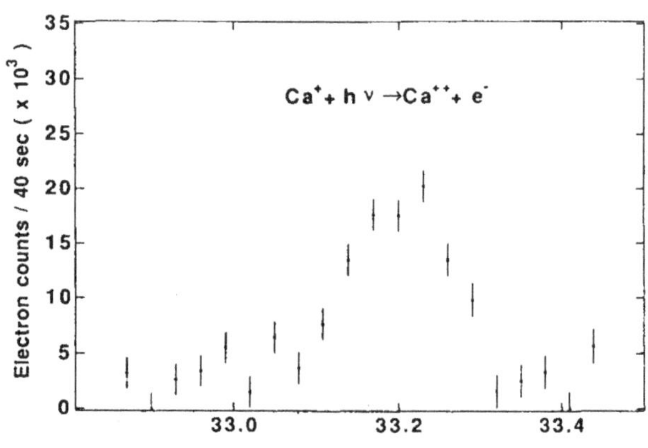

FIG. 4. Constant-ionic-state spectrum in the photon energy range of the resonance. The energy interval between each point is 0.03 eV. Error bars represent the statistical error.

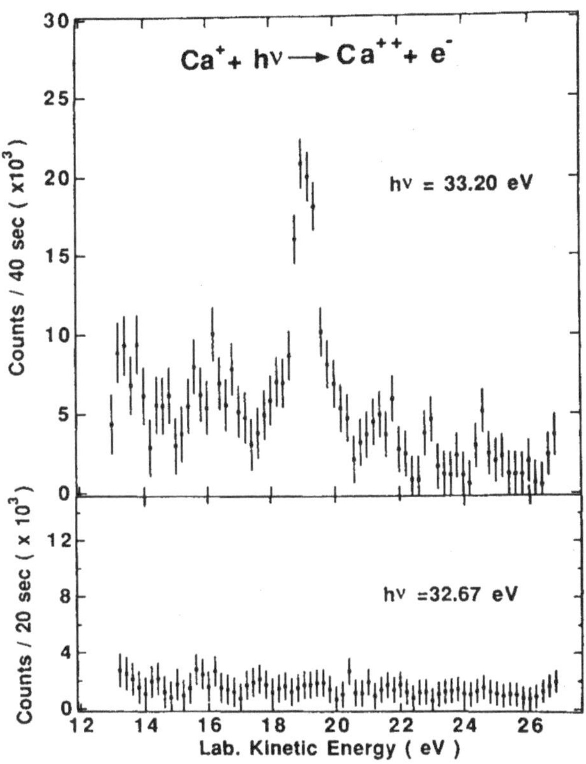

FIG. 5. Top: Photoelectron spectrum measured in the photoionization of a 20-keV, 35-μA Ca^{+}-ion beam with 33.20-eV (33.17 in the laboratory frame) photons. The Ca^{+} photoelectron line appears at 19.2 eV. The weaker electron line at 16 eV arises from photoionization of the $^{1}b_{1}$ state of H_2O. The averaged current in SuperACO was 200 mA. The error bars are larger than the statistical error on the count numbers, because the number of electrons in each channel results from the subtraction of two large numbers $[(1-2)\times 10^{5}]$. Bottom: Photoelectron spectrum obtained under the same experimental conditions, except for the photon energy, which is detuned from the maximum of the resonance by 0.5 eV. Only the H_2O photoelectron line is still visible around 16-eV kinetic energy, because the direct 4s photoionization cross section of Ca^{+} is very weak (1–2 Mb) at this photon energy.

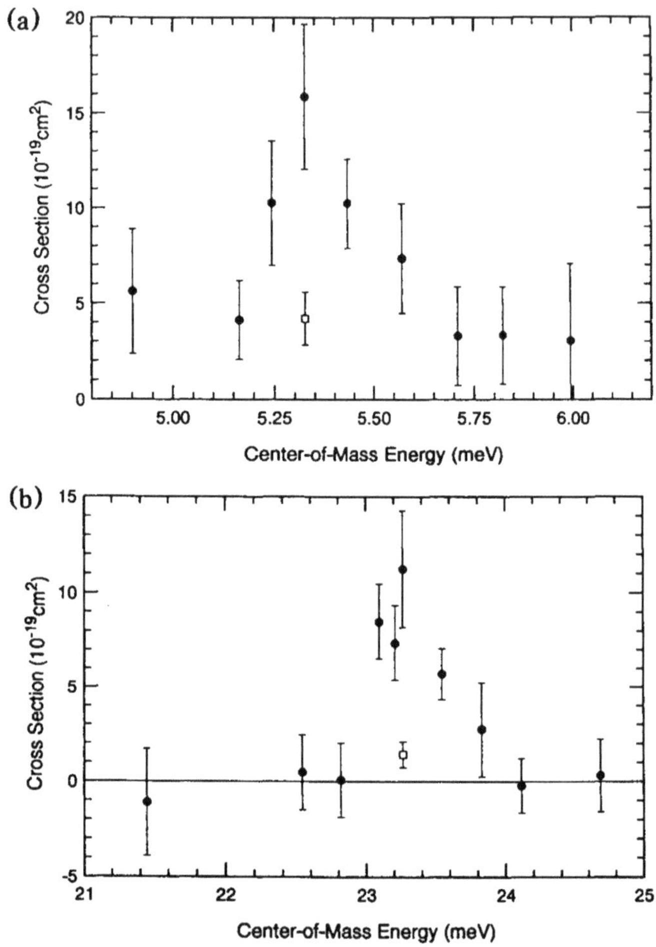

FIG. 6. Measured effective electron-proton recombination cross section with, ●, laser on and, □, laser off. (a) Peak at 5.33 meV corresponds to stimulated recombination to the $n=11$ level. (b) Peak at 23.3 meV indicates stimulated recombination to the $n=12$ level.

INVITED LECTURES

MATRIX EFFECTS IN ION DESORPTION FROM HEAVY-ION TRACKS IN SOLIDS

J. AI, R. NORD, and R. D. MACFARLANE
Department of Chemistry, Texas A&M University
College Station, Texas, 77843, U. S. A.

ABSTRACT

The deposition of energy by high energy ions (1 MeV/u) into matrices is dominated by electronic excitation processes. The electrical conductivity of the matrix plays an important role in the dissipation of energy from the heavy ion track with evidence that energy propagation in insulating solids is via phonon excitations leading to ion desorption at the surface and collapse of the surface structure. The results of three studies relating to the role of the matrix are presented in this paper. The first addresses the question of the role of electrical conductivity by examining ion emission from an electrically-conducting polymer where the conductivity can be continuously varied from a conducting to insulating state using electrochemical methods. The second study investigates the kinds of ions formed in heavy ion tracks where ions, molecules and atoms from multilayers mix in the heavy ion track and participate in ion-molecule reactions. The third study examines the question of whether the electronic structure at the surface of a heavy ion track is retained long enough to influence the ionization state of desorbed species.

1. INTRODUCTION

From the work that has been described in the previous lectures, a clear picture of the role of the incident ion in the electronic excitation of a solid matrix is now in place. More difficult to assess are the processes that occur as the energy deposited into the heavy ion track is dissipated into the surrounding matrix. For certain materials such as semiconductors, it is well known that a microplasma is produced and if the ionization occurs in the depletion layer of a reversed bias p/n junction, the electrons can be separated from the positive holes and the total electron charge can be collected. Solid state ion detectors operate on this principle. A heavy ion track generated in a scintillator populates electronic states which decay by photon emission but the density of these states is so high that exciton/exciton annihilation occurs resulting in reduced light output. These are examples of some of the dynamic properties of heavy ion tracks in selected matrices on time scales within a ns of the development of the heavy ion track. At much longer times (approaching infinity) the heavy ion track leaves a permanent record of its presence in the formation of a radiation-damaged volume leaving microscopic cylindrical damage tracks in certain insulating materials [1] and craters on the surface of other materials [2]. For both of these phenomena, the observation of the effects of heavy ion excitation are matrix dependent. So, it is not surprising that details of ion desorption triggered by fast heavy ion excitation are also matrix dependent. One of the difficulties in developing a universal model for ion desorption processes is the influence of the matrix; what might be perceived as a fundamental dynamic ion desorption property for a matrix of protein molecules adsorbed on the surface of nitrocellulose, for example, might not be observed for molecular ions ejected from a Langmuir-Blodgett film of amphiphilic molecules. For the subject of this paper, I have selected three studies carried out in our laboratory that give examples of specific ion desorption processes that are matrix dependent.

2. ION EMISSION FROM ORGANIC POLYMERS WITH VARIABLE ELECTRICAL CONDUCTIVITY.

2.1 Introduction

There is clear evidence that the excitation of insulating matrices by the fission fragments from ^{252}Cf results in the desorption of many more ions than for a metallic matrix. The current explanation is that in metals, energy transport from the fission track occurs very rapidly via electron conduction while for insulators, the energy transport is via phonon transport driven to shock wave intensity by an expansion of the material surrounding the fission track [3]. Some of this energy propagates to the surface resulting in ion desorption. A major problem in comparing metals with insulators is that too many variables are changed at one time so that if differences occur, it is not clear which of the many differences in the properties of the two materials are important. For example, the electrical conductivity and work functions both are quite different and both can influence ion emission. Ideally, one would like to study a matrix which can be switched reversibly from conducting to insulating state with minimal change in the physical chemical properties of the material. Such a class of materials exists among the organic conducting polymers that are of interest in materials science as electrochemical energy storage devices.

2.2 Properties of Poly(3-methylthiophene)

Poly(3-methylthiophene) (PMT) is an organic polymer that can be made in the form of a thin film by the electrochemical oxidation of the monomer. The polymer unit in the oxidized state is electron deficient and has a net positive charge that is balanced by negatively-charged counter ions residing in the matrix at sites along the polymer chain. In this form, the material is in an electrically-conducting state with an electronic structure characterized by a band gap of 2 eV containing 2 bipolaron states. The presence of these states gives the polymer a vivid blue color and it is through these bipolaron states that electron transport occurs. The polymer film can be transformed to an insulator by electrochemical reduction, destroying the bipolaron states resulting in an orders of magnitude increase in resistivity and a change in color from blue to red. The chemical structural changes that take place are minimal: the polymer chain is neutralized and the negatively-charged counter ions migrate from the film. The experiment carried out was to measure the ion emission properties of these films under fission fragment bombardment for various stages of oxidation/reduction or electrical conductivity to determine whether the intensity of ion emission was correlated with electrical conductivity.

2.3 Results and Discussion.

The ^{252}Cf-PD mass spectrum of the oxidized film (conducting form) was dominated by an intense counter ion in the negative ion spectrum ($CF_3SO_3^-$) and an SO_3^- fragment ion. The remainder of the peaks were identified as products of pyrolysis of the very strong polymer structure. An ion characteristic of the backbone thiophene sub-unit is the SH^- ion. A spectrum of this ion is shown in figure 1a.

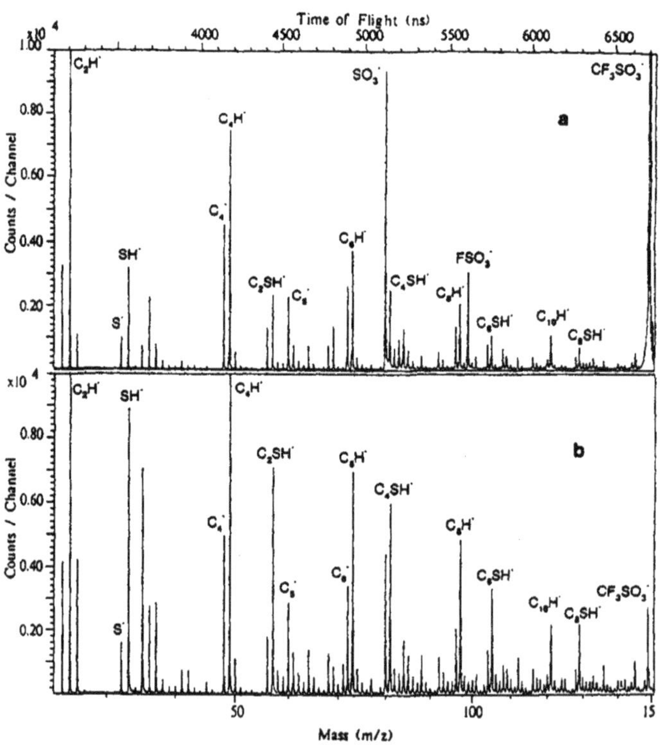

Figure 1: ^{252}Cf-PD mass spectra of poly(3-methylthiophene) in the a.) the oxidized (conducting) form and b.) reduced (insulating) form showing the influence of electrical conductivity on ion desorption yields.

When the polymer was reduced to the insulating form, the intensities of these pyrolysis ions increased by a factor of 2-5 as shown in figure 1b. When the film was electrochemically recycled to the conducting state the intensities of these ions diminished to their original values. We interpret these results as direct evidence that the electrical conductivity of the medium around the matrix has a direct influence on the dynamic conditions prevailing in the fission track.

3. ION-MOLECULE REACTIONS IN THE PLASMA REGION OF THE HEAVY-ION TRACK BETWEEN MULTILAYERS.

3.1 Introduction

A ^{252}Cf-PD mass spectrum is very complicated, composed of molecular and fragment ions of the molecules of interest in the matrix and a strong background of extraneous ions generally of unknown origin and little interest. Many of the species below 100 u, are ions that are observed in the high temperature pyrolysis of organic molecules. Examples include a series of C_xH_y positive and negative ions, CN^- and atomic ions. These ions are undoubtedly formed in the high energy density core of the heavy ion track. Some years ago, Tombrello made a curious observation that when a heavy ion track passes through two physically-different layers in intimate contact, they are welded together [4]. He later provided microscopic evidence for mixing between the two layers. This could only occur if at some stage of the development of the heavy ion track, the atoms and molecules freely moved between layers as if in a fluid state. This observation prompted us to to determine whether we could identify ion desorption products from surfaces where the material near the surface was a composite of several layers that were in intimate contact.

3.2 Experimental Details

The substrate consisted of three layers. The base layer was a film of the oxidized PMT. On top of this layer a 0.5 nm thick coating of Au was deposited. Electron micrographs of this surface showed the Au layer to be composed of islands of Au less that 0.1 micron in diameter. A third layer of CsI (1 ug/cm^2) also present in the form of small islands was deposited from an electrospray.

3.3 Results and Discussion

The positive ion mass spectrum of the composite is shown in figure 2a and the negative spectrum in figure 2b.

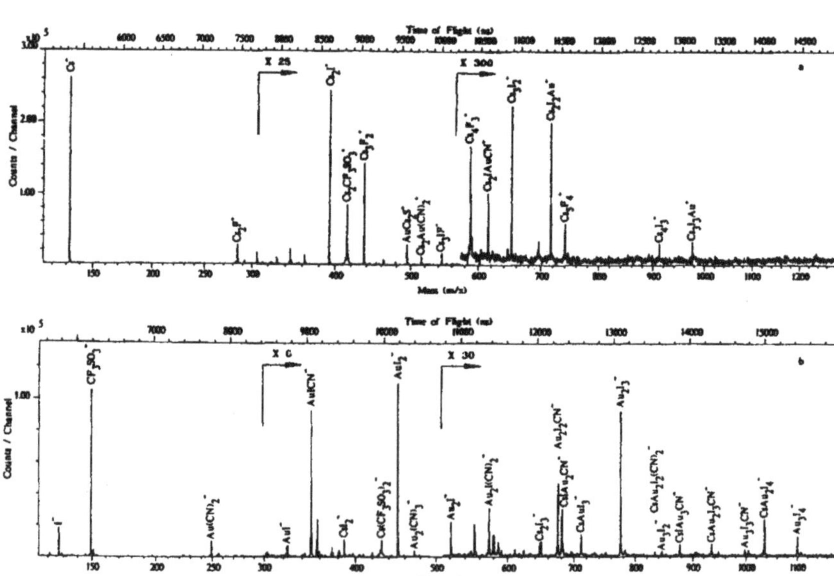

Figure 2: ^{252}Cf-PD mass spectra of a PMT/Au/CsI multilayer showing evidence of interlayer mixing in the fission track; a.) positive ions; b.) negative ions.

Each of the layers contributes ions and atoms to the composition of the plasma and all appear to have enough time to mix and produce the most stable molecular ion species based on the polarizability of the atoms comprising the molecular ion. For example, even though Cs^+ is associated with I^- in the matrix, the dominant Cs-containing molecular ion is Cs_2F^+. The F atom is generated as a decomposition product from $CF_3SO_3^-$ in the PMT layer and reacts with Cs atoms from the CsI layer to produce the CsF molecule, a very stable ionic species composed of two closed shell ions with low polarizability (hard/hard interaction). This molecule undergoes an ion molecule reaction with a Cs^+ ion to form the Cs_2F^+ product ion. Thus, in the short time that the plasma exists in the fission track, there is sufficient time for mixing and two chemically-reactive collisions to occur. There are also examples of chemical reactions between polarizable species in the same track. One of the strong peaks in the negative ion spectrum is AuI_2^- formed from the reaction of an Au atom with an I atom producing AuI followed by the attachment of an I^- ion to form a AuI_2^- molecular ion. Many other combinations of atoms and ions are present in the spectrum giving evidence of strong mixing between layers and multiple chemically-reactive collisions, as one might expect from a high temperature, high density plasma that rapidly cools to condense out complex multielement molecular ions.

4. DOES ION EMISSION OCCUR BEFORE OR AFTER THE ELECTRONIC STRUCTURE OF THE SURFACE IS DESTROYED?

4.1 Introduction

It is evident that when a fast heavy ion passes through an insulator there is an explosive release of energy and material at the surface. Craters are observed and most computer simulations present a scenario where surface layers erupt spewing molecules into vacuum in time scales on the order of tens of femtoseconds. But what we observe in ^{252}Cf-PDMS are ions and the question is, "At what stage in the eruption of this molecular volcano does ionic emission occur?" What is investigated in this study is whether there is any evidence that some ion emission occurs while the surface electronic structure is still intact.

4.2 Background Information for the Experiment.

CsI forms a classic ionic crystal in the solid state consisting of Cs^+ and I^- ions held together by Coulomb forces. When the CsI molecule in the gas phase is excited to dissociation energies, the two species separate first as ions but at a larger separation are converted to neutral Cs and I atoms by electron tunneling at a distance where the ionization potential and electron affinity match[5]. If dissociation occurs on the surface of a substrate, the final charge state of Cs and I can be influenced by the surface electronic states of the substrate if it is still intact and has not been destroyed by the explosive eruption of the surface surrounding the fission track. If the substrate is metallic, the relative energies of the Fermi level, ionization potential of Cs and electron affinity of I determines whether the final states of these species are neutral or charged. If the substrate is an insulator, the situation is a little more complicated. The ionization potential and electron affinity of Cs and I still play a role, but as each ion leaves the surface a vacancy level appears between the valence band and conduction band which changes in energy as the ion leaves the surface. The possibility for charge neutralization via resonant electron tunneling can occur for both species. The signature that the formation of gas phase Cs^+ and I^- ions is the result of dissociation of excited gas phase molecules desorbed in the plume of the fission track is the appearance of both ion in the emission spectrum while if only one is present, this would mean that dissociation took place within the field of the surface.

The experiment consisted of two parts. The first part examined ion emission from CsI adsorbed on a 2 micron thick polymer film. This configuration represents a matrix where if ion formation is due to dissociation of gas phase CsI molecules, both Cs^+ and I^- would be observed, or one or both are quenched due to ion/vacancy interactions. In the second part of the experiment, the polymer film was coated with a 0.5 micron thick deposit of a metal (Au, Cu, Pt) and the CsI was deposited as a thin film on this surface. The rationale was that the junction between the metal and CsI would perturb the surface electronic structure of the CsI and possibly influence its surface ionization properties. ^{252}Cf-PD mass spectra were recorded for both surfaces.

4.3 Results and Discussion

Figure 3a shows the negative ion ^{252}Cf-PD mass spectrum of CsI deposited on Mylar. The spectrum is very weak and there is no evidence for the formation of I^-. An intense Cs^+ peak was observed in the positive ion spectrum.

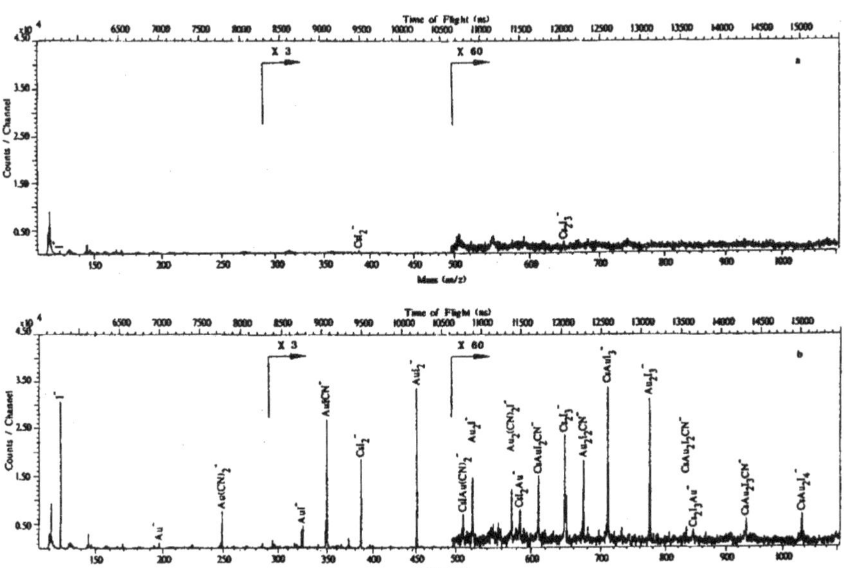

Figure 3: ^{252}Cf-PD negative ion spectra for a.) CsI deposited on an insulator and b.) on an insulator coated with 0.5 nm Au showing evidence for the retention of surface electronic states prior to ion formation.

These observations mean that the dissociation of CsI is occuring in the force field of the CsI microcrystalline surface. Many of the Cs^+ ions survive as ions as they are desorbed from the surface but the I^- ions are presumably neutralized by resonant tunneling interactions with its own vacancy level. Figure 3b shows the spectrum when a 0.5 nm thick Au layer is present under the CsI. The total negative ion emission is orders of magnitude larger and the I^- intensity is the dominant peak in the spectrum. We can only conclude from this major effect of the Au layer that the surface electronic structure of the matrix remains intact long enough to influence the charge state of the desorbing species.

5. ACKNOWLEDGEMENTS

I am grateful to the Organizing Committee, specifically, Drs. A. C. A. Souza and E. F. Silveira, for the opportunity to participate in this Summer School. I would also like to extend my appreciation to the Brazilian CNPq, the U.S. National Institutes of Health, and the Welch Foundation for their financial support of my research and my participation in this Summer School.

6. REFERENCES

[1] Fleischer, R. L., Price, P. B., and Walker, R. M., Nuclear Tracks in Solids: Principles and Applications; University of California Press, Berkeley, (1975).

[2] Chadderton, L. T., Nucl. Tracks Radiat. Meas., 15, 11 (1988).

[3] Bitensky, I. S., and Parilis, E. S., Nucl. Instrum. Meth. Phys. Res., B21, 26 (1987).

[4] Tombrello, T. A., Int. J. Mass Spectrom,. Ion Phys., 53, 307 (1983).

[5] Williams, P., Int. J. Mass Spectrom,. Ion Phys., 53, 101, (1983).

THE CLUSTER-SOLID INTERACTION: FUNDAMENTALS AND APPLICATIONS

M. A. PARK, M. E. INMAN, A. Y. YAU, B. D. COX, E. A. SCHWEIKERT

Center for Chemical Characterization and Analysis
Department of Chemistry
Texas A&M University
College Station, TX 77843-3144 USA

ABSTRACT

Bombardment of solids with clusters (e.g., CsI_2^-) results in high secondary ion (SI) yields. For projectiles of 5-30 keV, SI yields as high as 10% have been observed. Our data show that the SI yields increase with the mass of the projectile and its velocity. The SIs are identified by TOF-MS with event-by-event counting. The SIs emitted from a single projectile impact can be acquired in a coincidence mode. Chemical and spatial relationships of compounds desorbed from the same impact site are thus revealed.

1.0 Introduction

The interaction of a keV cluster (e.g., $Cs_3I_2^+$) with a solid occurs with some distinctive features: the atomic constituents of the projectile impact the solid coincidentally in time, in a spatially correlated manner and they deposit a high density of translational energy which is dissipated in nonlinear processes.[1] The latter result is an enhanced number of Secondary Ions (SIs) and in enhanced sputter yields. We have observed up to 50-fold enhancement in the SI yield for clusters vs. monoatomic projectiles of equal velocity.[2] The spatial correlation of the impacting projectile constituents applies also to the emission of the corresponding SIs and neutrals. We have recently shown that the SIs emitted coincidentally from an impact site reveal surface homogeneity at the submicron level.[3] The purpose of this paper is to review the rationale for using keV clusters as projectiles for probing solid surfaces and to describe the use of coincidence counting of the SIs emitted to maximize the physiochemical information from each projectile impact.

2.0 Experimental

The experiments discussed here were run in time-of-flight mass spectrometry and event-by-event counting mode. A diagram of the TOF-MS is shown in Figure 1. The multistop time-to-digital converter (TDC) used was of the type described by Festa.[4] Clusters were produced either by Cf-252 fission fragment induced desorption,[5] by spontaneous desorption (SD) or by laser desorption.[6] SD was first observed by Della Negra.[7] In this process (Figure 2), ions are field desorbed from the edge of an aperture in a steel plate and accelerated towards a negatively biased source foil. Ions striking the source foil cause the emission of neutrals, electrons and primary ions. The primary ions and electrons are accelerated by the electric field through the aperture and guard rings. Primary electrons are accelerated by a weak magnetic field (~ 3 Gauss) into a microchannel detector. This provides a "start" signal for the TDC. The potential difference between the source and aperture was about 10 kV to maintain the primary ions flux at about $10^4 s^{-1}$. The kinetic energy of the primary ions in the SD cluster SIMS experiments ranged from 18 to 30 keV. Figure 3 shows a TOF primary ion mass spectrum of CsI biased to 20 kV. The signal-to-noise ratios for the I^- and CsI_2^- peaks are 51 and 36, respectively. The SD cluster SIMS experiments were performed in a vacuum chamber at a pressure of about 10^{-6} torr. The primary ion source was prepared by grinding CsI crystals into a fine powder and then mounting the powder onto a holder with double-stick tape.

Large organic polyatomic ions can be produced by matrix assisted laser desorption (MALD). MALD-MS has seen recent and widespread use in the field of biomolecular analysis.[6] Direct laser desorption is limited to masses of only a few thousand Daltons due to extreme fragmentation of the molecular ion. However, in MALD the matrix most often an organic acid absorbs the majority of the laser energy and then transfers this energy to the surface layer.[8] It is therefore important to match the emission wavelength of the laser to the absorption wavelength of the acidic matrix. Figure 4 shows the TOF mass spectrum of the protein Gramicidin-S (M.W. = 1215 Daltons) obtained using a nitrogen laser (VSL-337 Laser Science Inc.). The laser was operated at a wavelength of 337 nm and a pulse rate of 20 Hz. The protein was mixed at a ratio of 1 part protein: 10 parts matrix (4-nitroanaline) in THF.

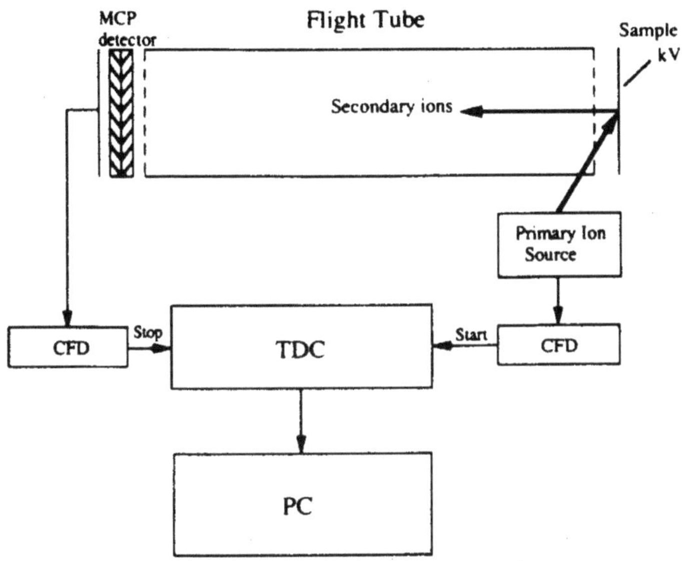

Figure 1. Generalized TOF-MS instrument.

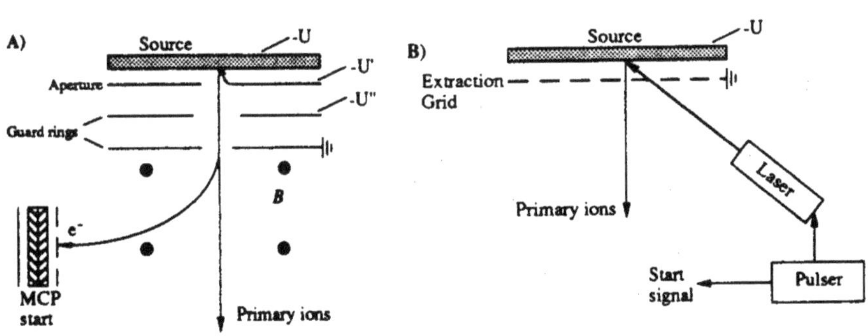

Figure 2. Diagram of a (A) spontaneous desorption based source and a (B) matrix assisted laser desorption source.

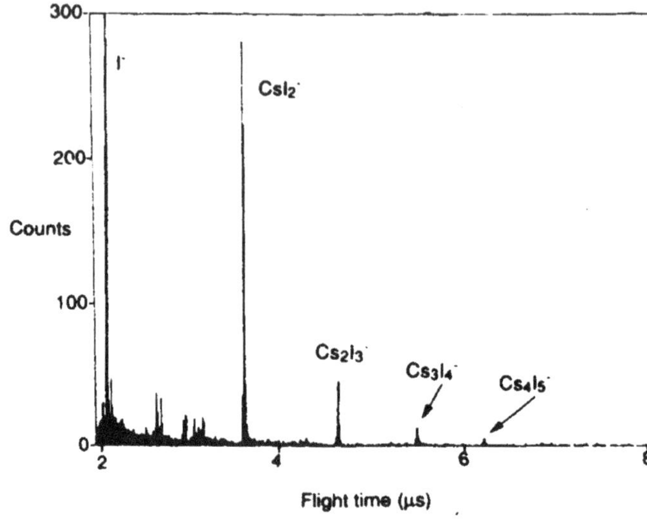

Figure 3. TOF mass spectrum of the primary ion produced from the spontaneous desorption source using a CsI covered source foil.

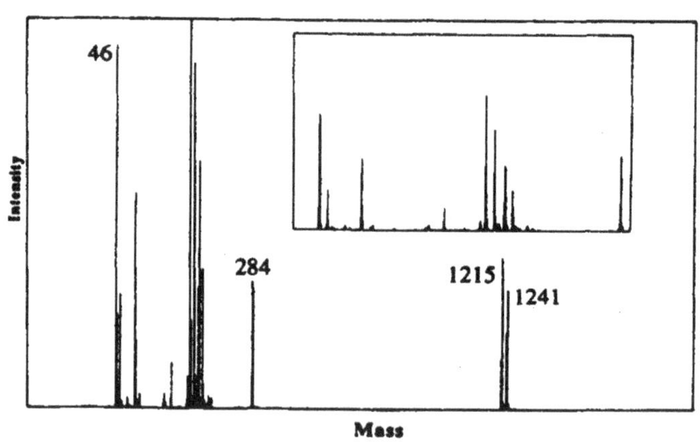

Figure 4. TOF mass spectrum of the primary ions produced from the matrix assisted laser desorption source using a Gramicidin-S covered source foil. Inset is an expansion of the low mass region.

3.0 Results and Discussion

3.1 *Secondary Ion Yields*

The yields are defined as the number of ejected ions per incident primary projectile. The data presented here (Figure 5) are for bombarding energies ranging from 15 to 30 keV. The yields are presented as a function of the energy per mass unit (E/A) in keV/amu. The secondary ion yields for different primary projectiles are obtained simultaneously at each target bias setting. Thus, the data can be readily compared since the experimental conditions (target surface condition, spectrometer transmission and detector efficiencies) are the same; indeed the relative error in the data is less than 10%.

The data (Figure 5) demonstrate the high SI yields generated from an organic target bombarded with polyatomic projectiles in the keV range. Similar observations have been made previously with several other projectile-target combinations. Our data confirm also earlier findings that the largest relative yield increases from mono- to polyatomic projectiles occur with organic targets. The effect of projectile complexity on SI yields becomes apparent when considering the yield enhancement, ϵ, that is the yield per incident atom of a polyatomic projectile vs. the SI yield of the monoatomic equivalent at equal velocity. Typically ϵ will have values in the 2-5 range for projectiles with 3-7 constituent atoms. Thus, at equal velocities, an atom "packaged" in a polyatomic ensemble may be 2-5 times more efficient for SI production than when the same species hits the target as a monoatomic ion. Figure 5 illustrates also the dependence of SI yields on the projectile velocity. For the velocity regime studied, the normalized yield of a polyatomic projectile (yield per constituent atom) increases faster than the yield of the monoatomic ion. There are no data yet to clarify at what velocity the SI yield will maximize, i.e., how the latter scales with the nuclear stopping power of the projectile. It may be noted that a similar trend in the normalized yield vs. the velocity of molecular projectiles has also been observed in the electronic stopping power regime.

To illustrate the scope of secondary ion mass spectrometry with cluster projectiles, we have compared the yields obtained with 21 keV $(CsI)I^-$ on four organic compounds with those generated on the same targets with 21 keV I^- and fission fragments. The same targets and spectrometer were used for these comparisons. The secondary ion yields from plasma desorption are shown in Table I. PDMS produced comparable but somewhat higher secondary ion yields than the clusters projectiles from the compounds studied. The molecular ions

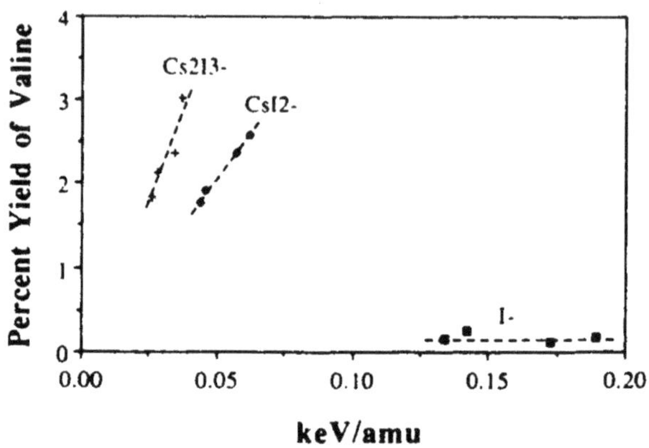

Figure 5. Plot of Y of molecular ions of valine versus primary ion kinetic energy.

Table I. Comparison of yields of organic compounds between fission fragments, monoatomic ions, and cluster ions.

Compound	Fission Fragments	I^- (20 keV)	CsI_2^- (20 keV)
GABA	6.1	0.03	0.5
Epinephrine	3.1	0.1	0.6
Phenylalanine	7.0	0.3	2.2
Valine	5.0	0.2	1.8

were easily identifiable in the PDMS secondary ion mass spectra. Little fragmentation was observed. The PDMS spectra were typically less noisy and had a higher signal-to-noise ratio than the cluster SIMS spectra. It should be noted that higher SI yields can be expected with higher energy and more massive clusters.[9] The limits to the trend depicted in Figure 5 have not yet been reached. Still in the realm of potentialities, it may also be recalled that keV ions can be readily focused and rastered by electrostatic means. This opens prospects for surface mapping which cannot be matched with MeV/amu ion beams.

3.2 *SI Coincidence Counting*

Coincidence methods have existed and been used extensively for many years in fields such as nuclear physics. Conventional TOF-MS takes advantages of the coincidences between primary ions and secondary ions to produce mass spectra. A primary ion striking a surface causes the emission of secondary ions. Thus, the emission of the secondary ions is coincident in time with the impact of the primary ion on the sample surface. Recording only those signals which are in coincidence with the primary ion impact reduces the background due to unrelated ions and signals. All the secondary ions in coincidence with a primary ion are recorded in the mass spectrum. By summing the results of many ($\sim 10^5$) primary ion impacts, a mass spectrum of statistical significance is produced.

A keV or MeV kinetic energy ion striking a surface may cause the desorption of several secondary ions. In addition to the relationship between the secondary ions and a primary ion, coincidence counting can be used to determine the relationship between the secondary ions themselves. When two or more secondary ions are observed as the result of the same primary ion impact, they are said to be in coincidence with one another. By observing these coincidences, it is possible to determine the relationships between the ions. If two ions are observed in coincidence more frequently than would be expected by statistics, then they are said to be correlated with one another. Ions are anticorrelated with one another if they are observed in coincidence less frequently than would be expected by statistics. Correlations and anticorrelations between ions may be the result of heterogeneities in the sample, or chemical or physical relationships between the ions.

In the experiments performed here, a low flux ($10^4\ s^{-1}$) of primary ions was used. This made it possible for a personal computer to examine and sort the results of primary ion impacts on an individual basis. The PC produces a conventional spectrum by recording all secondary ions which were observed in coincidence with a primary ion. The PC simultaneously produces coincidence spectra in which only the results of primary ion impacts producing a secondary ion of interest are recorded.

The relationships between the secondary ions are determined by comparing the relative intensities of the secondary ion peaks in the coincidence spectrum with those of the peaks in the conventional mass spectrum. This may be done by calculating either the percent coincidence of the ions with one another or the correlation coefficient of the ions. The correlation coefficient, Q, represents the actual divided by the expected probability of observing two ions in coincidence. This is calculated as $P(A^+,B^+)/P(A^+)P(B^+)$ where $P(A^+,B^+)$ is the observed probability of detecting the ions in coincidence, and $P(A^+)$ and $P(B^+)$ are the probabilities of observing ions A^+ and B^+, respectively. If the ions A^+ and B^+ are observed in coincidence only by random chance, $P(A^+,B^+)$ and $P(A^+)P(B^+)$ should be equal and Q should equal one. A Q greater than one would indicate a correlation between A^+ and B^+ and a Q less than one would indicate an anticorrelation.

The percent coincidence of one ion with another is defined as the percentage of the ions which are observed in coincidence with the ion of interest. For example, the percent coincidence of ion B^+ with A^+,$\%(A^+,B^+)$, is given by $[N(A^+,B^+)/N(B^+)] \times 100\%$ where $N(A^+,B^+)$ is the number of counts in the B^+ peak in the A^+ coincidence spectrum (i.e., the number of times A^+ and B^+ were observed in coincidence), and $N(B^+)$ is the number of counts in the B^+ peak in the conventional spectrum. The percent coincidence of ions with one another is proportional to their correlation coefficient. By comparing the percent coincidence of ions with the ion of interest, one can determine which ions are more correlated with the ion of interest. This type of comparison can be made readily via a percent coincidence plot such as that shown in Figure 6.

The plot of Figure 6 resulted from the bombardment of a poly(ethylene) surface with MeV kinetic energy primary ions. Here, the percent coincidence of secondary ions with H^+ is plotted versus the secondary ion mass. Several sets of related points are apparent. Each set of points corresponds to a given number of carbon atoms and a varying number of hydrogen atoms in the secondary ion. Within a set (neglecting n = 0), the type of ion which is most correlated with H^+ is that which contains the least hydrogen. This suggests a chemical relationship between H^+ and the hydrocarbon ions. The simplest explanation of the data is that H^+ ions are formed from hydrogen lost from the species from which the hydrocarbon ions are formed.

Another example of chemical relationships is observed in data obtained from a CsI surface bombarded by 21 keV CsI_2^- primary ions. The I^-

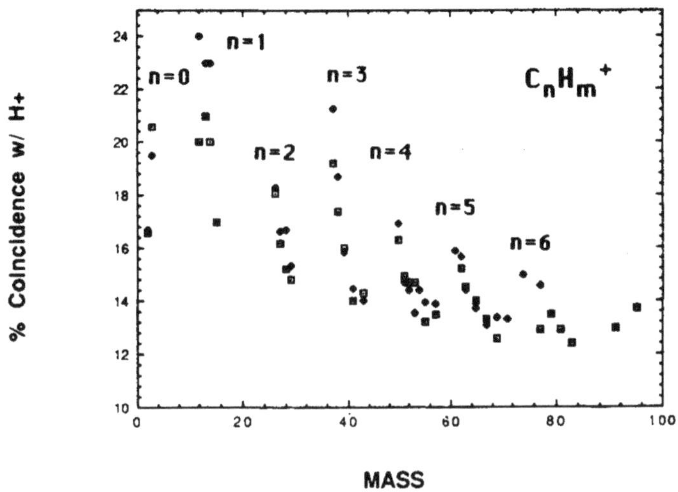

Figure 6. H^+ percent coincidence plot resulting from the analysis of poly(ethylene) via MeV kinetic energy primary ions.

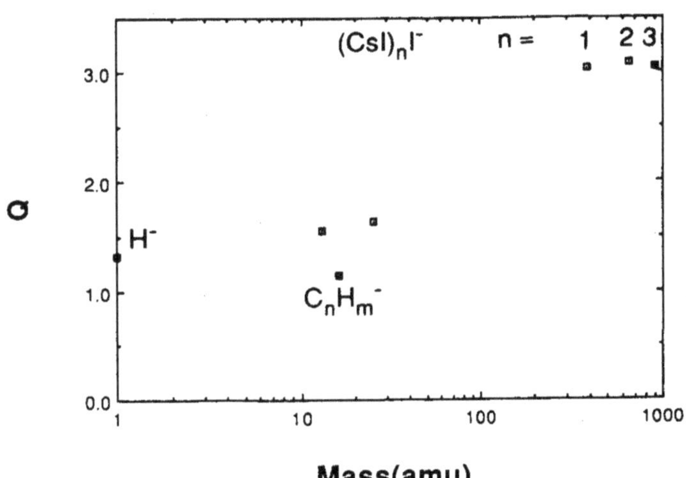

Figure 7. I^- correlation plot resulting from the analysis of CsI via keV kinetic energy primary ions.

correlation plot obtained in this way is shown in Figure 7. Here the correlation coefficient of the ions with I^- is plotted versus the mass of the secondary ions. Two sets of points are observed in this plot. The low mass set of points associated with hydrocarbon ions is seen to be relatively uncorrelated with I^- whereas the higher mass $(CsI)_nI^-$ ions are formed by breaking a chemical bond between the two.

Although these data are explained as due to chemical relationships, one must realize that a physical process is involved in the preservation of such information in the coincidence data. That is, the observed chemical correlations are indicative of process occurring during ion formation and desorption. In each of the above cases, the secondary ions are formed by breaking a chemical bond, however, one may question the circumstances under which such bonds are broken. Among these circumstances, one may consider whether the ions are formed before or after desorption from the surface. That is if the ions are formed by a "prompt" or "metastable" decay process.

In the case of the H^+ desorption from poly(ethylene), it is most likely that the ions are formed by a prompt decay. It is now apparent from the literature that H^+ is desorbed directly from the Coulomb explosion that results from the passage of a MeV ion through a surface.[10,11] This being the case, H^+ must be formed early in the desorption process (i.e., promptly). The observed correlation with H^+ implies that the low mass hydrocarbon ions are formed from the same precursor as the H^+ by a DIET type of mechanism.[12]

In the case of the I^- desorbed from CsI, a metastable decay mechanism may be more likely. Assuming this mechanism, larger doubly charged cluster ions would be desorbed from the surface (e.g., CsI_3^{2-}). Soon after desorption (perhaps 10^{-10} s) the ion decays to form smaller singly charged ions (e.g., $I^-$and CsI^{2-}) which are then observed in coincidence with one another.

Aside from chemical and physical phenomenon, coincidence counting can be used to observe the extent of homogeneity of a surface. keV and MeV ions are essentially nanoprobes of sample surfaces. Such a primary ion will cause secondary ion desorption from a region of about 10 nm in diameter. If a sample is chemically heterogeneous on a scale larger than 10 nm then a given primary ion may address a single chemical component. Secondary ions resulting from this primary ion impact would be indicative of the chemical component struck. In contrast, primary ions striking a homogeneous sample would produce secondary ions of all chemical components simultaneously. Thus, if one observes secondary ions indicative of the chemical components in coincidence then the sample should be homogeneous whereas if the such secondary ions are anticorrelated then the sample is chemically heterogeneous. The degree to which the sample components are mixed should be reflected by the degree to which the secondary ions are observed in coincidence.

To test a surface for chemical homogeneity, the primary ion must address only a small region of the surface at a time as mentioned above and each chemical component of the surface must produce a distinct mass spectral peak in order to be distinguishable from one another. Having identified peaks indicative of the chemical components one observes the coincidences of secondary ions with those peaks via the coincidence spectra. Again, percent coincidence plots can be generated based on the coincidence and conventional spectra. Considering a two component sample, a heterogeneous sample should produce two distinct sets of points in the percent coincidence plot. The plot in Figure 8 is the result of the analysis of a sample of ~ 10 μm NaF crystals on a poly(vinyl methyl ether) substrate. Based on a model discussed in a previous paper,[3] it is possible to estimate the difference between the sets of points if the size and coverage of the segregations on the surface are know. Alternatively, the average size of the segregations can be determined if the separation between the sets of points and the coverage is known:

$$b = 2Hc / 3(1 - \sigma)$$

where b is the average particle size, H, the homogeneity factor, is the average percent coincidence of the set of points associated with the segregation divided by the average percent coincidence of the set of points associated with the substrate, c is the diameter of the area addressed by the primary ion, and σ is the coverage of the segregations on the surface.

Clearly, the relationship between particle size and homogeneity factor is linear. This opens the possibility for the production of calibration curves and the determination of the size of chemical components on surfaces. To explore this possibility, we examined samples of NaF crystals on poly(styrene) and poly(vinyl methyl ether) and rhodamine doped PS spheres on a PVME surface. The resulting plot of H versus particle size (see Figure 9) shows two curves: one based on NaF segregations and the other based on PS segregations. As expected from the model, the curves are indeed linear. However, the value of the y-intercept of the two sets of points differ by an order of magnitude. From the model, when the size of the segregations are about the diameter of the area addressed, the homogeneity factor should be approximately one. That the NaF data are much higher than one may indicate a chemical correlation. Such a chemical correlation would not affect the slope of the line between NaF type ions but only the y-intercept.

The usefulness of coincidence counting in TOF-MS has yet to be fully explored. As discussed above, methods based on coincidence counting are useful for the determination of the size of segregations on surfaces, and for

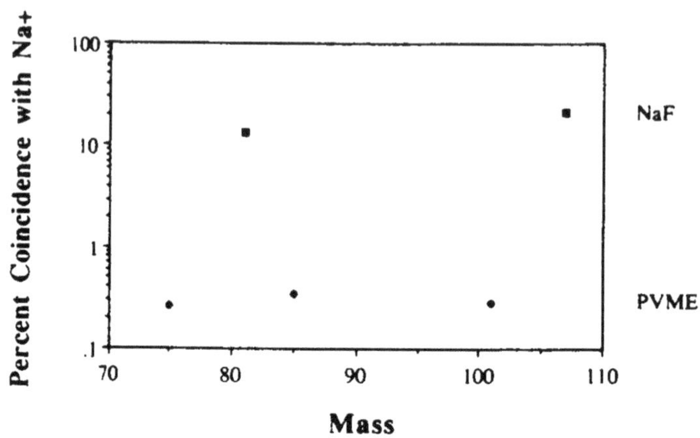

Figure 8. Na^+ percent coincidence plot produced from 10 μm NaF crystals on a poly(vinyl methyl ether) substrate. The large difference in the coincidence of NaF type peaks and PVME type peaks with Na^+ shows that the sample is segregated.

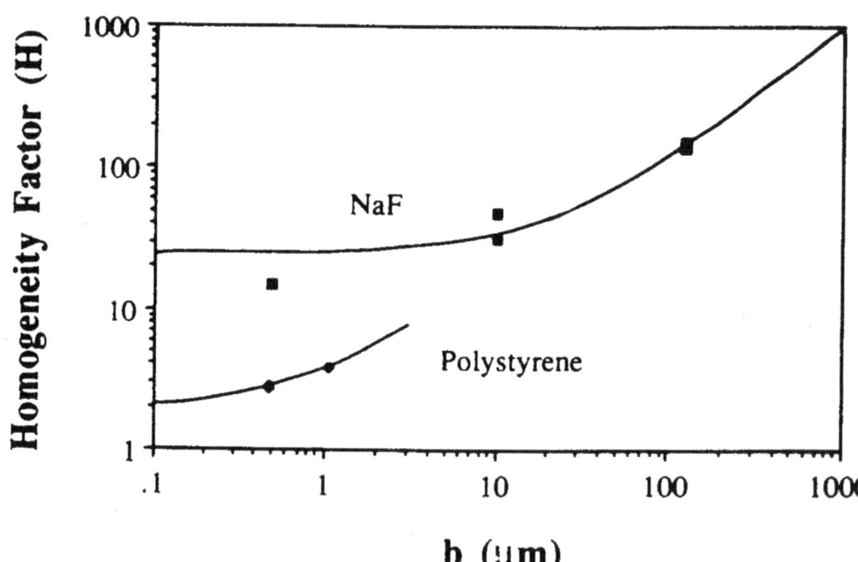

Figure 9. Plot of the homogeneity factor, H, versus particle size, b, for NaF crystals on a poly(styrene) or poly(vinyl methyl ether) surface and rhodamine doped poly(styrene) spheres on a PVME surface.

the examination of chemical and physical relationships between secondary ions. Though not discussed here, coincidence counting can also be used for the determination of the yield of secondary ions even when the primary ion flux is not known.[13]

4.0 Conclusion

The fundamental aspects of energy dissipation and ion formation involved in keV cluster-solid interactions remain to be fully understood. Further cluster bombardment with coincidence counting of the SIs emitted provides an approach for studying chemical and spatial relationships on domains of tens to hundreds of nm. Thus cluster-SIMS could provide "chemical vision" at a scale where at present our understanding is mostly based on physical and morphological characterization.

5.0 References

1. R. J. Beuhler, L. Friedman, *Chem Rev.* **86** (1986) 521.
2. M. G. Blain, S. Della-Negra, H. Joret, Y. Le Beyec, E. A. Schweikert, *Phys. Rev. Lett.* **63** (1989) 1625.
3. M. A. Park, K. A. Gibson, L. Quinones, E. A. Schweikert, *Science* **248** (1990) 988.
4. E. Festa, R. Sellem, L. Tassan-Got, *Nucl. Instr. Meth.* **A234** (1985) 305.
5. M. G. Blain, E. A. Schweikert, E. F. da Silveira, *J. de Physique* **50**, C2 (1989) 85.
6. M. Karas, A. Ingendoh, U. Bahr, F. Hillenkamp, *Biomed. Environ. Mass Spectrom.* **18** (1989) 841.
7. S. Della-Negra, Y. Le Beyec, P. Hakansson, *Nucl. Instr. Meth.* **B9** (1985) 103.
8. F. Hillenkamp, M. Karas, A. Ingendoh, B. Stahl, in *Biological Mass Spectrometry,* A. L. Burlingame and J. A. McCloskey, eds. (Elsevier Science Publishers, 1991).
9. M. Benguerba, A. Brunelle, S. Della-Negra, J. Depauw, H. Joret, Y. Le Beyec, M. G. Blain, E. A. Schweikert, G. Ben Assayag, P. Sudreaud, *Nucl. Instr. Meth. Phys. Res.* **B** (1991) 000.
10. W. Ens, B. U. R. Sundqvist, P. Hakansson, D. Fenyo, A. Hedin, and G. Jonsson, *J. de Phys.* **50(2)** (1989) C2-9.
11. S. Della-Negra, Y. Le Beyec, B. Monart, K. Standing, K. Wien, *Phys. Rev. Lett.* **58** (1987) 18.

12. E. F. da Silveira, M. G. Blain, E. A. Schweikert, *J. de Phys.* 50(2) (1989) C2-79.
13. M. A. Park, B. D. Cox, E. A. Schweikert, *J. Vac. Sci. Technol. A* 9(3) (1991) 1300.

HEAVY-PARTICLE INDUCED ELECTRON EMISSION FROM SOLIDS

RAUL A. BARAGIOLA

Dept. of Nuclear Engineering and Engineering Physics,
University of Virginia, Charlottesville, VA 22901, USA

ABSTRACT

We discuss properties of the electron emission induced by ion bombardment of solids and delineate basic theoretical principles.

1. Introduction

Ionization events in solids can be caused in primary interactions with photons, electrons, or heavy particles, or in the cascade of secondary electrons. The observables of ionization events include the inelastic energy loss of particle beams, and temporary or permanent changes in the target, like the ejection of electrons or atomic particles, generation of defects, induced electrical conductivity, changes in the chemical composition and heating. Since changes in the target will be discussed by several authors in this volume, I will restrict this article to electron emission.

This phenomenon is one of the most conspicuous effects of the bombardment of solids with energetic atomic particles (neutral and charged atoms and molecules).[1-4] The basic measure of electron emission is γ, the average number of electrons emitted per primary projectile. Other quantities of interest include the energy and angular distributions of the electrons and the emission statistics. Usual experimental variables are the type, energy and angle of incidence of the projectile and the type and condition of the target surface.

Electron emission can result from two essentially different mechanisms, driven by the potential or by the kinetic energy of the projectile. In potential electron emission (**PEE**),[5] electrons are excited by the energy released in the Auger neutralization or de-excitation of an incoming ion or excited atom, and can occur even in the limit of zero projectile velocity v. A conceptually different mechanism is kinetic electron emission (**KEE**), where the *dynamic* perturbation set up by the moving projectile transfers kinetic energy from the projectile to an electron.

2. Potential Electron Emission

The Auger de-excitation process can result in electron emission if the potential energy brought by the projectile, E_i, is higher than twice ϕ, the work function of the surface. PEE, being exothermic, has no kinetic threshold, i.e., it can occur even if the incoming particle has zero kinetic energy.[6] The magnitude of the PEE yield γ_p is governed by the overlap of electron states in the solid and the atom, and by the availability of initial and final states. Fig. 1 shows different electron transfer processes for an ion near a surface, most of which can result in electron emission.

Fig. 2 shows a compilation of PEE yields γ_p of clean metals induced by single charged rare gas ions. The yields fall within a narrow band of the line given by the relation:[1]

$$\gamma_p = 0.032\ (0.78E_i - 2\phi), \tag{1}$$

where E_i is the ionization potential. For a typical metal with $\phi \sim 4$ eV, only ions with $E_i > 10$ eV will eject electrons.

The absorption of gases generally *lowers* the PEE yields; electron emission at eV energies does not occur at the surface of electron multipliers or other surfaces that have been exposed to the atmosphere, even if the relation $0.78E_i > 2\phi$ is satisfied. A related finding is that PEE yields are also small for molecular ions, even light ones, not only because of a small E_i, but possibly due also to the existence of molecular decay channels competing with the electron emitting Auger process.[7] An important aspect in PEE is the decrease in the effective ionization potential near the surface due to the image interaction. This explains the factor of 0.78 (instead of 1) in Eq. 1 and causes the probability of PEE to depend on the distance between the ion and the surface. Since faster ions tend to undergo Auger neutralization closer to the surface than slow ions, PEE yields show a dependence on the velocity of the projectile. At sufficiently high velocities, $\gtrsim 5 \times 10^7$ cm/s, Auger decay will occur inside the solid; this problem has received theoretical attention recently in connection with the charge state and energy loss of low energy ions in solids.[8]

A different kinetic effect in PEE arises because of the Heisenberg broadening of the energy levels due to the finite time the ion spends near the surface. This is responsible for part of the changes in the energy distribution of electrons as a function of impact velocity (Fig. 3).[9] A different effect is produced by the dynamic population of states above the Fermi level produced by the ion as it approaches the surface. This is caused by the time-dependent Coulomb interaction between the ion and the surface.[10] This effect is conceptually the same mechanism as KEE. An additional effect in metals is the shake-up near the Fermi level produced by the sudden annihilation of the core hole and electron ejection which occurs as a result of AN.

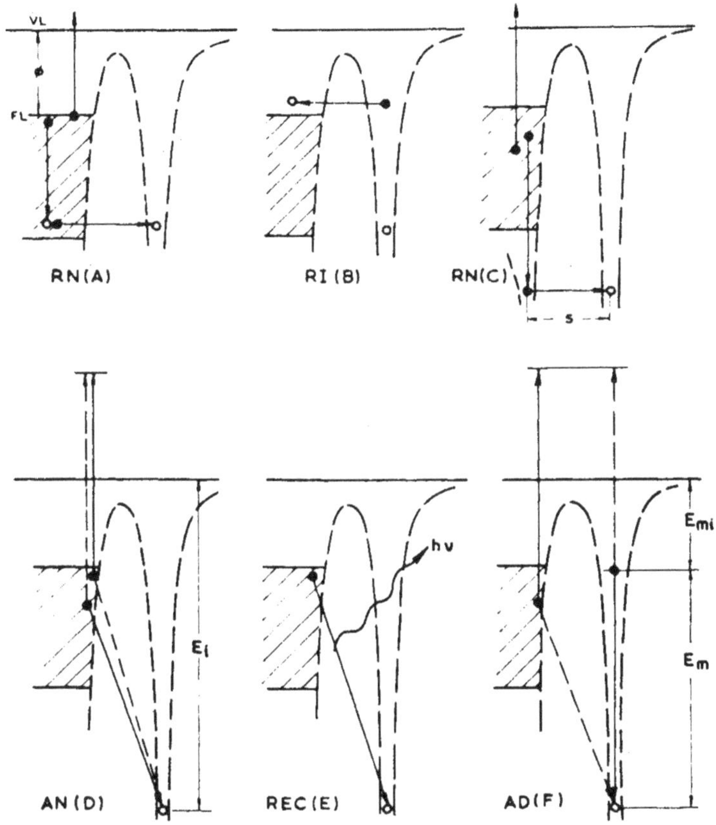

Fig. 1. Electron energy diagrams appropriate for (A) Resonance neutralization, RN; (B) Resonance Ionization, RI; (C) RN from a core level; (D) Auger Neutralization, AN; (E) Radiative electron capture, REC; (F) Auger de-excitation (AD). VL - vacuum level, FL - Fermi level, ϕ - work function. Arrows indicate electronic transitions. Initially filled (empty) electron states in atoms or the solid are represented by filled (open) circles. Light arrows show secondary transitions following RN, and dashed lines in AN and AD, exchange transition.

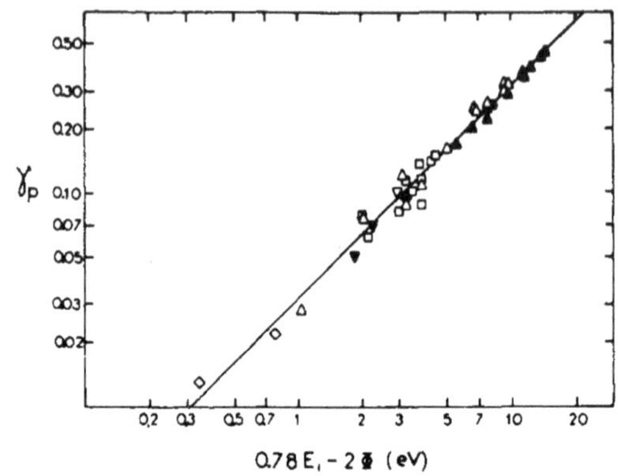

Fig. 2. PEE yields from metals versus 0.78 E_i - ϕ. For references to the data, see Baragiola et al.[1]

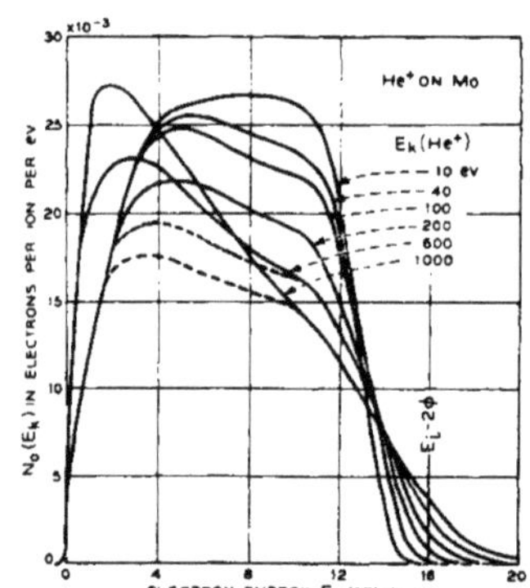

Fig. 3. Electron energy distributions showing the transition from PEE to KEE.[9]

The energy distribution of electrons resulting from AN (Fig. 3, low energies) is a self convolution of the *local* density of states at the surface, weighted by energy and symmetry dependent transition probabilities. The broadening due to interactions with other electrons from the solid[11] is much smaller than in photoemission, since AN occurs outside the surface. The study of the energy distributions of electrons in PEE has been developed by Hagstrum into a technique of probing the electronic states of surfaces called Ion Neutralization Spectroscopy (INS).[12] INS is different from more common surface physics techniques, like photoelectron or electron induced Auger spectroscopies in that the information obtained pertains to electronic states *outside* the solid, the most important region for surface chemical processes.

At low work function surfaces, ions can neutralize to an excited state before de-exciting by an Auger process. This Auger decay of excited atoms into target ionization, also termed Penning Ionization, can occur also with incident metastable atoms, e.g., He^* and Ne^*, when the excitation energy exceeds 2ϕ. Metastable de-excitation differs from Auger neutralization of an ion in a different initial state not only of the projectile, but also of the target, since the initial state of the surface prior to the transition is perturbed differently by the potential set up by an ion than by an excited neutral atom. More details on PEE by single and multiply-charged ions can be found in review articles.[5, 13]

3. Kinetic Electron Emission from Metals

Unlike PEE, KEE can result from the impact of ground state atoms or of ions with low neutralization energy (like heavy molecules or alkali ions). Several reviews of KEE cover the topic from different perspectives.[1-4,14-16] We will now discuss some properties of KEE for clean metal surfaces.

KEE yields rise from a threshold at projectile velocities v in the range 10^6 - 10^7 cm/s and, after a transition region where $\gamma \propto v$ (Fig. 4), they fall into a curve $\gamma(v)$ similar to $S_e(v)$ = dE/dx, the electronic stopping power of the projectile in the surface region of the target. (Fig. 5)

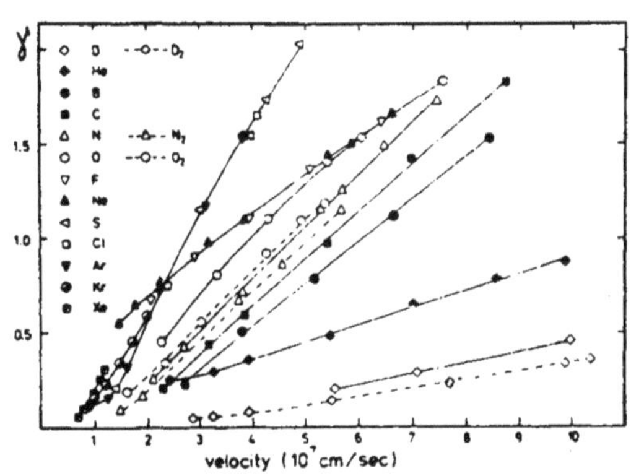

Fig. 4. Electron emission yields for different ions on clean Al as a function of the velocity of the ions.[17]

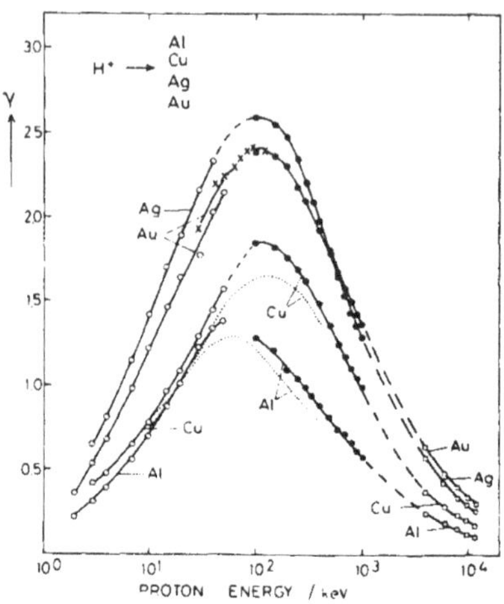

Fig. 5. Electron emission yields vs. ion energy for H^+ on different metal targets. For keys to the symbols, see Ref. 18.

The proportionality $\gamma \propto S_e$ holds surprisingly well for *light* ions over a wide range of energies.[18] Moreover, we found that for clean surfaces bombarded at normal incidence by slow ions not too close to the threshold of KEE, the material constant $\kappa = \gamma/S_e = 0.1$ Å/eV to within 30%, for different projectile-target combinations. This finding allows easy prediction of γ, since S_e values are readily available in most cases.[19] This property is similar to the well known fact that the average energy to produce an ion-pair in gases, W, is independent, in first order, of the type and energy of the projectile, and depends weakly on the type of gas.[20]

The material constant is strongly affected by gas adsorption, and yields for surfaces exposed to the atmosphere are usually several times larger than yields from clean metal surfaces, especially near threshold. The degree of cleanliness of the surface is a *crucial* factor to consider when evaluating data from the literature.

For single crystal targets, electron yields depend sensitively on the crystal face bombarded and on the alignment of the ion beam with respect to major crystallographic axes. This is related to channeling and blocking effects that limit the range of impact parameters in collisions with target atoms.

Thresholds

The threshold behavior of KEE has recently received much attention, since it is hoped that studies will assist in developing more efficient detectors for mass spectrometry of large molecules.[21-24] Thresholds seem governed by several mechanisms. For light ions on metals, we found that a good description of the excitation is in terms of binary collisions between the projectile and a free target electron.[25] The maximum energy transfer T between a heavy projectile of velocity v and an electron of velocity v_e and mass m is $T = 2mv(v+v_e)$. Therefore, the threshold velocity v_{th} occurs when the collision involves the fastest free electrons (those with the Fermi velocity, v_F) and T equals the work function ϕ. Thus, $v_{th} = -(v_F/2)[(1 + 2\phi/mv_F^2)^{1/2} - 1]$. The threshold velocities obtained with this formula are around 1.5 - 3 x 10^7 cm/s for most metals and consistent with extrapolation of the experimental results for light projectiles.

Heavier projectiles have thresholds at velocities one order of magnitude lower, indicating that the free electron model is not applicable to them. For instance, measurements of KEE induced by Xe^+ ions on atomically clean metals[26,27] showed no indication of a threshold down to velocities of 1.2 x 10^6 cm/s. A possible reason for these lower thresholds is emission from electronic states of the projectile, since binding forces allow more momentum exchange during the collision at a given velocity. The minimum possible threshold for KEE is when the center-of-mass energy equals the minimum energy needed to eject an electron (the work function for metals, the electron affinity plus the band gap for non-metals).

Electron emission can be enhanced by non linear or "hot spike" effects, produced when a large and energetic molecule is stopped, depositing most of its energy in a very small region of a solid. Due to multiple collisions between moving atoms during impact, some of these atoms will acquire velocities larger than the beam velocity, making emission more likely. Recent molecular dynamics simulations[28] have shown high energy tails in the energy distribution of atoms in large clusters bombarding solids. These non-linear effects produce microplasmas in hypervelocity impact of small dust particles in solids, which provide a means of detection in space science.[29]

Energy Distributions

Typical energy distributions peak at a few eV and decay rapidly with energy, like in the case of PEE (Fig. 3). They are essentially independent of v except near the threshold. If plotted in a log scale, long high energy tails are apparent in the electron energy distributions of KEE. Sometimes, Auger peaks appear superimposed on these tails. They result from inner-shell excitations in violent collisions between recoils or between the projectile and a stationary target atom. The Auger spectra consist of broad bands caused by core-excited atoms decaying in the bulk of the solid, and sharp (~ 1 eV) atomic lines caused by excited atoms and ions decaying in vacuum. The research on ion-induced Auger emission is very rich and vigorous.[530-38]

The origin of the continuum tails is still obscure. At high impact velocities, they have been explained to arise from binary collisions between the projectile and a target electron. At low velocities, the explanation is still a matter of debate,[39] models proposed include emission of Auger electrons and direct coupling to the continuum *during* the collision. Our measurements for very low energy (0.5-6 keV) ions in solids have shown that the continuum electron energy distributions extend up to a surprisingly large fraction of the center-of-mass energy (Fig. 6). This shows that the high electron energies result from nearly head-on collisions between the projectile and a target atom. The high energy tails are strongly dependent on the type of target, they decay generally slower with electron energy the higher the atomic number of the target.

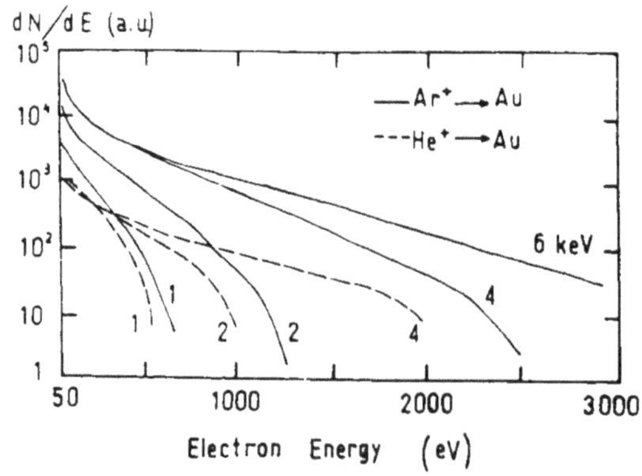

Fig. 6. High energy continuum tails in the electron energy distributions from an Au target bombarded by He^+ and Ar^+ at different energies.

Effect of adsorbed gases

Electron yields are very sensitive to the state of the surface.[1] Our study of the effect of oxidation showed that KEE yields can increase or decrease accompanying changes in surface composition but in a way that is not directly related to work function changes.[40]

KEE from thin oxide layers may be enhanced by a lowering of the inner potential from the high value for metals to the low electron affinity of oxides; this enhancement appears to decrease with bombarding energy.[41] On the other hand, to explain experiments near the threshold, Amme[42] found sufficient to propose that electron emission from gas-covered surfaces bombarded with slow atoms is dominated by binary encounters between the projectile and the outermost adsorbed gas layer. He assumed that the surface layer in normal vacuum condition is made of water molecules, and obtained good correlation with gas-phase ionization cross sections.

4. Theoretical Concepts in KEE

So far no theory of KEE can calculate electron yields from first principles,[43] and we have to rely on a heuristic approach. It is customary to describe KEE with a model that divides the process in three steps: electron excitation, transport, and escape. This model, which has been quite successful in explaining most observations, can be framed in the following general picture.

Excitation

Heavy particles moving through matter transfer energy to electrons as a result of a time-varying Coulomb interaction; this can lead to excitation or ionization of the projectile and/or the target. The projectiles undergo multiple collisions, set recoiling target atoms in motion, and finally stop in the solid or are reflected into vacuum. On their paths, the projectiles and the recoils create many excited electrons (sources) in ionization events, which move initially with a range of energies and directions. The source of electrons is generally peaked in the direction of the projectile (forward). This explains the anysotropy in the forward/backward emission ratio from thin foils[44] (i.e., more electrons are emitted in the downstream than in the upstream side) and contributes to the dependence of KEE on the incidence angle of light projectile.[45] Another source of the electrons comes from the Auger decay of inner-shell holes. It is delayed with respect to the primary electrons, and generally isotropic.

Electron cascades

Due to strong electron-electron interactions in solids, the electrons from primary excitations degrade rapidly in energy exciting other electrons in the solid, within constraints imposed by existing band gaps. This continuous energy degradation and multiplication in the number of electrons continues until a large number of electrons end at the bottom of the conduction band, and their kinetic energy eventually dissipates as heat.

The primary electrons emitted in the ionizing collision interact with other electrons of the solid and initiate a cascade that degrades to the bottom of the conduction band. The presence of the surface interrupts this cascade, and a sample of it crosses the surface and appears in vacuum. At the time of ejection, the spatial extent of the low energy part of the electron cascade is limited, due to the strong electron-electron interactions, and elastic scattering of low energy electrons with ion cores, to distances of the order of the mean electron attenuation length L (typically 10 - 30 Å in metals and 30 Å - 1 μ in insulators).

If electron escape from the surface can be separated from the ionization events and the development of the electron cascade, we can write the electron yields as:

$$\gamma = \int dE\, N(E)\, P(E)\, dE, \quad (2)$$

where E is the energy with respect to the vacuum level, N(E) the energy distribution of electrons arriving to the surface, and P(E) the escape probability through the surface barrier. N(E) can in turn be approximated by:

$$N(E) = f(E)\, L\, R, \quad (3)$$

where the function f accounts for the energy distribution of the electrons and includes angular factors. R is the production rate of electrons excited above the vacuum level per unit path length near the surface. It includes ionizations caused by the projectile, recoiling target atoms and secondary electrons.

For fast particles, and in analogy with ionization processes in semiconductors, R is approximated by S_e/J, where $S_e = dE/dx$ is the electronic stopping power of the projectile, and W a mean energy to produce an excited electron. In this high velocity approximation (electronic energy deposition constant over L) we obtain:

$$\gamma = \int dE\, f(E)\, L\, S_e/W\, (E/(E+I))\, dE \quad (4)$$

Recoil effects are particularly important at low velocities, and their contribution to ionization has to be added to S_e. However, at slow velocities it is not longer valid to assume that the number of excited electrons is proportional to a deposited energy divided by W. The lower the energy, the larger the proportion of the inelastic energy deposited that goes into sub-threshold events, with final electron energies below the vacuum level. This can be accounted for by using in Eq. 4 a modified stopping power that includes only processes that end with an electron above the vacuum level. This approach succeeds in describing KEE by heavy projectiles and energetic target recoils.[46]

Electron escape through the surface barrier

Once the excited electrons reach the surface, they must cross the surface barrier and escape into vacuum to be detected. Approximating the surface by a classical planar potential barrier, and neglecting band-structure effects, the probability of escape is given by:[47] $P_e = E/(E + I)$, where E is the electron energy outside the solid and I the height of the surface barrier. For clean metals, I is equal to the inner potential, or of the order of 10 eV. For insulators, E_i is equal to the electron affinity of the surface, of the order of 1 eV for most insulators. This lower barrier partially explains why insulators are more efficient electron emitters. Some materials, like solid Ar, have negative electron affinities and exceptionally large electron yields.

Total Electron Yields

Total electron yields are obtained by integrating Eq. 4 over electron energies. One obtains:

$$\gamma = K <P> S_e = \kappa S_e , \tag{5}$$

where K and κ are material constants and $<P>$ is the average escape probability. As mentioned above, not too close to the threshold it is found[1] that electron yields for light ions can be estimated to within about 30% by the expression:

$$\gamma = 0.1\ S_e(\mathrm{eV/Å}), \tag{6}$$

a result that also holds for backward electron emission from thin foils, if they have a clean surface.[48] Variations around the value of Eq. 6 come from target dependences[1] and from shell effects in the projectiles[17] but they are small, supporting the general proportionality between electron yields and electronic stopping powers. The yields for forward emission from foils by fast ions are larger than predicted by Eq. 6 due to the preferential energy deposition of fast primary electrons (δ electrons) in the forward direction.[36]

Eq. 6 can be extended easily to non-normal incidence by considering that the important quantity is the inelastic energy deposited in a thin slab of the solid parallel to its surface. In the absence of recoil and backscattering effects, source anysotropy, and energy degradation of the projectile, the yields scale with the secant of the incidence angle.[37]

5. Electron Emission from Insulators

Insulators are hard to study due to the charging of their surfaces by the primary beam and by the emitted electrons. During positive ion impact, the surface of a good insulator will tend to charge to the beam voltage, where it will repel the primary ions and stop the charge accumulation. For keV ions, breakdown will usually occur first, with the surface discharging to the metallic sample holder through the bulk of the insulator, across its surface, or by breakdown to nearby electrodes. Even for neutral beams, the surface will charge positive due to electron emission. As the surface potential increases due to irradiation, only the most energetic electrons will escape. For negative ions, the surface potential will stabilize if the electron yield is higher than one, to a value that will allow the escape of enough electrons to balance the charge input from the ion beam.

The way to study insulators without altering them is to use very small beam currents, to avoid significant sample charging or to use very thin insulators, to ease conduction paths to a metallic sample holder. In the latter case, a thickness dependence is found for the yields. Very few experiments have achieved this condition and at the same time ensuring a clean surface. Electron yields from insulators are usually several times larger than for metals or semiconductors. There are two reasons for it: 1) the larger escape probability through the surface barrier due to the small electron affinity, and 2) the larger electron escape depth due to the band-gap which hinders collisions with energy losses smaller than the band-gap energy.

6. Outlook

The field of electron emission has evolved more slowly than other topics of particle-solid interactions. Part of the reason is the surface sensitivity, which demands controlled experiments under ultra-high vacuum conditions, and part is due to the complexities brought by the very strong electron-electron interactions in solids. Challenging areas where the understanding is presently lacking are insulating targets, adsorbed layers, threshold behavior, molecular and cluster impact and high density ionizations, to name a few.

References

1. R. A. Baragiola, E. Alonso, J. Ferrón and A. Oliva, Surface Sci. **90**, 240 (1980)
2. J. Schou, Scan. Microsc. **2**, 607 (1988)
3. B. A. Brusilovsky, Appl. Phys. **A50**, 111 (1990).
4. W. O. Hofer, Scan. Microsc. **S4**, 265 (1990)
5. R. A. Baragiola, Radiation Eff. **61**, 47 (1982).
6. There is a minimum impact energy for ions due to their acceleration by the image charge.
7. D. W. Vance, Phys. Rev. **169**, 252 and 263 (1968); F. M. Propst and E. Luescher, Phys. Rev. **132**, 1037 (1963); J. B. French and R. H. Prince, in *The Structure and Chemistry of Solid Surfaces*, G. A. Somorjai, Ed. (Wiley, NY, Pt. 48, 1969).
8. P. M. Echenique, F. Flores and R. H. Ritchie, Sol. St. Phys. **43**, 229 (1990)
9. H. D. Hagstrum, in *Inelastic Ion-Surface Collisions* (Academic, 1977) p. 7
10. H. D. Hagstrum, Y. Takeishi and D. D. Pretzer, Phys. Rev. **139**, A256 (1965)
11. H. D. Hagstrum and Y. Takeishi, Phys. Rev. **137**, A304 (1965)
12. H. D. Hagstrum, J. Vac. Sci. Technol. **12**, 7 (1975)
13. P. Varga, Comments Atom. Molec. Phys. **23**, 111 (1989)
14. K. H. Krebs, Vacuum **33**, 535 (1983)

15. D. Hasselkamp, Comments At. Mol. Phys. **21**, 241 (1985)
16. N. Benazeth, J. Microsc. Spectrosc. Electron. **12**, 235 (1987)
17. E. V. Alonso, R. A. Baragiola, J. Ferrón, M. M. Jakas and A. Oliva-Florio, Phys. Rev. **B22**, 80 (1980)
18. D. Hasselkamp, Comments. At. Mol. Phys. **21**, 241 (1988)
19. J. F. Ziegler, *Stopp[ing and Ranges of Ions in Matter* (Pergamon, NY 1977)
20. *ICRU Rept. 31*, International Commission on Radiation Units and Measurements (1979)
21. L. A. Dietz and J. C. Sheffield, J. Appl. Phys. **46**, 4361 (1975).
22. R. J. Beuhler and L. Friedman, Nucl. Instr. Methods **170**, 309 (1980); R. J. Beuhler, J. Appl. Phys. **54**, 4118 (1983).
23. M. Salehpour, P. Hakansson, B. Sundqvist and S. Widdiyasekera, Nucl. Instr. Meth. **183**, 278 (1986).
24. P. W. Geno and R. D. Macfarlane, Int. J. Mass Spectrom. Ion. Phys. **92**, 1955 (1989).
25. R. A. Baragiola, E. V. Alonso and A. Oliva, Phys. Rev. **B19**, 121 (1979).
26. E. V. Alonso, M. Alurralde and R. A. Baragiola, Surface Sci. **166**, L155 (1986).
27. G. Lakits, F. Aumayr, M. Heim and H. Winter, Phys. Rev. **A42**, 5780 (1990)
28. M. I. Haftel, Bull. Am. Phys. Soc. **36**, 1334 (1991); M. Hautala, Z. Pan and P. Sigmund, submitted to Phys. Rev. A.
29. J. A. M. McDonnell, J. Phys. **E 20**, 741 (1987).
30. R. A. Baragiola, Springer Ser. Chem. Phys. **17**, *Inelastic Particle Surface Collisions*, ed. E. Taglauer and W. Heiland (Springer, Berlin: 1981) p. 38; L. de Ferrariis, O. Grizzi, G. E. Zampieri, E. V. Alonso and R. A. Baragiola, Surf. Sci. **167**, L175 (1986); O. Grizzi and R. A. Baragiola, Phys. Rev. **A35**, 135 (1987); R. A. Baragiola, L. Nair and T. E. Madey, Nucl. Instr. Meth. **B** (1991, in press)
31. E. W. Thomas, Vacuum **34**, 1031 (1984)
32. J. Mischler and N. Benazeth, Scan. Electron Microsc. **II**, 351 (1986); N. Benazeth, J. Mischler and M. Negre, Surf. Sci. **205**, 419 (1988); C. Benazeth, P. Hecquet, C. Mayoral and N. Benazeth, Rad. Eff. Def. Solids **108**, 227 (1989)
33. K. Saiki, I. Rittaporn and S. Tanaka, Japan. J. App. Phys. **26**, 45 (1987)
34. M. H. Shapiro and J. Fine, Nucl. Instr. Meth. **B44**, 43 (1989); Nucl. Instr. Meth. (1991, in press)
35. M. Herteux, W. Hink, J. P. Biersack and E. Schreck, Nucl. Instr. Meth. **B48**, 621 (1990)
36. A. Koyama, H. Ishikawa, K. Maeda and Y. Sasa, Nucl. Instr. Meth. **B48**, 608 (1990)
37. S. Valeri, R. Tonini and G. Ottaviani, Phys. Rev. **B38**, 13282 (1989); S. Valeri and R. Tonini, Surf. Sci. **220**, 407 (1989); S. Valeri and R. Verucchi, Nucl. Instr.

Meth. **B** (1991, in press)

38. F. Xu, R. Siciliano,, M. Camarca and A. Oliva, Surf. Sci. **209**, L133 (1989); A. Bonnano, F. Zu, M. Camarca, R. Siciliano and A. Oliva, Nucl. Instr. Meth. **B48**, 371 (1990)
39. P. Clapis and Q. Kessel, Phys. Rev. **A 41**, 4766 (1990)
40. J. Ferrón, E. Alonso, R. A. Baragiola and A. Oliva Florio; Surface Sci. **120**, 427 (1982).
41. E. V. Alonso, R. A. Baragiola, J. Ferron and A. Oliva-Florio, Rad. Effects **45**, 119 (1979).
42. R. C. Amme, J. Chem. Phys. **50**, 1891, 1969.
43. P. Sigmund and S. Tougaard, Springer Ser. Chem. Phys. **17**, *Inelastic Particle-Surface Collisions*, ed. by E. Taglauer and W. Heiland (Springer, Berlin: 1981) 2
44. E. J. Sternglass, Phys. Rev. **108**, 1 (1957)
45. J. Ferrón, E. Alonso, R. A. Baragiola and A. Oliva Florio; Phys. Rev. **B24**, 4412 (1981)
46. J. Ferrón, E. Alonso, R. A. Baragiola and A. Oliva Florio; J. Phys. **D14**, 1707 (1981).
47. M. S. Chung and T. E. Everhart, J. Appl. Phys. **45**, 707 (1984)
48. H. Rothard, K. Kroneberger, A. Clouvas, E. Veje, P. Lorenzen, N. Keller, J. Kemmler, W. Meckbach, and K.-O. Groeneveld, Phys. Rev. **41**, 2521 (1990)

PRINCIPLES OF ELECTRON STIMULATED DESORPTION

RAUL A. BARAGIOLA

Department of Nuclear Engineering and Engineering Physics
University of Virginia, Charlottesville, VA 22901, U.S.A.

ABSTRACT

We review the mechanisms for desorption from surfaces under electron bombardment, emphasizing the importance of multiple excitations in creating a localized, repulsive state, sufficiently long lived to escape bond healing an lead to desorption.

1. Introduction

Energetic electrons can desorb neutral and ionized species from surfaces by stimulating direct electronic transitions in surface bonds. This effect of electron bombardment, called Electron Stimulated Desorption (ESD), is different from desorption induced by heating or by direct momentum transfer to an atom. Other forms of desorption induced by electronic transitions (also known as DIET) include photon stimulated desorption, some forms of heavy particle induced desorption and electric field desorption.

Basic interest in ESD arises because it is one of the richer test grounds for our understanding of fundamental processes in atomic physics, and of the evolution of excited states in solids on the femtosecond scale. Strong interest also results from positive and negative influences of ESD. For instance, ESD complicates measurements of ultra low pressures, alters samples in surface analysis and shortens the lives of electron multipliers and vacuum tubes. On the other hand, ESD is useful in electron beam lithography, in mass spectrometry and in the determination of the structure of adsorbed molecules.

We will focus here on the main physical processes involved in ESD and how they can be applied to understand observations; we refer the reader to recent reviews[1-8] for applications and historical aspects, and to the proceedings of the topical DIET workshops.[9-12]

2. General Principles

It is important to realize that although most reported ESD experiments have observed light desorbed ions (H^+, F^+, O^+, etc.), the *vast majority of desorbed particles are neutral.* The reason for the distorted view given by published data is that it is easier to detect ions than low energy neutral particles. In usual experiments, a collimated, monoenergetic electron beam bombards a surface and data is collected in the form of yields (desorbed particles per incident electron), mass spectra, angular and energy distributions of desorbed species. Typical results are: 1) yields rise from a threshold in the range 5-40 eV, pass through a maximum in the 10^2 eV range and then decay with energy; 2) neutrals are ejected with average energies below 1 eV, ions with several eV; 3) emission is most intense normal to the surface and is suppressed in off-normal directions; 4) desorbed species from ordered adsorbed layers on single crystal form angular patterns which reflect initial bond geometries. Although most extant data follow these "typical" lines, wide variations continue appearing as research broadens to include more materials and more elaborate detectors. Among these are desorption at eV incident energies, thresholds in yield-energy curves of hundreds of eV, ion energies above 20 eV, desorption of multiply charged ions, negative ions and large clusters.

a. Comparison with gas-phase processes

ESD results from the electronic excitation of a bound state to a repulsive configuration where an atom or molecule can gain sufficient kinetic energy to desorb. The repulsive state can result from a one-electron excitation of bonding levels, or from more complex multiple excitations or ionizations. One-electron excitation is conceptually the same as the Franck-Condon transition to repulsive molecular states usual in photon or electron dissociation of free molecules.[13] This transition is so sudden (~0.1 fs) that the nuclei cannot adjust to the new potential produced by the change of the electron distribution and so evolve from the same instantaneous position and velocity of the initial state. In the absence of other electronic transitions, the molecule dissociates if it lies energetically above the dissociation continuum. In this simple picture, the energy distribution of the dissociation fragments results from mapping the initial distribution of interatomic distances through the repulsive potential-energy curve (Fig. 1).

This idea does not apply directly to ESD from surfaces; *desorption cross sections are found to be two to more than nine orders of magnitude smaller than typical gas-phase dissociation cross sections.* There are also differences in the distributions in charge, mass, and energy of the fragments, and in the dependence of cross sections on electron energy. The general rule is that cross sections for rupture of an internal bond in an adsorbed molecule (e.g., the C-O bond in CO adsorbed on a metal) are high, and may be comparable to gas phase processes. Cross sections are smaller for breaking the bond between a molecule and a surface (e.g. that between CO and a metal). Very small cross sections apply to breaking internal bonds in insulators (e.g., desorbing an O or a Si atom from SiO_2), while desorption from pure metals by individual electrons is not observable.

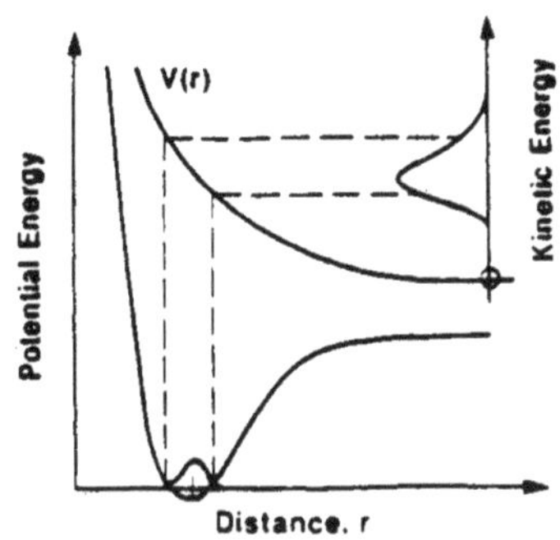

Fig. 1. Schematic potential energy curves for dissociation or desorption.

b. Quenching of the Excitations

The reason for the differences between gas phase and surface cross sections is that *electronic excited states are very short lived near a metal surface*. Initial excitations are quenched by electron transfer to the solid which "heals" the ruptured bond. Electron transfer is allowed by a large density of empty final states in the solid with the same energy (i.e., resonant) as the dissociation channel. At typical adsorption distances ($\sim$2 Å), one-electron, resonant electron transfer (RT) is usually very fast (< 1 fs). A likely sequence is that RT first ionizes the excited molecule to an empty state in the metal; then the ground state of the molecule is repopulated by a slower, two-electron Auger neutralization (AN) process[14] where an electron from the metal tunnels to the molecule and another is excited. Thus, a state results with two excited electrons in the solid, and the molecule in its ground state.

Since electron transfer times are much shorter than the time it takes an atom to escape from the solid (10 - 200 fs), practically all excitations will be quenched before significant motion occurs, frustrating desorption. For desorption to occur, the one-electron de-excitation channel needs to be inhibited in most cases. This can happen when the excited state lies opposite to a forbidden energy region of the solid, like a band-gap or below the valence band; this explains the larger desorption yields for adsorbates in insulators.[15] Examples are self trapped excitons in insulators which are long lived since they can only diffuse out by two-electron transfer: one electron drops to the ground state and an electron is excited in an adjacent atom. Since the exciton distorts the lattice, it diffuses like a polaron, i.e., the electron transfer is accompanied by a phonon field.[16]

Excited states can drop below the valence band and escape de-excitation if there is additional excitation or ionization, e.g., in the form of two valence holes or one valence and one core hole. Two valence holes may migrate by resonant electron transfer, with hopping times of ~10 fs. Thus even if a repulsive state is formed, an individual atom may not remain in it sufficiently long to gain enough kinetic energy for desorption. On the other hand, a core-hole can localize the excitation since its migration is hindered by the very small overlap between core states. This excitation will be localized during the lifetime of the core hole. Shallow core holes in solids decay very fast, usually in less than 1 fs, by Auger emission. If the outer electrons needed for the Auger transition are missing, like in the case of cations in insulators, the lifetime can be very long, of the order of desorption times. Furthermore, core holes in insulators excert an essentially unscreened Coulomb force on their neighbors, distorting the lattice. This accumulated strain could conceivably relax in desorption, when the core excitation finally decays. Such an effect occurs in free molecules.[17]

Desorption is more likely when the molecular fragment spends less time close to the surface so that electron exchange with the solid is less probable. One can write the probability of desorption as $P_d = e^{-t/\tau}$, where τ is the mean lifetime of the repulsive electronic state and t is the time taken by the nucleus to escape from the quenching effect of the surface. Ejection times to distances of 3 Å are $t \sim 40(M/E)^{1/2}$ fs, where E is the final energy of the ejected species in eV, and M its mass in amu. Appropriate times for most desorption studies range from 10 fs for 20 eV H to 840 fs for 0.1 eV N_2O.

The population N(s) of the excited state decays like $dN(s)/dt = -N(s)\ R(s)$, where R(s) is the deexcitation rate. Due to the exponential fall of electron density outside atoms and solids, $R \sim A\ e^{-as}$ at large s, with a saturation at small distances. Recent theories[18] confirm this behavior with $a \sim 1.5\ Å^{-1}$ for atoms near metal surfaces. Values of $A = 1/\tau$ depend strongly on the work function of the surface and the energy and extent of the atomic orbitals.

The desorption probability for an atom of velocity v(s) is then given by:

$$P_d = \exp\left(-\int_{s_o}^{\infty} R(s)/v(s)\ ds\right) \qquad (1)$$

The velocity of the desorbing particle increases from a low value at s_o due to the force $F_r = dU/dr$ produced by the repulsive potential U(r) along the direction of motion. Since the integrand in Eq. 1 falls very fast over distances of the order of 1/a, we obtain, neglecting $v(s_o)$, a mean normal velocity, $v_m \sim [F_r(s_o)/(aM)]^{1/2} \cos\Theta$ in the range of interest. Here M is the mass of the desorbing particle and Θ the angle between the initial velocity vector and the surface normal. With this approximation, the desorption probability becomes:

$$P_d \sim \exp[-M^{1/2} A a^{-1/2} e^{-as_o} / |F_r|^{1/2}(s_o) \cos\Theta] \qquad (2)$$

Due to the lack of good estimates of A, it is not possible to obtain reliable values of P_d from Eq. 2. We can, however, make many predictions for the dependence of P_d on the different parameters. The strongest dependence is on s_o., thus, desorption is strongly favored for species adsorbed far from the surface, explaining why internal bonds of adsorbed molecules are easier to break than metal-adsorbate bonds. The mass term in Eq. 2 favors the emission of light species (like H^+) and produces and isotope effect.[19] The angular term in Eq. 2 explains why desorption is suppressed in off-normal directions. Transition probabilities A are larger for species bound to several atoms, due to stronger wavefunction overlap, so desorption generally occurs more readily from atop adsorption sites than from bridge-bonded sites.[20]

3. The Repulsive State

The repulsive state responsible for desorption has been identified in only few cases. In principle, one can learn about it by studying the threshold energy for desorption. The lowest ESD thresholds are usually in the range of valence excitations, i.e., below 15 eV for most adsorbed gases. A case of ESD by single valence excitations where some of the potential energy curves are known is condensed rare gases. Here, excitation leads to the formation of a molecular exciton, which decays radiatively to the repulsive part of the ground state potential energy curve, as shown in Fig. 2, producing ejected atoms of characteristic energies.[21] The formation of an intermediate excited state which moves the nuclei closer before relaxing to a repulsive state is often known as an Antoniewicz mechanism.

Another example of ESD by valence excitation is the emission of negative ions by dissociative attachment (DA). Here, low energy electrons attach to a molecule (A-B) to form a transient negative ion $(A\text{-}B)^-$ which later dissociates into $A+B^-$. Threshold energies for DA are usually < 5 eV, smaller than those for production of negative ions via dipolar dissociation leading to $A^- + B^+ + e^-$. Fig. 3 shows an example of desorption by DA for O_2 on Pt covered with insulating spacer layers of condensed Ar.[22] For very thin spacers, the substrate quenches DA, and ESD of negative ions is mainly given by bipolar dissociation. Increasing spacer thickness weakens the interaction with the metal and causes emission to resemble the gas phase process.

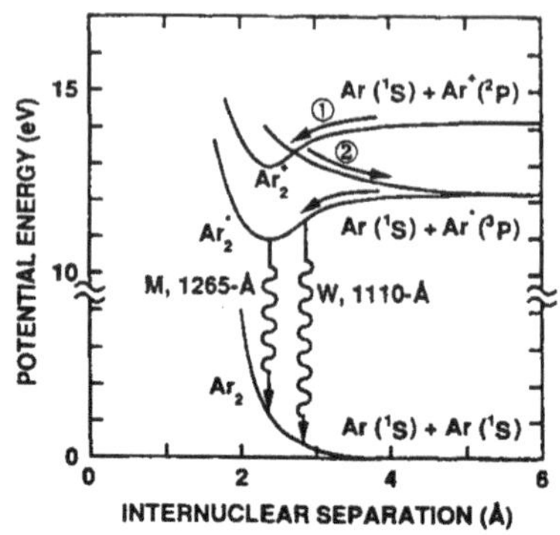

Fig. 2. Solid-state potential for two Ar atoms. ① is the self-trapping of a hole. ② is the repulsive recombination of an electron and a self-trapped hole. M and W are luminescence lines.[21]

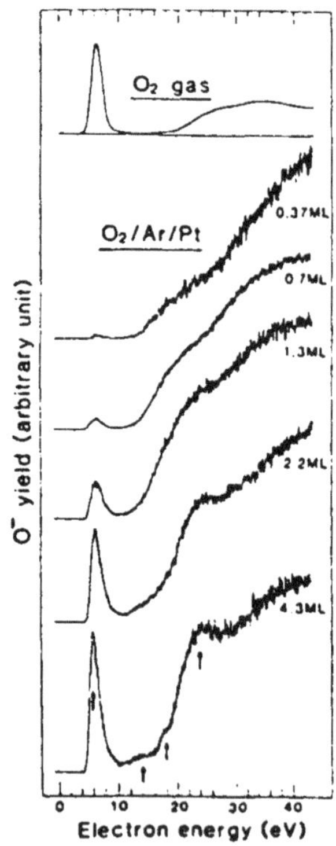

Fig. 3. O^- desorption yields as a function of electron energy for the gas-phase and for 0.1 monolayer (ML) of O_2 on Pt.[22]

Low energy excitations can be caused by secondary electrons created by the primary beam,[23] when the energy required for the production of the repulsive state is a few eV, or comparable with the energies of the secondaries. Electron with insufficient energy to be emitted (so-called "hot electrons") may nevertheless induce desorption by attaching to long lived negative-ion states.[24] The image attraction will pull the adsorbed ion closer to the surface, where a subsequent relaxation to the repulsive part of the neutral state can produce desorption by a variation of the Antoniewicz mechanism. One-electron valence excitation was the first mechanism proposed for ESD,[25] but now appears to be relatively unimportant for *ion* desorption in most cases. This follows from the appearance of sharp onsets in desorption yields versus electron energy at the thresholds for excitation of core levels. This correlation points to bond breaking induced by the Auger decay of the core hole. This form of desorption, called ASD, for Auger Stimulated Desorption, is usually more efficient than that caused by one-electron excitations.

In the ASD model of Knotek and Feibelman,[26] interatomic Auger decay of cation core holes in oxides cause desorption when the stoichiometry of the oxide is that of "maximal valency", meaning that the cation is ionized to the noble gas configuration. In this model, after a core hole is created in either the cation or the anion, the subsequent Auger decay will ionize the valence electrons, localized at the anion. Thus, in TiO_2, a Ti core hole can decay by a high order Auger process transforming an adjacent O^- into O^+. This reversal of the Madelung potential at the anion site produces a strong repulsion from the neighboring cations (Ti^{4+}), which can drive desorption. Although this model played historically an important role, its predictions are not accurate. For example, the model predicts desorption from MgO, but it is found that this oxide is very hard to decompose under electron bombardment.[27] Even for the case of TiO_2 the model has difficulties: little desorption is observed from a maximum valency surface, while strong emission results for an oxygen-defficient surface.[28]

The ASD model has been extended to covalent systems[29] (e.g., SiO_2). Decay of the core hole by an Auger process may localize two holes in a bonding orbital. If the lifetime of the final two-hole state is long enough, it may relax by Coulomb repulsion into atomic displacements. This extended ASD model successfully predicts that the energy dependence of the oxygen loss from SiO_2 should follow that of the cross section for excitation of the Si*2p* level.[30] The model can be also applied to covalent bonds of atoms to metal surfaces.

More complex excitations are even more effective than two-hole final states for stimulating desorption. They are signalled by sharp rises in the desorption yields occurring not at the precise core excitation threshold but shifted to higher energies by tens of eV. For instance, the compilation in Fig. 4 shows that in ESD from adsorbed CO and NO, the onsets for the emission of O^+ and O^{++} are delayed tens of eV from the O*1s* core threshold.

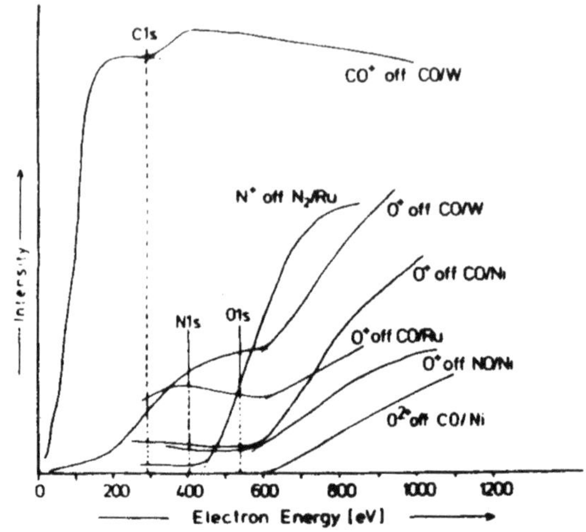

Fig. 4. Compilation of ion yields vs electron energy for desorption from metals.[31]

Our recent results suggest that multi-electron excitations are very important in ion emission from insulators where they can cause the ejection of multiply-charged ions.[32] In SiO_2, the emission of Si^+ ions has an onset delayed ~30 eV with respect to that for Si2*p*excitation; Si^{2+} emission is even more delayed. Since the Si charge state is ˜1 in the lattice,[29] the emission of Si^{3+} ions requires the excitation of no less than four valence electrons, two from Si, and two from an adjacent O^- to convert it to O^+ so that Coulomb repulsion can take place. For this reason, and in contrast with the desorption yield for atoms, the desorption yields for *ions* do not follow the core ionization cross section.

4. The Desorbed Flux

a. Energy distributions

Energies of desorbed neutrals peak at < 1 eV (Fig. 5),[33] while those for ions and excited neutrals usually peak at a few eV.[34] Coulomb repulsion from multiply charged ions can produce energy distributions with long tails extending to tens of eV (Fig. 6).[35]

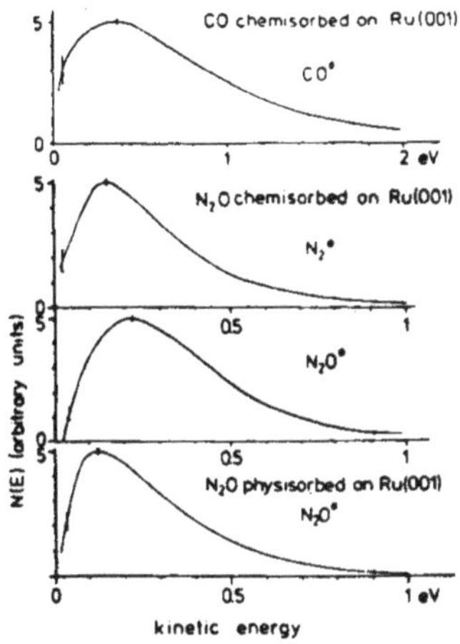

Fig. 5. Energy distributions of neutrals desorbed by 250 eV electrons.

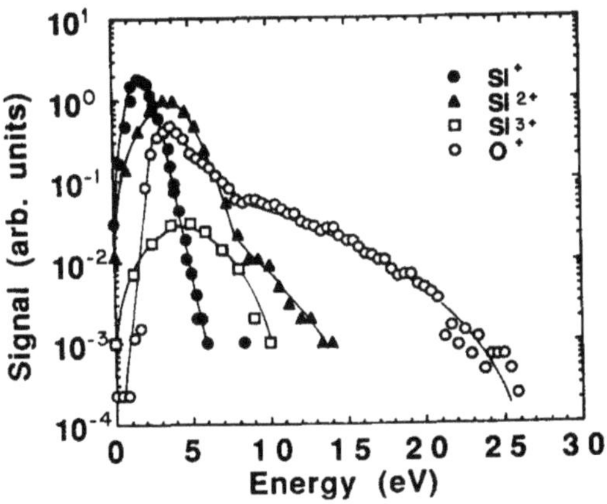

Fig. 6. Energy distributions of ions desorbed from SiO_2 by 2 keV electrons.[35]

The repulsion energy during dissociation is shared between the fragments inversely proportional to their masses, in order to conserve momentum. This tends to make light ions, like H^+, more energetic. However, the trend shown by measurements is that energies are not inversely proportional to mass, but rather, they are of similar magnitude. This suggests that the surface as a whole absorbs all momentum making the desorption "recoilless."[36] Maximum energies occur for adsorbed atoms; when an atom embedded in a solid is being ejected, the other lattice atoms adjust their positions to their final relaxed configuration thereby slowing down the ejected species. This rearrangement can be severe for ionic or partially ionic lattices, due to the long range nature of the interatomic potentials,[37] and can inhibit direct desorption.

b. Angular distributions

The appearance of angular patterns in the desorbed flux from adsorbed layers on single crystals forms the basis of the technique called Electron Stimulated Desorption Ion Angular Distributions (ESDIAD).[38,39] For singly coordinated bonds, a transition to an antibonding state should produce a repulsive force directed, by reasons of symmetry, along the internuclear axis. In other cases, relaxation of nearby atoms can deflect the desorbed species away from the initial bonding direction. This is seldom observed since ion emission is unlikely from multiple coordinated sites. In general, the desorbed beams are not narrow but are broadened by bending, vibrational and rotational modes in the surface molecules.[40] Factors which affect the angular distributions of *ions* are electron transfer, which depletes the intensity for emission at small angles to the surface, and deflection of ion trajectories by the long-range image charge[41] and Coulomb interactions with adsorbed ionic species.[42]

ESDIAD has proved to be valuable for the study of structures of adsorbed layers. Unlike low-energy electron diffraction (LEED), the information is direct, and since the imaging forces are local, long range order is not necessary for directed ion emission. Fig. 7 shows an example ESDIAD for H^+ from NH_3 adsorbed on a Ni(111) surface.[43] The halo shows that molecules are bonded via the nitrogen atom with the hydrogens pointing away from the surface. The hydrogen atoms have random azimuthal orientations, probably due to free rotations around the Ni-N bond. When a small quantity of oxygen is adsorbed, the O-H interactions lock the ammonia molecule to a fixed orientation, producing a pattern with three peaks.

This technique gives useful symmetry information, but it is not yet possible to derive accurate information of bond angles, taking due account of the interactions of the ejected ions with the surface. Electron transfer dominates the emission of ions in ESDIAD, favoring the emission of singly-coordinated species, of atoms at edges or steps, of atoms more distant from the surface, which more likely escape neutralization. It is important to take this into account in interpreting ion emission data, since *the observed ions may not be representative of the majority of the surface under study.* The observation of angular patterns of desorbed species is a unique way to study local bonding in ordered adlayers; its usefulness will increase when some of its problems are circumvented by detecting emitted neutrals atoms, e.g., by post-ionization with an intense laser beam.

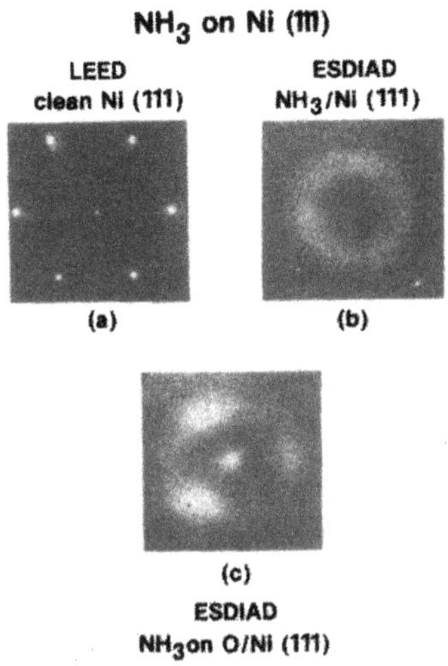

Fig. 7. LEED and ESDIAD patterns of NH_3 adsorbed on Ni(111). (a) clean LEED at 110 eV. ; (b) H^+ ESDIAD of NH_3 on Ni(111); (c) H^+ ESDIAD of NH_3 on oxygen predosed Ni(111).[43]

c. Charge distributions of desorbed particles

As mentioned above, the vast majority of desorbed species are neutral, due to electron capture by RT or AN. Neutralization rates are lower for species like H, O, and F, which have a high ionization potential, so the states fall below the bottom of the valence band. Neutralization is also unlikely for alkali ions, whose ionization potentials are smaller than the work function of most materials so that the states fall above the valence band. In insulators, neutralization rates are greatly reduced since there are usually no occupied states opposite empty states in the desorbing ion.

Electron bombardment of SiO_2, can eject ions with charges up to 3+ (Fig. 8). The final charges observed result from the initial charge states following Auger decay plus AN during ejection. AN of mulitply-charged ions is slower[44] than RT because the smaller screening shrinks the orbitals, causing less overlap with neighboring atoms and therefore a lower probability of hole hopping and inter-atomic Auger decay (AN).[45]

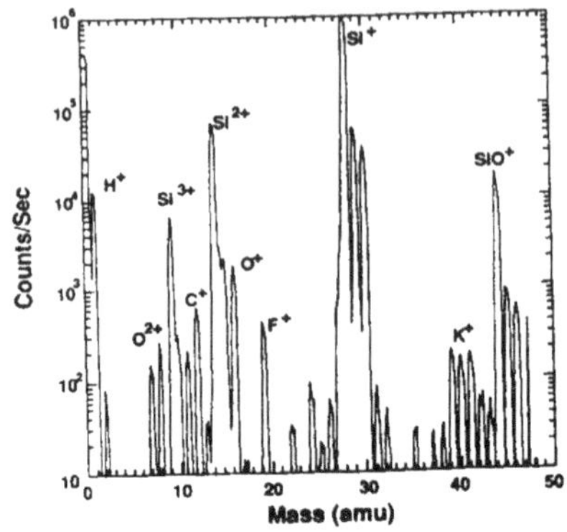

Fig. 8. Mass spectra of ions from 65 Å SiO_2 films under 2 keV electron bombardment.[46]

Insight on the importance of electron transfer comes from the finding[46] that negative ion yields can also be correlated with inner-shell excitations cross sections (Fig. 9). Since Auger processes make ions more positive, the emission of Si^- and O^- from SiO_2 means that excitation is decoupled from processes which determine the final charge. In this case, negative ions more likely result from electron capture when the ejected particle crosses the surface into vacuum.

Neutralization processes are one of the least known steps in the ejection of particles from surfaces. Some surprising findings are the general absence of N^+ emission,[47] and the very strong impurity F^+ signals observed from most surfaces. In fact, F^+ signals are observed from surfaces which no detectable F Auger signal ($<$ 1% surface concentration of F). We propose that this unusual sensitivity results from excitation to doubly-excited, autoionizing states. These states are extended, so we expect Pauli repulsion from their overlap with the solid. Due to relatively long lifetimes, autoionizing atoms will decay into ions far away from the surface, where reneutralization is unlikely. It is possible that the observed threshold for F^+ emission from Si,[48] corresponds to the onset for double excitation of F to an autoionizing state. This autoionization mechanism for ion emission has recently been identified and used to explain anomalous high ion yields for kev neon scattered from magnesium surfaces.[49]

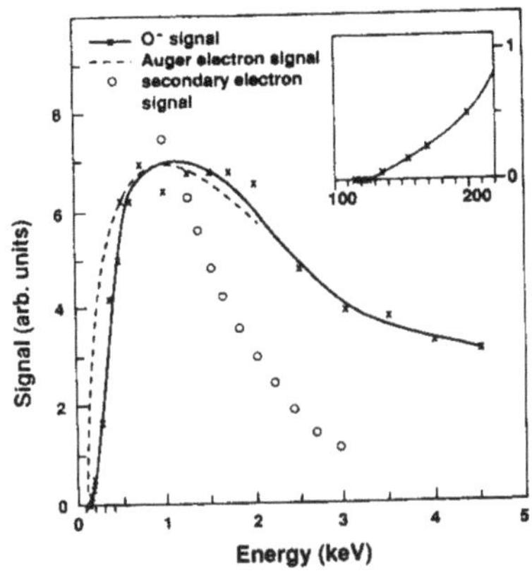

Fig. 9. O^- yield versus electron energy for 65 Å SiO_2 films bombarded with electrons at 70° incidence. Also shown are the Auger emission cross section, which approximately correlates with the ESD yields, and the secondary electron yield, which does not. The inset shows that the onset of the ion yield is delayed with respect to the ~110 eV Si2p binding energy.[46]

5. Conclusions

The knowledge of ESD is qualitative, mainly because there is a limited understanding of the details of electron transfer processes at surfaces, which determine the principal aspects of the observations. Important factors in ESD are the lifetime of the excited state, which needs to be of the order or larger than 10-1000 fs, depending on the velocity of the desorbed species. Lifetimes are enhanced by multi-electron excitations and by the presence of a core-hole. Many trends in desorption can be qualitatively explained by the ratio between lifetimes and desorption times and by the variation of lifetimes with properties of the system.

References

1. M. L. Knotek, Rep. Prog. Phys. **47**, 1499 (1984)
2. T. E. Madey, D. E. Ramaker, and R. Stockbauer, Ann. Rev. Phys. Chem. **35**, 215 (1984)
3. D. Menzel, Nucl. Inst. Meth. **B13**, 507 (1986)
4. T. E. Madey, Science **234**, 316 (1986)
5. V. N. Ageev, O. P. Burmistrova, and Yu. A. Kuznetsov, Sov. Phys. Usp. **32**, 588 (1989)
6. P. Avouris and R. E. Walkup, Ann. Rev. Phys. Chem. **40**, 173 (1989)
7. R. D. Ramsier and J. T. Yates, Surface Sci. Repts. **12**, 243 (1991)
8. R. A. Baragiola and T. E. Madey, in *Interactions of Charged Particles with Solids and Surfaces*, ed. by F. Flores et al. (Plenum, 1991) 313
9. *DIET-I*, ed. by N. H. Tolk, M. M. Traum, J. E. Tully, and T. E. Madey (Springer Verlag, Berlin, 1983)
10. *DIET-II*, ed. by W. Brenig and D. Menzel (Springer Verlag, Berlin, 1985)
11. *DIET-II*, ed. by R. H. Stulen and M. L. Knotek (Springer, Berlin, 1988)
12. *DIET-IV*, ed. by G. Betz and P. Varga (Springer Verlag, Berlin 1990)
13. H. S. Massey, E. H. S. Burhop, and H. B. Gilbody, *Electronic and Ionic Impact Phenomena*, Vol II (Oxford Univ. Press, London, 1969)
14. R. A. Baragiola, Rad. Effects **61**, 47 (1982). See also article by the author on electron emission, this volume.
15. T. R. Hayes and J. F. Evans, Surface Sci. **159**, 466 (1985)
16. D. R. Jennison and D. Emin, Phys. Rev. Lett. **51**, 1390 (1983).
17. G. G. de Souza, P. Morin and I. Nenner, Phys. Rev. **34**, 4770 (1986)
18. P. Nordlander, *DIET-IV*, p. 12
19. T. E. Madey, J. T. Yates, D. A. King, and C. J. Uhlaner, J. Chem. Phys. **52**, 5215 (1970)
20. F. P. Netzer and T. E. Madey, J. Chem. Phys. **76**, 710 (1982)
21. C. T. Reimann, W. L. Brown, M. J. Nowakowski, S.-T. Cui and R. E. Johnson, *DIET-IV*, p.226.
22. H. Sambe, D. E. Ramaker, L. Parenteau, and L. Sanche, Phys. Rev. Lett. **59**, 236 (1987).
23. D. E. Ramaker, T. E. Madey, R. L. Kurtz, and H. Sambe, Phys. Rev. **B33**, 2099 (1988)
24. W. Ho, Comments Cond. Matt. Phys. **13**, 293 (1988)
25. D. Menzel and R. Gomer, J. Chem. Phys. **41**, 4411 (1964); P. A. Redhead, Can. J. Phys. **42**, 886 (1964).
26. M. L. Knotek and P. J. Feibelman, Phys. Rev. Lett. **40**, 964 (1978)
27. J. Dresner and B. Goldstein, J. Appl. Phys. **47**, 1038 (1976).

28. R. L. Kurtz, R. Stockbauer and T. E. Madey, in *DIET-II*, p. 89.
29. D. E. Ramaker, J. S. Murday, N. H. Turner, G. Moore, M. G. Lagally and J. Houston, Phys. Rev. **B19**, 5375 (1979)
30. L. Calliari, M. Dapor, L. Gonzo and F. Marchetti, in *DIET-IV*, p. 373.
31. P. Feulner, R. Treichler, and D. Menzel, Phys. Rev. **B24**, 7427 (1981)
32. R. Baragiola, T. Madey and A-M. Lanzillotto, *DIET-IV*, p. 322.
33. P. Feulner, D. Menzel, H. J. Kreuzer, and Z. W. Gortel, Phys. Rev. Lett. **53**, 671 (1984)
34. R. H. Stulen, in *DIET-II*, p. 130.
35. R. A. Baragiola, T. E. Madey, A-M. Lanzillotto, Phys. Rev. **B41**, 9541 (1990).
36. J.I.Gersten and N.Tzoar, Phys. Rev. **B16**, 945 (1977)
37. R. E. Walkup and P. Avouris, Phys. Rev. Lett. **56**, 524 (1986)
38. T. E. Madey, Science **234**, 316 (1986)
39. M. D. Alvey and J. T. Yates, Jr., J. Am. Chem. Soc. **110**, 1782 (1988)
40. W. Riedl and D. Menzel, Surf. Sci. **207**, 494 (1989)
41. Z. Miscovic, J. Vukanic, and T. E. Madey, Surface Sci. **169**, 405 (1986)
42. C. Z. Dong, P. Nordlander and T. E. Madey, in *DIET-IV*, p. 34.
43. F. P. Netzer and T. E. Madey, Surface Sci. **119**, 422 (1982)
44. P. Varga, Comments At. Mol. Phys. **23**, 111 (1989)
45. P. J. Feibelman, Surf. Sci. **102**, L51 (1981)
46. A. M. Lanzillotto, T. E. Madey and R. A. Baragiola, Phys. Rev. Lett. **67**, 232 (1991)
47. M. L. Knotek and J. E. Houston, J. Vac. Sci. Technol. **20**, 544 (1982)
48. M. J. Bozack, M. J. Dresser, W. J. Choyke, P. A. Taylor and J. T. Yates, Surf. Sci. **184**, L332 (1987)
49. O. Grizzi, M. Shi, H. Bu, J. Rabalais and R. Baragiola, Phys. Rev. **B41**, 4789 (1990).

DEPENDENCE OF SECONDARY ION EMISSION FROM ORGANIC MATERIAL ON THE ENERGY LOSS OF THE IMPACTING HEAVY ION

Jerry E. Hunt
Chemistry Division, Argonne National Laboratory, Argonne (U.S.A.)

and

Karl Wien
Institut für Kernphysik der Technischen Hochschule Darmstadt (FRG)

ABSTRACT

At the Argonne Dynamitron accelerator samples of the aminoacid valine were irradiated by 2.5 MeV-Ar, 0.8 MeV-Kr and 1.0 MeV Xe beams, in order to study the energy distributions of ejected secondary ions. In case of Kr and Xe the nuclear stopping power exceeded the electronic stopping power by a factor 2 or 3, respectively, but the functional shape of the energy distributions and the mean ejection energies (0.9 - 1.4 eV) indicated that the molecular ions $(M{\pm}H)^{\pm}$ are desorbed by an electronic sputter process. Contributions of atomic collision cascades were observed for the H^- ion. In the second part of the article, the overall dependence of molecular ion yields on the electronic energy loss is discussed in the frame work of recent desorption models and the structure of the nuclear track.

1. Introduction

This article presents first experimental studies of heavy ion induced sputtering in an energy regime, where nuclear and electronic stopping have the same order of magnitude, and second a discussion of the electronic sputter process in context with nuclear track structure. The subject of interest is sputtering of organic material.
In the domain of electronic energy loss - i.e. at projectile energies between 0.02 and 30 MeV/n - sputtering from organic material certainly originates in the electronic excitation of the solid by the swift heavy ion. At lower projectile energies nuclear stopping becomes the dominant component of the energy loss, but its role in the sputter process of organic material is not clear. Regarding the dependence of secondary ion yields on the energy loss, Salehpour et al.[1] concluded that at projectile velocities

around 0.1 cm/ns the ejection of the $(M\text{-}H)^-$ ion of valine is an electronic sputter process. As shown in Fig. 1, an influence of nuclear stopping on the $(M\text{-}H)^-$ yield seems not to be present. Such an influence is, however, observed for certain small fragment ions like OH^-.

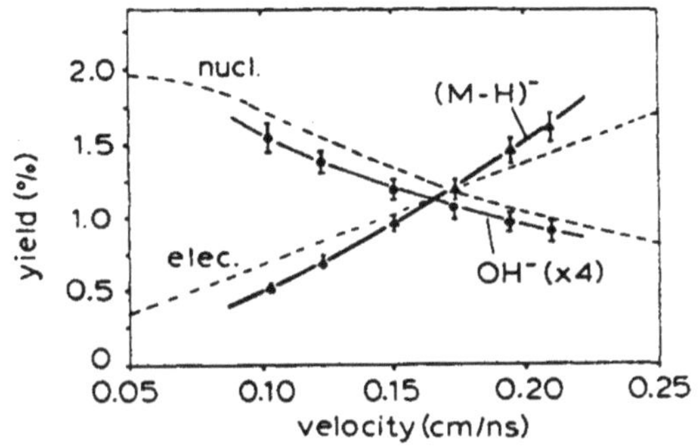

Fig. 1: Yield of secondary ions ejected from a valine sample by Xe ions as function of the projectile velocity. The dashed curves represent the nuclear and electronic energy loss. At their intersection point they are normalized to the yield curve of the valine parent ion $(M\text{-}H)^-$. The data were measured by Salehpour et al.[1]

The question arises, if the trend of yields as displayed in Fig. 1 continues down into the keV regime. Albers et al.[2] published $(M+H)^+$ yields of valine measured at projectile velocities between 0.05 and 1.2 cm/ns, but - as discussed in reference [3] - below 0.2 cm/ns the results are questionable. Reliable data obtained with projectile energies between 1 keV and 18 keV have been presented by Standing et al.[4]. They irradiated the aminoacid alanine with various alcali ions and studied the yields of the $(M+H)^+$ ion and other positive ions. The essential result of their experiments was that in the low keV regime ion emission from organic samples is related to nuclear stopping. The main argument is the large difference in ion yields obtained with Li, Na, K and Cs projectiles. Going from Li to Cs at constant energy, the $(M+H)^+$ yield increases by almost 3 orders of magnitude, whereas the electronic energy loss differs only by 30 % and does not exceed 10 eV/Å.

This unsatisfying situation gave the impetous to study the energy distributions of organic molecular ions ejected by heavy ions in the velocity regime of about 0.1 cm/ns. Electronic sputtering is generally associated with smaller ejection energies than sputtering via direct atomic collisions[5]. Also the functional form of the energy distributions is different. In the next chapter energy distributions taken in the high electronic energy loss region are compared with those, where the nuclear energy loss is dominant.

The second part of this article deals with the functional dependence of ion yields on the electronic energy loss $(dE/dx)_e$. Reviewing the large body of data collected during 18 years of Plasma Desorption Mass Spectrometry (= PDMS) it turns out that only in case of valine the data cover a reasonable part of the $(dE/dx)_e$ dependence. Models developed for heavy ion induced desorption describe in general only fractions of the total dE/dx dependence of yields. Considering the picture of the desorption process, it will be shown that the missing link of a consistent theory is the description of the nuclear track sturcture and its variation with projectile velocity.

2. Experimental Investigation of Energy Distributions of Secondary Ions Ejected from Valine

The Argonne 4.5 MeV Depossitron accelerator was used to produce beams of 2.8 MeV - ^{40}Ar ions, 0.80 MeV - ^{84}Kr ions and 1.0 MeV - ^{131}Xe ions. As illustrated in Fig. 2 the beam penetrated the target from the backside and ejected the secondary ions of interest from the frontside. The target had an effective diameter of 1.5 mm and consisted of a 27 μg/cm^2 C foil covered with a vacuum evaporated film of 1000 Å valine. The beam energy at the target exit - i.e. in the uppermost surface layers of the sample - was 2.5 MeV for Ar, about 0.35 MeV for Kr and about 0.55 MeV for Xe.

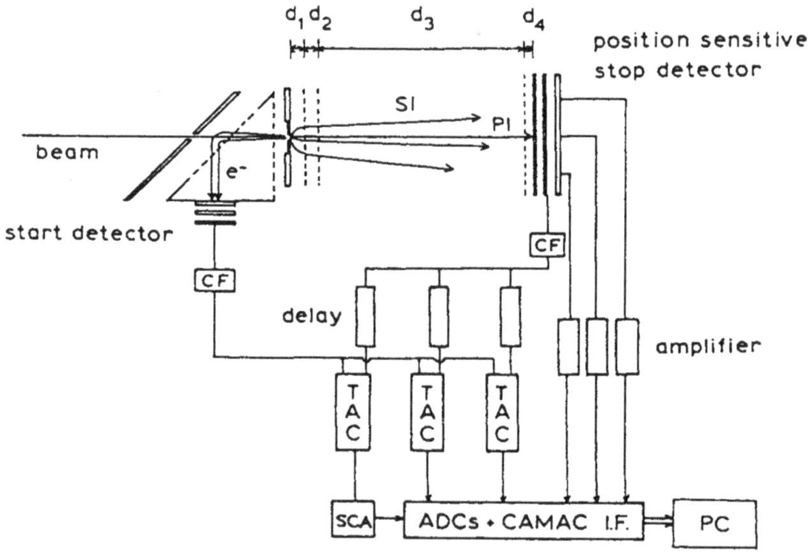

Fig. 2: Experimental set up for measuring energy distributions of secondary ions ejected by beam ions, which penetrate the target at the backside through a 1.5 mm apperture. SI = secondary ions, PI = primary ions, CF = constant fraction discriminator, TAC = time-to-amplitude converter, SCA = single channel analyzer, ADC = analog to digital converter. Only the most important part of the electronic equipment is shown. For details see text and reference 7 concerning the position sensitive stop detector.

The time-of-flight (TOF) set up sketched in Fig. 2 was equiped with two channel plate assemblies - each containing a tandem of 2 plates. The start signal for the various TOF measurements was allways generated by secondary electrons being ejected from the target backside by the beam ions and then after acceleration deflected by 90° onto the start-detector. The beam ions as well as the secondary ions reached the first channel plate of the stop detector after a total flight distance of 29.98 cm. In accordance to the three time-of-amplitude converters (TAC) shown in Fig. 2 three TOF measurements were performed simultanously:

1. Measurement of the projectile velocity behind the target in order to determine the actual beam energy.

2. Generation of a gate signal for the third measurement: The second TAC delivered the total TOF spectrum of particles reaching the stop detector-including beam ions and secondary ions. Because this TAC could handle only one stop event per start event, secondary ions are registered only, when the proceeding beam ion was stopped in one of the 3 grids inserted in the flight path. The total grid transmission was 32 %. The gate signal was taken from the mass line of interest by means of a single channel analyser behind the TAC. The purpose of this gating is, to analyse in the third measurement only secondary ion events without a precursor, in order to avoid pulse pile up.

3. Measurement of energy distributions: In order to evaluate the axial (in direction of the surface normal) and radial (perpendicular to the surface normal) energy distributions of a certain ion, two numbers had to be determined for this ion, when it hits the stopdetector: the time of flight and the distance r between the spectrometer axis and the point of impact at the planar stop detector. The corresponding time and spatial distributions $\Delta n/\Delta t$ and $\Delta n/\Delta r$ can be transformed into energy distributions by

$$\frac{\Delta n}{\Delta E_a} = \frac{\Delta n}{\Delta t} \cdot \frac{dt}{dE_a} \quad \text{and} \quad \frac{\Delta n}{\Delta E_r} = \frac{\Delta n}{\Delta r} \cdot \frac{dr}{dE_r}$$

with

$$t = \sqrt{m} \cdot f(d_i, U_i, E_a) \quad \text{and} \quad r = \sqrt{\frac{2E_r}{m}} \cdot t$$

Here, d_i are the four distances between the sample, the 3 grids and the stopdetector and U_i the corresponding electric potentials.

In order to allow broadening of the mass lines due to the axial energy E_a together with sufficient mass separation, the secondary ions leaving the sample were accelerated first by a low potential (89 Volt) and then by a high potential (4966 Volt). TOF mass spectra as shown in Figs. 3 and 4 delivered the flight time t and the time distribution $\Delta n/\Delta t$ for a certain mass m.

The distance r of impact and the corresponding spatial distribution $\Delta n/\Delta r$ were determined by means of a position sensitive stop detector. It consisted of two channel plates ($\Phi = 45$ mm) in series and a stripped „backgammon" anode. Details of this detector type and the methode are described elsewhere[5]. The coordinates of the point of impact are determined with help of three charge pulses taken from the three types of stripes on the anode. The amplitudes of the charge pulses and the TAC signal were stored in list mode for each detected ion in coincidence to the mentioned gate signal. The spatial resolution of the position sensitive detector was about 0.5 mm.

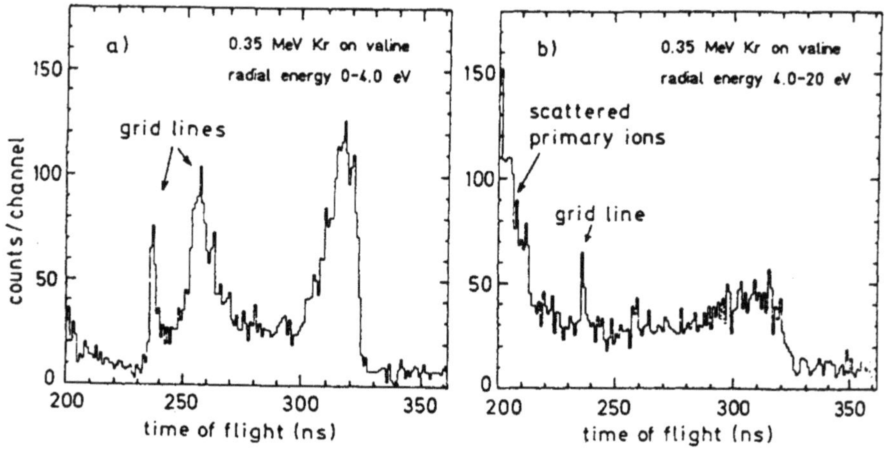

Fig. 3: TOF spectrum of the H^- ion ejected from valine by a Kr beam (300 ions/s). The spectrum a) corresponds to ions having a radial energy between 0 and 4.0 eV, the spectrum b) to ions between 4.0 eV and 20 eV. The narrow line at 237 ns is associated to H^- ions, which are ejected from the front side of the first grid or the walls of the grid holes. The broader line at about 255 ns has been produced by H^- being emitted at the backside of the first grid, then reflected by the low potential between sample and this grid and sucked off through the grid. Of interest are the broad distributions in the right part of the spectra.

Because of rotational symmetry the spectrometer axis intersects the plane of the stop detector at the gravity centre of the spatial distribution of stop events. More difficult is the determination of the time $t_o = \sqrt{m} \cdot f(d_i, U_i, E_a = 0)$ under a mass line. As explained in reference 5 one has to calculate t_o with help of very precisly measured values of d_i and U_i. We used a microscope to measure d_1 and d_2, the distances between the sample and grid 1 and grid 2, respectively. In general, radial energy distributions are more accurate than axial distributions because of uncertainties of t_o.

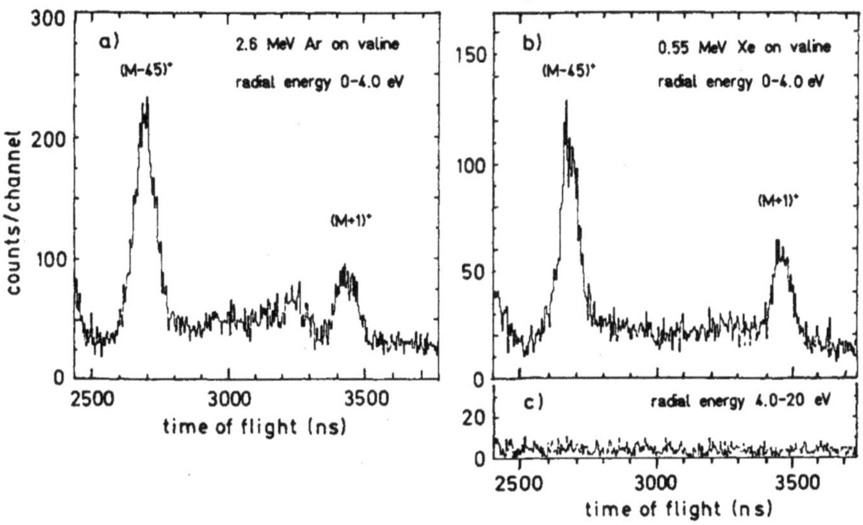

Fig. 4: TOF spectra of the $(M+H)^+$ and the $(M-45)^+$ ions ejected by a) 2.5 MeV-Ar and b) and c) 0.55 MeV-Xe ions. The spectra a) and b) correspond to secondary ions having a radial energy between 0 and 4.0 eV, the spectrum c) to ions between 4.0 and 20 eV. Spectrum c) contains only background events, because $(M+1)^+$ and $(M-45)^+$ ions with radial energies above 4.0 eV were hardly emitted.

We investigated the secondary ions H^-, $(M-H)^-$, $(M+H)^+$ and $(M-45)^+$ of valine. Examples of energy distributions are presented in Figs. 5 and 6. The mean energies of the measured distributions are given in Table 1 - together with the corresponding energy loss values in valine. For comparison Table 1 contains also data of Moshammer[7] recently measured by means of a 1.4 MeV/n - ^{131}Xe beam at the GSI/Darmstadt/FRG and ^{252}Cf-fission fragments. The method and the experimental conditions were quite similar to the present work: irradiation from the target backside and vacuum evaporated valine samples (5000 Å and 1500 Å, respectively).

As illustrated in Fig. 4, allready the TOF spectra taken in the high mass region exhibited that the differences of ion ejection by Ar and by Kr or Xe projectiles are small. In case of 2.5 MeV-Ar the contribution of nuclear stopping to the total energy loss is only 8.3 %, whereas for 0.35 MeV/Kr and 0.55 MeV-Xe $(dE/dx)_n$ exceeds $(dE/dx)_e$ by a factor 2 and 3, respectively. Comparing the energy disstributions - examples are presented in Figs. 5 and 6 - it becomes evident that in case of Ar and Kr or Xe the emission of the molecular ions and the $(M\text{-}45)^+$ fragment ion are caused by the same final ejection mechanism. Combining the yield data of Salehpour et al.[1] and the energy distributions of this work, the essential result is that down to projectile energies of 0.005 MeV/n heavy ion induced sputtering of molecular ions from organic material originates in the electronic excitation of the solid.

Table 1: Mean energies of secondary ions (SI) ejected from valine by heavy ions (PI) above 0.35 MeV. The samples were irradiated from the backside under normal incidence. The error of the radial energies is about 10% , of the axial energies about 15% . Results obtained with Xe* and f.f.* (Cf-fission fragments) were taken from the thesis of Moshammer[7]. The number in brackets given in the first column is the thickness of the valine sample in Å.

PI	$(dE/dx)_n$ [eV/Å]	$(dE/dx)_e$ [eV/Å]	SI	$<E_a>$ [eV]	$<E_r>$ [eV]
Ar	12	133	H^-	1.4	1.15
(1000)			$(M\text{-}H)^-$	0.37	0.56
			$(M+H)^+$	0.58	0.75
			$(M\text{-}45)^+$	0.87	0.76
K	83	40	H^-	1.39	1.21
(1000)			$(M\text{-}H)^-$	0.50	0.68
			$(M+H)^+$		0.74
			$(M\text{-}45)^+$		0.66
Xe	132	46	$(M\text{-}H)^-$	0.47	0.80
(1000)			$(M+H)^+$	0.57	0.82
			$(M\text{-}45)^+$	0.74	0.75
Xe*	<10	1100	$(M\text{-}H)^-$	0.7	0.8
(5000)			$(M+H)^+$	1.6	1.6
f.f.*	<10	950	H^-	1.3	1.8
(1500)			$(M\text{-}H)^-$	0.8	0.6
			$(M+H)^+$	1.5	0.9
			$(M\text{-}45)^+$	1.9	0.6

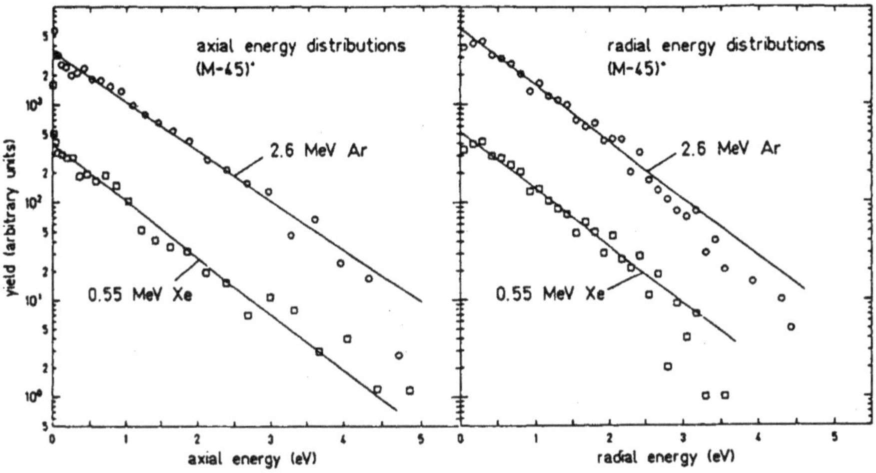

Fig. 5: Axial and radial energy distributions of the $(M\text{-}45)^+$ ion being ejected from valine by Ar and Xe ions. The solid lines are expontial fits to the data points, the corresponding mean energies are given in table 1. At high energies several data points have been averaged.

Regarding the body of data we want to add the following comments:

1. The energy distributions measured in this work can be reasonably fitted to exponential functions. The statistics of counts is, however, rather low and a steeper decay of the radial distributions then exponential can not be excluded as,for instance, indicated in Fig. 5. Moshammer[7] observed for molecular ions at high energies a E_r^{-3} dependence, his axial distributions followed always an exponential law.

2. The mean total energies ($= < E_a > + < E_r >$) of the $(M\text{-}H)^-$ ion range from 0.9 eV to 1.5 eV, the slight increase going from Ar to Xe is within the experimental errors.

3. The mean axial energies $< E_a >$ of $(M+1)^+$ are about a factor 2.7 smaller than the corresponding values of Moshammer[7]. The essential differences between the two experiments are the projectile energy and the energy loss as listed in Table 1. An increase of the axial energy with projectile energy or energy loss, respectively, has also been reported by Becker et al.[5]. Our observation that $< E_r >$ is larger than $< E_a >$ could indicate that the emission probability of $(M+H)^+$ is not forward peaked. That is in contradiction to the studies of Moshammer[7],

who found an angular distribution close to cosϑ in the high energy regime.

4. In order to detect small contributions of nuclear sputtering, TOF spectra were extracted from the list mode data in coincidence to certain windows on the radial energy spectrum. Ions, which impacted the stopdetector within a distance r = 9 - 20 mm, had a radial energy between 4.0 eV and 20 eV. As shown in Fig. 4c, the corresponding stop events are not correlated with definit masslines in the TOF spectrum. The origin of such ions are chance coincidence events with H^+ ions or true coincidence events with primary ions after being scattered and degraded at the grids.

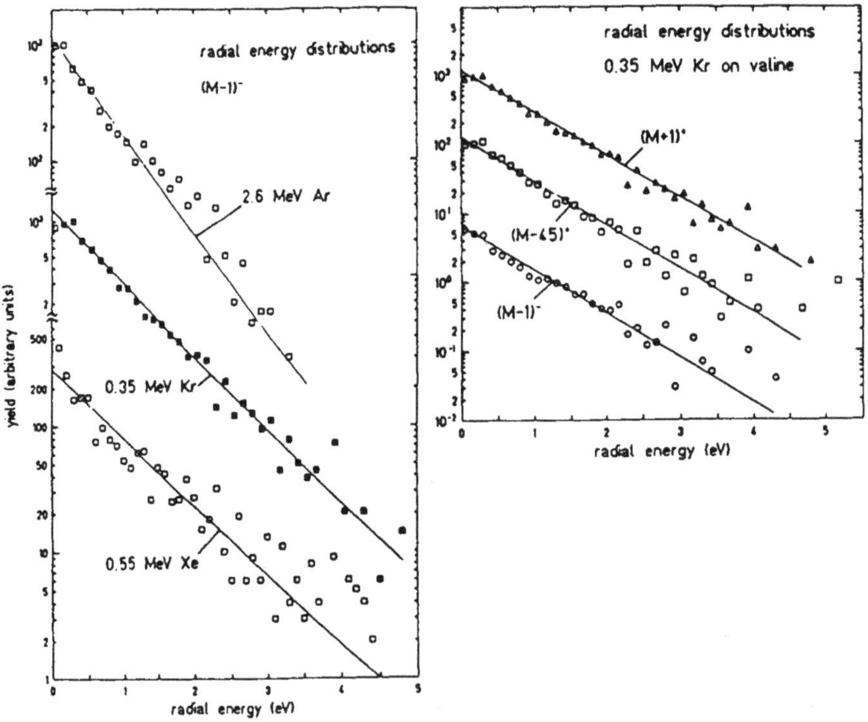

Fig. 6: Radial energy distributions of secondary ions ejected from valine. In the left figure distributions of the $(M\text{-}H)^-$ ion produced by Ar, Kr and Xe are presented. In the right figure the three ions $(M+H)^+$, $(M\text{-}4)^+$ and $(M\text{-}H)^-$ were emitted always from the same sample by Kr ions. The solid lines represent exponential fits, the corresponding mean energies are listed in Table 1.

5. The TOF spectra of the H^- ions presented in Fig. 3 demonstrate clearly the coexistence of nuclear and electronic sputtering. Of interest in both spectra is the peak at large flight times. The other peaks are explained in the figure caption (see also Ref. 5). The broad distribution in the right spectrum measured in coincidence to E_r = 4.0 - 20 eV is mainly generated by H^- ions ejected by atomic collision cascades. The mean energy $< E_a >$ of this distribution has been roughly estimated to be 25 eV. The left spectrum consists of both sputter components. The nuclear component was subtracted with help of the right spectrum. The remaining spectrum was almost identical with the H^- spectrum produced by the Ar beam. The mean axial and radial energies listed in Table 1 are equal within the error bars and agree also with values reported by Moshammer[7].

3. Desorption Yields of Molecular Ions in Relation to $(dE/dx)_e$

Regarding the proceeding chapter, the desorption yields of molecular ions level down with decreasing projectile energy. In case of valine a threshold energy or a threshold energy loss have not been observed so far. At projectile energies above

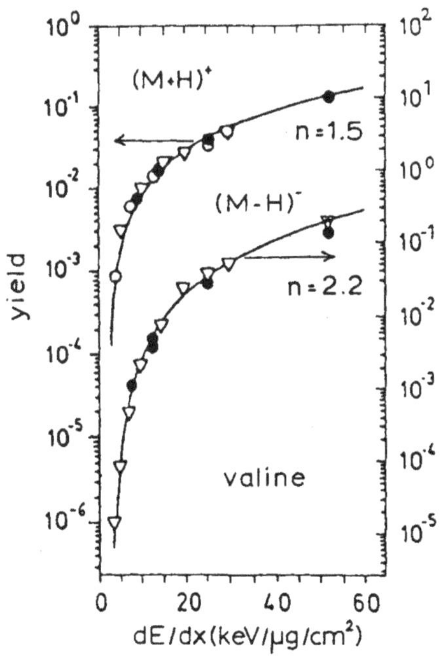

Fig. 7: Yield of positive and negative parentions of valine plotted versus primary ion energy loss. The solid circles represent data obtained with primary ions of same velocity, the open points data measured with primary ions having a velocity v = 1.39 cm/ns ± 13%. The solid lines corresponed to the function $(dE/dx - S)^n$. The data were measured by Brandl et al.[8].

the maximum of $(dE/dx)_e$, however, Brandl et al.[8] reported a destinct threshold behaviour. The two yield curves of valine $(M+H)^+$ and $(M-H)^-$ presented in Fig. 7 were fit by the authors to the expression

$$Y = A(dE/dx - S)^n \tag{1}$$

The threshold energy loss S has for $(M+H)^+$ the value 25 eV/Åand for $(M-H)^-$ 42 eV/Å. These values are considerably higher than the electronic energy loss of the keV ions used by Standing et al.[4] to desorb the parentions of valine. That seems to be an other argument for nuclear sputtering in the keV regime, but one has to be aware that in the high energy regime only a fraction of $(dE/dx)_e$ - probably 1/3 of $(dE/dx)_e$ - contributes to the sputter process. The main part of the deposited energy is carried away by long range δ-electrons[9], wich are of minor or of no importance for molecular ion desorption as will be discussed next.

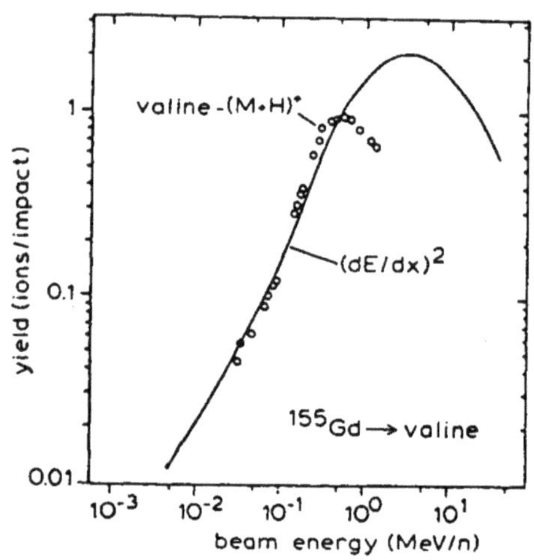

Fig. 8: The yield of the valine $(M+H)^+$ ion compared with $(dE/dx)^2$. The valine sample was irradiated by ^{155}Gd ions.

The yield curve of the $(M+H)^+$ ion in Fig. 8 illustrates that the maximum desorption yield does usually not coincide with the maximum of $(dE/dx)_e$. This phenomenon leads to peculiar „hooks" in yield-dE/dx diagrams[10]. The most complete hook can be constructed from the various data sets being published for the $(M+H)^+$ ion emitted from valine by ^{16}O beams[2,8,10]. In Fig. 9a four data sets were adjusted to each other in the overlapping regions and then plotted versus dE/dx taken from the Northcliffe-Schilling tables[11]. For comparison, in the neighbouring figure an other type of hook is presented: Here, the inactivation cross section of living cells[12] is plotted versus the energy loss LET. Regarding the two radiation effects one should notice, that in both cases molecular bonds have to be broken - either the bonds between the valine molecule and its neighbours or bonds between parts of the DNA. The remarkable difference between the two hooks in Fig. 9 is, that they are rounded in opposite directions. In other words, at the same value of dE/dx the desorption yields are higher in the low energy branch of dE/dx as in the high energy branch, whereas the inactivation cross sections exhibit the opposite behaviour. The two plotts in Fig. 9 demonstrate the significance of electronic excitation and its spatial distribution in the surrounding of the nuclear track: In case of heavy ion induced desorption the energy density within a small cylinder around the track is essential for the desorption process, - in case of living cells, the long range δ-electrons beyond the central core play the major role in cell destruction.

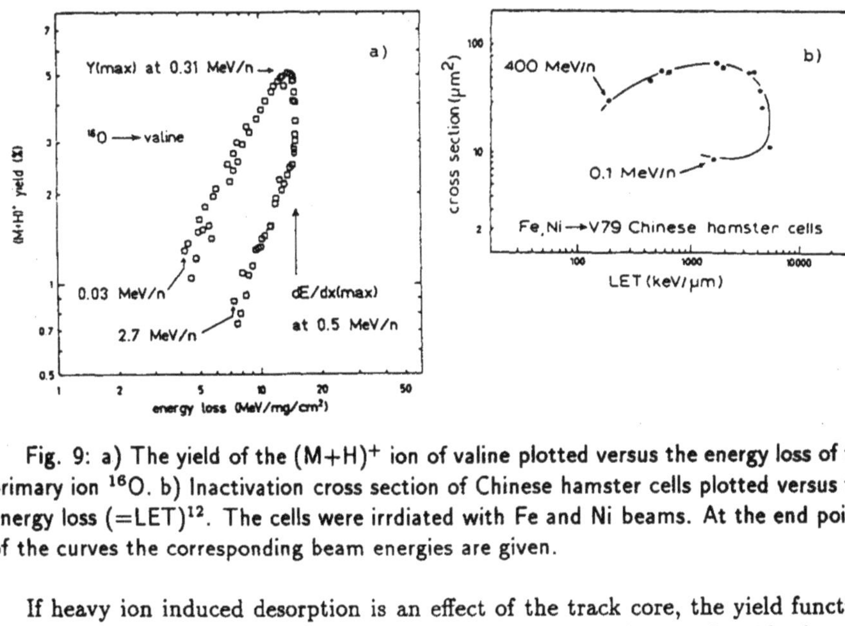

Fig. 9: a) The yield of the $(M+H)^+$ ion of valine plotted versus the energy loss of the primary ion ^{16}O. b) Inactivation cross section of Chinese hamster cells plotted versus the energy loss (=LET)[12]. The cells were irrdiated with Fe and Ni beams. At the end points of the curves the corresponding beam energies are given.

If heavy ion induced desorption is an effect of the track core, the yield function is probably better characterized by that part of the energy deposited inside the core volume. In order to calculate this partial energy loss $(dE/dx)_p$ we used the radial energy density

$$\epsilon(r) = \epsilon_o(r^{-2} - R_e^{-2})/(L^{-2} - R_e^{-2}) \qquad r \geq L \tag{2}$$

$$\epsilon(r) = \epsilon_o \qquad r < L \tag{3}$$

r is the radial distance from the track, L the radius of a central zone, in which the energy density has a constant value ϵ_o, and R_e the mean range of δ-electrons perpendicular to the track. A similar expression has been used in several desorption modells, in order to account for the variation of the energized core with projectile velocity[13,14,15]. Essential is the $1/r^2$ dependence of $\epsilon(r)$ being verified experimentally for gases[16]. The range R_e is estimated by means of the semiempirical expression[17]

$$R_e = (3.91 * 10^5/\rho)(v/c)^2 \ \mathring{A} \tag{4}$$

Here, ρ is the sample density in g/cm^3 and v/c the ratio of projectile velocity and light velocity. The central energy density ϵ_o is evaluated by means of the condition, that the spatial integral over $\epsilon(r)$ equals dE/dx. The result is

$$\epsilon_o = (dE/dx)(L^{-2} - R_e^{-2})/(2\pi * lnR_e/L) \tag{5}$$

Because the varation of the core radius r_o with projectile velocity is not known, we investigated two extreme cases:

1. r_o is kept constant. The partial energy loss is then the integral over $\epsilon(r)$ up to r_o. L is choosen to be 2 Å.

2. The mean energy per atom inside the core volume is kept constant. The corresponding radius r_o is then determined again by intgration over $\epsilon(r)$.

In accordance to case 1 and 2, Fig. 10 persents the yield curves of the $(M+H)^+$ ion as function of the partial energy loss with $r_o = 20$ Å and of the core radius with $\epsilon_o = 10$ eV/atom (here, L was assumed to be identical with r_o). As seen, the hook character of the yield curves is reduced, but the low and high energy branches of the curves follow different power laws of $(dE/dx)_p$ or r_o, respectively. That means, only certain sections of the curves can be parameterized by such simple power laws.

The values of $r_o = 20$ Å and $\epsilon_o = 10$ eV/atom were choosen with respect to the desorption models of Johnson et al.[18] and Bitensky et al.[19]. Both models are based on the assumption, that a pressure pulse expands from the energized central core into the surrounding solid and that desorption occurs at the surface, when this pressure exceeds a critical value in the uppermost surface layers. Regarding the pressure pulse model of Johnson et al., at high excitation energies a certain volume is ejected characterized by a radius

$$r_c \sim n_M^{-2/3} U^{-1} (dE/dx)_{eff} \tag{6}$$

n_M is the molecular number density, U the cohesive energy per molecule and $(dE/dx)_{eff}$ the partial energy loss inside the core volume. The desorption yield is then proportional to $(dE/dx)^3_{eff}$. In the model of Johnson et al. a variation of the core radius r_o is not included, a comparison with experimental data is therefore restricted to measurements at constant projectile velocity. In Fig. 10a the low energy branch of the yield curve scales like $(dE/dx)^3_p$. That means in context with the pressure pulse model of Johnson et al., that the effective core radius is constant in this energy regime, whereas above the maximum of $(dE/dx)_e$ the core radius and the partial (or effective) energy loss varry with projectile velocity.

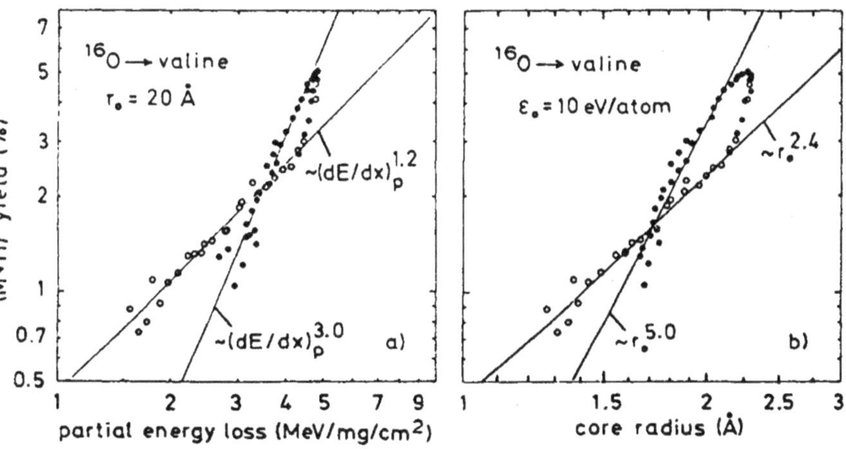

Fig. 10: The $(M+H)^+$ yield of valine measured with ^{16}O beams as function of a) the partial energy loss inside a core radius of $r_o = 20$ Å and b) of the radius r_o of a cylindrical zone around the track, where the mean energy per atom is allways 10 eV. The dark and open circles represent data from different experiments in the low and the high energy regime with respect to the maximum of dE/dx.

In the model of Bitensky et al.[19] the pressure pulse forms a shock wave, which causes ablation at the surface. Again a volume is ejected having a radius

$$r_c = r_o P_o / P_{cr} \tag{7}$$

P_{cr} is a critical pressure, above which surface molecules are ejected into vacuum. P_o is the initial pressure in the track core, from where the shock wave originates, and r_o the corresponding core radius. To initiate a shock wave, the mean energy per atom inside the core volume has to reach a value of typical ϵ_{cr} = 10 eV/atom. Bitensky et al.derived the following relation between r_o, the constant ϵ_{cr} and dE/dx :

$$r_o^2 = 3(dE/dx)/(4\pi n\epsilon_{cr}) \tag{8}$$

n is the number density of atoms. The desorption yield is proportional to r_o^3 or $(dE/dx)^{3/2}$, respectively. Using the energy density $\epsilon(r)$ as defined above, r_o was evaluated with $\epsilon_{cr} = \epsilon_o$ = 10 eV/atom. The yield curve plotted as function of r_o has again two branches, which are more separated than in Fig. 10a. The exponent n = 5 of the low energy branch is clearly larger than predicted, the slope of the high energy branch being proportional to $r_o^{2.4}$ is in reasonable agreement with the model. As seen in Fig. 10b, the condition for shock wave formation can be achieved only within rather small radii r_o (< 2.3 Å). These radii have the dimensions of an atom and it seems questionable, if a macroscopic model like the shock wave model of Bitensky et al. is capable to describe processes within the atomic scale. Our conclusion is, that at such low excitation energies as they are accomplished with ^{16}O ions the shock wave mechanism can hardly account for molecular ion desorption.

4. Conclusions

The yield measurements of Salehpour et al.[1] and the energy distributions of secondary ions presented here have clearly shown, that down to heavy ion energies of 5 keV/n ejection of molecular ions from organic material is mainly an electronic sputter process. Nuclear sputtering might play a role in the low keV regime, because the energy deposited in the electronic system is too low with respect to the threshold energies observed at the high energy branch of yield curves. Essential for electronic sputtering is the track core. It turned out that the energy content of the core volume or its effective radius are better suited to parameterize desorption yields than the total energy loss. Both parameter are ingredients of desorption models recently developed[18,19]. Their role in the desorption process is, however, only partially understood. So far, a consistent description of the generation and evolution of the track core is not available. A promising first step is the theoretical approach of Kammer[20], who investigated the propagation and motion of heavy ion created secondary electrons in solids. Important are his calculations concerning the transient confinement of electronic energy in the track core. The next step, transfer of electronic energy to the atomic system, is the missing link in the hierarchy of processes leading finally to particle desorption at the surface.

References

1. M. Salehpour, J. E. Hunt, D. L. Fishel and J.C. Tou, Int. J. Mass Spec. Ion Proc. 88 (1989) 211
2. A. Albers, K. Wien, P. Dück, W. Treu and H. Voit, Nucl. Instr. and Meth. 198 (1982) 69
3. K. Wien, Nucl. Instr. and Meth. (1992), in press
4. K. G. Standing, B. T. Chait, W. Ens, G. McIntosh and R. Beavis, Nucl. Instr. and Meth. 198 (1982) 343
5. O. Becker and K. Wien, Nucl. Instr. and Meth. B16 (1986) 456
   O. Becker, thesis at the Institut für Kernphysik, Technische Hochschule Darmstadt, FRG, 1985
6. I. S. Lapington, R. Kessel and D.M. Walton, Nucl. Instr. and Meth. A273 (1988) 663
7. R. Moshammer, thesis at the Inst. für Kernphysik, Techn. Hochschule Darmstadt, FRG, 1991
8. D. Brandl, Ch. Schoppmann, R.Schmidt, B. Nees, A. Ostrowski and H. Voit, Phys. Rev. B43 (1991) 5253
9. H.G. Paretzke, GSI report 39 (1988), Darmstadt/FRG
10. P. Dück, H.Fröhlich, W.Treu, H.Voit, Nucl. Instr. and Meth. 189 (1989) 615
11. L.C. Nortcliffe and R.F. Schilling in: Nuclear Data Tables A, ed. K. Way, 7/3-4 (1970), Acad. Press
12. G. Kraft, W. Kraft-Weyrather, E.A. Blakely and R. Roots, Adv. Space Res. 6/11 (1986) 141
13. C.C. Watson and T.A. Tombrello, Rad. Eff. 89 (1985) 26
14. A. Hedin, P. Hakansson, B. Sundqvist and R.E. Johnson, Phys. Rev. B31 (1985) 1780
15. H. Voit, E. Nieschler, B. Nees, R. Schmidt, Ch. Schoppmann, P. Beinig, and J. Scheer, Journal de Physique C2/50 (1989) 237
16. M. Varma, Nucl. Tracks Radiat. Meas., Vol. 16 (1989)
17. E. J. Kobetich and R. Katz, Phys. Rev. 170 (1968) 391
18. R.E. Johnson, B.U.R. Sundqvist, A. Hedin and D. Fenyö, Phys. Rev. B40 (1989) 49
19. I.S. Bitensky and E.S. Parilis, Nucl. Instr. and Meth. B21 (1987) 26
20. F.E. Kammer, thesis at the University of Oldenburg/FRG, Fachbereich Physik, 1991 (in english)

DESORPTION INDUCED BY MOLECULAR AND ATOMIC IONS

C.V.BARROS LEITE, E.F.DA SILVEIRA, J.M.F.JERONYMO AND R.R.PINHO

Pontifícia Univ.Católica do Rio de Janeiro
Rio de Janeiro, Brazil

E.A.SCHWEIKERT AND M.A.PARK
Texas A&M Univ. ,College Station, Texas, USA.

Abstract -

The ion desorption induced by MeV molecular and atomic projectiles on a phenylalanine target is studied. The effects of the incident ion charge state and of its velocity on the desorption yield are analyzed.

Introduction -

This work refers to some of our results in the analysis of the desorption process. Relative yield of positive and negative ions were measured for charged atomic and molecular MeV ions of C^+, O^+, CO^+, and CO_2^+ impinging on a phenylalanine target. The secondary ions were identified by a time of flight mass spectrometer. The yield of $[M-H]^-$, $[M+H]^+$, and $[M-COOH]^+$, secondary ions shows a strong dependence on the incident ion stopping power .The desorption yield dependence on the projectile charge state was measured and compared with experimental data obtained by other groups.

Experimental -

Time of flight (TOF) mass analysis was employed to separate and identify the molecular ions sputtered from the target irradiated with the fast ion beams. The desorbed ions were detected in coincidence with primary ions or with the secondary electrons resulting from the beam impact.

Incident ion beams were obtained from a 4MV Van de Graaff accelerator at PUC/RIO. An analyzing magnet was used to select the ion mass, the energy and the initial charge state. The switching magnet enables a second analysis of the beam by selecting different charge states resulting from collisions with a carbon foil used as stripper. The target was placed in the chamber forming an angle of 45° with the direction of the incident ions. Positive and negative molecular secondary ions were measured for projectile ions with energies ranging from 300 keV up to 3000 keV. An excitation curve was obtained for each single charge state. The study of the yield dependence with the incident ion charge state was performed using the stripper carbon foil to generate the charge state distribution. The desorbed ions were accelerated towards a 20cm TOF tube by an electrical field applied between the target (biased at 2.0 kV) and a grounded grid 5mm away. The transmission coefficient of the mass spectrometer was not determined but it was the same for all experiments. Two different arrangement were used for the coincidence measurements: the usual set-up, making use of the time to amplitude converter (TAC) , and another one using a time to digital converter (TDC). In the first case only two correlated events can be measured, that is, to each start signal only one stop can be in coincidence. In the second case, to each start signal, several correlated events can be measured. The results obtained with both arrangements indicate that the emission of just one secondary ion species for each impact is the dominant process.

Samples of phenylalanine evaporated on a thin film of carbon or formvar coated with aluminum or on a metallic grid were used as targets. The thin films were required to allow the primary ions to traverse the sample and to produce the start signal for the measurement of positive SI. The target deposited on the grid was used for bombardment with molecular ion beams and only negative SI were measured. In this situation the secondary ions detected on the MCP detector were uncorrelated to ions traversing the target. The silicon detector placed after the target was used for monitoring the incident beam. To minimize the dissociation of the molecular ions of the beam in the line a cold trap was used in order to maintain the pressure below 10^{-6}torr. Figure 1 shows the diagram of the electronic set-up.

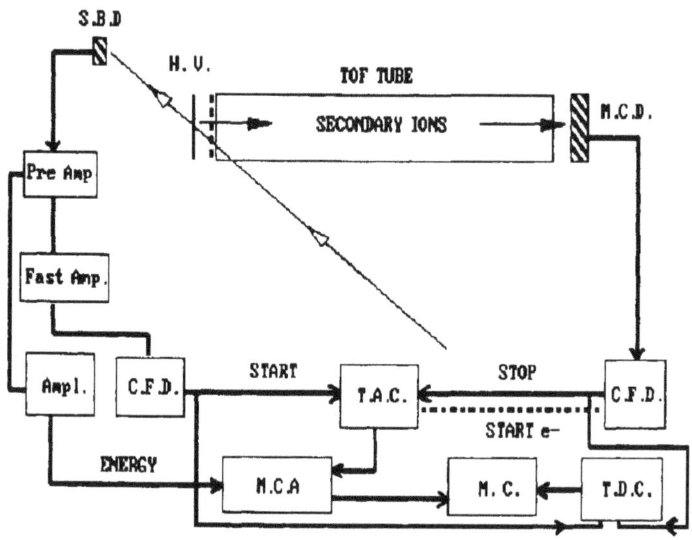

Figura 1 - Diagram of the eletronic set-up

A typical time of flight spectrum is shown in figure 2. This figure presents a positive SI spectrum for C^+ ion beam of 1.5 MeV.

The charge state study was performed with the same set-up, except for the fact that it was used a stripper carbon foil to produce high charge state.

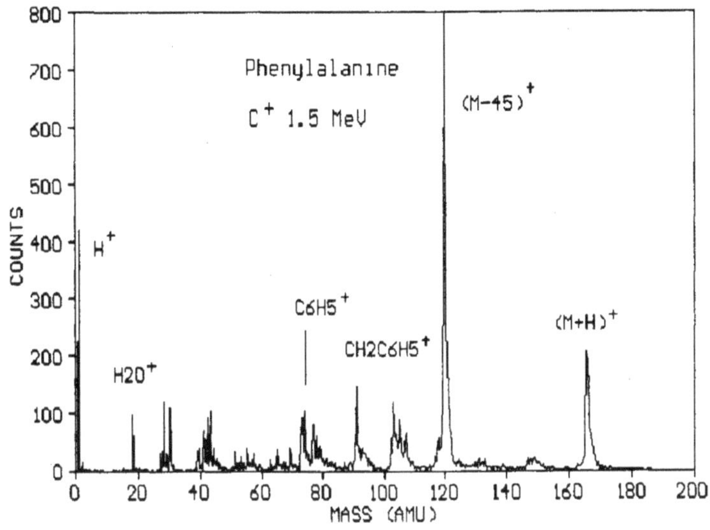

Figure 2 - Spectrum of desorbed positive ions produced by 1.5 MeV C^+ ions incident on a target of phenylalanine. The peak labelled $(M-45)^+$ is the phenylalanine fragment $(M-COOH)^+$ of mass 120u. Others fragments and the water contaminant are also present.

The charge of the incident ions was selected by the switching magnet. The neutral particles were obtained after the deflection of the single charge state beam by the switching magnet through the collision of these particles with the residual gas inside the line. The charged particles were deflected away from the target by means of a magnet placed near the scattering chamber. It was observed that the fraction of neutral particles in the mono charged beam was less than 1% . The negative beam charge state was obtained by reversing the current of the switching magnet for approximately the same value of the single positive charge state.

Results and Discussion

Samples of phenylalanine (molecular mass M=165) were irradiated with beams of C^+ and O^+ with energies ranging from 300 up to 3000 keV and with beams of CO^+ and $CO_2{}^+$ with incident energies of 300 up to 900 keV. The measured relative yields as a function of the ion velocity for positive secondary ions of $[M+H]^+$ and of M=120 $[M-COOH]^+$, are shown in figure 3.

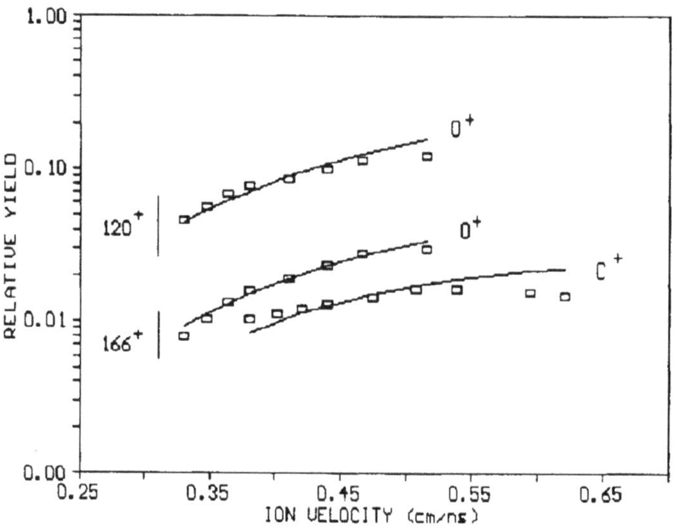

Figure 3 - Relative yield of positive SI of mass 120 and 165 induced by incident ions of $C^+$and O^+ with velocity ranging from 0.3 to 0.65 cm/ns. - ▫ experimental points, __ normalized stopping power to the third power.

The line is a curve proportional to the third power of the stopping power (SP) of the incident ions on the phenylalanine. The SP values were obtained by using the code Trim developed by Ziegler et al [1] and normalized to fit with the experimental data.

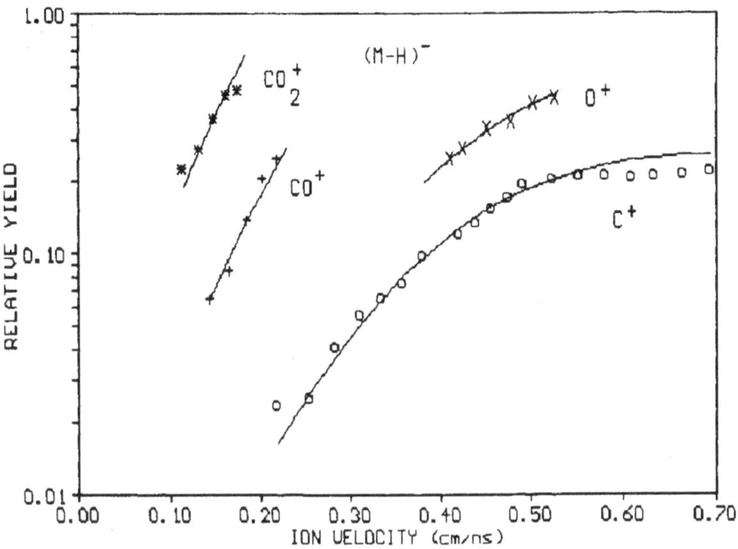

*Figure 4 - Relative yield of the (M-H) SI induced by ions of C^+, O^+, CO^+ and $CO_2{}^+$ with velocities varying from 0.1 to 0.7 cm/ns. o - experimental points for C^+ ions , X - O^+ ions ,+ - CO^+ and * - $CO_2{}^+$ ions, ___ - are the normalized stopping power to the third power.*

The yields of the negative secondary ions of mass M=164 $[M-H]^-$ for the different beams are shown in figure 4. Again the dashed line represents the normalized stopping power to the third power. The stopping power for the molecular ions like CO^+ and $CO_2{}^+$, was calculated using the Bragg's rule [2],i.e., by adding the stopping power for the atomic components of the molecule corresponding to the projectile velocity. Data presented in figure 3 show that the desorption yield of $M=166^+$ induced by O^+ beam was roughly twice as large as the corresponding yield of the desorption induced by the C^+ beam. The same ratio is found in figure 4 for the yield of $M=164^-$ desorbed by O^+ and C^+ beams.

In fact, M.Salehpour et al (1988) [3] have found the same ratio (2) for the de-protonated $M=116^-$ valine ion desorbed by C^+ and O^+ beams at same energy range . The calculated stopping power ratio for oxygen and carbon (S_o/S_c) beams at v = 0.4 cm/ns was approximately 1.3. This means that $(S_o/S_c)^2 = 1.7$ and that $(So/S_c)^3 = 2.2$, suggesting that the desorption yield would scale with stopping power having a power between 2 and 3. When molecular ions are used as projectiles, similar results were obtained.

Table 1

Beam	S/S_C	$(S/S_C)^2$	$(S/S_C)^3$	Y / Yc	
				ANL	OURS
O^+	1.3	1.7	2.2	2	2
O_2^+	2.6	6.8	18	11	-
C_2^+	2.0	4.0	8.0	8	-
CO^+	2.3	5.3	12	9	14
CO_2^+	3.6	13	47	22	~ 70

Table 1 summarizes the results obtained in the present work (for phenylalanine de-protonated peak, $M=164^-$) and for some results of the Argonne group [3] (for the valine de-protonated , $M=116^-$). It is also presented the electronic stopping power values calculated for the beams at v = 0.4cm/ns . The stopping powers, S=dE/dx, and the desorption yields, Y, are relative to the C^+ beam.

Starting with the assumption that $Y=kS^n$, the result of table 1 indicates that n would lie between 2 and 3, being even slightly larger than 3 in some cases. Salehpour et al [3] had proposed n=2 based on their results with C^+, O^+ and Ar^+ beams, as well as from measurements performed by other groups [4-7]. They have attributed the larger values of the experimental results, when compared with KS^2 , to the enhancement of the electronic stopping power due to collective effects produced by molecular projectiles. This enhancement has also been discussed by other authors [8-10].

It is interesting to note that the observed third power dependence of the stopping power is a prediction of the molecular expansion model for the desorption of large neutral molecules. In the study by Fenyö et al [11,12] , a molecular dynamics numerical simulation was used to evaluate the yield of desorbed molecules for incidence angles of 0° and 45° . No distinction was made between charged or non charged species and the results exhibit a third power dependence on the stopping power.

Charge State -

The study of the desorption yield dependence on the charge state of the incident ion was performed using a carbon beam with charge states C^0,C^+,C^{2+},C^{3+},C^{4+},C^{5+} and C^{6+}. The charge state distribution was obtained by accelerating ions of C^+ and passing the ion beam through a carbon foil used as stripper before the switching magnet. The switching magnet selected the charge state of the projectile. Samples of phenylalanine were used in this study.

The yield of the negative SI $(M-H)^-$ as a function of incident charge state for carbon beam of 2.5 MeV is presented in figure 5. The figure also shows the normalized data for Ne, Ar and Kr incident ions obtained from Monart [13]. The general behavior for the de-protonated SI $M=164^-$

shows a linear dependence on the charge state for charge states greater than 4+ and it was practically constant for lower charge states.

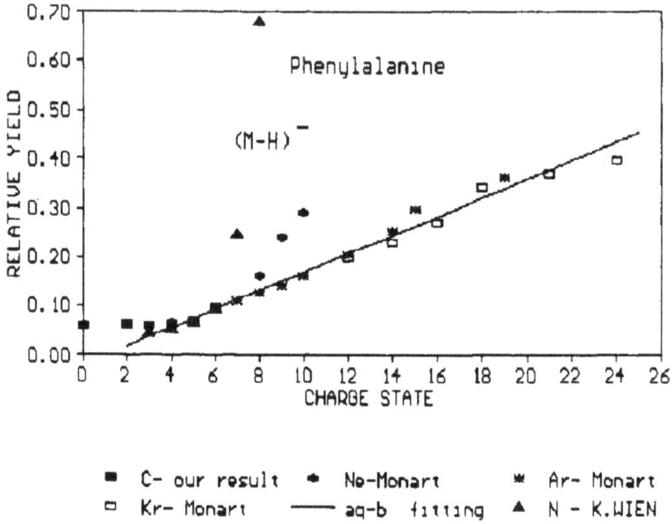

Figure 5 - Relative yields of desorbed phenylalanine de-protonated negative ion $(M-H)^-$ produced by 2.5 MeV carbon beam with incident charge varying from 0 to 6^+. Data of Ne, Ar and Kr ions at 1.0 MeV/u [13] and N (1.4 MeV/u) [14] were normalized to fit with our data.

The same charge dependence was observed for the yield of $(M+H)^+$, that is, constant for incident charge smaller than 4^+ and varying linearly with the incident charge for charge greater than 4^+. Figure 6 shows the $(M+H)^+$ yield for carbon beam of 1.5 MeV. The figure presents also normalized data of S and Kr obtained from Monart [13].

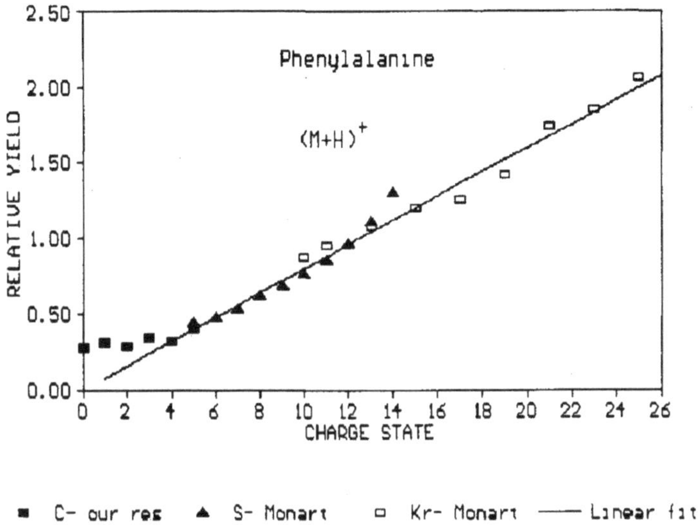

Figure 6 - Relative yields of the protonated phenylalanine SI $(M+H)^+$ induced by C (1.5MeV). Data of S and Kr at (1.0MeV/u) [13] were normalized to fit with our data. The line represents the linear fitting.

The yields of negative and positive hydrogen secondary ions for beams of carbon at 1.8 MeV and 1.5 MeV show a suggestively behavior with the incident charge. The yield of H^- SI has a dependence with the charge of the incident ion which is very similar to the yield of heavy secondary ions like $(M-H)^-$ or $(M+H)^+$. That is, it is pratically constant for incident charge smaller than 4^+ varying with q^n for charge higher than 4. Figures 7a and 7b show the yield as a function of the incident charge for ions of C (this work) and Ne, Ar and Kr (1.0 MeV/u), obtained from reference [13] and N (1.4 MeV/u) from [14]. The data show that the dependence of the H^- yield on incident charge varies with the target material.

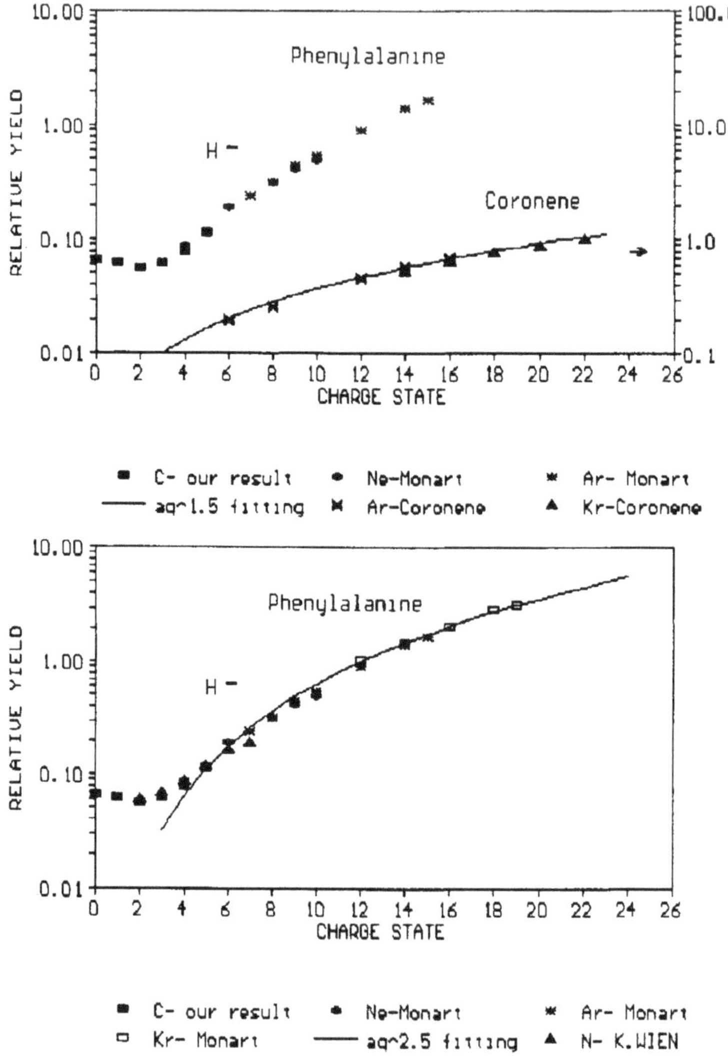

Figure 7 - Relative yield of the H^- SI induced by C (1.8MeV) on phenylalanine. Data from Monart [13] and K.Wien [14] were nomalized to fit with our results. The lines are $q^{2.5}$ fitting for phenylalanine. and $q^{1.5}$ for coronene target.

In fact, the behavior for a target of phenylalanine was not the same as for a coronene target. The yield for the phenylalanine was proportional to $q^{2.5}$ and for coronene, $q^{1.5}$. This may be explained by assuming that the H^- SI are produced by capture of target electrons.

The H^+ SI yield has a charge behavior indicating to be the hydrogen desorbed from the contaminant surface layer. Figure 8 shows the H^+ yield as a function of the ion charge state. The data obtained from Monart [13] for S, Ne, AR, and Kr and from K.Wien [14] for N were normalized to fit with our data. The H^+ yield shows a strong dependence on the incident charge state ($\propto q^3$) for ion charge greater than 4+.

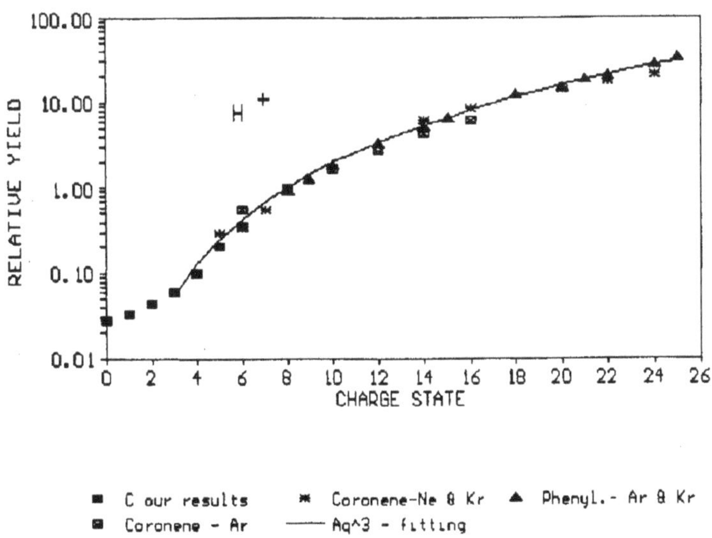

Figure 8 - Relative yield of the H^+ SI induced by C (1.5MeV) on phenylalanine. and of data from [13]. Ne and Kr on coronene target are included in the figure. The line is the q^3 fitting.

The assumption that the hydrogen comes from the contaminant layer is supported by the slight dependence of the H^+ yield on the target material, as can be observed by the fitting of the data obtained with targets of phenylalanine, coronene and CsI. All targets present the same charge dependence, that is, $Y \propto q^3$ (figure 9).

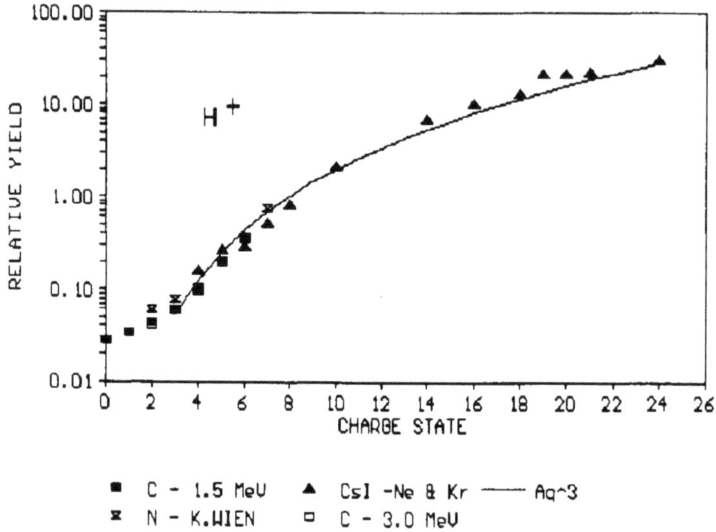

Figure 9 - Relative yield of H^+ SI induced by C (1.5 and 3.0 MeV) on phenylalanine. Ne and Kr [13] on CsI and N (1.4 MeV/u) [14] on phenylalanine. The line is the q^3 fitting.

Based on the yield curves, it is possible to consider that there are two different regimes for the desorption process: the low charge state regime and the high charge state regime. The low charge state regime may be described as an effect of the ion interaction at the depth region of the target. In this case the ion is supposed to be at the equilibrium charge state. At this regime the desorption yield would be dependent on the primary ion velocity and on the stopping power. At the high charge state regime the desorption process is due to a surface interaction and the dependence on the state of charge of the incident ion is the main factor that contributes to the secondary emission.

The behavior of the H^+ yield as a function of the incident charge state for ions of carbon for different energies is shown in the figure 10. The figure indicates that increasing the velocity of the incident ion the yield of the H^+ decreases, keeping the same dependence on the ion charge state.

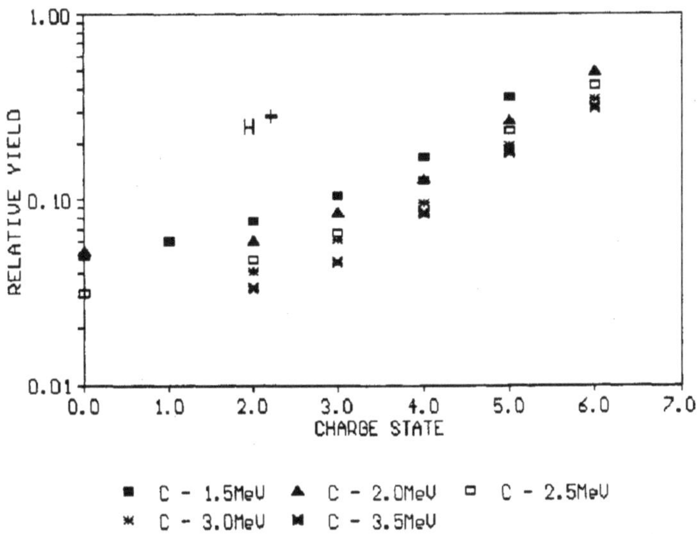

Figure 10 - Relative yield of H^+SI versus incident charge of C beam at energies of 1.5, 2.0, 2.5, 3.0 and 3.5 MeV.

This indicates that for high velocity the behavior of the yield of the H^+ with the ion charge tends to be the same. Data from K.Wien [14] using N at 1.4 MeV/u show the same behavior as our data for C 1.5 MeV (figure 9).

In addition,the dependence of the H^- yield on the ion charge does not depend very much on the ion velocity (figure 11).

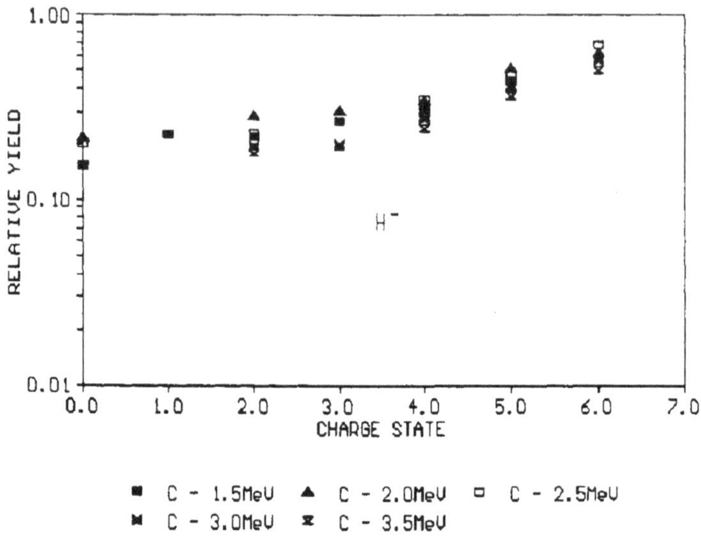

Figure 11 - Relative yield of H^- SI induced by C beam for charge states from 0 to 6^+ with energies of 1.5, 2.0, 2.5, 3.0 and 3.5 MeV.

Supposing that the H^+ emission is a result of the projectile ion interaction on the first superficial layers of the target, the dependence on the equilibrium charge of the incident ion is less important. The H^- emission has a behavior that indicates to be dependent of the depth region of the target and does not have a pronounced dependence on the ion charge state. The equilibrium charge state of C, in the range of velocity studied, was calculated using the Montenegro's relations [15] and varies from 2.5 to 3.6. It should be remarked that the dependence of the H^- yield on the charge of the ion is very weak in the charge range studied in this work.

Conclusions -

The understanding of the desorption process is still far from being complete and many experiments have to be performed to obtain enough data in all range of velocity and charge states. The analysis of the H^+ or H^- secondary ions is a very important subject to be studied because of their dependence on the charge state of the incident ion.

Other target materials and energies have to be analyzed to get a better idea of the emission process. Information about this mechanism may be obtained by analyzing the energy or the velocity distribution of the emitted particles.

The authors acknowledge the financial support of SCT, FAPERJ, CNPq and NSF.

References -

1 - J.F.Ziegler, Handbook of Stopping Cross Section for Energetic Ions in All Elements, Pergamon Press, New York, 1980 also J.P.Biersack and J.F.Ziegler, Trim Code

2 - Bragg, W.H.,and Kleeman,R. (1905), Phil. Mag. 10, S318

3 - M.Salehpour, D.Fishel and J.Hunt, Phys. Rev. B38 (1988) 12320

4 - P.Dück, H. Fröhlich, W.Treu and H.Voit, Nucl. Instr. and Meth. 191 (1981) 245

5 - S.Della-Negra , D.Jacquet, I.Lorthiois, Y.Le Beyec, O.Becker and K.Wien, Int. J. Mass Spectr. Ion Phys. 53 (1983) 215

6 - O.Becker, S.Della-Negra, Y. Le Beyec and K.Wien, Nucl. Intr. and Meth. B16 (1986) 321

7 - S.Della-Negra, O. Becker, R.Cotter, Y. Le Beyec, B. Monart, K. Styandi ng and K.Wien, J. Physique 48 (1987) 151

8 - W.Brandt, A.Ratkowski and R.H.Ritchie, Phys. Rev. Lett. 33 (1974) 1325325

9 - J.W.Tape, W.M.Gibson, J.Remillieux, R.Laubert and H.E.Wegner, Nucl. Instr. Meth. 132 (1976) 75

10 - N.R.Arista, Phys. Rev. B18 (1978) 1

11 - R.E.Johnson, B.Sundqvist, A.Hedin and D.Fenyö, Phys. Rev. B40 (1989)

12 - D.Fenyö, B.Sundqvist, B.R.Karlsson and R.E.Johnson, Phys. Rev. B42 (1990) 1895

13 - Monart,B., Thesis Université Paris Sud (1988)

14 - K.Wien, Radiation Effects and Defects in Solids, 109, 137 (1987)

15 - E.C.Montenegro,S.A.Cruz and C.Vargas Aburto, Phys. Lett. 92A 4 (1982) 195

APPLICATIONS OF THE GENERALIZED MULTISTRUCTURAL (GMS) WAVE FUNCTION TO PHOTOELECTRON SPECTRA AND ELECTRON SCATTERING PROCESSES

Marco Antonio Chaer do Nascimento
Instituto de Química, Departamento de Físico-Química, Universidade Federal do Rio de Janeiro, RJ, 21910, Brazil

ABSTRACT

A Generalized Multistructural (GMS) wave function is presented which combines the advantages of the SCF-MO and VB models, preserving the classical chemical structures but optimizing the orbitals in a self-consistent way. This wave function is particularly suitable to treat situations where the description of the molecular state requires localized wave functions. It also provides a very convenient way of treating the electron correlation problem, avoiding large CI expansions. The final wave functions are much more compact and easier to interpret than the ones obtained by the conventional methods, using orthogonal orbitals. Applications of the GMS wave function to the study of the photoelectron spectra of the trans-glyoxal molecule and to electron impact excitation processes in the nitrogen molecule are presented as an illustration of the method.

1. INTRODUCTION

An accurate description of the electronic structure of atoms and molecules requires the incorporation of the effects of electron correlation. However, the rigorous description of the electronic correlation effects is a difficult problem from the computational point of view, even for simple systems. Although formally known, the exact solution of the correlation problem is seldom reached because of the very complex nature of the calculations, regardless of the adopted approach.

In the usual treatment of the electronic correlation effects, a Hartree-Fock (**HF**) or multi-configurational (**MCSCF**) wave function is used as zero-order function for a perturbation calculation (**MBPT**), or as a reference wave function either for a configuration interaction (CI) or else for a coupled-cluster (CC) calculation.

Much less used, the valence-bond (**VB**) approach starts from classical chemical structures, where the individuality of the atoms is preserved and the molecule is "visualized" as a collection of atoms held together by chemical bonds. In the **VB** approach configuration interaction is introduced by the superposition of chemical "structures" representing different distributions of electrons among the various atoms (Lewis structures). Consequently, the correlation effects are introduced without destroying the picture provided by the classical chemical structures. This approach is conceptually simpler, and the results are easier to interpret, contrary to the ones obtained from large scale **CI** or **CC** calculations.

Unfortunately the **VB** method turns out to be computationally even more complex than the methods based on the **MO** model, because of the large number of chemical structures needed in the expansion of the wave function and also because of the non-orthogonality of the orbitals. Another serious problem of the original **VB** model is that the atomic orbitals cannot readjust their shapes to take into account the formation of the chemical bonds. Orbital optimization within the **VB** approach has been considered by several authors,[1-9] and it is now recognized that it can drastically reduce the number of structures needed in the **VB** expansion.

In spite of its tremendous success, the MO theory is faced with serious problems in certain situations where the correct description of the state of the molecule requires localized wave functions. One of the best known examples is the case of core-ionized states of homonuclear diatomic molecules. In the case of the O_2 molecule, for instance, solving for the HF wave function with a 1s core hole localized on one oxygen atom yields an energy more than 10 eV lower than if the wave function is forced to retain the full molecular symmetry, with the singly-occupied core orbital delocalized onto both oxygens. Many other examples of this phenomenon, generally known as "symmetry breaking", can be found in the literature. From the mathematical point of view this effect can be related to the more general problem of instabilities of the Hartree-Fock (HF) equations. Although the broken symmetry solution gives a much better energy than the symmetric solution, it does not possess the full symmetry of the nuclear geometry, as it should for the correct electronic wave function. However, if molecular symmetry is imposed during the self-consistent procedure, we obtain a quite poor description of the ion-molecule.

On the other hand, the VB model can very easily deal with these situations because symmetry breaking can be viewed as a tendency for the wave function to achieve the chemically intuitive localized structure. Nevertheless the VB wave function for the ionized molecule with a localized hole presents the same symmetry problems as the MO function.

From either point of view, the correct way of dealing with such a situation is to construct a wave function with the proper symmetry, by taking a linear combination of the wave functions representing localized holes on the left and on the right oxygen atoms. From the VB point of view, combining the two localized-hole structures poses no new problem to the model since the superposition of chemical structures is inherent to the theory. However, the process of combining two localized MO functions brings the non-orthogonality problem deep into the framework of the MO theory.

There might be situations where the hole state is not completely localized, in which case one could add the delocalized wave function to the linear combination of localized structures, and associate its coefficient to the degree of hole delocalization.

In this paper we describe a generalized multi-structural wave function which combines the advantages of the two models (**SCF-MO** and **VB**), preserving the classical chemical structures but optimizing their orbitals in a self-consistent way. Each "structure" can be represented by a **HF**, **MCSCF** or **CI** wave function which contains the most important correlation effects clearly described. The final wave function is obtained variationally and consists of a superposition of the wave functions representing the different "structures" considered. Because each structure can be represented by a correlated wave function containing the dominant (and clearly identifiable) effects, a very small number of structures will be needed in the final wave function, in order to describe the less important effects. The recovery of the residual, but not negligible effects, by the more usual approaches, usually requires an extremely large number of configurations in a **CI** or **CC** calculation or a high order **MBPT** treatment. On the other hand, the use of more than one bonding structure drastically reduces the number of configurations needed in conventional **CI**, **MBPT** or **CC** calculations. With the generalized multi-structural wave function the correlation effects are therefore introduced without loosing the simple picture provided by the chemical structure and at a moderate computational effort.

2. THE GENERALIZED MULTI-STRUCTURAL WAVE FUNCTION

Only a brief description of the generalized multi-structural (**GMS**) wave function will be presented here. A full discussion of the theory, with a detailed description of all the computational aspects involved in obtaining the solution for this kind of wave function, can be found elsewhere[10-12].

The generalized multi-structural wave function can be defined as

$$\Psi_{GMS} = \sum_{I=1}^{NSTRUCT} a_I \psi_I = \sum_{I=1}^{NSTRUCT} \sum_{i=1}^{NSEF} c_i \psi_i^{\,I} \tag{1}$$

where ψ^I represents the ith spin-eigenfunction (NSEF) of the Ith bonding structure (NSTRUCT) and c its weight in the expansion (1). In principle there are no restrictions whatsoever on the form of the wave functions ψ . Each of the ψ's can be individually optimized at the **HF or MCSCF (GVB, CASSCF,** etc.) level, followed or not by a CI treatment, or else can be taken simply as **VB**-type functions. In this last case the ψ_{GMS} is obviously identical to the **VB** function for the molecule.

Each one of the ψ's is represented in a basis of orthonormal orbitals, $\{ \varnothing^I \}$, properly optimized for the Ith structure. Although the orbitals of a given structure are taken to be orthogonal to each other,

$$< \varnothing^I_\alpha \mid \varnothing^I_\beta > = \delta_{\alpha\beta} \tag{2}$$

no such restriction exists for the orbitals belonging to different structures,

$$< \varnothing^I_\alpha \mid \varnothing^{II}_\beta > = S_{\alpha,\beta}^{\,I,II} \tag{3}$$

The coefficients c_i are obtained variationally by solving the equations,

$$<\delta\Psi_{GMS} \mid \mathcal{H} - E \mid \Psi_{GMS} > = 0 \tag{4}$$

$$(\mathbb{H} - \mathbb{S}\, E) \,\mathbb{C} = 0 \tag{5}$$

where $\mathbb{H}$ and $\mathbb{S}$ are interaction supermatrices containing the diagonal (same structure) and the inter-structural matrix elements. The matrix elements involving orbitals belonging to different structures are computed using the biorthogonalization technique proposed by Amos and Hall[13] and extended by King et al[14]. The structure of those supermatrices is very important in defining the programming strategy, as will be discussed in a forthcoming publication.

Once matrices $\mathbb{H}$ and $\mathbb{S}$ have been constructed eq.(5) can be diagonalized using Cholesky's decomposition[15], followed by a standard diagonalization. The matrix of coefficients, $\mathbb{C}$, is then obtained by a back transformation between the eigenvectors of the standard diagonalization and the Choleski transformation matrix.

The **GMS** wave function can also be used to study systems with mixed character. For instance, one could combine localized and delocalized structures or valence-like and Rydberg-like wave functions, at any level of description. In the former case the coefficients could then be associated to the degreee of localization of the state being considered as well as, in the latter case, to the amount of diffuse character of the state under investigation. Examples of the different situations where **GMS** wave functions apply can be found elsewhere[11].

3. Illustrative Examples

In this communication we discuss the application of the GMS wave function to the study of the photoelectron spectra of the trans-glyoxal molecule and to the electron impact excitation of $1\sigma_g$ and $1\sigma_u$ electrons of the N_2 molecule, which are examples of systems which exhibit symmetry breaking solutions.

3.1 The photoelectron spectra of trans-glyoxal

The photoelectron spectrum of the trans-glyoxal molecule was recorded by Turner et al.[16], who assigned the first four bands to the

$a_g(n)$, $b_u(n)$, $b_g(\pi)$ and $a_u(\pi)$ ionization potentials. This spectrum has been the subject of several theoretical calculations but in spite of that there seems to be considerable disagreement between the ordering predicted by the calculations and the experimental assignement. The results of semi-empirical calculations[17] indicate the ordering n, π, n, σ while some ab-initio calculations[18-20] predicted the ordering n, π, n, π. On the other hand a Green's function calculation[21], in third order perturbation expansion and including renormalization contributions, predicted not only the ordering but also ionization potentials in good agreement with the experiment[16]. However this last calculation did not use optimized geometries and ground state results surprisingly predicted the cis isomer to be more stable than the trans isomer. More recently Osamura and Schaefer[22], using an optimized geometry also obtained the same ordering of states suggested by the experiments. The ionization potentials, calculated at the ΔSCF level, showed fair agreement with the experimental results although the separation found between the second and third bands (0.49 eV) was much smaller than the observed value (1.8 eV).

None of the above mentioned calculations reveal that similarly to the $n \longrightarrow \pi^*$ excited states[23-24], the ion states of the trans-glyoxal molecule also exhibit symmetry-breaking solutions. Therefore, as in the case of the electronic excitations, the ionization should be better viewed as a localized process.

The generalized multistructural (GMS) wave function[10-12] is particularly suitable to describe systems which exhibit symmetry-breaking solutions. Localized solutions can be obtained at any level of calculation and recombined to recover the full symmetry of the molecular wave function.

The calculations were carried out with the Dunning[25] valence double-ζ (DZ) contraction of Huzinaga's[26] DZ (9s/5p,4s) gaussian basis set. Due to limited computational capabilities we have not considered polarization functions, but it is resonable to expect that they will affect the ionization potentials to the same extent as the $n \longrightarrow \pi^*$ transition energies[27].

The ground state geometry was optimized[24] at the Hartree-Fock (HF) level and used in all the calculations.

Using the optimized ground state geometry we solved self-consistently for the ^2A_g, ^2B_u, ^2B_g and ^2A_u ion states at the HF level and used the respective optimized orbitals to obtain CI wave functions. The CI calculations were performed allowing all single and double (SD) excitations in a space formed by 37 (occupied + virtuals) orbitals (14 a_g, 13 b_u, 5 a_u and 5 b_g) while keeping frozen the first seven occupied (4 a_g and 3 b_u) orbitals. Those calculations served not only to identify the important correlation effects but also as a basis for comparison with the GMS results.

In order to investigate the symmetry-broken solutions, revealed by HF calculations using the C_s symmetry point group, we constructed GMS wavefunctions represeting the ion states with localized (n^{-1} or π^{-1}) holes, at HF, GVB and CI levels of calculation. This can be achieved by recombining the delocalized π and n orbitals as:

$$\pi_{l,r}=\frac{1}{\sqrt{2}}\left\{1a_u \pm 1b_g\right\} \tag{6}$$

and

$$n_{l,r}=\frac{1}{\sqrt{2}}\left\{7a_u \pm 6b_u\right\} \tag{7}$$

and solving self-consistently, for the different ion states, using the procedure suggested by Bagus and Schaefer[28]. In equations (6) and (7), π_l and n_r represent the π and n orbitals localized on the left and right C=O groups of the glyoxal molecule. Correlation effects within each C=O group are classified as intra-moiety while those involving electrons from both C=O groups are called inter-moiety effects.

The GMS wavefunctions for all the ion states were constructed as a superposition of configurations (or better, spin-eigenfunctions) associated with three different structures. Using eq. (6), the wavefunction for the ^2B_g state ($[1a_u^2\ 1b_g^1]$), for instance, can be written, in terms of localized orbitals, as :

$$\psi(^2B_g) = \mathcal{A}\ [(core)(\pi_l+\pi_r)(\pi_l+\pi_r)(\pi_l-\pi_r)\ \ldots\alpha\beta\alpha] \quad (8)$$

$$\psi(^2B_g) = \mathcal{A}\ [(core)\{\pi_l\pi_l\pi_r-\pi_r\pi_r\pi_l\}\ \ldots\ \alpha\beta\alpha\ ] \quad (8')$$

which can be translated to a two localized structures GMS wavefunction:

$\Psi_{GMS}(^2B_g) = C$ 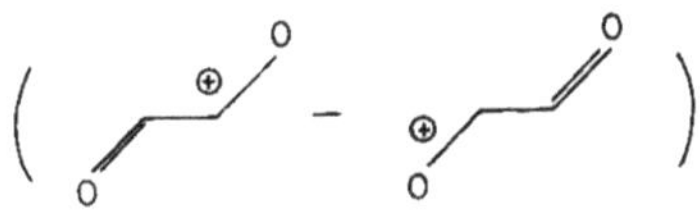 (9)

where C is a normalization constant. In the GMS wavefunction of eq. (9), structures (I) and (II) can be represented by HF, MCSCF or CI wavefunctions. In order to add some flexibility to the GMS wavefunctions we have also considered the delocalized structure obtained with C_{2h} symmetry. For the ^2B_g state, for instance, we have:

$\Psi_{GMS}(^2B_g) = c_1$ 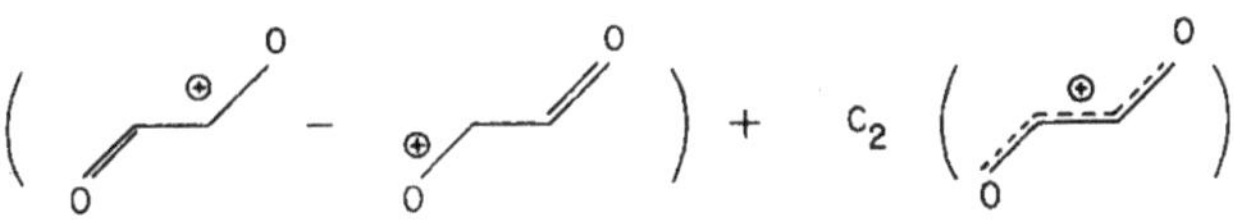 (10)

The coefficients c_1 and c_2 being determined variationally will fix the degree of localization of the ion states.

At the HF level of calculation, the localized structures were represented by the solutions obtained in the C_s symmetry point group while the delocalized structure resulted from calculations performed with C_{2h} symmetry. From these wavefunctions we constructed, for each ion state, GVB-PP (3/6) localized and delocalized wavefunctions, which were used to build the GMS wavefunctions at the CI level.

The reference state to compute the ionization potentials at the GMS-HF level is simply the ground-state HF wavefunction. However, the choice of the reference state at the GMS-CI level of calculation deserves special consideration.

It would be impractical to construct GMS wavefunctions for the ion states including as many configurations as the CI(SD) wavefunctions, because of the large amount of computational work due to the nonorthogonality of the molecular orbitals belonging to the different structures. Besides, the phylosophy behind the GMS method is to construct more compact and easily interpretable wavefunctions. Therefore, as reference state for the GMS calculations, one should try to represent the ground state through a more compact wavefunction which includes the most important correlation effects.

The GVB method[3] is known to provide quite simple wavefunctions containing, in general, the most important correlation effects. Inter-pair correlation effects can be easily taken into account through a small CI calculation in the space defined by the GVB natural pairs, and other minor effects can also be incorporated allowing the CI space to contain some of the virtual orbitals.

Several exploratory GVB-PP(4/8) calculations, correlating the two oxygens' lone pairs (n) and the two π bond electron pairs, have been performed in order to establish the ground state wave function to be used as the reference state. From these calculations it was concluded that the dominant correlation effects could be accounted for with a full closed shell CI calculation in the GVB space (A), what implies up to octuples excitations in the space formed by the eight natural orbitals. Using those configurations (A) in a multireference CI(S+D) calculation in the GVB space we observed that only the π $[(\pi_1)^1 (\pi^*_1)^1(\pi_r)^1(\pi^*_r)^1]$ doubly excited open shell configuration contribute with a significant energy lowering. Therefore we chose the reference state to be represented at the A+ S+ π [1111] level of excitation, using the orbitals obtained at the

GBV (4/8) level of calculation. This wavefunction comprises 151 configurations in contrast to the 2437 terms HF+CI(S+D) wavefunction previously described.

Table 1 shows the results obtained at various levels of calculation compared to the experiments. The comparison between the HF results obtained with C_{2h} and C_s symmetries clearly reveals the existence of broken-symmetry solutions, similarly to the case of the $n \rightarrow \pi^*$ states of glyoxal. Of course, when using the C_s symmetry one cannot distinguish, at the HF level, the $A_{g(u)}$ from the $B_{u(g)}$ state. The solutions in the C_s symmetry describe the ionization processes as localized in one of the carbonyl groups.

The full molecular symmetry can be easily recovered at the GMS(HF) level of calculation. The combination of the two degenerated localized (π^{-1}) structures gives rise to the ^2A_u and ^2B_g states while the two degenerated (σ^{-1}) structures furnish the ^2A_g and ^2B_u states.

The CI(SD) results for the ion states were obtained as described before, using the respective orbitals from HF calculations performed in the C_{2h} symmetry point group. Much larger CIs would be necessary to recover the symmetric description of the ion states if one uses the localized (C_s) orbitals. The results of the CI(SD), except for the ^2B_g state, agree well with the experiments and confirm the ordering (n, n, π, π) of the ion states.

To construct the GMS(CI) wave functions we made use of the GVB-PP wave functions for the localized (I and II, eq. 9)and the symmetric (III) structures. Each one of the localized wavefunctions already contain most of the intrapair (n or π) correlation effects and the superposition of the GVB functions representing the two localized structures allow us to describe the inter-moiety $(n_1, n_r$ and $\pi_1, \pi_r)$ correlation effects.

The interpair correlation $(n_1, \pi_1$ and $n_r, \pi_r)$ and the inter-moiety $(n_1, \pi_r$ and $n_r, \pi_1)$ correlation effects are more conveniently introduced through the symmetric structures (III), by performing CI calculations in the GVB space, similar to the ones

Table 1 Ionization potentials (eV) for the trans-glyoxal molecule, compared to experiment.

	Present Calculation					Exp.[a]
State	HF[b] (C_{2h})	HF[b] (C_s)	GMS (HF)[b]	CI (SD)[c]	GMS (CI)[d]	
^2A_g	11.38	10.38	9.90 (3)[e]	10.79 (2529)	10.26 (38)	10.6
^2B_u	13.40	(10.38)	11.12 (3)	12.06 (2529)	12.40 (38)	12.19
^2B_g	13.50	13.02	12.44 (3)	13.51 (2491)	13.34 (34)	14.0
^2A_u	15.74	(13.02)	14.58 (3)	15.17 (2491)	15.33 (34)	15.4

a. Ref 16

b. Relative to a ground state energy of -226.5171 h

c. Relative to a ground state energy of -226.6557 h

d. Relative to a ground state energy of -226.6132 h

e. Number of configurations in the GMS or CI wave functions.

described for the ground state reference wave function. The GMS(CI) wavefunctions (eq. 1) contain therefore configurations belonging to the GVB-PP wavefunctions of the two localized structures and the ones associated with the GVB+CI wavefunctions of the symmetric structure.

Because of the presence of the configurations related to the localized structures, only a few configurations of the symmetric structure contributed a significant energy lowering and the final GMS wavefunctions contain just 34 configurations for the ^2A_g and ^2B_u states and only 38 for the ^2A_u and ^2B_g states. From table 1 one sees that the results obtained at the GMS(CI) level are quite remarkable considering the compactness of these wavefunctions compared to the respective CI(SD) wavefunctions. It is clear from these results that the consideration of the localized structures reduce dramatically the number of configurations needed in CI calculation using the symmetric orbitals, and that the GMS wave functions provide a much simpler way of attaining the correct balance of the different correlation effects.

3.2 Electron Impact excitation of the N_2 Molecule

Electron impact excitation of the N_2 molecule plays an important rọle in the understanding of auroral and ionospheric phenomena since N_2 is the major atmospheric constituent to altitudes of about 200 Km. This process is also important in astrophysics, for instance, in connection with the chemistry of the interstellar medium, the composition of comets, and the opacity of stars.

Two of the most studied excitation processes in the N_2 molecule are the $X\ ^1\Sigma_g^+ \longrightarrow a^1\Pi_g$ $(3\sigma_g \longrightarrow 1\pi_g)$ transition known as Lyman-Birge-Hopfield (LHB) band and the so-called pre ionization peak formed by two experimentally unresolved transitions, $X^1\ \Sigma_g^+ \longrightarrow {}^1\Pi_g$ $(1\sigma_g \longrightarrow 1\pi_g)$ and $X\ ^1\Sigma_g^+ \longrightarrow {}^1\Pi_u$ $(1\sigma_u \longrightarrow 1\pi_g)$. Despite the large amount of theoretical and experimental effort spent in the characterization of these processes, some discrepancies in the generalized (GOS) and optical (OOS) oscillator strenghts results still remain.

The LBH transition has been experimentally studied by several authors[29-34] at impact energies ranging from 0.3 keV to 25 keV. Theoretical calculations of the GOS as a function of the transferred momentum and making use of a Born-type approximation for the electron excitation cross section have also been published[35-36]. Relaxation effects have not been considered, i.e., the frozen-core approximation was used, and correlation effects were only partially included in one of the calculations[35]. As a result of these simplifications, the agreement with experiments is mostly qualitative and no calculation describes correctly the behavior of the GOS over the full range of momentum transfer known experimentally. The a $^1\Pi_g$ state does not exhibit symmetry breaking solutions and therefore only one structure is needed for its description. However, in order to investigate relaxation and correlation effects on the GOS behavior one can make use of the ability of the GMS program to compute matrix elements between non-orthogonal molecular orbitals. The ground state and the a $^1\Pi_g$ excited state were solved at the HF and CI levels using a (12s, 7p, 1d) / [9s, 5p, 1d] gaussian basis set[37]. All the calculations were performed at the equilibrium internuclear distance[38] of 2.068 a.u. Two different calculations were performed to obtain the excited-state wave function, one using the frozen core approximation and the other solving self-consistently for the excited-state wave function, allowing all the orbitals to fully relax. Correlation effects were introduced via CI calculations allowing double and quadruple excitations from all the respective occupied HF orbitals to a virtual space formed by improved virtual orbitals[39] of σ_g, σ_u, π_g and π_u symmetries. The introduction of single and /or triple excitations in the CI expansion induces a mixing of the a $^1\Pi_g$ ($3\sigma_g \longrightarrow 1\pi_g$) LBH transition with the resonably intense ($2\sigma_g \longrightarrow 1\pi_g$) transition as discussed in detail in another publication[37]. A generalization of the analytical expression obtained by Chung and Lin[36] was used to compute the GOS values in the Born-Ochkur-Bonham (BOB) approximation. Table 2 shows the effect of relaxation and correlation effects on the GOS results at selected k^2 values. Those effects are also shown in figure 1, where the GOS values are compared to experiments and other theoretical calculations.

Table 2: Relaxation and correlation effects on the generalized oscillator strengths for the LBH transition.

Calculation	Generalized Oscillator Strength k^2=0.066	k^2=0.202	k^2=0.803	k^2=1.425
HF (frozen)	0.0130	0.0330	0.0610	0.0530
HF (unfrozen)	0.0115	0.0294	0.0539	0.0461
CI (DQ)	0.0110	0.0279	0.0514	0.0450

It is quite clear from the results shown in table 2 and figure 1 that relaxation and correlation effects are very important for the correct description of the GOS behavior. In fact, this behavior can be well reproduced, in the BOB approximation, once relaxation and correlation effects are properly taken into account.

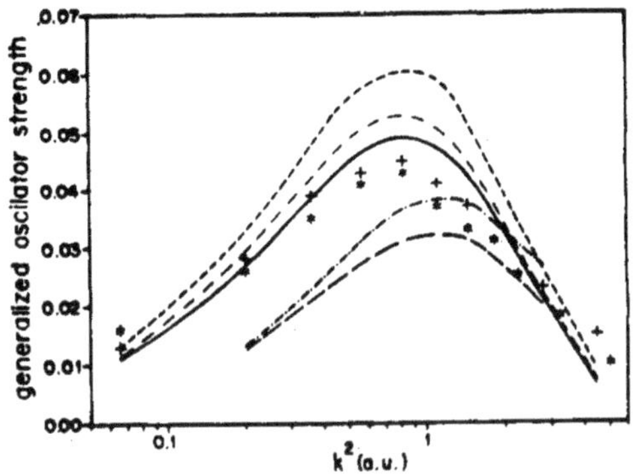

FIG. 1 - GOS results of different theoretical calculations compared to the experiments: ----, HF frozen core; – –, HF relaxed core; —— CI; -.-. and — —, TDA and RPA, Ref. 41; + Ref. 31; * Ref. 30

The agreement between our theoretical and the experimental values, in the region of small and medium k^2, is good even for the limited, but well-balanced, basis set used in the calculations. For larger k^2 values the discrepancy observed between theory and experiments should be ascribed much more to the failure of the BOB approximation to describe this region than to the quality of the wave functions employed. For large k^2 values higher-order terms in the Born series might be needed to correctly describe the stronger interaction between the incident electron and the target.

The so-called preionization band comprises two experimentally unresolved transitions ($1\sigma_g$, $1\sigma_u \longrightarrow 1\pi_g$). Since they are both inner-shell transitions, we can antecipate a strong relaxation effect on the shape of all occupied orbitals and consequently on all the properties associated with the excitation process, such as OOS, GOS and transition energies (ΔE). Besides, in this case both transitions exhibit symmetry breaking solutions and the effects associated with the hole localization process must be taken into account. As discussed before, the GMS wave function is particularly suitable to describe localized processes. Moreover, the different effects - orbital relaxation, hole localization and electronic correlation - can be independently incorporated to the GMS wave function, making possible a systematic study of the relative contribution of each of these effects to the value of a given property. In order to allow a better description of the relaxation process following inner-shell excitation, the basis used in the LBH study[37] was partially uncontracted to a [10s, 6p, 1d] basis set. To discuss the influence of the relaxation and correlation effects on the OOS, GOS and ΔE values, different calculations at the HF level and at the GMS level including up to three structures, have been performed[40]. The GOS values were obtained from these wave functions using the BOB approximation. In the first calculation both the ground and the excited states are descibed in a basis of molecular orbitals self-consistently optimized for the ground state, at the HF level.

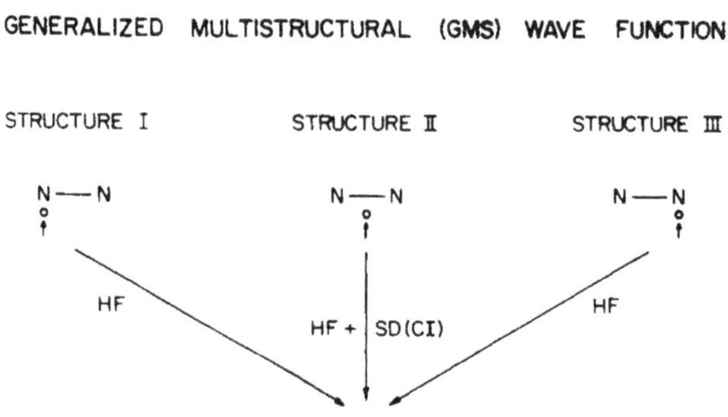

$$\Psi^{CI}_{GMS} = \sum_{I=1}^{NSTRUCT} \sum_{i=1}^{NSEF} C^I_i \varphi^I_i$$

FIG. 2 - The generalized multistructural wave function for the core excited N_2 molecule.

This is called the frozen-core approximation (HF-FC) in which none of the above mentioned effects is included. In calculation 2 the molecular orbitals for the excited state are also obtained self-consistently and therefore orbital relaxation effetcs are directly considered. This is called the HF relaxed calculation (HF-R). In calculation 3 the excited state is described at the GMS level, by a two-structure wave function (GMS-2S), one describing a hole localized on the left atom and the other a hole on the right nitrogen atom (see figure 2). At this point relaxation effects associated to the hole localization are also taken into account. In calculation 4 localization is relaxed and the system is allowed to choose in the variational sense, the degree of hole localization. This is accomplished by solving for a GMS three-structure (GMS-3S) wave function, containing the same two localized structures of the GMS-2S wave function plus a third structure which describes a delocalized hole (structure II in figure 2). The GMS-3S wave function is also a three configuration wave function because each structure is described at the HF level. Finally a GMS-CI wave function is constructed by adding to the GMS-3S wave function singly and doubly excited configurations describing core-valence and valence-valence correlation effects. These additional configurations were obtained using the HF wave function of structure III as the reference configuration. Table 3 shows the ΔE and OOS results obtained at the five different levels of calculation. Figure 3 shows the sum of the GOS values for both transitions at the HF-FC and GMS-CI levels of calculation, compared to experiments and other theoretical results.

Table 3: Transition energy and optical oscillator strength for the preionization band, at different levels of calculation, compared to experiments.

Calculation	ΔE (eV) $^1\Pi_u$	ΔE (eV) $^1\Pi_g$	OOS
1. HF-FC	413.64	413.70	0.26
2. HF-R	410.56	410.65	0.21
3. GMS - 2S	401.13	412.20	0.19
4. GMS - 3S	400.95	401.02	0.18
5. GMS - CI	402.55	402.62	0.20
Exp.	401.20[a]		0.23[b], 0.20[c,d] 0.14[e]

a. Ref. 42 b. Ref. 43 c. Ref. 44
d. Ref. 45 e. Ref. 46

From table 3 it is clear that hole localization is by far the most important effect to consider when computing ΔE, being responsible for a decreasing of ~ 9 eV relative to the delocalized hole picture. On the other hand, for the OOS value orbital relaxation is the predominant effect. The results shown in figure 3 also clearly indicate a large influence of correlation and relaxation effects on the GOS behavior. As in the case of the LBH transition, the GOS values show good agreement with the experiments for small and medium k^2 values. For large k^2 values the observed discrepancy should once more be attributed to the failure of the BOB approximation.

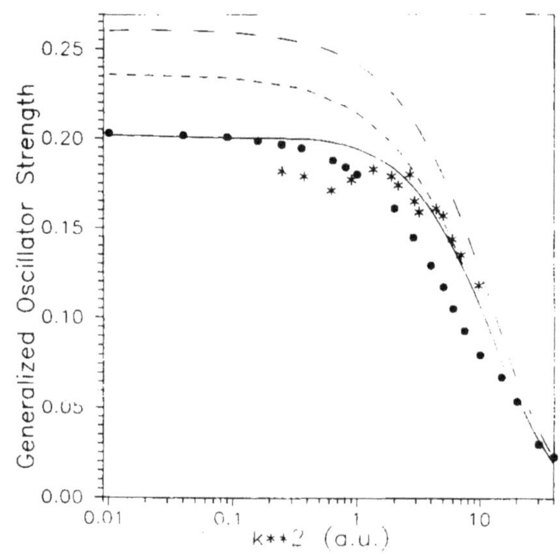

FIG. 3 - Total Generalized Oscillator Strength for the preionization peak compared to the experiments and other theoretical results: solid line, GMS-CI; — —, HF-FC; ----, Ref. 47; • Ref. 48; * Ref. 49.

4. CONCLUSIONS

The results obtained clearly indicate that the **GMS** wave function is an useful toll to investigated systems which exhibit symmetry-broken solutions. It provides a simple way of recovering the full symmetry of the molecule and the most important correlation effects.

The photoelectron spectra of the trans-glyoxal molecule is quite well described using a localized description of the ionization process. For the case of the N_2 molecule, the transition energies are largely affected by hole localization while for the **GOS** and **OOS** orbital relaxation dominates.

ACKNOWLEDGEMENTS

It is a pleasure to acknowledge Eduardo Hollauer, who wrote the **GMS** code and performed the calculations for the trans-glyoxal molecule, and Dr. Carlos Eduardo Bielschowsky, who adapted his Born-scattering program to accept **GMS** wave functions and did most of the calculations for the N_2 molecule.

The author also acknowledges CNPq and Finep for financial support.

REFERENCES

1. W.A. Goddard III, Phys. Rev., 157 (1967) 73, 81
2. W.A. Goddard III, J. Chem. Phys., 48 (1968) 450
3. F.W. Bobrowickz and W.A. Goddard III in: *Modern Theoretical Chemistry, vol. 3, Methods of Electronic Structure Theory*, ed. H.F. Sehalfer III (Plenum Press, New York 1977), Chp. 6 .
4. J. Gerratt, Adv. At. Mol. Phys., 7 (1971) 141
5. N.C. Pyper and J. Gerrat, Proc. Roy. Soc., A355 (1977) 407
6. I.G. Kaplan *Symmetry of Many-electron Systems*, (Academic Press, New York, 1973)
7. D.L. Cooper, J. Gerratt and M. Raimondi, Adv. Chem. Phys., 69 (1987) 319
8. J. Gerratt and M. Raimondi, Proc. Roy. Soc., A371 (1980) 525
9. R. McWeeny, Int. J. quantum Chem., 34 (1988) 25
10. E. Hollauer and Marco A.C. Nascimento, Chem Phys Lett., 184 (1991) 470
11. E. Hollauer and Marco A.C. Nascimento, J.Chem Phys (submitted)
12. E. Hollauer and Marco A.C. Nascimento, Chem. Phys. Lett.(submitted),
13. A.T. Amos and G.G. Hall, Proc. Roy. Soc. (London), A263 (1961) 483

14. H.F. King, R.E. Stanton, H. Kim, R.E. Wyatt and R.G. Parr, J. Chem. Phys. 47 (1967) 1936
15. J.H. Wilkinson, *The Algebraic Eigenvalue Problem*, (Clarendon Press, Oxford, 1965)
16. D.W. Turner, C.Baker, A.D. Baker and C.R. Brundle, *Molecular Photoelectron Spectroscopy* (Wiley Interscience, New York, N.Y., 1970) p. 252
17. H.Kato, H. Konishi, H. Yamabe and T. Yonezawa, Bull. Chem. Soc. Jpn, 40 (1967) 2761
18. C.E. Dykstra and H.F. Schaefer, J. Am. Chem. Soc., 97 (1975) 7210
19. V. Pincelli, B. Cadioli and D.J. David, J. Mol. Struct., 9 (1971) 173
20. T. Ha, J. Mol. Struct., 12 (1972) 171
21. W. von Niessen, J. Am. Chem. Soc., 99 (1977) 7151
22. Y. Osamura and H.F. Schaefer, J. Chem. Phys., 74, (1981) 4576
23. L.E. Nitzsche and E.R. Davidson, Chem. Phys. Lett., 58 (1978) 171
24. E. Hollauer and Marco A.C. Nascimento, Chem. Phys. Lett. 181 (1991) 463
25. T.H. Dunning Jr. and P.J. Hay in: *Modern Theoretical Chemistry: Methods of Electronic Structure Theory*, vol 3, ed. H.F., Schaefer (Plenum Press, New York, 1977), p.1
26. S. Huzinaga, J. Chem. Phys., 42 (1965) 1293
27. C.E. Dykstra, R.P. Lucchese and H.F. Schaefer III, J. Chem. Phys., 67 (1977) 2422
28. P.S. Bagus and H.F. Schaefer III, J. Chem. Phys., 56 (1972) 224
29. T.C. Wong, J.S. Lee, H.F. Wellenstein and R.A. Bonham, J. Chem. Phys. 63 (1975), 1538
30. E. Feinelli, R. Camilonni, G. Petrocelli and G. Stefani, Nuovo Cimento, 9 (1987), 33
31. C.A. Lucas, M. Sc. Thesis, Universidade Federal do Rio de Janeiro, 1984; G.G.B. de Souza and C.A. Lucas, in *Abstracts of Contributed Papers*, Fourteenth International Conference on the Physics of Electronic and Atomic Collisions, Palo Alto, 1985, edited by M.J. Coggiola, D.L. Heutis, and R.P. Saxon (North-Holland, Amsterdam, 1986) p. 252
32. E.N. Lassetre and M.E. Krasnow, J. Chem. Phys., 10,(1984) 1248
33. A. Skerbele and E.N. Lassetre, ibid. 53 (1979) 3806
34. N. Oda and T. Osawa, J. Phys. B 14 (1981) L563

35. A. Szabo adn N.S. Ostlund, Chem. Phys. Lett **17** (1972) 163
36. C. Chung and L. Lin, Phys. Rev. A6 (1972) 988
37. C.E. Bielschowsky, M.A.C. Nascimento and E. Hollauer, Phys. Rev. A42 (1990) 5223
38. G.H. Hezberg, *Spectra of Diatomic Molecules* (Van Nostrand, New York, 1950)
39. W.J. Hunt and W.. Goddard III, Chem. Phys. Lett., 3 (1989) 414
40. C.E. Bielschowsky and Marco A.C. Nascimento, Phys. Rev. **A** (submited for publication)
41. A. Szabo and N.S. Ostlund, Chem. Phys. Lett. 17 (1972) 163
42. R.N.S. Sodhi and C.E. Brion, J. Electron. Spectrosc., 34 (1984) 363
43. A. Bianconi, H. Petersen, F.C. Brown and R.A. Bacharach, Phys. Rev. A17 (1978) 1907
44. R.B. Kay, E. van der Leerew and M.J. van der Wiel, J. Phys. B: At. Mol. Phys., 10 (1977) 2513
45. R. Camilloni, E. Fainelli, G. Petrocelli and G. Stefani, J. Phys. B: At. Mol. Phys., 20 (1987) 1839
46. N. Oda, T. Nishimura and T. Osawa, *Pro . 11th Int. Conf. on Physics of Electronic and Atomic Collisions*, eds. K. Tatayanagi and N. Oda (1979) p. 408
47. T.N. Rescigno and A.E. Orel, J. Chem. Phys., 76 (1979) 3390
48. R. Camilloni, E. Fainelli, G. Petrocelli, G. Stefani, F. Maracci and R. Platania, in: *Lectures Notes in Chemistry*, eds. F.A. Gianturco and G. Stefani (Springer, Berlin, 1984), vol. 35, p. 172
49. R.S. Barbieri and R. Bonham (submitted to Phys. Rev. A)

MAIN FEATURES IN THE ATOMIC IONIZATION BY ION IMPACT

C.R. GARIBOTTI and R. PREGLIASCO

Centro Atómico Bariloche, 8400 Bariloche, Argentina
and Consejo Nacional de Investigaciones Científicas y Técnicas

ABSTRACT

We consider the relevant characteristics of the energy and angular spectra of the electrons ejected in the collision of ions with atomic targets. We analyze recent results relative to the projectile charge and impact energy dependence of these features. We point out problems which merit further experimental and theoretical research.

1. Introduction

In the collision of a fast particle with an atom or a molecule one or more electrons are ejected. The production of electrons by ion impact is of relevant interest in many branches of physics. Is is possible to study atomic structures and dynamic excitation mechanisms by electron detection, and electron spectroscopy has received a great deal of attention.

As early as 1933 Bethe[1] used the Born approximation for the calculation the ion impact excitation and ionization of electrons bound in atomic states. Experimental data come much later and in the 1960s complete energy and angular distribution of emitted electrons were measured for light atoms and projectiles[2]. Since then many laboratories were involved in the measurement of the ejected electron spectra[3,4]. Namely, the double differential cross section (DDCS) in the angle and energy of emission ($d\sigma/dEd\Omega$).

In this paper we will discuss recent results relative to the main characteristics of the electron spectra, which are described in the Figure 1. This figure represents in the electron velocity space ($\vec{v}$) the region where large densities of emitted electrons are observed, these are:

SE: soft collision peak, which contains the electrons ejected with low velocity relative to the target. They are produced by the projectile through a small momentum transfer collision.

BE: binary collisions: originate from a head on collision between the projectile and the electron, with a large momentum transfer. In the simple picture of a rest free electron, the energy and momentum conservation laws, gives for the centroid energy E_B of the binary ring:

$$E_B = \frac{E_P}{M} \cos^2\vartheta$$

where E_P and M are the energy and mass of the projectile (atomic units), ϑ is the scattering angle and we will call v_p the ion velocity. The transversal profile of that ring is determined by the velocity distribution of the electron in the target. At scattering laboratory angles larger than $\pi/2$ there is not binary peak.

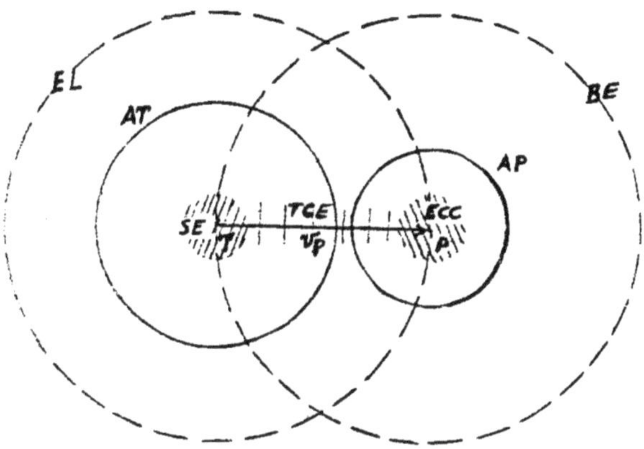

Fig. 1: Zones in the electron velocity space where the DDCS for electron emission is large.

AT: Auger electrons from the target produce a relatively narrow peak. At large collision energies the peak position does not changes with ϑ, indicating a small recoil energy transference to the target atom. The angular dependence of the Auger emission is determined from the initial excited state, in particular for K-shell excitation the Auger peak intensity is found to be isotropic.

ECC: These are target electrons captured into low-energy continuum states of the projectile, which are preferentially ejected at forward angles with a velocity equal to the projectile velocity, producing a peak at that point of the velocity space. They overlap with the electron loss electrons which give in general a larger yield than the ECC process. Therefore for the observation of the ECC peak is convenient to use bare ions as projectiles.

The electron spectra which comes from electrons in the projectile is similar to that discussed for the target, except that the reference system must be changed.

EL: This is the electron loss ring which is produced by outer-shell electrons ejected from the projectile through elastic scattering in the screened Coulomb field of the target. The EL peak is centered in an energy E_L which corresponds to a velocity equal to the projectile velocity (v_p) and E_L results almost independent of the observation angle[5]. Soft and binary collision electrons from the projectile are both found to contribute to the electron loss peak.

AP: The Auger electrons from the projectile are centered at a fixed velocity relative to the projectile. In the laboratory system they show large variation in energy and intensity as the observation angle is changed. These variations come from kinematical effects due to the moving emitter. Usually the Auger peaks are composed of several lines.

Finally we have a recently discovered feature where the electron seems to be shared by both the target and projectile ions[6].

TCE: Two centers electrons refers to a large electron density observed in the velocities comprised between the target and projectile velocities.

2. Recent Developments

Now we will discuss recent progress in the analysis of these features of the electron spectra and its dependence with the physical parameters, e.g., the charge Z_p and energy E_p of the projectile. We will only consider the target electrons spectra.

Binary Encounter Electrons.

The DDCS behaves as Z_p^2 in any first order perturbative method. When the BE are interpreted as due to the collision between the projectile and a "quasi free" electron is valid the Rutherford cross section (in au):

$$\frac{d\sigma}{dE\, d\Omega}(E', \Theta', Z_p) = \frac{Z_p^2}{16\, E'^2 \sin^4\left(\frac{\pi - \vartheta'}{2}\right)} \tag{1}$$

Where E' and ϑ' are the energy and angle of the electron in the projectile frame. Allowance for the momentum distribution of the target electrons $|\Psi(\vec{v})|^2$ can be considered in the impulse approximation (IA) by writing:

$$\left.\frac{d\sigma}{dE\, d\Omega}\right|_{BE} = \int \frac{d\sigma}{dE\, d\Omega}\left(E', \vartheta', Z_p\right) \cdot |\Psi(\vec{v})|^2\, d\vec{v} \tag{2}$$

where variables in the projectile and target frames have a simple relationship. Therefore the resulting DDCS for BE will be proportional to Z_p^2. For bare projectile the projectile charge is well defined and experiments show a good agreement with a Z_p^2 - law. In figure 2. we show recent experimental results[7] for the BE peak at 0° emission angle for bare ions with energy 1.5 MeV/amu on H_2 and the excellent comparison with Eq. 2.

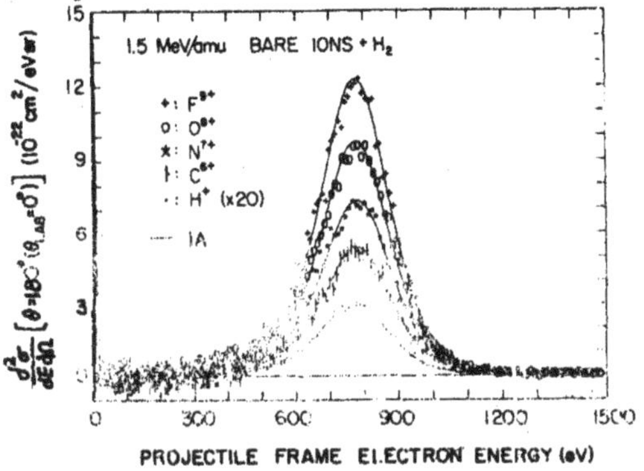

Fig. 2: BE profile in the projectile frame, at θ = 0°, for different bare projectiles[7]. The solid lines are theoretical results from Eq. 2. For H^+, data and theory have been multiplied by 20.

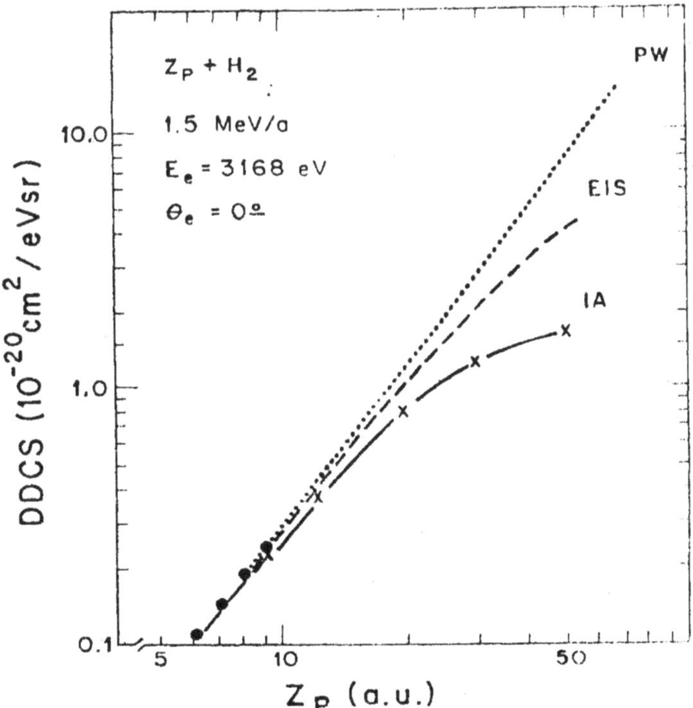

Fig. 3: Value of the DDCS in the top of the BE peak[8], in the forward direction, for bare ions on H_2. The impact energy is 1.5 meV/amu.

Experiment:● ● ●; Theories: plane wave; · · ·; CDW-EIS:- - -; impulse approximation:- x -

However deviations must be expected for large Z_p/V_p, ie. the strong distortion region, where perturbative methods are not valid. Experimental results are no available there, but theoretical calculations for the BE peak yield shown in Fig. 3 that a saturation effects occurs[8], i.e. the cross section becomes stationary at large values of Z_p.

For electron carrying projectiles the electrons are expected to screen the nuclear charge and therefore the BE peak should be lower than that for the bare ions. In Fig. 4 we display experimental results for F^{+q} ions[9]. Surprisingly, experiments give a quite different result, showing that the height of the BE peak increases with the screening. It should be noted that the effective charge Z_{eff} of the dressed ion is not such a well-defined quantity as Z_p. Recently some authors introduce an non-Coulombic potential with an energy dependent Z_{eff}[10]. They evaluate the potential from Hartree-Fock calculations which accounts for the electronic structure of the projectile. The theoretical results are also shown in Fig. 4.

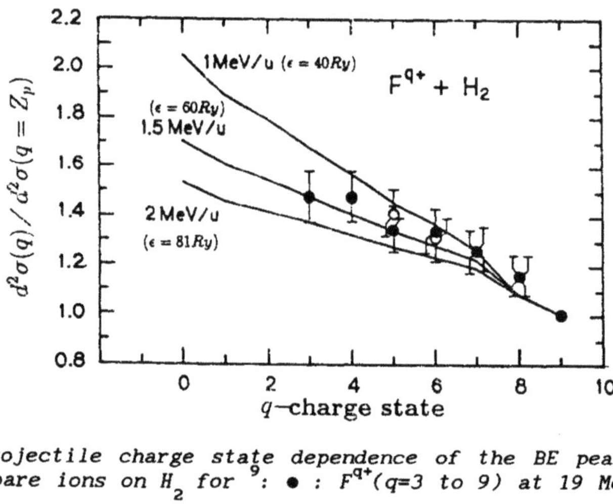

Fig. 4: Projectile charge state dependence of the BE peak relative to that for bare ions on H_2 for [9]: • : F^{q+}(q=3 to 9) at 19 MeV; o: F^{q+}(q=5 to 9) at 28.5 MeV. Lines: calculations from Ref. 10, for different energies..

Soft Electrons

The first Born approximation considers that the ionized electron is described by a Coulomb wave centered at the target ion. This gives a DDCS divergent in the limit $v \to 0$ and independent of the emission angle θ, i.e. an isotropic peak. The main difficulty to check that isotropic characteristic resides in the large uncertainties presented in the measurement of very low velocity electrons. However recent measures[11] show that the DDCS, at fixed angles and energies down to 1 eV (v = 0.27 a.u.), does not seems to converge to a common limit suggesting that the soft electron peak could be not symmetric. In Fig. 5 we shown some recent results for H^+-Ne collision at 100 keV impact energy.

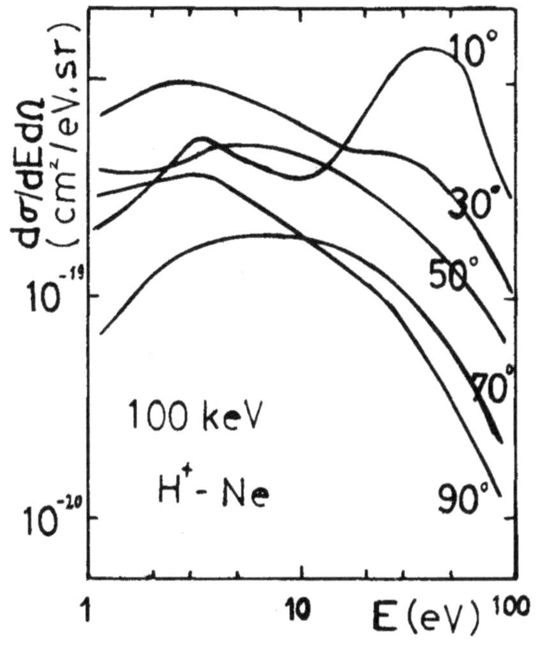

Fig. 5: Ionization DDCS for H^+-Ne collision, at 100 keV/amu, for each fixed emission angle as a function of the electron energy[11].

Electron Capture to the Continuum

In a capture formalism the final ECC state can be described by a continuum Coulomb wave function centered at the projectile[12]. The peak results from the divergent normalization factor of that function, which gives a singular electron distribution in the projectile frame. However, the experiments exhibit in the forward direction a number of electrons larger for $v<v_p$, than for $v>v_p$ as show in Fig. 6 [13]. The shape of the cusp reflects the dynamics of the collision process and, in particular, that asymmetry of the cusp has been ascribed to the contribution from high order perturbative terms[14]. Different theoretical formalisms have been introduce for the inclusion of those terms: second-order Born, distorted wave and semiclassical models.

The DDCS is singular in the ECC peak and the shape of the observed cusp depends critically on the experimental conditions[13]. Theoretical DDCS must be convoluted with a function that correctly describes the analyzer-detector system to obtain values for comparison with experimental data. Every experimental device has its own view of the interaction region from which the electron fluxes flow. The observed electron distribution $Q(v,\vartheta)$ is related to the DDCS through a

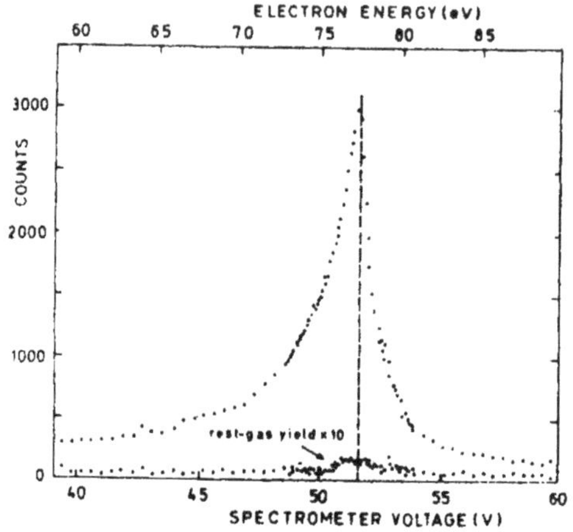

Fig. 6: ECC cusp in the forward direction for H^+-He at 141keV/amu, with $\theta_o = 1°$

transmission function of the spectrometer $S(E,\vartheta)$ which depends of the angular and velocity resolution of the equipment[13]. This relationship is given by:

$$Q(v,\theta) = \int S(E,\vartheta) \frac{d\sigma}{dE\, d\Omega} dE\, d\Omega \tag{3}$$

The accuracy of the results will depend on how well the S function is described. It is usual to assume a separable form, in angle and energy:

$$S(E,\vartheta) = R(E)\, G(\vartheta) \tag{4}$$

and write analytical expressions for the R and G functions which are derived from simple modelization of the spectrometer. For forward emission the S is independent of the angle ϕ, due to the azimuthal symmetry of the process.

The R(E) is taken as a narrow step function, or a gaussian, and $G(\vartheta)$ is considered either a step or a trapezoidal function[13], as shown in Fig. 7. Accordingly, 2 or 3 experimental parameters should be introduced: the velocity resolution of the spectrometer $R = \Delta v/v$ and the angular resolution ϑ_0 (or ϑ_1 and ϑ_2). With these simple approximations for the transmission function every theoretical DDCS can be convoluted with Eq. 3 for comparison with data. However comparison between experiments from different laboratories is not clear. In 1981 we proposed a way for analyzing the data which has become the standard method[13]. This method reveals the detailed shape of the ECC cusp in an equipment independent manner, so providing a straight-forward comparison between the work of different groups, both experimental and theoretical.

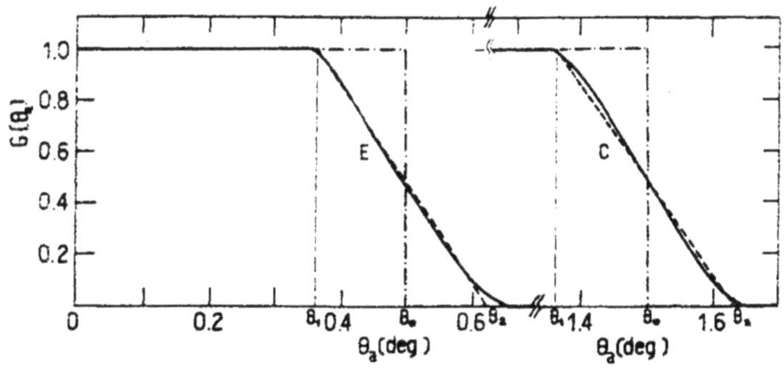

Fig. 7: Model angular acceptance G (θ) for two experimental set ups[13]*. The trapezoidal (θ_1,θ_2) and rectangular (θ_0) approximations are also shown.*

The DDCS is expressed as a series expansion:

$$\frac{d\sigma}{dE\ d\Omega} = \frac{v}{v'} \sum_{n,l=0} B_{nl}\ v'^{\,n}\ P_l\ (\cos\vartheta') \tag{5}$$

where P_l are the Legendre polynomials and the prime labels the quantities in the projectile frame. Then it is:

$$Q(v,\vartheta) = \sum_{n,l=0} B_{nl}\ U_{nl} \tag{6}$$

where:

$$U_{nl} = \int (v')^{n-1}\ P_l\ (\cos\vartheta')\ S\ (v,\Omega)\ dE\ d\Omega$$

can be exactly evaluate. The parameters B_{nl} are independent from the experimental conditions and are obtained by fitting data with Eq (6) to be compared with those resulting from other experiments or from a theoretical approach expanded according to Eq. 5.

The terms l=0 describes the isotropic part of the DDCS, l=1 the dipole content, l=2 the quadrupole content and so on. The shape of the first terms are shown in Fig. 8.

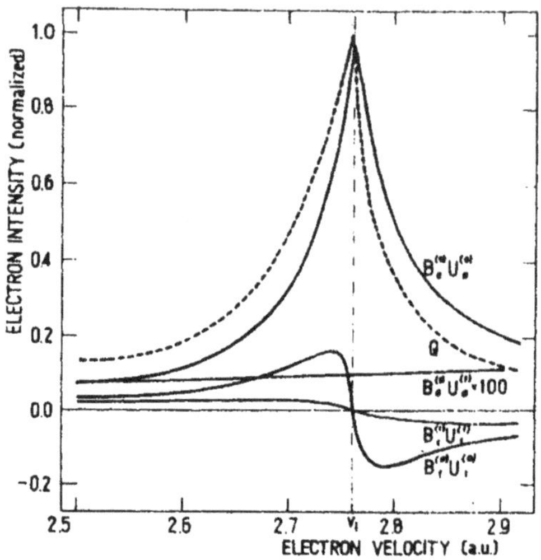

Fig. 8. Contribution of each terms in the Eq. 6 to Q (v), near the ECC peak.

A large number of authors have used Eq. 6 for the interpretation of experimental and theoretical results, for different projectile charges and energies. In Figs. 9 and 10 we show results for the coefficient B_{00} and the ratio B_{01}/B_{00} respectively, the sources of the values are indicate in the figure.

The projectile field is only a perturbation when it corresponds to a small charge and large velocity ion In this case the first Born approximation predicts a DDCS scaling as $Z_P^3\ V_P^{-12}$ and the second order gives $Z_P^3 V_P^{-11}$ [19]. In Fig. 9 the experimental B_{00}/Z_p^3, for H^+ and He^{2+} + He collisions, are compared with the OBK approximation for charge exchange[20]. The data show a dependence $v^{-11.3\pm0.2}$ in the range 1-2.6 MeV/amu. and in low energies are observed deviations from Z_P^3 law[16].

In Fig. 10 we compared the ratio $\beta = B_{10}/B_{00}$ obtained by different experimental and theoretical groups. The β measures the forward to backward asymmetry of the cusp and reflects the dynamics of the transfer process. The results assign a negative value for β, but show a very large spread. The precise determination of the parameters in Eq. 5 remains an unsettled problem and require an exact identification of:

a) the spectrometer transmission function $S(E,\vartheta)$

b) impact energy

c) uncertainties due to the truncation of the series (Eq. 6)

It has been shown that a small variations of these quantities (of the order of 0.5%) produces larger uncertainties in the values of the B_{nl}[21].

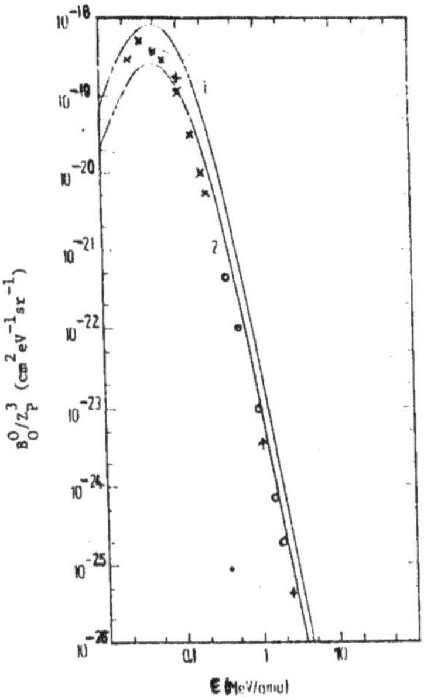

Fig. 9: B_{oo}/Z_P^3 as a function of the impact energy E, for H^+ and He^{2+} colliding with He. Data: x: H^++He (Ref.15); +: H^+ + He and o: He^{2+} + He (Ref.16). Curves 1 and 2: two versions of OBK approximation[20].

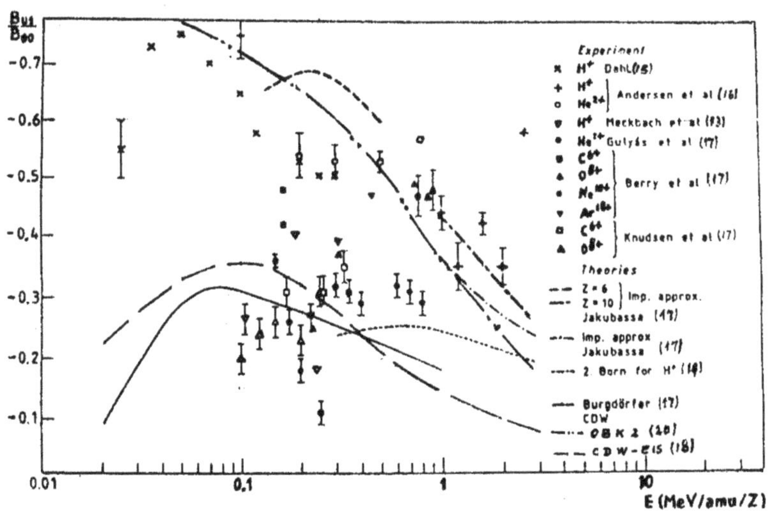

Fig. 10: Experimental and theoretical values of B_{o1}/B_{oo} obtained with different projectiles on He, as function of E/Z_P. References are indicated in the figure

Furthermore we note that the cusp from a non coincidence measurements may be composed of contributions from other processes, in particular transfer plus ionization (TI) and double ionization (DI). The total cross section for DI is found to be much smaller than that for single ionization at the whole impact energy range, and up to now there is not information about the spectra of the DI electrons. Meanwhile it has been found that for Au ions on He, H_2 and Ar, the TI exhibits a cusp for $\vec{v} = \vec{v}_p$ comparable to the ECC peak[22]. Therefore in these cases TI must be separate before the interpretation of the data.

Two center electrons (TCE)

The DDCS exhibits a high and wide saddle for small scattering angles and electron velocities comprise between the SE and ECC cusps.

In Fig. 11 we display "transverse distributions" represented as a function of $v_T = v$ sen θ for two electron velocities v = 1.43 and 1 au, for H^+ and He^{2+} at 100 keV/amu.

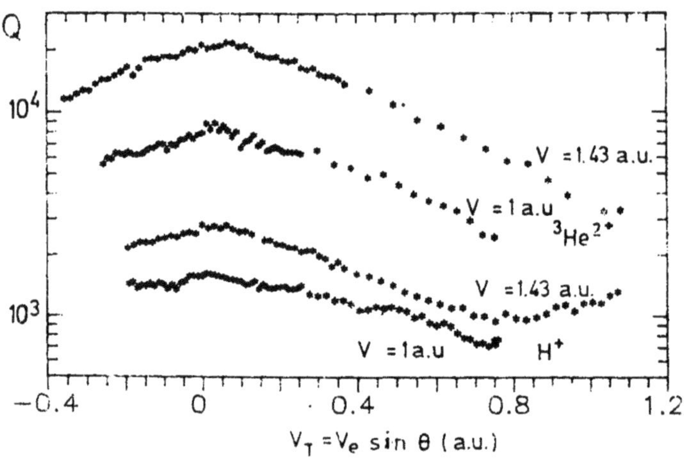

Fig. 11: DDCS for He^{2+} and H^+ on He at 100keV/amu, represented for two fixed electron energies as a function of the transverse velocity $v_T = v$ sen θ. Data ar relatively normalized[6].

Previous measurements indicate an additional narrow, high and step ridge for very small angles[1]. However recent experimental analysis shows that this sharp ridge is mainly produced by an extended gas target originated in the expansion of the incoming gas stream[23], and in fact remains only the broader ridge.

A classical interpretation considers the ridge as a consequence of the permanence of the electron on the saddle potential conformed, in the

0^0 direction, from the superposition of the projectile and target ions electrostatic potentials. This predicts a large fraction of electrons moving with the velocity of the potential saddle point (v_s) and leads to a shift of the saddle towards the target, and hence a decrease of v_s, when Z_p is increased. This effect would results in a shift in the maximum of $d\sigma/dv\, d\Omega$, i.e., the "velocity cross section". This shift was no confirmed by accurate measurements[24].

Quantum approximations have been considered for the description of the ridge. In Fig. 12 we compare some current approximations with data for H^+ on He at 50, 100, 150 and 200 keV/amu[25]

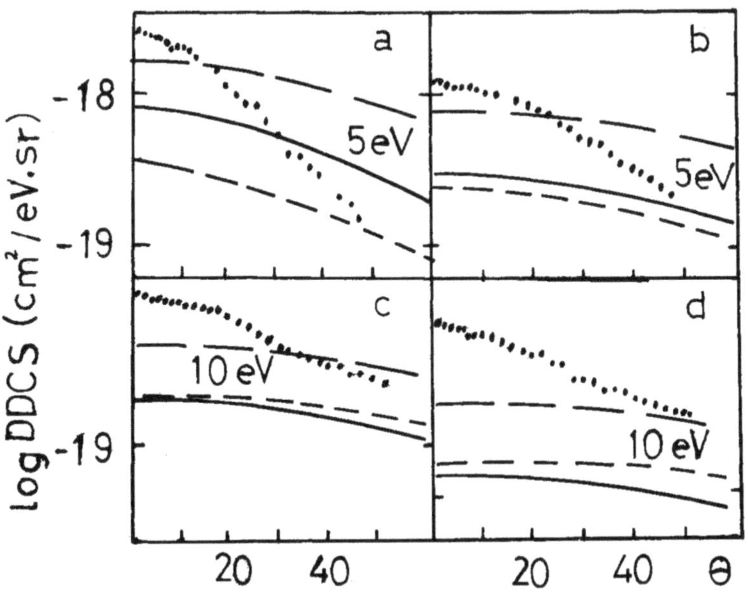

Fig. 12: DDCS for electron emission m H^+-He collision at: a) 50keV b) 100keV, c) 150keV, d) 200keV as a function of the angle at the electron energies indicated on the curves. Experimental values from reference 25. Theories: —: effective change; — —: multiple scattering; - - - -: CDW-EIS.

The data gives the angular distributions of the electrons with energy of 5 and 10 eV, which correspond to v = 0.61 and 0.86 au. respectively. We have considered three usual theoretical approaches[26]: the effective charge, the multiple scattering and the CDW-EIS. We observe that these methods give a rather flat DDCS.

From Figs. 12 we note that the hump smooths for increasing proton energies. This suggests that its origin must be found in the formation of a dynamic molecule produced by the projectile and the residual ion during the collision process. Otherwise we can say that the ejected electron moves in a two-center potential and consequently it should be described by a three-body Coulomb continuum state. The theories mentioned above relays on the requirement the initial and final scattering states must have correct asymptotic conditions. It seems that this condition is not enough for the description of two-center electrons. The full impulse approximation[8] gives a better description of wave function near to the colliding partners and in Fig. 13 we show the good comparison with our experimental results.

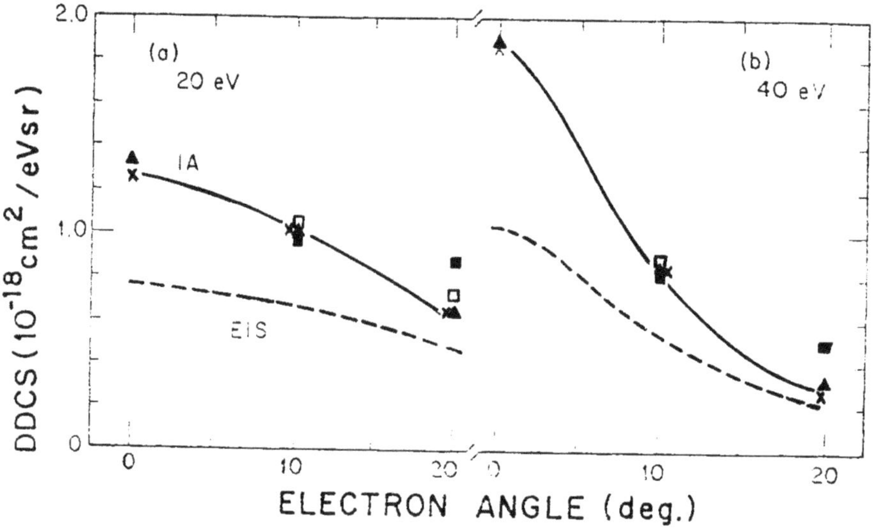

Fig. 13: DDCS for electrons emitted in H^+-He collision at 100keV/amu as a function of the ejection angle for E_e = 20 and 40eV. Experiment: Δ from Ref. 25; □ and ■ from Ref. 2. Theory[8]: ——: impulse approximation; - - - -: CDW-EIS approach.

3. Conclusions

A great amount of information is known about spectra of electrons emitted in ion-atom collisions. However, we can say that a systematic knowledge about the dependence of the main features of the spectra with the projectile charge, the impact energy and the target properties is not available yet.

The SE peak has not been measured and nearby data do not confirm the angular symmetry predicted by the first Born approximation.

We have measured angular distribution of the TCE for H^+ and He^{++} on Ne, at 100keV/a, but further study is necessary to obtain the quantitative dependence of the ridge with Z_p/V_p.

The BE hump is the best known part of the spectra, but most of the studies were done in the forward direction and, focusing effects in the angular distribution have not been considered.

The ECC cusp is the most studied constituent of the spectra, however, 20 years after its discovery, there is not a precise separation between physical and instrumental contributions to its shape. Even when correlation effects between two or more electrons is now the hot topic in ion-atom collisions, we have to conclude that a proper understanding of the single electron emission has not been attained and will require a large quantity of work.

References

1. M. Bethe: *Handbuch der Physik*, (1933 Vol.**24**,1,273.
2. M.E. Rudd and T. Jorgensen, *Phys.Rev.* (1963) **131**,666.
   C.E. Kuyatt and T. Jorgensen, *Phys.Rev.*(1963) **130**, 144.
   H.W. Berry, *Phys.Rev.*(1961) **121**, 1714.
3. L.H. Toburen. *Phys.Rev.*(1971) **A3**, 216.
   N. Stolterfoht, Z. *Physik* (1971) **248** 81,
4. N. Stolterforht, *Topics in Current Physics*, Ed. I. Sellin, (Springer 1978) **Vol.5**, 155.
5. F. Drepper and J.S. Briggs, *J.Phys.* 1976) **B9**, 2063.
6. S. Suarez, C. Garibotti, W. Meckbach, G. Bernardi, P. Focke, R. Pregliasco and G. Sigaud, *Lectures Notes in Physics* (1990), (Springer-Verlag) **376**, 88.
7. D. Lee, P. Richard, T. Zouros, J. Sanders, J. Shinpaugh and H. Hidmi, *Phys.Rev.*(1990) **A41**, 4816.
8. J.E. Miraglia and J. Macek, *Oak Ridge Nat.Lab. preprint*, (1991), *Phys.Rev.* (to be published).
9. T. Zouros, P. Richards and D. Lee, *Lectures Notes in Physics* (1990), (Springer-Verlag) **376**, 61.
10. C.O. Reinhold, D. Schultz and R. Olson, *J. Phys.***B** (1991) to be published.
11. W. Cheng, M.E. Rudd and Y.Y. Hsu, *Phys.Rev.* (1989) **A39**, 2359.
12. A. Salin, *J. Phys.* (1969) **B2**, 631.
13. C.R. Garibotti and J. Miraglia, *J. Phys.* (1981) **B14**, 863.
    W. Meckbach, I. Nemirovsky and C. Garibotti, *Phys.Rev.* (1981) **A24**, 1793.
14. R. Shakeshaft and L. Spruch, *Rev.Mod Phys.* (1979) **51**, 369.
15. P. Dahl, *J. Phys.* (1985) **B18**, 1181.
16. L.H. Andersen, K.E. Jensen and H. Knudsen, *J. Phys.* (1986) **B19**, L161.
17. L. Gulyas, A. Kover, G. Szabo, T. Vajni and D. Berenyi, *Lectures Notes in Physics* (1990), (Springer-Verlag) **376**, 56.
    L. Gulyas, G. Szabo, D. Berenyi, A. Kover, K. Groenfeld, D. Hoffmann and M. Borkhard, *Phys.Rev.* (1986) **A34**, 2751.
    S.D. Berry et al. *Phys.Rev.* (1985) **A31**, 1392.
    H. Knudsen, L.H. Andersen and K. Jensen, *J.Phys.* (1986) **B19**, 3341.
    D.M. Jakubassa, *J.Phys.* (1983) **B16**, 1768; *Phys.Rev.* (1988) **A38**, 70.
    P. Burgdörfer, *Phys.Rev.* (1986) **A33**, 1578.
18. D.S. Crothers and J. McCann, *J.Phys.* (1987) **B20**, L19.
19. K. Dettmann, K.G. Harrinson and M. Lucas, *J.Phys.* (1974) **B7**, 269.
20. B.L. Moiseiwitsch, *J. Phys.* (1991) **B24**, 983.
21. H. Atan, W. Steckelmacher and W. Lucas, *J. Phys.* (1990) **B23**, 2579.

22. Z.H. Andersen, M. Frost, P. Hvelplund and H. Knudsen, *J.Phys.* (1984) **B17**, 4701.
23. G. Bernardi, S. Suarez, P. Focke and W. Meckbach, *Lectures Notes in Physics* (1990), (Springer-Verlag) **294**, 295.
24. G. Bernardi, P. Fainstein, C. Garibotti and S. Suarez, *J. Phys.* (1990) **B23**, L139.
25. G. Bernardi, S. Suarez, P. Fainstein, C. Garibotti, W. Meckbach and P. Focke, *Phys.Rev.* (1989) **A40**, 6863.
26. C. Garibotti, M. Martiarena and D. Zanette, *Lectures Notes in Physics* (1987), (Springer-Verlag) **294**, 262.

HEAVY ION COOLER RINGS AND THEIR APPLICATION IN STUDIES OF ION - ELECTRON - PHOTON INTERACTIONS

R. Schuch
Manne Siegbahn Institute, Frescativ 24, S-10405 Stockholm, Sweden

ABSTRACT

After an introduction to the basic principles of storing fast moving ions by the magnetic elements of a ring, the schemes of ion beam cooling will be introduced by the examples of laser cooling and electron cooling. Results obtained with both cooling schemes at one of the presently operating storage rings (TSR) will be shown. In the last part of the lecture the use of the high-quality electron beam in the cooler for electron-ion collision experiments is discussed. The excellent resolution and luminosity of this device will be demonstrated with examples of electron impact radiative, dielectronic, and laser stimulated recombination.

I. Introduction

Heavy-ion cooler rings are new types of experimental facilities for collisional and spectroscopical studies in atomic physics, plasma physics and nuclear physics. The technology of these machines originates in the sophisticated circular particle accelerators used presently in elementary particle physics. One of the modivations for the development of these so called heavy ion cooler rings for atomic physics, plasma physics and nuclear physics is the possibility of cooling the fast moving stored ions to a narrow momentum spread, i.e. to low temperatures. Another reason is the high circulating ion current of a stored beam and the repeated passage through an interaction region such as a laser, an electron- or ion- beam, or an internal target. This gives an enormeous increase of the reaction rates (luminosity) compared to conventional accelerators where a beam acts in a single passage through a target. Therefore studies of collisions with small cross sections, and spectroscopy of unprecedented resolution and quality should be possible.

There are additionally many new effects appearing if a beam of ions is cooled to a low internal temperature (velocity spread). Collective motion of the ions appear like in a strongly coupled one-component plasma. At very low beam temperatures the beam is supposed to crystalize and phase transitions of this new form of matter, provided by a cooled stored ion beam, can be investigated. These new experimental opportunities in atomic physics, plasma physics, and nuclear physics have become evident to a number of groups and lead to the construction of several heavy ion cooler rings. At present these rings are in particular built in Europe. The cooler storage ring TSR in Heidelberg (1), ESR at GSI Darmstadt (2), ASTRID in Arhus (3) and CRYRING in Stockholm (4). Cooler storage rings for light ions used for medium energy nuclear physics are partly operating since some time or are presently built. Those are CELSIUS in Uppsala, Sweden (5), COSY in Juelich, FRG, IUCF cooler ring in Bloomington, USA, LEAR at CERN, EC, TARN II in Tokyo, and a cooler ring in Osaka, Japan. The phase space compression of the stored ion beam by laser cooling and electron cooling is applied in all storage rings. For both methods a repetitive interaction of the stored ions with the cooling medium in the ring is essential. Low temperatures of highly charged ions were reached by electron cooling in the TSR (6) and even much lower temperatures of singly charged ions have been obtained by laser cooling both in the TSR (7) and in ASTRID (8). Some atomic physics experiments which make use of the laser and the electron cooler are presented below.

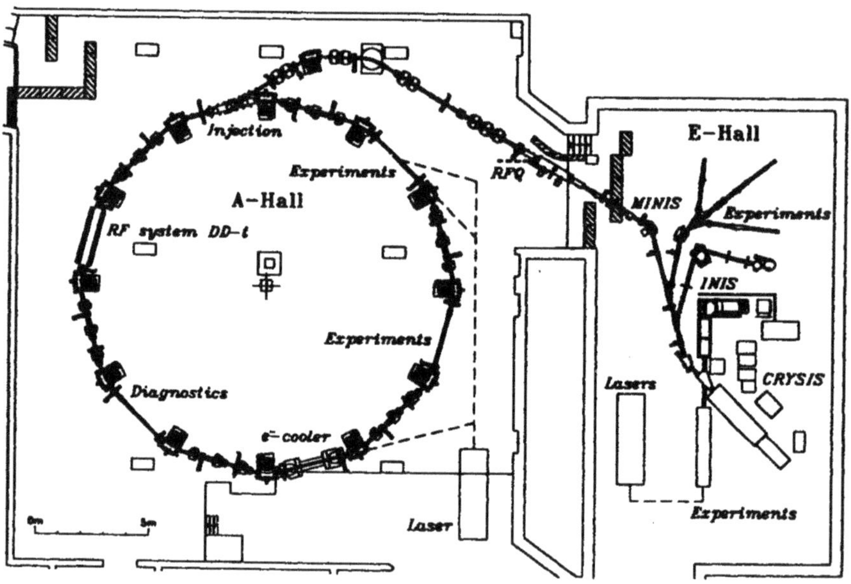

Fig. 1 CRYRING layout.

An ion storage ring has generally a periodic structure of magnetic elements: dipole magnets for bending (B), focussing and defocussing quadrupoles (F, D),and possibly higher order multipoles. Between the magnets there are drift lengths (O). Since a quadrupole is always focussing in one and defocussing in the other plane, a combination of F and D has to be used in order to have overall focussing in both planes. A superperiod of the CRYRING (4) consists e.g. of a BFDFBO structure. CRYRING (whose layout is shown in Fig. 1) is a storage ring for heavy ions with beam cooling and synchrotron acceleration capabilities. The highly charged ions are obtained from a cryogenic electron beam ion source (CRYSIS), preaccelerated by an RFQ to medium energies and then injected into the CRYRING. The fast acceleration mode of the ring brings the ions quickly (100ms) to high enough energy (up to $\sim 24MeV/amu$) for storage and cooling or decelerates them after cooling for certain experiments. Here we do not want to discuss all the conditions for storing charged particles in a storage ring. For the interested reader we refer to some of the excellent introductory courses in the CERN accelerator schools (see ref. 9) In short, the motion of ions without momentum spread in an ideal storage ring is described by a homogeneous differential equation of Hill's type. The solution of it gives a pseudoharmonic oscillation (betatron oscillation), and the number of betatron oscillations per revolution is the so called tune. For a stable operation of the machine, a working point has to be found, where the tune is far from low valued rational numbers, otherwise the betratron oscillation hits a resonance.

Besides the reduced momentum spread of a cooled beam, one can also have a largely enhanced luminosity in a storage ring, compared to using the beam directly from the same source in a single pass through the target. This enhancement is particularly high for a pulsed source. It has its optimum when the pulse width and duty cycle can be matched with the injection, cooling, and storage times in the ring. It reaches values of 10^4 and more, depending on certain parameters, e.g., loss rate of the stored beam. This gain of luminosity in a storage ring is very important for the use of a low-density target, such as an electron target with a low energy spread, or an internal atomic beam target.

II. Beam Cooling

Two cooling methods appear to be at present most effective for heavy ions: electron cooling and laser cooling which will be explained below. First of all we want to mention the main reasons for the interest in cooling heavy ion beams:
- the preparation of a beam with extremely low momentum spread and low emittance for experiments.
- the possible observation of collective effects or even crystallization of the stored beam.
- the compensation of emittance growth due to beam-target or residual gas interaction,
- the possibility of accumulating and cooling beams of rare exotic ions.

Electron cooling is based on the exchange of energy between the ion beam and a low-temperature beam of electrons having the same velocity as the ions (10,

11). The beams are merged over a distance of typically $1-2$ m. A magnetic field with a strength B in the order of 0.1 T is applied in the direction of the electron beam, preventing it from diverging. The electrons are emitted with a homogenous charge density from a hot cathode ($T \approx 1300\ K$) having a radius of around 2 cm. The thermal emission results in a transverse electron temperature of about 0.1 eV while the acceleration reduces the longitudinal temperature by a large factor ($10^{<} - 4 >$). To these temperatures one has to add, however, contributions from ripple in power supplies, heating during acceleration, electron-beam space charge, misalignment between ion and electron beams, etc., which can increase in particular the longitudinal energy spread.

The Coulomb scattering between ions and electrons can be seen as a friction force experienced by an ion moving with velocity v_i through a gas of electrons with density n_e giving: $-\frac{dE}{dx} = 4\pi\frac{q^2 e^4 n_e L_C}{m_e v_i^2}$. Here $L_C = \ln(b_{max}/b_{min})$ is the Coulomb logarithm, typically having a value between 5 and 10. In order to obtain cooling times the force has to be integrated over the velocity distributions of both ions and electrons (12). For the example of Ar^{18+} ions stored at 10 MeV/amu in CRYRING a cooling time in the order of 50 ms has been estimated (13). The first electron cooling for heavy ions has been demonstrated in the TSR (1) where for example a 1.6 mA C^{6+} beam of 73 MeV energy has been cooled from an initial momentum spread of $\frac{\delta p}{p} = 1.8\ 10^{-2}$ to $\frac{\delta p}{p} = 2\ 10^{-4}$ in 3 seconds. Schottky noise spectra which reflect the momentum spread of such a cooled beam will be shown below.

The cooling by laser photons is mediated by the momentum transfer to the ion in the absorption process. Because the photon is emitted in a random direction in a spontaneous decay. The repeated action of the excitation/decay-cycle will give a net momentum change in the direction defined by the laser light (14). Only a specific laser frequency which corresponds to the excitation energy will cause momentum transfer. For an ensemble of atoms with a particular velocity distribution, only atoms of a specific velocity class (a specific Doppler shift) will be in resonance with a fixed laser frequency. So this process depletes the population of this specific velocity class in the distribution (burning a Bennet hole in the distribution) moving this population to an adjacent velocity class (the velocity difference corresponding to the photon momentum). By tuning the laser frequency through the velocity distribution it is thus possible to narrow (cool) the distribution (15).

For a working laser-cooling scheme a two-level system is required in order to make it possible to repeat the excitation many times. The number of such atomic systems accessible for cw lasers are very limited. Ions which have been discussed are for instance Li^+, Be^+, Be^{++} and Mg^+. In the Li^+andBe^{++} ions the fraction of the beam in the metastable $1s2s\ ^3S$ state could be cooled by the $1s2s\ ^3S - 1s2p\ ^3P^o$ transition. A set-up which has been successfully applied for laser cooling of Li^+ in the TSR (7) is shown together with the Li^+ level scheme and the fluorescence spectra in Fig. 2. The narrow peak in the frequency spectrum was obtained by a fixed frequency Ar^+ ion laser counterpropagating the ion beam and being in resonance with the low velocity part of the distribution. A tunable dye laser was merged in direction of the ion beam and its frequency was changed to higher values in order to decelerate the ions which are in resonance. In such scans only the metastable fraction of the ions are cooled. The obtained velocity spread was in the order of 10^{-6} which is still much above the theoretical value.

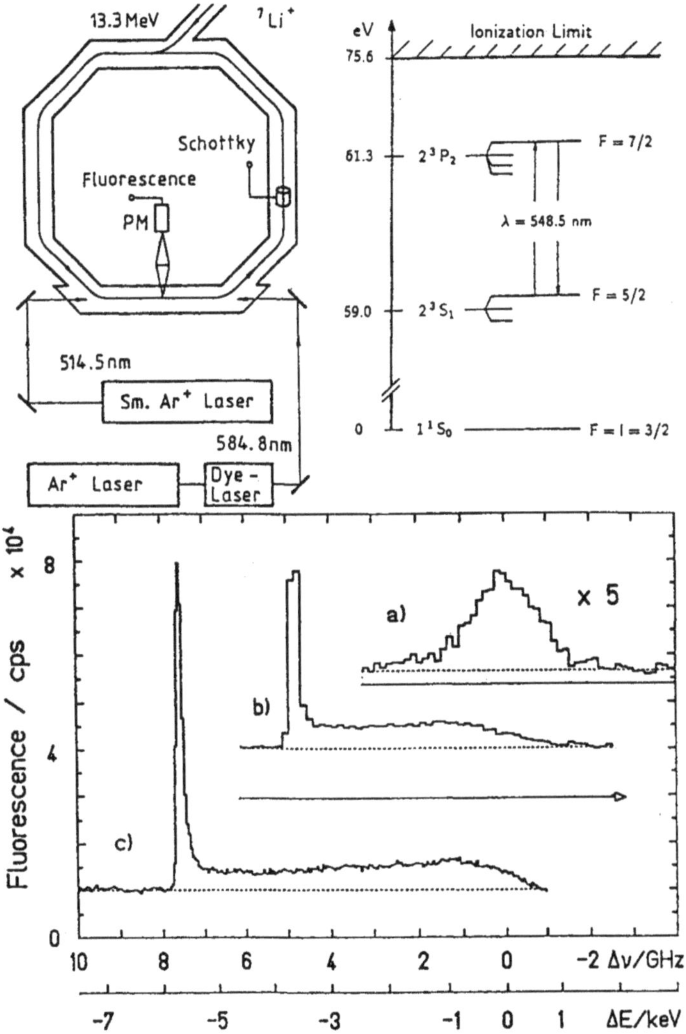

Fig. 2. Laser cooling scheme for metastable Li^+ ions (see level scheme) and the momentum compression measured by the fluorescence detection at the TSR (from ref. 7).

The revolution frequency of ions in the ring can be measured by a frequency analysis of a signal from a charge sensitive pick-up. The according noise peaks appear in harmonics (bands n) of the revolution frequency. One calls this the longitudinal Schottky scan in contrast to the transversal Schottky spectrum, which is caused by the betatron oscillations (ref 9). The longitudinal Schottky noise can be directly related to the momentum profile of the beam. Modifications of the longitudinal Schottky noise have been observed for very cold and dense ion beams first at NAP-M (16), but also at LEAR, TSR, and ESR. Fig. 3 shows such an example from electron cooling of C^{6+} ions in the TSR (6).

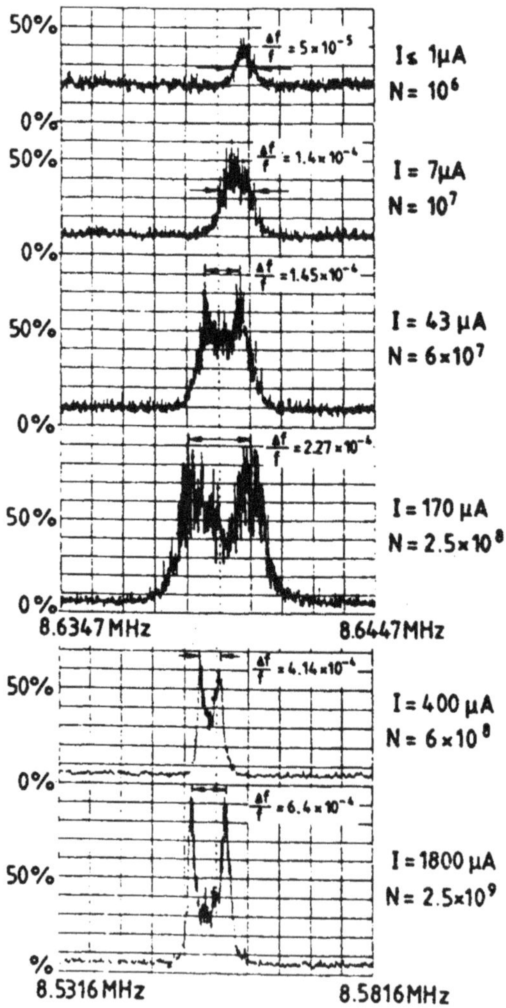

Fig. 3 Longitudinal Schottky noise spectra of different intensities of 73.3 MeV C^{6+}

ions stored in TSR after electron cooling (from ref. 6).

Two effects are observed: the signal strength decreases with the beam temperature and the noise spectrum splits into two peaks around the harmonic center frequency. The decrease of the signal power is explained by an increasing correlation between the particles due to the increased importance of the Coulomb potential relative to the internal thermal energy. This reduces the random component of the density fluctuations and thereby the Schottky noise power. The splitting of the central line into two peaks is explained by back coupling of particle density fluctuations via the walls of the vacuum chambers on the following particles in the beam. This leads in the case of cold dense beams, were the random component is surpressed, to the excitation of collective modes for the particle motion: i.e. waves of density fluctuations propagate and counterpropagate in the stored beam. These waves modulate the revolution frequency and show up as a peak splitting in the longitudinal noise spectrum. From the splitting seen e.g. in Fig. 3 the "velocity of sound" in the stored beam can be infered (6).

Further cooling could lead to a phase transition of a strongly coupled one-component plasma to an ordered structure (crystallization of the stored beam) (16). A simple criterium for obtaining an ordering effect in the beam is given by the condition that the potential energy between the ions should be much larger than their kinetic energy (temperature). Or, a so called Γ-value $\Gamma = \frac{q^2}{ak_BT} \geq 170$ (q is the ion charge state, $a = (\frac{3}{4\pi n_0})^{\frac{1}{3}}$, k_B is the Boltzmann constant, n_0 is the ion density). In computer simulations, applying a molecular dynamics code, of an idealized storage ring (17, 18), one has obtained helixes of three dimensional crystals of the one-component ion plasma. In these calculations one solves the equations of motion for the Hamiltonian:

$$H = \sum_i \frac{p_i^2}{2m} + \sum_i \sum_{j \neq i} \frac{Z^2}{r_{ij}} + k_x(z_i)x_i^2 + k_y(z_i)y_i^2$$

where the second term represents the electrostatic interaction between all the ions and the harmonic potentials induced by the focussing elements in the ring. The temperature of such an assemble of identical particles is calculated from an averaged square of the particle velocities. Uniform cooling of such a beam is then equivalent to an appropriate scaling of the particle velocities. Longitudinal and transverse cooling can also be simulated by similar scaling of the longitudinal and transverse velocities respective to the moving frame.

In figure 4 the succesive ordering of the 400 Ar^{18+} ions moving at the velocity of 5 MeV/a.m.u. is displayed. The amount of Ar^{18+} ions corresponds to the selected 2mm long slice of the beam used together with periodic boundary conditions in the calculations (19). In a simulation with the realistic periodic focusing and defocusing structure of the ring (see figure 1) it was found (19) that a crystallized beam would retain its order after 4 revolutions in the simulated ring. Similar results have been obtained also for the ESR at GSI in Darmstadt(18). The bending magnets introduce shearing forces on the three-dimensional ordered beam and the passage through bending magnets could destroy the order in the beam if the forces between the particles do not compensate this shearing force. Simulations of the shearing force in the dipole magnets are now in progress. One does not have such a shearing force in the case of a one-dimensional crystall (a chain) which could occur for less than about 10^6 particles in a ring of the size of CRYRING or about 100 particles in a 2mm slice. It is in fact believed that this kind of ordering was observed in the Novosibirsk NAP-M ring (16).

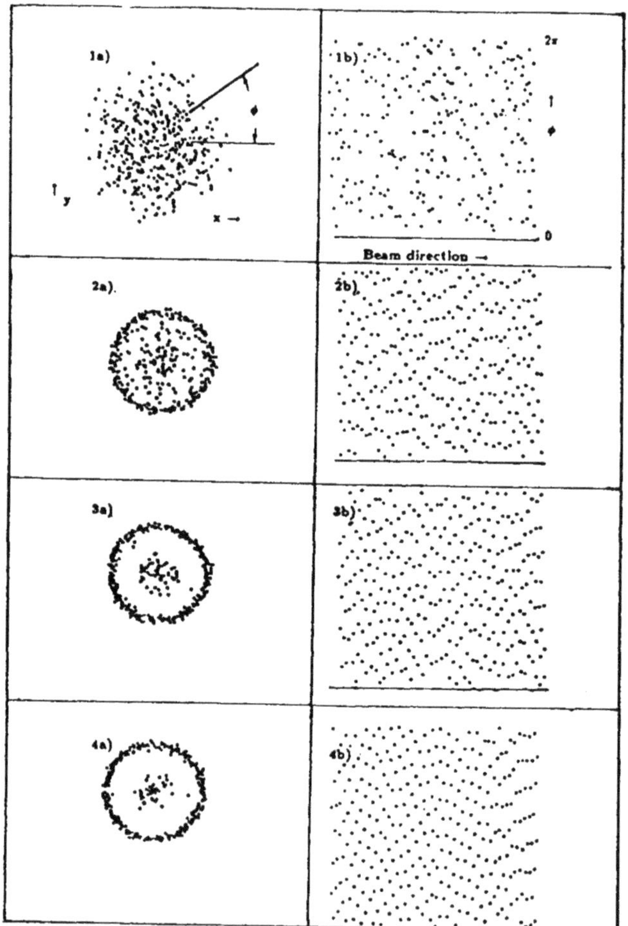

Fig. 4 Cooling of a stored Ar^{18+} ion beam, simulated by a molecular dynamics calculation with a 2mm long slice which contains 400 ions (from ref. 13). Sections a) of the graphs are cuts through the beam cross section. Sections b) of the graphs are $\varphi - z$ projection of the beam slice. 1a) and 1b) uncooled beam at 800 K. 2a

and 2b) after 1 time unit and T = 20.5 K. 3a) and 3b) after 2 time units and T = 1 K. 4a) and 4b) after 4 time units and T = 0.5 K.

The phase-transition to a liquid or even a crystalline phase of a stored ion beam in a ring could be of great importance for the physics of one-component plasmas, i.e. large-size interactive Coulomb systems, for luminosity enhancement in future colliders, and for spectroscopy in rings by means of reduction of the Doppler widths. The crystallization of a stored ion beam would eliminate the intra-beam scattering, and improve the obtainable velocity resolution.

III. Electron-Ion Collision Experiments

The interest in electron impact phenomena such as excitation, ionization, radiative, dielectronic and collisional recombination of electrons with bare or few electron ions is due to several practical applications and also to very fundamental questions. The applications are in plasma physics or astrophysics where electron-ion recombinations and their spectroscopy play an important role, but also in electron cooling itself, where recombination could cause significant ion beam losses and is expected to be useful for the diagnosis of the beam properties. On the other hand, studies of electron-ion impact phenomena can be very useful for studying the structure and the decay modes in few electron highly charged ions. They can serve e.g. as testing grounds for calculations of energy levels in few electron ions which require a proper treatment of relativistic, QED, and correlation effects.

A very high resolution in centre-of-mass energy can be obtained in a cooler used as an electron target and also small collision energies can be reached. If the electron energy (in the laboratory frame) is changed by a small amount ΔE from its value E_e where electron and ion velocities are equal, the resulting centre-of-mass energy is, non-relativistically, $E_{\rm cm} = \Delta E \frac{\Delta E}{4E_e}$, which is much smaller than ΔE. For instance, if a 10 keV electron beam is detuned by 500 eV, the resulting collision energy is only 6 eV.

The energy resolution that can be achieved is limited by the energy spread of the two beams. For a cooled ion beam even of relatively high intensity one might reach a relative energy spread in the 10^{-4} range, both for the ion and the electron beam. At small collision energies the transverse energy spread will usually dominate since it remains unaffected by the transformation to the centre-of-mass system. In a storage ring, furthermore, the transverse velocity spread of the ion beam is small compared to that of the electron beam. By the cathode temperature a transverse energy spread of the order of 0.1 eV is introduced. But the main part in the resolution is due to space-charge effects in the electron beam. There one has to take into account that the cooled ion beam has a diameter considerably smaller than the electron-beam diameter. This fact improves the resolution since the space-charge-induced potential drop across the electron beam has a parabolic shape. In practice one can achieve an energy resolution of better than 1 eV at center of mass collision energies of some 100 eV. A value of that order was obtained in the dielectronic recombination experiments at the TSR (see section III.2).

III.1. Spontaneous and induced radiative recombination

The rate for radiative recombination of free electrons with ions can be obtained from the time reverse of photoionization which was first derived by Stobbe (20). Within the low energy approximation, valid for cooling conditions and recombination into the lower bound states, i.e. the binding energy in the final state ($E_{nl} = Ry \cdot \frac{Z^2}{n^2}$) is large compared to the relative energy between electrons and ions (E_r) an analytical formula for the recombination cross section can be derived (21). An analysis of the validity conditions for this approximation shows that it is valid for $n \leq Z$. The cross section for emitting a photon of energy $E_\gamma = E_{nl} + E_r$ is then:

$$\sigma_{rr}^{nl} = \frac{\pi^2}{3}\alpha^3 a_0^2 \frac{E_{nl}}{E_r}[(l+1)c_{l+1}^2(n,l) + lc_{l-1}^2(n,l)]$$

with α the fine structure constant, a_0 the Bohr radius, and $c_{l+1}(n,l)$ and $c_{l-1}(n,l)]$ as reduced electric dipole matrix elements (21) which are a combination of confluent hypergeometric functions. This cross section contains the dependence on the l state which is populated, in contrary to the very widely used Bethe Salpeter formula (22) which does only contain the n-state dependence. Fig. 5 shows the population of the different l states versus l/n relative to the cross section in the maximum for values of n$\leq$ 30.

The recombination rate is obtained by averaging the cross section over a temperature distribution of the electron beam. If one assumes a flattened Maxwell temperature distribution with $T_\perp$ transverse and $T_\parallel$ longitudinal temperature, respectively, one obtaines (21):

$$\alpha_{rr}^{nl} = \frac{E_r \sigma_{rr}^{nl}}{(2\pi m k_B T_\perp)^{1/2}}\left(\frac{\frac{(T_\perp - T_\parallel)}{T_\parallel} + 1}{\frac{(T_\perp - T_\parallel)}{T_\parallel}}\right)^{1/2} arctan\left(\frac{(T_\perp - T_\parallel)}{T_\parallel}\right)^{1/2}$$

In low-temperature high density plasmas, besides radiative recombination the collisional (three-body) recombination or a combination of both can become the dominant process (23). This is due to the fact that the collisional recombination coefficient has a very different dependence on the electron temperature ($\sim T^{-4.5}$) than the radiative recombination coefficient ($\sim T^{-0.5}$). Even in an electron cooler, where one deals with a quite dilute plasma ($n_e = 10^8 cm^{-3}$), the collisional recombination can become a noticeable contribution to the total recombination because of the low electron beam temperature. The relative contribution increases also with the ionic charge, as they have different dependences on the ionic charge,

$\sim Z^3$ and $\sim Z^2$ for collisional recombination and radiative recombination, respectively. Taking the electron density of $n_e = 10^8 cm^{-3}$ the two would be equal for Ar^{18+} for $k_B T \simeq 5 \cdot 10^{-2}$ eV; whereas for U^{92+} they are equal for $k_B T \simeq 8 \cdot 10^{-2}$ eV.

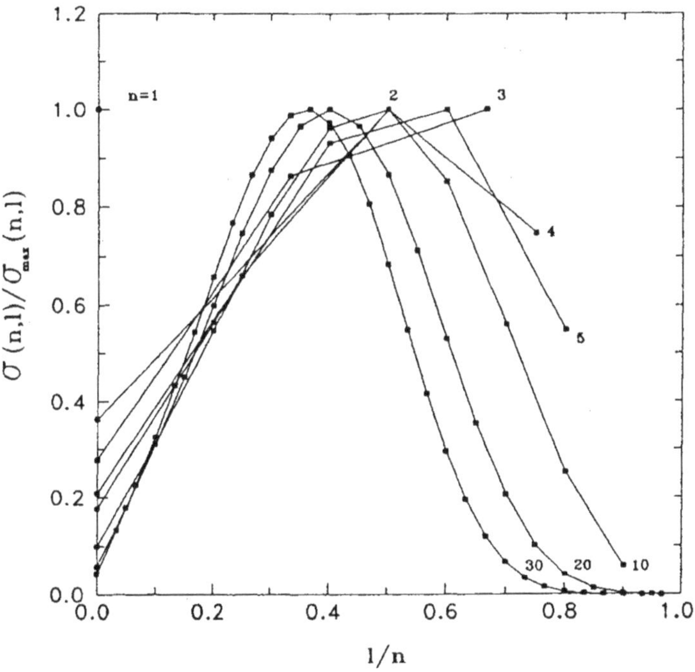

Fig. 5 Cross section for the population of the different l states by radiative recombination versus l/n relative to the cross section in the maximum for certain values of n$\leq$ 30. (from ref. 21).

Only very high n states are populated by collisional recombination. Most of the recombined electrons are again ionized by collisions with free electrons. This leads to a distribution of the population of the high n-states, and the formation of a "bottleneck" for recombination (23). The effect of field ionization in the toroid magnet at the exit of the cooler or the dipole magnets of the ring can also influence the n-value of this bottleneck. So the formation of these states and their interplay with magnetic and electric fields could well influence the cooling process and the beam losses.

Laser induced radiative recombination can be understood from the concept of induced transitions between bound states as applied to the radiative recombination of a free electron into a bound state induced by a laser photon. The laser being merged with the electron and ion beam can stimulate radiative recombination into a state E_n if it fits within the level width to the laser energy (24). Using a optical tunable laser, such as a dye laser in the energy range of 2 to 4 eV, a wide band of states from e.g. n=20 to 100 can be populated in a highly-charged heavy ion. Laser induced recombination is of interest for measurements of energy structures in the electron continuum. It is speculated (25) that the electron density fluctuations (Debye-spheres) resulting from the screening of the ionic charge can be observed by laser induced transitions.

The enhancement factor for stimulated radiative recombination is:

$$G = g(\nu)\frac{B_{if}}{A_{if}}\frac{\alpha_{rr}^{n}}{\alpha_{rr}}$$

The spectral energy density of the laser $g(\nu)$ can be written as:

$$g(\nu) = \frac{P}{c}A\Delta\nu$$

where P is the laser power, A the geometrical area, $\Delta\nu$ the laser bandwidth. Since $A_{if} = 8\pi h(\frac{\nu}{c})^3 B_{if}$ from the Einstein relation, one gets for G from the above relation. As one example we consider stimulated recombination into n=33 of Ar^{17+} by a laser power of about 25 MW/cm^2 (pulsed) this gives a gain value of about 20. With a total pulse length per sec of presently available lasers, $\Delta t = 10^{-5}$ s, one has a time averaged induced rate of 300 s^{-1}.

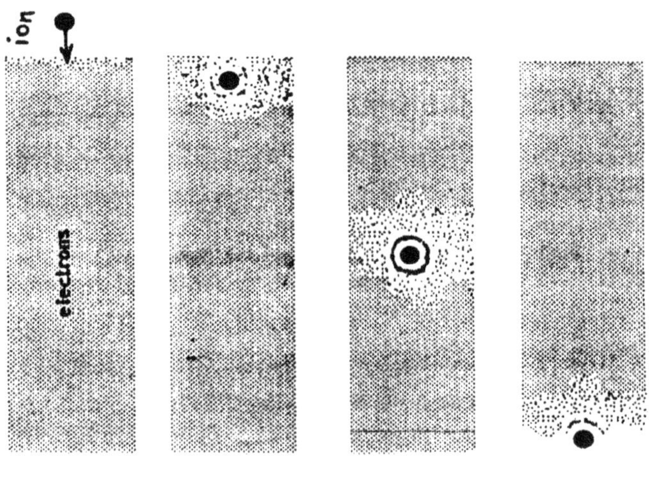

Fig. 6. Schematic picture of Debye screening spheres formed when a highly charged ion moves within a cold gas of electrons(25).

A first observation of laser-induced radiative recombination was obtained with merged beams of protons, electrons, and laser photons in the TSR (26). For this experiment intense laser light pulses at the Balmer series limit were send into the overlap region of electrons and protons (of an energy of 21 MeV) in the cooler, and the formation of hydrogen in the quantum state n=2 was measured. When the electrons are cooling the protons, the relative velocity between them is essentially determined by the electron velocity in the ion rest frame. Mainly based on the dielectronic recombination measurements (see above), a transverse velocity spread corresponding to the cathode temperature of 1220 K (0.105 eV) and a longitudinal temperature of about 1 10^{-3} eV is expected. The recombined ions were seperated from the circulating protons in the bending magnet downstream of the electron cooler and detected in coincidence with the laser pulses. The observed spontaneous recombination rate was in good agreement with the theoretical calculated ones using the relations given above. The light pulses of 450.46 nm from a tunable dye laser, pumped by an excimer laser, were 20 ns long and had a peak power of 1 to 3 MW.

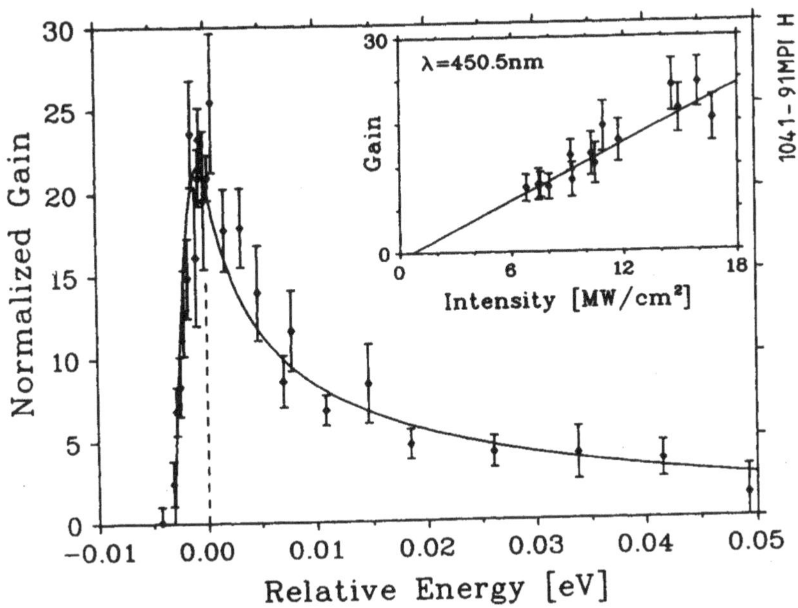

Fig. 7. Normalized gain factor for laser-induced recombination into the hydrogenic n=2 state as a function of the c.m. photon energy minus the n=2 binding energy (26). The full line was calculated for the external field of 20 V/cm. The insert shows the maximum gain factor as a function of the laser intensity with a straight line to guide the eye.

The time spectrum from the coincidences between hydrogen atoms reaching the detector and the laser pulses has a flat background of spontaneous recombination and a peak from the induced process. The width of the peak is given by the laser pulse length and the overlap time between laser photons and protons in the cooler. The ratio between the counts in the time window of the peak and the background represents the gain factor. In Fig. 7 the gain factor, normalized to a laser intensity of 17.5 MW/cm^2, is shown as a function of the photon energy in the c.m. system (the n=2 binding energy substracted). A surprising result was that the gain factor starts to rise already at -25.5 cm^{-1}, clearly below the threshold given by the n=2 binding energy. Above the maximum it falls off slowly, which is very well described by the electron velocity distribution in the moving frame, as explained above.

The calculated gain factor for recombination into the 2p state is 59, for a laser power of 17.5 MW/cm^2. This value is nearly a factor 3 above the measured value (shown in Fig. 7). A possible photo re-ionization of the recombined ions by the same laser pulse is negligible for the 2p state, but nearly 100% for the 2s state. We explain the deviations, both in the maximum gain and at the threshold, by the influence of external static electrical fields on the electronic states at the hydrogen ionization limit (Stark states at the ionization limit) (26). From a measured small displacement between electron and proton beam an electric field of (20+-4) V/cm from the electron beam space charge was deduced. The linear potential of this field exceeds the Coulomb potential at a distance of $\sim 1\mu m$ from the proton. It leads to a saddle point of the potential at $\epsilon = -27.4 + -3cm^{-1}$ which reduces the ionization threshold of the hydrogenic n=2 level by this amount. This value is in good agreement with the shift towards energies below threshold seen in Fig. 7. In a model, which combines the classical motion of the electrons in the Stark potential with the quantum mechanical description of radiative recombination, the full line in Fig. 7 was calculated. This calculation also explaines the reduction of the gain by binding and anti-binding Stark states, i.e. the Stark potential allows only certain electron trajectories to recombine with the protons.

III.2 Dielectronic recombination

For dielectronic recombination (DR) to occur an electron of the projectile ion is excited and at the same time a free electron is captured to form a doubly excited state in the ion, which stabilizes by emitting photons. Due to energy conservation, the binding energy plus the kinetic energy of the captured electron must equal the excitation energy of the bound electron. For the case of isolated resonances in the intermediate state $(N, L; n, l)$ the DR cross section can be approximated as:

$$\sigma_{DR} = \sigma_0 \Gamma(N, L) \Sigma_l 2(2l+1) \frac{A_a(n,l) A_r(n,l)}{A_a(n,l) + A_r(n,l)}$$

Where σ_0 is a cross section containing all the constants including the energy bin, Γ is the Lorentz profile for the decay of the state (N,L;n,l), $A_a(n,l)$ and $A_r(n,l)$ are the autoionizing and radiative decay rates of this state. For high n-states (Rydberg states) the radiative rate $A_r(n,l)$ is independent of n,l and determined only by N,L. The autoionization rate $A_a(n,l)$ varies with n^{-3}. For low n,l one has $A_a \gg A_r$ and $\sigma_{DR} \sim \Sigma_l 2(2l+1) A_r(n,l) \sim n^2$. Therefore, σ_{DR} for e.g. the KLn series should increase for low n and start to decrease for higher n. Since $A_r \sim Z^4$

this n-dependence of the DR cross section should vary with the ion nuclear charge. One can see the n dependence in the shape of the measured DR excitation function. A result from a first DR measurement at the TSR (27) with O^{7+} ions is shown in Fig. 8 for the KLn series. The cross section increases for O^{7+} up to $n \sim 3$, from there on it starts to decrease. The increase at above $n \sim 7$ is due to a pile up of unresolved states. Actually, one can fit the shape of the spectrum very well at high n with a n^{-3} scaling, where we find (27) that only states up to $n \sim 35$ contribute to the measured intensity. This makes clear that field ionization of high n states in the ring dipoles can not influence the data because the field ionization limit was estimated to at $n \sim 69$.

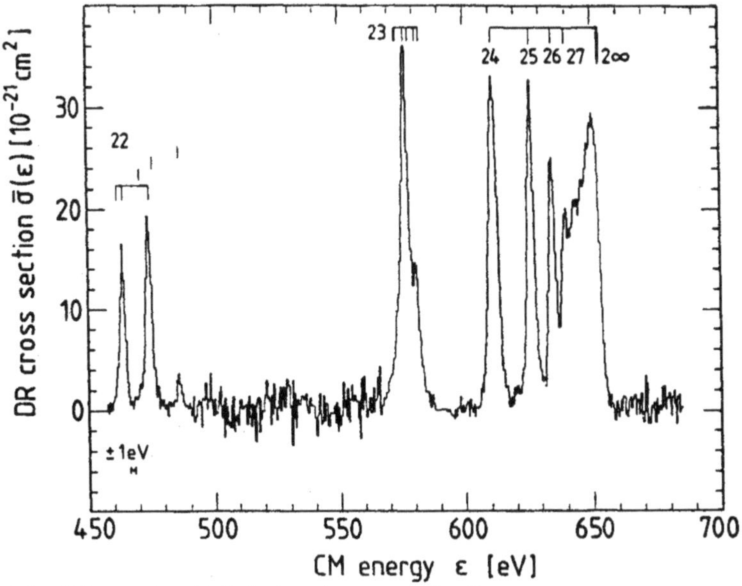

Fig. 8. The KLn serie of dielectronic recombination resonances measured with 143 MeV O^{7+} ions stored in the TSR (27).

A quite different behaviour is found for the He-like ion C^{4+} which was also measured at the TSR (28). The presence of the additional inner shell electron causes four Rydberg series corresponding to the four possible configurations (see fig. 9). The ^1P series dominates the spectrum (fig. 9) because the stabilizing radiative transition 1s2p(^1P) to $1s^2(^1S)$ is favoured. It is, however, remarkable that the cross section drops off when it approaches the series limit, whereas it continues increasing for the C^{5+} ion. It was found in a theoretical analysis that this decrease for C^{4+} is due to the interaction of the 1s2p(^1P) resonance states with continuum states of the other Rydberg series, which enhances the autoionization probability of the intermediate states and reduces thereby the DR cross section.

The selective formation and spectroscopy of a doubly excited state in a highly ionized ion, which is formed as an intermediate state, is the scope of these experiments. The resolution in these measurements is in the order of one eV from the size of the ion beam (5 mm) in the space potential of the electron beam. For inner-shell states with lifetimes of $\sim 10^{-15}$ s the resolution is therefore somewhat above the width of these lines. Thus a measurement of the yield of recombined ions as a function of the electron-ion relative energy results in a series of peaks reflecting these doubly excited states. Both the measurement of the *areas* of these peaks and the *energy positions* of the peaks tests our understanding of the electron-electron interaction in doubly-excited states. In fact, DR can serve as the basis for a new spectroscopy, one in which emitted radiation need not be directly observed, and which can compete favorably with traditional x-ray and electron spectroscopy.

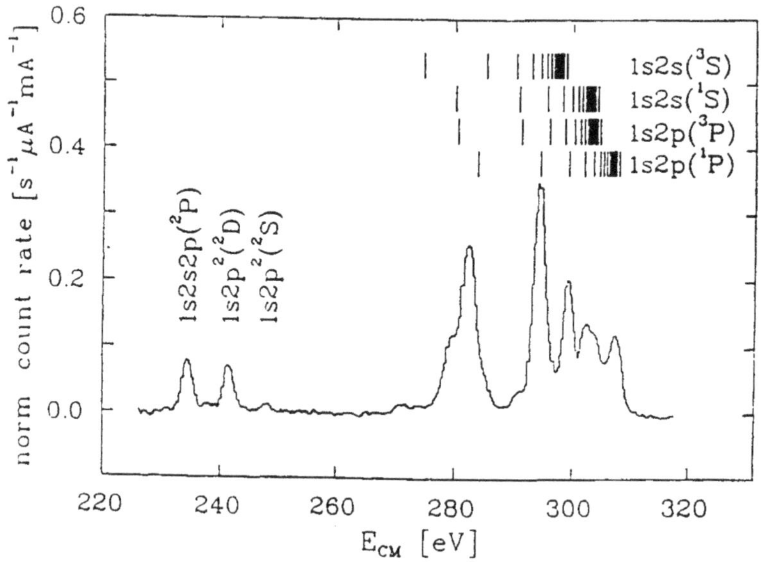

Fig. 9. The KLn serie of dielectronic recombination resonances measured with C^{4+} ions in the TSR (28).

The author acknowledges many fruitful discussions and enjoyable experiments with his colleagues at the MSI atomic physics department and CRYRING group, the TSR group at Heidelberg, and the ESR group at GSI, Darmstadt.

References

1 P. Baumann, M. Blum, A. Friedrich, C. Geyer, M. Grieser, B. Holzer, E. Jaeschke, D. Kraemer, C. Martin, K. Matl, R. Mayer, W. Ott, B. Povh, R. Repnow, M. Steck, E. Steffens, W. Arnold, Nucl. Instr. Meth. A 268 (1988) 531.

2 K. Blasche, D. Bohne, B. Franzke, and H. Prange, IEEE Trans. Nucl. Sci. NS-32 (1985) 2657, and B. Franzke, Nucl. Instr. Meth. in Phys. Res. B 24/25 (1987) 18.

3 R. Stensgaard, Physica Scripta T22 (1988) 315.

4 C.J. Herrlander, A. Bárány, K.-G. Rensfelt, J. Starker, Phys. Scripta T22 (1988) 282.

5 C. Ekstroem et al. Phys. Scripta T22 (1988) 256.

6 D. Habs, W. Baumann, P. Blatt, J. Berger, A. Faulstich, C. Geyer, M. Grieser, B. Holzer, E. Jaeschke, D. Kraemer, C. Martin, K. Matl, M. Dutta, T. Kuehl, D. Marx, S. Schroeder, M. Gerhard, G. Huber, R. Mayer, W. Ott, B. Povh, R. Repnow, M. Steck, E. Steffens, W. Arnold, R. Schuch, J.F. Babb, L. Spruch, A. Noda, Nucl. Instr. Meth. B43, (1988) 390.

7 S. Schroeder, R. Klein, N. Boos, M. Gerhard, R. Grieser, G. Huber, A. Karafillis, M. Krieg, N. Schmidt, T. Kuehl, R. Neumann, A. Balykin, M. Grieser, D. Habs, E. Jaeschke, D. Kraemer, M. Kristensen, M. Music, W. Petrich, D. Schwalm, P. Sigray, M. Steck, B. Wanner, A. Wolf, R. Schuch, Phys. Rev. Lett. 64, 737 (1990).

8 O. Poulsen, private communications.

9 See, e.g., CERN Accelerator School, CERN 84-15 (edt. by P. Bryant and S. Newman, 1984), CERN 85-19 (edt. by P. Bryant and S. Turner,1985), and CERN 87-03 (edt. by S. Turner, 1987).

10 C.I. Budker, N.S. Dikansky, V.I. Kudelainen, I.N. Meshkov, V.V. Parkomchuk, D.V. Pestrikov, A.N. Skrinsky, and B.N. Sukhina, Part. Accelerator 7 (1976) 197.

11 A. Wolf, Phys. Scripta T22 (1988) 55., and H. Poth et al, Z. Phys. A332, (1989), 171.

12 L. Spitzer, Physics of Fully Ionized Gases, Interscience, N.Y.,(1956).

13 R. Schuch, A. Barany, H. Danared, N. Elander, S. Mannervik, Nucl. Instr. Meth. B43, (1988) 411.

14 J. Opt. Soc. Am. B 2 (1985) 11, Phys. Today, June 1987, and ref. therein.

15 J. Javanainen, M. Kaivola, U. Nielsen, O. Poulsen and E. Riis, J. Opt. Soc. Am. B2 (1985) 1768.

16 N.S. Dikansky, D. Pestrikov, KfK Report 3846, Proceedings of the ECOOL, Karlsruhe 1984

17 A. Rahman and J.P. Schiffer, Phys. Rev. Lett. 57 (1986) 1133.

18 I. Hoffman, Proceedings of the Workshop on Beam Crystallization, Wertheim, 1988, and private communication.

19 N. Elander, J. Beebe-Wang, R. Schuch and R. Cauchau, work in progress.

20 M. Stobbe, Ann. der Physik 7 (1930) 661.

21 M. Pajek and R. Schuch, to be published

22 H.A. Bethe and E. Salpeter, Handbuch der Physik, Vol. 35 (1957) 88.

23 S. Stevefelt, J. Boulmer, and J.F. Delpech, Phys. Rev. A12 (1975) 1246.

24 R. Neumann, H. Poth, A. Winnacker, and A. Wolf, Z. Phys. A313 (1983) 253.

25 N. Angert, F. Bosch, B. Franzke, K. Finnlayson, D. Liesen, T. Kuehl, R. Mann, D. Marx, P. Mokler, R. Neumann, J. Palinkas, H. Poth, S. Schroeder, R. Schuch, H. Schulte, GSI Proposal # 121 (1990)

26 U. Schramm, J. Berger, M. Grieser, D. Habs, E. Jaeschke, G. Kilgus, D. Schwalm, A. Wolf, R. Neumann, and R. Schuch, Phys. Rev. Lett. 67, 22 (1991).

27 G. Kilgus, J. Berger, P. Blatt, M. Grieser, D. Habs, B. Hochadel, E. Jaeschke, D. Krmer, R. Neumann, G. Neurether, W. Ott, D. Schwalm, M. Steck, R. Stockstad, E. Szmola, A. Wolf, R. Schuch, A. Mueller, and M. Wagner, Phys. Rev. Lett. 64, 737 (1990).

28 G. Kilgus, D. Habs, E. Jaeschke, D. Krmer, D. Schwalm, M. Steck, A. Wolf, R. Schuch, A. Mueller, and M. Wagner, MPI Heidelberg, Annual Report 1990.

ELECTRON-MULTIPLY CHARGED ION COLLISIONS*

N.V. de Castro Faria
Departamento de Física
PUC-Rio - Cx. Postal 38071
Rio de Janeiro, 22453 RJ, Brazil

We present a method for the study of ionization and recombination of multiply charged ions by fast electrons. The method is based on the channeling of ions through single crystals and was tested in the case of fast ions ($v \sim 33v_o$) of χe^{Q+} $(35 \leq Q \leq 53)$ incident on [110] silicon axis. In the rest frame of the ion this experimental situation is equivalent to an electron beam of $\sim$ 15 keV and $\sim 10^{14} A/cm^2$ colliding with this ion.

INTRODUCTION

Multiply charged ions are very interesting objects of study. They can be employed to test fundamental concepts of quantum electrodynamics (QED), as, for example, it was presented in a recent publication of Schweppe et al[1]. An experiment with fast U^{89+} ($v \sim 0.42c$) was done to test QED effects in this few electrons and high charged system, i.e., the relativistic coulomb interaction between electrons, the screening of the self-energy and the vacuum polarization. Briefly, the U^{89+} beam impinges an aluminum foil that collisionally excites a fraction of the ions, initially in their ground state ($2\,^2S_{1/2}$), into the lowest excited state $2\,^2P_{1/2}$, which then decays downstream the foil with a decay length of about one centimeter. The $2\,^2P_{1/2} - 2\,^2S_{1/2}$ energy splitting was measured and the value so obtained can be used as a guide for new improvements of the current theory[2].

In fact, there are many reasons to use highly charged ions as a way of studying field theoretical effects. These effects are enhanced, when the nuclear charge is large, by a factor of the order of Z^4. Another reason is that as Z gets higher, the energy in the Coulomb field begins to approach the large amount of energy required to excite an electron-positron pair from the vacuum (two electron masses) and the structure of the QED vacuum begins to play an important role.

Other situations where multiply charged ions are important are the high temperature plasmas, such as those found in stellar atmospheres or fusion devices. In these plasmas, the collisions between ions and electrons are the most significant particle interactions. This statement can be better understood if one remembers that usually the atomic processes are parameterized in the current plasma models by rate coefficients $\alpha(T)$ rather than cross sections $\sigma(v)$. These coefficients take in account the velocity v of the particle and its distribution $f(v)$, i.e., $\alpha(T) = \int_0^\infty v\sigma(v)\ f(v)dv$ and correspond to an equilibrium temperature T. Since electrons are much less massive than ions, their velocities will be larger than that of an ion of the same energy. If the cross sections are comparable, processes involving electrons will have much higher rates than those involving only ions. Otherwise, if the total charge of the plasma is almost zero, the density of electrons will be higher than that of ions by a factor roughly proportional to the average ion charge, because to each ion with charge Ne corresponds N unbound electrons. While some care must be taken with this explanation mainly because it supposes comparable cross sections, neutral plasma and thermal equilibrium, conditions not satisfied, for example, near the injection site of a Tokamak or near the walks of a plasma container vessel[3], it is very general and includes much practical situations.

In high temperature plasmas, the collisions between electrons and ions determine the mechanisms of such processes as electric conduction and diffusion and they give place to excitation, ionization and recombination of the ions. They give the major contribution for the ion charge distribution and for the composition and flux of the radiation emitted by the plasma. Consequently, the measurement of cross sections of electron-ion collisions is fundamental for a better understanding of plasmas. These cross sections are also of fundamental interest by themselves[3], because these collisions are dominated by the long range attractive Coulomb potential, rather than by the short range interaction of electron and neutral particles. This difference leads to the finite cross section at threshold for electron-impact excitation which is unique to electron-ion interactions.

Although techniques exist for measuring the rate coefficients from the plasmas, where the motion is random and the energy distribution is very large, the direct measurement of cross sections provides the more complete knowledge about the system. In fact, a cross section can be convoluted with any electron velocity distribution in order to produce the respective process rate, even if the system is not in thermal equilibrium and/or the distribution is non-Maxwellian.

Since the pioneering interesecting-beams experiment of Dolder et al[4] and of Dance et al[5], considerable progress has been made in electron-ion collisions, particularly by the group of Müller et al[6,7] which has developed a powerfull electron gun[8]. In these experiments the particles are in a well defined state and the relative velocity is known so that detailed information can be obtained. The main difficulty comes from the normalization of the data, essentially given by a form factor which takes into account the inhomogeneities of both beams. In fact, in order to obtain absolute values of the cross sections, it is necessary to measure the density profiles of both beams. This difficulty was transposed by Müller et al[8] with a modified version of the animated beam method of Defrance et al[9]. The electron beam is swept mechanically through the ion beam and, in this case, the normalization does not requer the knowledge of the density profiles, so that their measurement is no more needed.

The animated beam method can also be applied to merging beam experiments, including the ones involving storage ring[10,11], where form factor measurements are impossible. Although the interest in storage rings for multiply charged heavy ions was initially to study problems in nuclear physics, their applicability to atomic physics has become evident very early. Today, there are various projects atomic physics oriented and some have already started their operations. While the main interest in storage rings came with the development of methods of phase space cooling, the electron beam cooling section of the storage rings offers an interesting place for the study of electron-multiply charged ion collisions.[10,11]

However, for high energy electron-multiply charged ion collision regime, the cross sections are very small and consequently very high currents are needed. In that case, the methods discussed until now are of limited application. In a recent work[12], a new method for the measurement of electron impact ionization cross sections of multiply charged ions was proposed, a method that can be extended to radiative recombination. The method is an applications of channeling of heavy charged ions through single crystals and was tested in the case of fast ions ($v \sim 33v_\circ$) of χe^{Q+} $(35 \leq Q \leq 53)$ incident on [110] silicon axis. The technique is relatively simple and will be presented in the next sections.

METHOD

The technique of channeling is well know; a complete review is presented on reference 13. For axial channeling and incident positive ion (the relevant conditions for our method) we can say, that the well collimated ion beam, with its direction making a small angle with one axial direction of the crystal, could be guided by the transverse electric field produced by the screened nuclear charges and go further and loses less energy than if it was incident in a random direction. It is these steering forces and consequently the trajectory stability of the channeled ion, in opposition to a simple crystal transparency phenomenon, that make channeling a major and peculiar effect.

To simplify the description of the method, we present it on the framework of a particular tested case, i.e, fast ions ($v \sim 33v_o$) of χe^{Q+} ($35 \leq Q \leq 54$) incident on [110] silicon axis.[12]

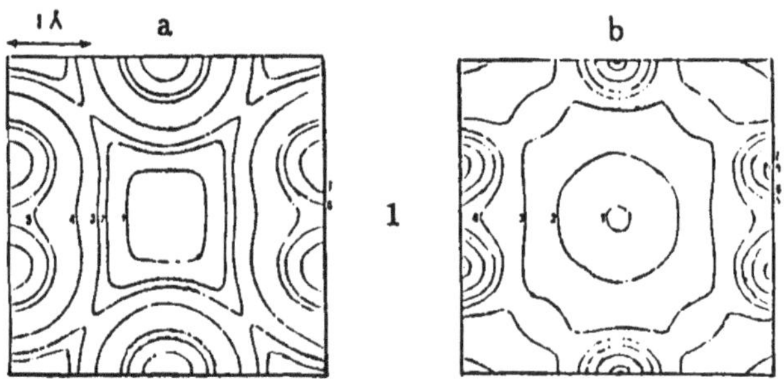

Figure 1a and 1b represent, respectively, the map of the potential energy V for a unit charge and the map of the electronic densities ρ_e, both averaged along the [110]-axis of Si. Suppose that a positive ion (mass M, charge Q and energy E) is incident upon the crystal at a point P, its direction making a small angle φ with the axis direction (normal to plane of figure 1). We define a transverse kinetic energy $E_k = \frac{1}{2}mv_T^2 = E\text{sen}^2\,\varphi \sim E\varphi^2$. To the point P corresponds a potential energy V_P and consequently, a total transverse energy $E_T = E\varphi^2 + V_P$. Suppose now that V' is an equipotential corresponding to E_T. Channeling implies that the ion will be trapped inside the equipotential V', if we can neglect the small scattering angle and energy transfer between the ion and the electrons of the "channel" limited by V'. The critical value E_n for normal axial channeling ($V' > E_n$ corresponds to dechanneling) is usually much larger than the value of the potential at the saddle point between channels. Thus, the majority of channeled particles have energy to transpose the low potential barrier between adjacent channels. When V' is lower than these saddle point potentials, the particle will be confined to one axial channel. If the majority of particles are trapped in one channel, one say that there are hyperchanneling. In that case, the ion sees a beam composed by the valence electrons of the crystal. For an equipotential like the number 3 of figure 1a, the equivalent average "beam current" could be as high as tenths of one Ampere. That was the idea used to obtain the ionization cross sections (next section) and that could be used to get radiative electron capture cross sections.

IONIZATION

In the experiment described in reference 12, a beam of 27MeV/u χe^{35+} was used. The outcoming charge state distribution after crossing $20.7\mu m$ of Si along the [110] direction extended from $Q_{out} = 35$ to $Q_{out} = 53$. In the random crystal orientation case, the equilibrium distribution was also obtained, with an average charge state of $Q \sim 50$. Both curves are shown in figure 2.

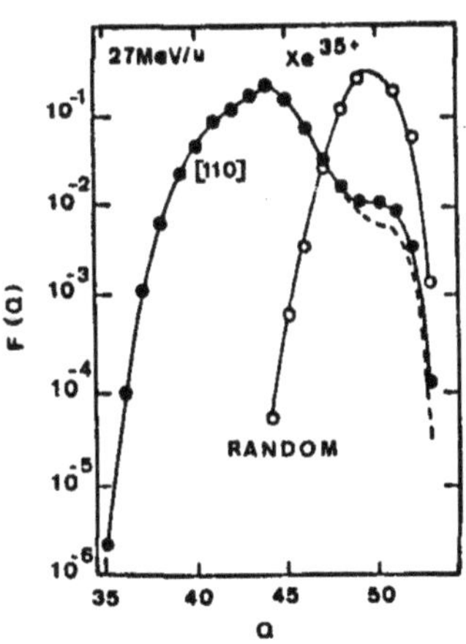

2

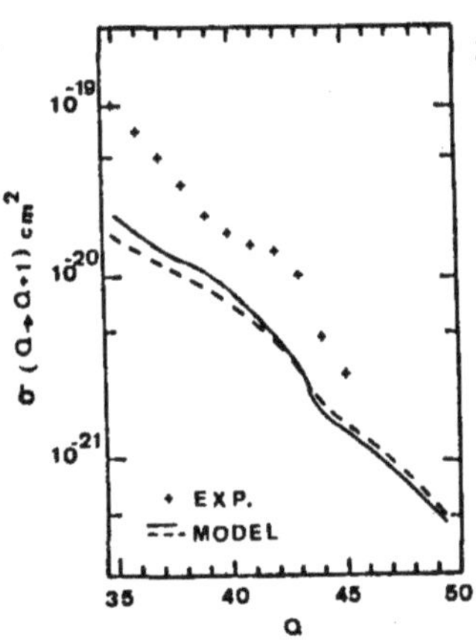

3

The charge state distribution in [110] geometry ($Q \leq 45$) has been fitted using Monte Carlo simulation, with the electron impact ionization cross sections being the free parameters. As a result values for $\sigma(Q \to Q+1)$ are obtained (fig. 3). The main approximations were to consider the xenon ions as point charges, because even the M-orbitals are rather well localized (within $0.15\AA$) around the χe nucleus, and also as trial charges. We may think that highly charged ions significantly polarize the medium. However, this is a delayed effect which has a minor influence on the potential felt by the ion. We have not considered electron capture, not only because the charge 35 is far from equilibrium, to the lower side, and consequently these cross sections are very small in respect to the loss cross sections, but also because the valence electrons are almost free and mechanical capture is inhibited (see nest section). This last statement can be better understood when we analyse the distribution of paths of the ions along the channel. Particles coming out with low Q_{out} are expected to have experienced a low electron density in the center of the crystal as it is indeed confirmed by the simulation. In fact, the majority of the particles emerging with charges lower than ~ 40 were hyperchanneled. We have also neglected double ionization.

The experiments just described correspond to impact of 14.7keV electrons. A sharp decrease is observed between $\sigma(43 \to 44)$ and $\sigma(44 \to 45)$ (see fig. 3), that corresponds to a shell effect, because χe^{44+} has an empty M-shell. The results are compared to the predictions of the Lotz empirical formula and to a simple calculation using the Thomson formula, with a cut off for energy transfers below the binding energy. Although these models are very simple and do not include such as probable mechanisms as excitation-autoionization, our cross sections are 2 to 4 times larger than these predictions. There is the possibility that the electron flux is too high to induce multistep ionization. The analysis of this effect is under study. Recently, Moores and Pindzola[14] have presented a calculation of the electron impact ionization of highly charged ions in a level-to-level multiconfiguration Dirac-Fock approximation. Numerical results were obtained for several highly charged ions, but χe^{Q+} was not included in these calculations.

RECOMBINATION

The results that correspond to channeling of χe^{Q+} ($50 \leq Q \leq 53$) were not published yet; only a preliminary analysis was presented[15]. As an example, we show in figure 4 the outcoming charge state distribution in aligned and random orientations for incident χe^{52+}. The set up is essentially the same used for ionization measurements. The equilibrium charge state in random orientation is $Q \sim 50$.

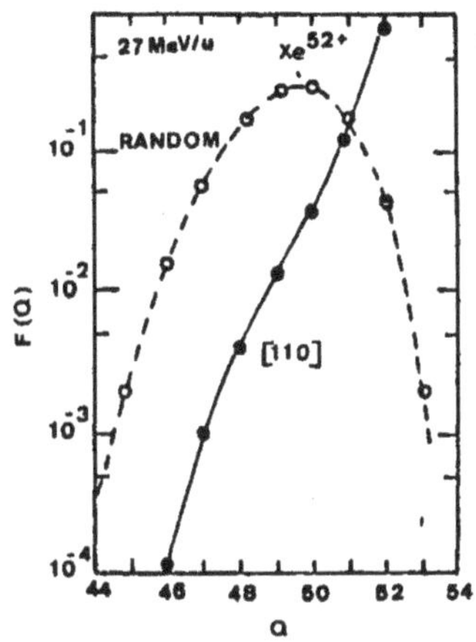

4

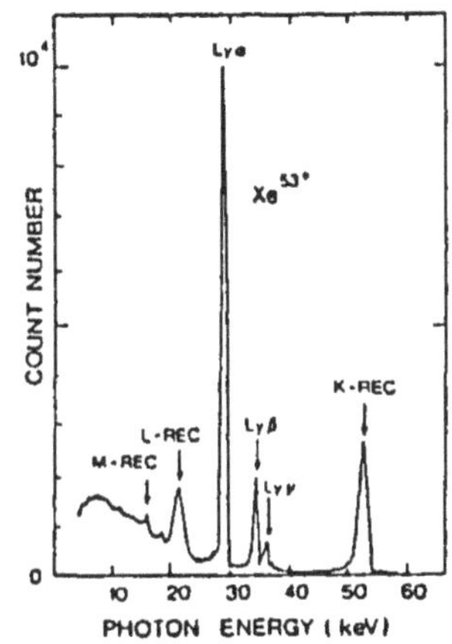

5

The figure 4 represents a typical result for $Q > 50$. We can observe that almost all ions stay "frozen" in their incident charge state. The reason is, first, that the removal of K and L electrons in a collision with a "free" target electron is, respectively, impossible and unlikely due to the maximum energy transfer in such collisions. The second, we have already advanced in the last section, i.e., the mechanical capture (usual non radiative capture) is suppressed for channeled trajectories that prevent close ion atom collisions (momentum and energy conservation).

While the charge distribution allows to get much information on other probable recombination processes, paralleling the ionization case (last section), the most complete study lies in the analysis of the photon energy spectra (direct and in coincidence with transmitted ions). An example is given in figure 5, where we show the spectrum obtained at 90° for χe^{53+} incident ions. The spectrum is composed of a continuum due to primary and secondary bremsstrahlung and of sharp peaks, which belong either to the Lyman series or to the radiative electron capture (REC). In this last process, a capture of a free electron by the ion is possible because one photon with energy equal to the kinetic electron energy as seen in the ion frame, plus the binding energy of the electron in the level where the capture happened, is emitted.

Monte Carlo simulation that fits the charge distribution curve and photon spectra could be done and cross sections extracted. This analysis of the data is just beginning. Finally, some theoretical work has just appear in the litterature trying to interpret new data of REC using channeled ions[16].

CONCLUSION

The main conclusion is that channeling offers the possibility for studying the interaction of multiply charged ions with a very intense electron "beam" in very precise conditions of ion state and relative velocity. The experiment, although relatively simple, has some important constraints. The first is the quality of the beam. In the experiments described in reference 12, the emittance was very small (spot size $\sim 3mm$, angular divergence $<< 1$ mrad). All target movements were driven by stepping motors (angular precision 10^{-3} degree, translation precision 0.1 mm). The crystal was a $17\mu m$ thick (111) silicon layer expitaxially grown on a highly doped Si substrate. After deposition, the central part of the substrate is dissolved over $\sim 1cm^2$.

Another important feature of the experiments just described is that they provide information on the impact parameter dependence of the stopping power. In fact, the choice of Q_{out} coresponds to a selection in the transverse energy distribution of the ions and on their accessible transverse space. Detailed information is thus obtained on "local stopping power", i.e., in a given point of the transverse [110] space, and on its relation with the corresponding local electronic density. For hyperchanneled ions the energy loss is very close to the energy loss corresponding to a homogeneous electron gas with density equal to the mean density of valence electrons. That is a very important conclusion of all the work.

* The french collaboration that prepared the experiments at GANIL, analyse the data and proposed the new method was composed by the physicists, engineers and technicians listed by alphabetic order, in reference 12.

The author acknowledges with thanks the warm hospitality extended by his colleagues of the collaboration and the finantial support ("Professeur Associé" at Lyon) provided by the Ministère des Universités of the French Government. This work was supported in part by Secretaria de Ciência e Tecnologia and CNPq, Brazilian Governmental Agencies.

REFERENCES

1. J. Schweppe, A. Belkacem, L. Blumenfeld, N. Claytor, B. Feinberg, H. Gould, V.E. Kostroun, L. Levy, S. Misawa, J.R. Mowat and M.H. Prior, Phys. Rev. Lett. **66**, 1434 (1991).
2. K.T. Cheng, W.R. Johnson and J. Sapirstein, Phys. Rev. Lett. **66**, 2960 (1991).
3. D.W. Mueller, Nuc. Instr. Methods **B42**, 467 (1989).
4. K. Dolder, M.F.A. Harrison and P.C. Thoneman, Proc. Roy. Soc. **A264**, 367 (1961).
5. D.F. Dance, M.F.A. Harrison and A.C.H. Smith, Proc. Roy. Soc. **A290**, 73 (1966).
6. A. Müller, G. Hoffman, K. Tinschert, E. Salzborn and G.H. Dunn, Phys. Rev. Lett. **61**, 70 (1988).
7. A. Müller, G. Hoffman, K. Tinschert and E. Salzborn, Phys. Rev. Lett. **61**, 1352 (1988).
8. A. Müller, K. Huber, K. Tinschert, R. Becker and E. Salzborn, J. Phys. **B18**, 2993 (1985).
9. P. Defrance, F. Brouillard, W. Claeys and G. van Wassenhove, J. Phys. **B14**, 103 (1981).
10. S. Datz, Nucl. Instr. Methods **B24**, 3 (1987).
11. R. Shuch, Nucl. Instr. Methods **24**, 11 (1987).
12. S. Andriamonje, R. Anne, N.V. de Castro Faria, M. Chevallier, C. Cohen, J. Dural, M.J. Gaillard, R. Genre, M. Hage-Ali, R. Kirsch, A. L'Hoir, B. Farizon-Mazuy, J. Mory, J. Moulin, J.C. Poizat, Y. Queré, J. Remillieux, D. Schmaus and M. Toulemonde, Phys. Rev. Lett. **63**, 1930 (1989).
13. D. Gemmel, Rev. Mod. Phys. **46**, 129 (1974).

14. D.L. Moores and M.S. Pindzola, Phys. Rev. **A42**, 5384 (1990).

15. (same authors of ref. 12), Rad. Effects and Defects in Solids, **110**, 71 (1989).

16. J.M. Pitarke, R.H. Ritchie and P.M. Echenique, Phys. Rev. **B43**, 62 (1991).

PHENOMENA IN LOW ENERGY ELECTRON SCATTERING BY ATOMS AND MOLECULES STUDIED BY MEANS OF A MULTI-ANGLE PARALLEL DETECTION ELECTRON SPECTROMETER

D. ROY and D. TREMBLAY
LPAM, Département de Physique, Pav. Vachon, Université Laval, Québec, Qc, G1K 7P4, Canada

and

D. DUBÉ
Département d'Informatique, Université du Québec à Hull, Qc, J8X 3X7, Canada

ABSTRACT

Observations of spectacular resonance phenomena are reported here, in measurements of differential scattering of low energy electrons. The instrument used is a multidetection electron spectrometer capable of parallel measurements giving extensive angular distributions of a specific scattering event. This is achieved by means of a truncated spherical mirror coupled to 19 discrete electron multipliers. The two examples discussed in the present work deal with the angular dependence of the decay of Feshbach resonances into inelastic channels. The first case involves the triply excited resonances $(2s^2 2p)^2P$ and $(2s2p^2)^2D$ in helium, decaying with unusual strength in the excitation cross sections of the $n=2$ (singly excited) bound states. In the second case, a possible anomaly is observed in the strength of the $^2\Sigma_g^+$ resonance of N_2 at approximately 11.5 eV. The first two vibrational peaks of this resonant state, in the $v=1$ ground-state exit channel, exhibit unexpected relative variations, mostly at large scattering angles. This phenomenon is interpreted as an interference effect between the resonant and the non-resonant amplitude.

1. Introduction

The presence of resonances in low energy electron scattering cross sections has turned out to be more the rule than the exception during the last two decades. Still now, with the increasing sensitivity of modern electron spectrometers, the investigation of some resonances through their decay channels in extended ranges of scattering angles can reveal unexpected features. This work presents two examples of Feshbach resonances exhibiting such characteristics. The first case involves the triply excited resonances $(2s^22p)^2P$ and $(2s2p^2)^2D$ around 58 eV in helium which appear with unusual strengths in the excitation cross sections of the low-lying $n=2$ bound states. The second case is in molecular nitrogen, where the decay of the first two vibrational levels of the $^2\Sigma_g{}^+$ state of $N_2{}^-$, into the $v=1$ ground state of N_2, exhibits unexpected relative variations, mostly at large scattering angles.

2. Experimental

The measurements discussed here were carried out with our electron scattering spectrometer called MAPDESS (multi-angle parallel detection electron scattering spectrometer). It allows simultaneous differential measurements at 19 scattering angles, separated by 8° intervals, from 18° to 162°.

Figure 1 presents a schematic view of the instrument.

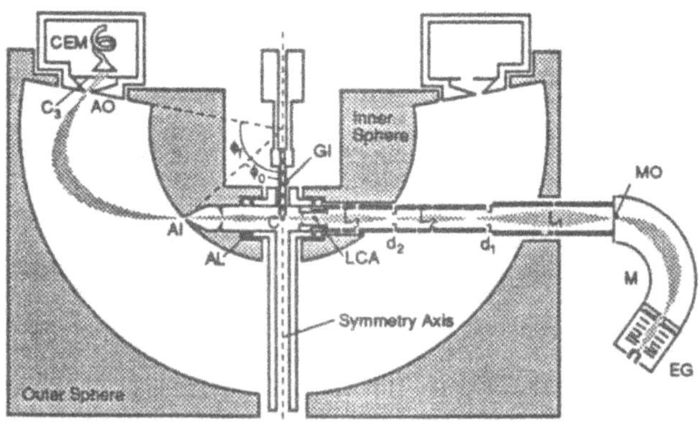

Fig. 1. Schematic view of the MAPDESS. Starting from the symmetry axis, the right part is in the plane of the incoming beam. The left part is symmetrical about the axis and represents one of the 19 scattering planes.

The unusual configuration of this spectrometer is determined by the special symmetry of the truncated spherical mirror analyzer surrounding the collision center. Starting from the symmetry axis, the right part is in the plane of the incoming electron beam. The left part is symmetrical about the axis and represents one of the 19 scattering planes. The various parts shown are the electron gun (EG), the monochromator (M), the monochromator exit slit (MO), the three electron lenses (L_1, L_2, L_3), the electron collimator (LCA), the collision center (C), the annular lens (AL), the analyzer input aperture (AI), the analyzer output aperture (AO), the exit aperture collimator (C_3), one of the 19 channel electron multipliers (CEM), and the gas inlet (GI).

The theoretical energy resolution of the 127° cylindrical deflector used as a monochromator is $(\Delta E/E_0)_m = 0.012$, where ΔE is the width at half maximum and E_0, the pass energy. For the analyzer, one has $(\Delta E/E_0)_a = 0.0028$; actually the measured value is 0.0022-0.0026, depending on the angle. In practice, an overall absolute energy resolution in the range 25-30 meV has been routinely obtained for the MAPDESS. The angular resolution was shown to be ±3°. More details are available elsewhere.[1,2]

3. Theoretical Background

When studying resonances, Fano and Cooper[3,4] are always referred to for the description of the appearance of this phenomenon in cross sections. Starting from the expression of the scattering amplitude for an incident electron of energy E, in the neighborhood of a resonance at E_r, the cross section can be expressed as[4,5]

$$\frac{\partial\sigma}{\partial\Omega} = \left|\frac{a}{E-E_r+i\Gamma/2} + be^{i\phi}\right|^2, \tag{1}$$

where a is a parameter independent of energy, Γ the resonance width; $be^{i\phi}$ is the complex amplitude of the non-resonant cross section, which varies only slowly with energy, but varies with the scattering angle θ. This equation can take the form of the so-called Fano-Cooper formula

$$\sigma(\varepsilon) = \sigma_a\left[(q+\varepsilon)^2/\left(1+\varepsilon^2\right)\right] + \sigma_b\,, \tag{2}$$

where ε is the reduced energy $2(E-E_r)/\Gamma$, $\sigma(\varepsilon)$ represents the differential cross section of Eq. (1), σ_a is the portion of the cross section affected by the resonance, σ_b is the nonresonant portion , and q is a line profile index which can be expressed in terms of the other parameters.[4] Off-resonance, the cross section is

$$\sigma_{or}=\sigma_a+\sigma_b. \tag{3}$$

According to the Fano-Cooper formula, the "resonant" cross section σ_a is modulated within a narrow energy interval by a special function possessing two extrema. This function has its minimum (zero) at $\varepsilon = -q$ and its maximum at $\varepsilon = 1/q$. The total excursion through the resonance is $1+q^2$. In absolute units, without the background, the total strength of the resonance σ_r is then given by

$$\sigma_r=\sigma_a(1+q^2), \tag{4}$$

while in relative units it is given by this quantity divided by the off-resonance value of the cross section σ_{or}. This effective strength of the resonance has also been represented[4] by the expression $\rho^2(1+q^2)$, where $\rho^2 q^2$ represents the fractional rise above the off-resonance value and ρ^2, the fractional trough. Actually ρ^2 has been defined as the ratio $\sigma_a/(\sigma_a+\sigma_b)$.

4. The Case of Helium

With the MAPDESS, we have studied these ^2P and ^2D resonances in the elastic and five inelastic channels at several angles θ in the range 18° to 162°. A report of this work was presented at XVI ICPEAC.[8] Previous measurements have been limited to smaller angular ranges in both elastic and inelastic channels.[4,6,7,9,10] Figure 2 shows the presence of both resonances in the inelastic channel $(1s2s)^3S$, and Fig. 3 in channel $(1s2p)^3P$. In each case, the plot shows 10 cross section curves taken at scattering angles between 18° and 162°, for incident electron energies 56.5 to 59 eV. For the normalization, the average of end points of each curve is fixed at one. Due to the very weak excitation cross section of the 2^3P state, no such measurement had been reported before for this channel. In these two channels it is surprising to what extent the cross section may be perturbed by the decay of these short lived states. Such perturbations are normally observed in the vicinity of the excitation threshold, but

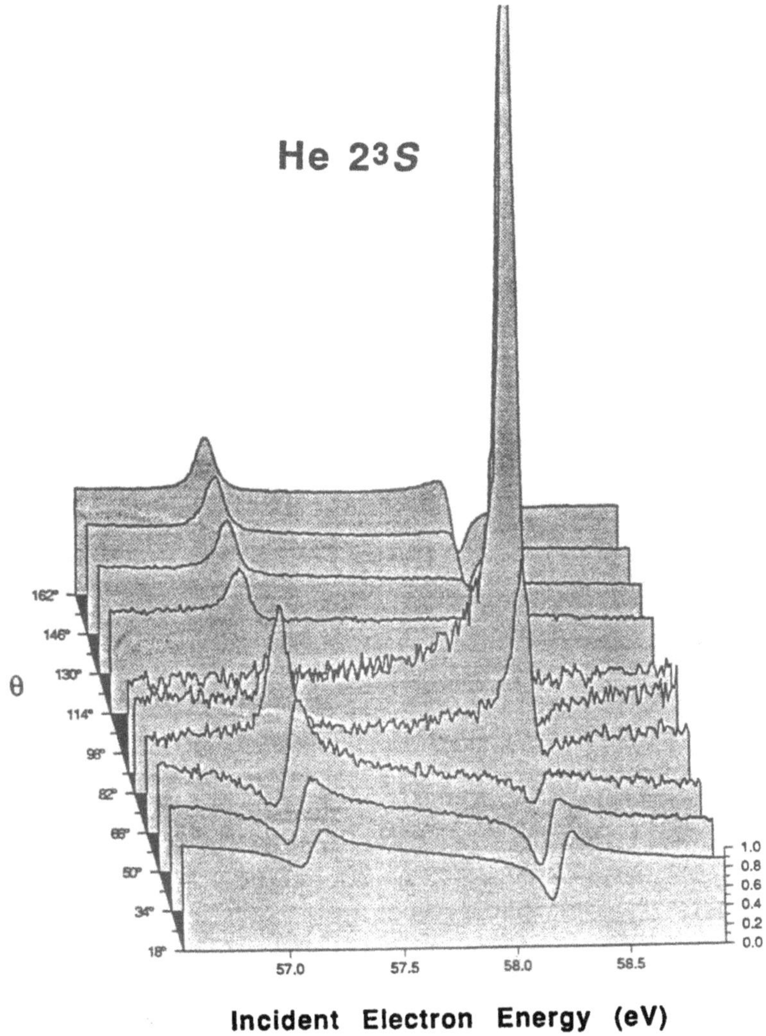

Fig. 2. The ^2P and ^2D resonances in the differential excitation cross section of the 2^3S state of He measured at different scattering angles θ. For the normalization of each curve, the average between the ends is fixed at 1.

here it occurs at almost 40 eV above the threshold.

For the six channels studied, the experimental curves have been fitted with Fano profiles, following Eq. 2, taking into account the broadening effects caused by the limited resolution of theinstrument and by the thermal velocity of the target gas. From

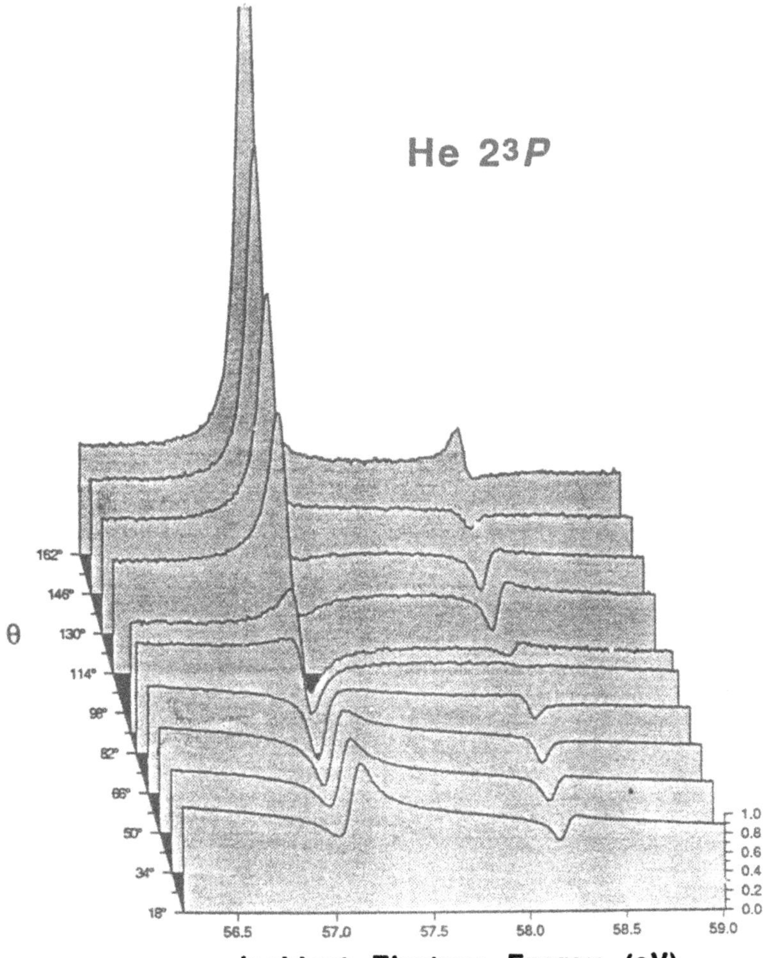

Fig. 3. The ^2P and ^2D resonances in the differential excitation cross section of the 2^3P state of He measured at different scattering angles θ. For the normalization of each curve, the average between the ends is fixed at 1.

the numerical analysis of the measurements involving a very high signal to noise ratio, the widths Γ of the resonances have been determined as 70 meV for the ^2P resonance and 40 meV for the ^2D resonance, with an estimated error of 25%. These values are in agreement with previous results from other works (e.g. Ref. 9),except for a recent determination[11] of 63±3 meV for the ^2D resonance.

Using these Γ values, the profile indices q have been obtained as a function of the scattering angle θ for all the channels studied: the elastic one and the inelastic ones 2^3S, 2^1S, 2^3P, 2^1P, and 3^3S. Figures 4 and 5 present the q values for the ^2P and ^2D resonances, respectively. Large variations of the profile indices are then observed for specific channels, showing special couplings and interferences. Since the data reported by Simpson *et al.*[4] for three of these channels were essentially limited to the range $\theta \leq 30°$, all that can be said is that they are roughly in agreement with the present data.

The next figures present the total strengths of the resonances, both in relative (σ_r/σ_{or}) and absolute units (σ_r). Figures 6 and 7 are for the ^2P resonance while Figs. 8 and 9 are for the ^2D resonance. The relative strengths were directly obtained from the measurements, taking the ratio of the total excursion of the resonance to the background. The conversion to absolute strengths was done by means of absolute experimental values of the involved cross sections available in the literature. Typically the accuracy on the absolute values is about ±25%. Again these results are roughly in agreement with the previous results[4], but clearly the extensive angular range gives a better overview of the phenomenon.

The variations of the relative strengths can easily reach a factor of ten in some channels, particularly in the triplet channels. In the case of the elastic channel, the strength is about 100 times smaller; but in absolute units (Figs. 7 and 9) the phenomenon appears to be as important as in the other channels. Thus for all of the channels the absolute variations of the strengths can reach two decades with the presence of deep minima at specific angles, particularly in the cases invoving S final states. Only one theoretical work has directly dealt with the decay channels of these resonances[12]. However, only the elastic and the inelastic 2^3S and 2^1S channels were considered, and the contributions of the ^2P and ^2D resonances were calculated for the integral (total) cross sections. According to that work, both resonances are present in the inelastic channels, but not in the elastic one (as thought at the beginning[4] but refuted later[10]). Clearly more theoretical investigation is needed to exploit all of the information contained in the results of the present work.

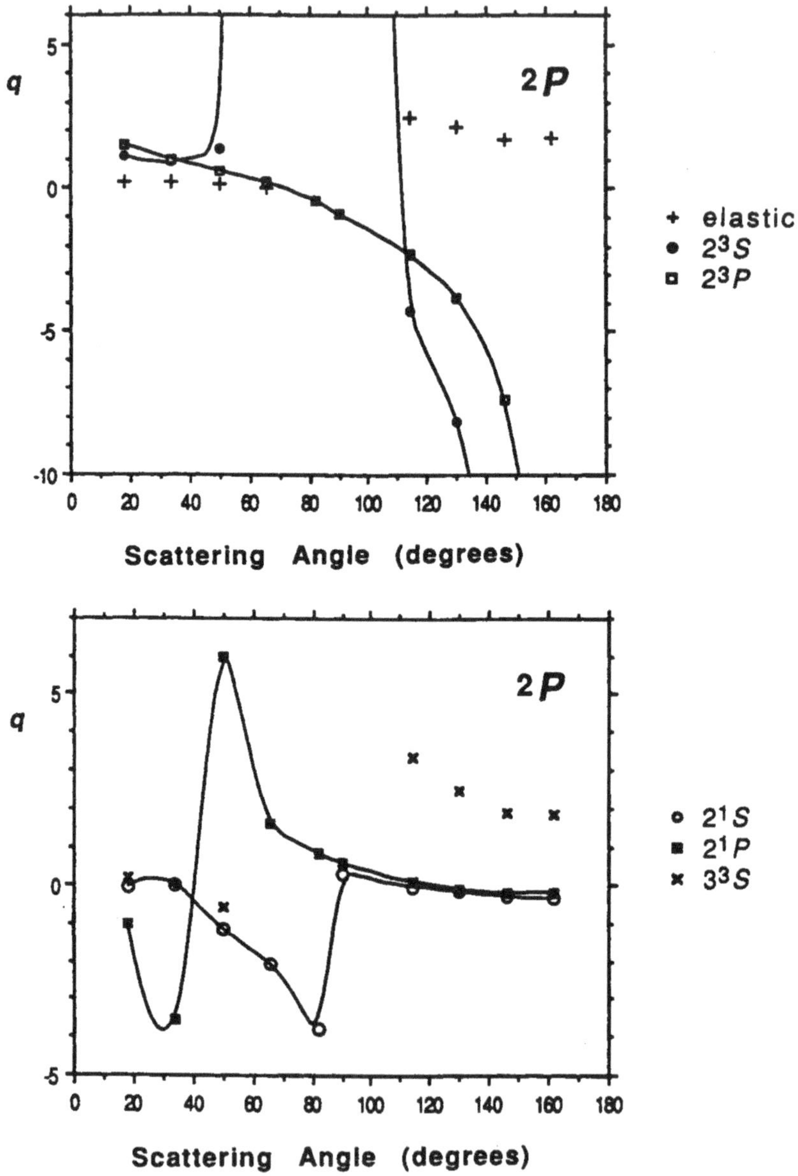

Fig. 4. Profile indices q of the ^2P resonance as a function of the scattering angle θ for the different decay channels.

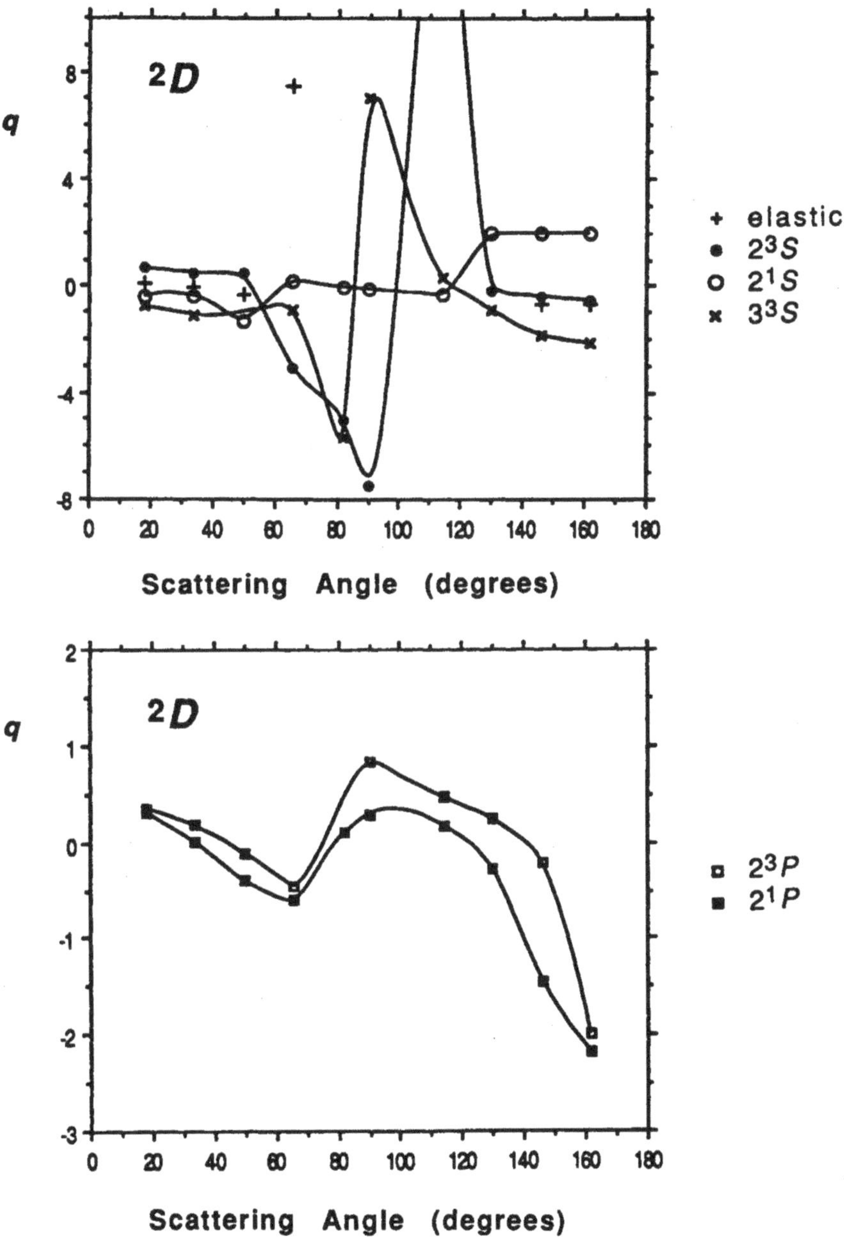

Fig. 5. Profile indices q of the ^2D resonance as a function of the scattering angle θ for the different decay channels.

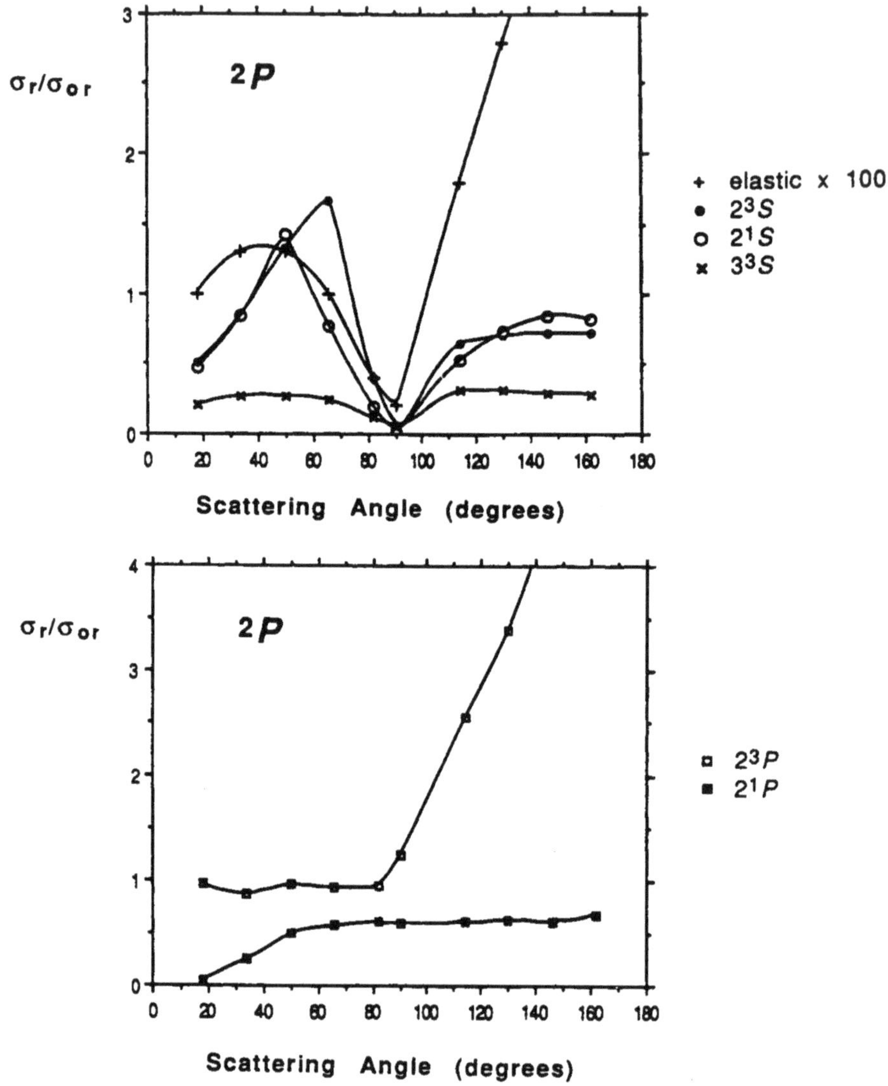

Fig. 6. Relative strength of the ${}^2P$ resonance, corresponding to σ_r/σ_{or}, as a function of the scattering angle θ, in the different decay channels. The strength in the elastic channel is multiplied by 100.

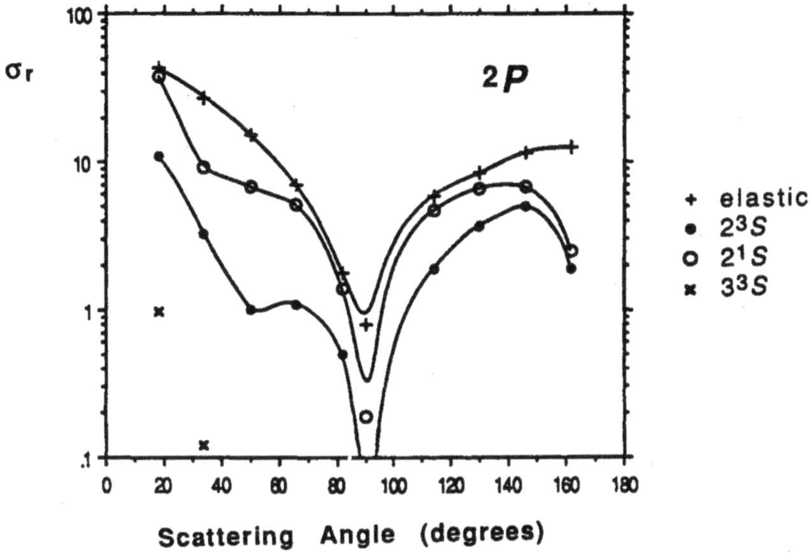

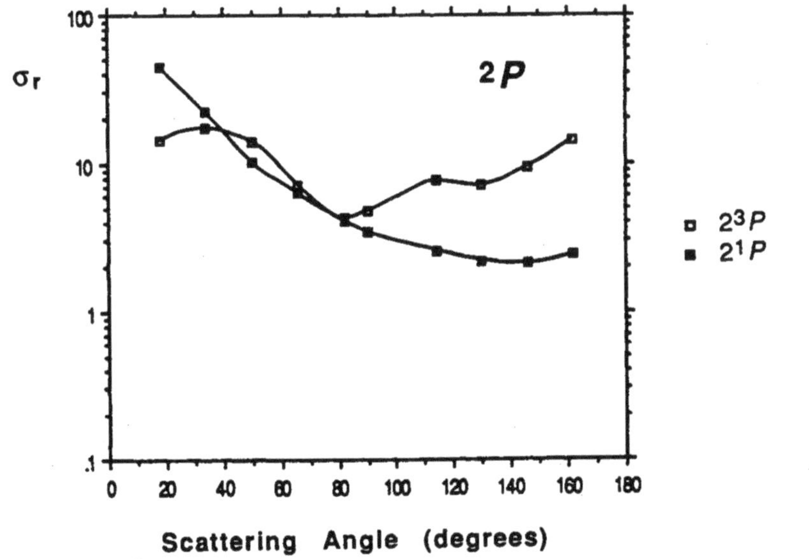

Fig. 7. Absolute strength of the ^2P resonance, corresponding to σ_r, as a function of the scattering angle θ, in the different decay channels.

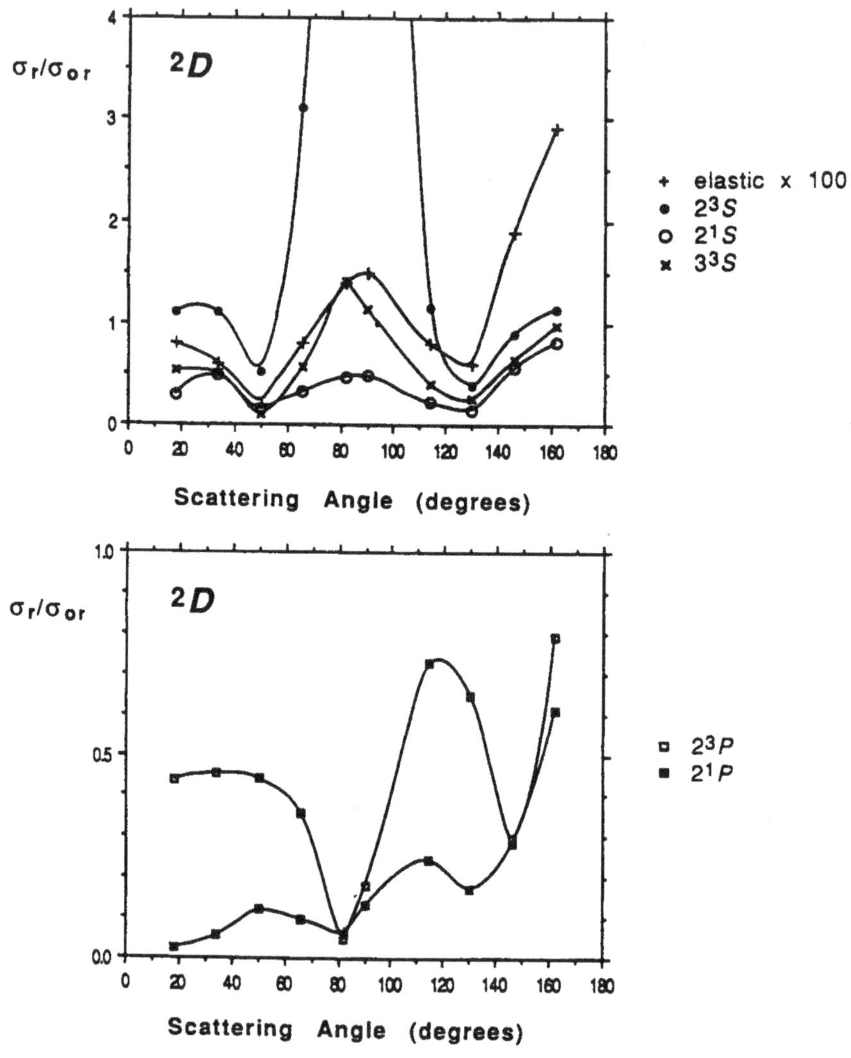

Fig. 8. Relative strength of the ^2D resonance, corresponding to σ_r/σ_{or}, as a function of the scattering angle θ, in the different decay channels. The strength in the elastic channel is multiplied by 100.

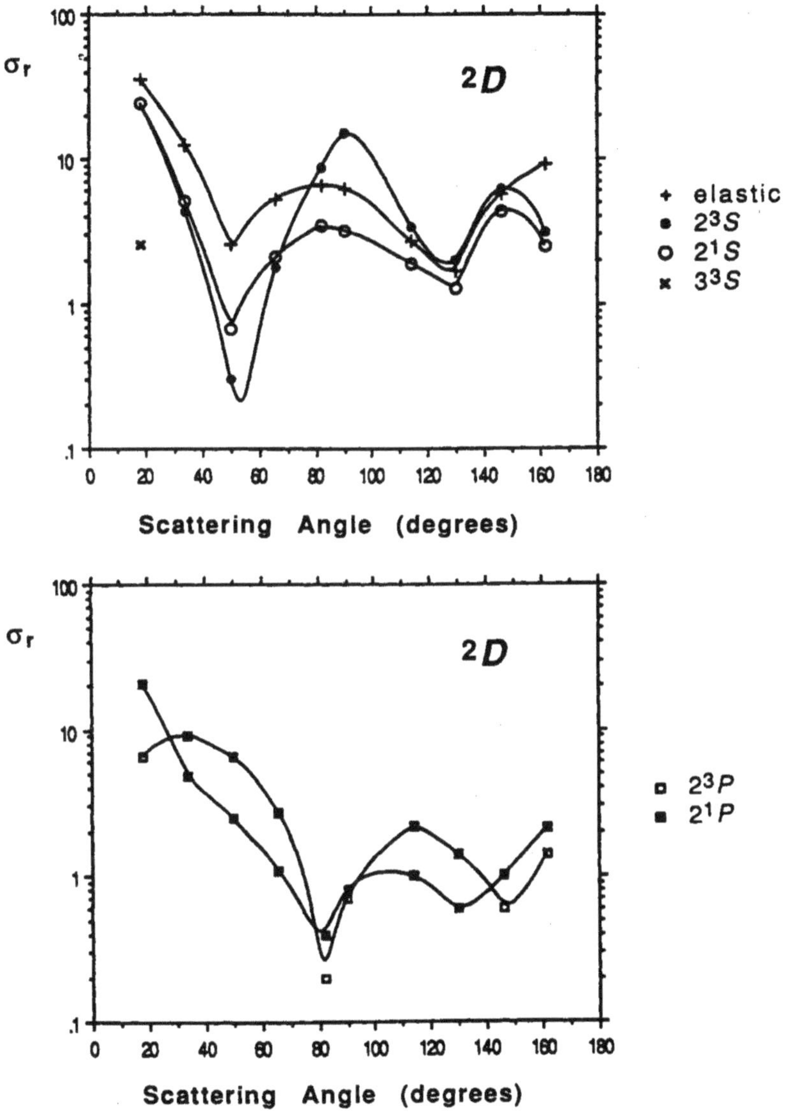

Fig. 9. Absolute strength of the ^2D resonance, corresponding to σ_r, as a function of the scattering angle θ, in the different decay channels.

5. The Case of Nitrogen

The first Feshbach resonance of N_2 appears at 11.48 eV; it is assigned to the state $^2\Sigma_g^+$ linked to the parent state $E\,^3\Sigma_g^+$. It was first observed by Heidemann *et al.*[13] and later studied by Comer and Read[5] who observed the decay of the first vibrational levels of the negative ion in the v=0, 1 and 2 levels of the N_2 ground state. Given the isotropic angular distribution (up to 85°) indicating an *s* wave in both the entrance and exit channels, the latter proposed the assignment and estimated its width Γ at 0.6 meV. According to Newman *et al.*[14] the energies of the v=0 and 1 vibrational levels of the resonant state are 11.499 and 11.770±0.010 eV.

As reported recently[15], we have reinvestigated this resonance by means of the MAPDESS, in an extended range of scattering angles, up to 162°. Figure 10 shows a sample of measurements in the v=1 ground state exit channel. In this

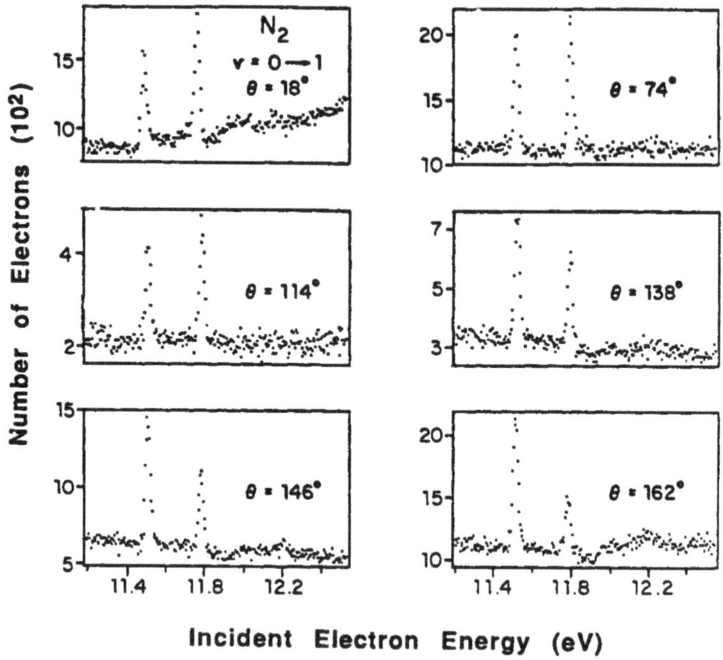

Fig. 10. Observation of the v=0 and v=1 vibrational levels of the $^2\Sigma_g^+$ resonant state of N_2^-, in the measurement of the differential cross section of the v=0→1 vibrational excitation of the ground state of N_2, at six different scattering angles.

inelastic channel where the non-resonant process is much smaller, it is possible to study the resonant process more easily, with fewer interference effects than in the case of the elastic ($v=0$) exit channel. The first two vibrational levels of the negative ion appear clearly at 11.50 eV and 11.77 eV on the scale of the incident electron energy; the energy width of the peaks is about 30 meV. At angles below 122°, the relative intensities of the two peaks are found to be nearly constant but at larger angles the relative intensities change quite dramatically, the first peak becoming increasingly larger than the second peak.

This change of relative intensities is surprising. It implies that the angular distributions are not symmetrical with respect to 90°, as should be the case[16] and, even more puzzling, that the angular distributions of the two peaks are not the same, even though they are thought to be relevant to the same electronic state, differing only on the individual vibrationnal level. What is the origin then of this unexpected behavior? As reported in our previous paper,[15] many tests were made in order to confirm the physical origin, the reproductibility and the reliability of the phenomenon.

The cross sections involved were determined absolutely through calibration from the experimental data for helium obtained by Register *et al.*[17] Figure 11 shows the resonant cross sections

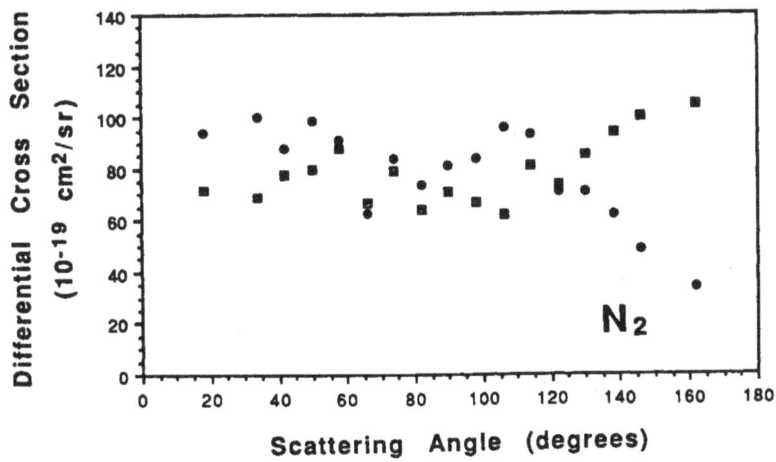

Fig. 11. Resonant cross sections, corresponding to the absolute strengths σ_r, involved in the features at 11.50 eV (squares) and 11.77 eV (full circles) of the N_2^- resonance, observed in the inelastic channel $v=0\rightarrow1$ of the ground state (*cf.* Fig. 10).

for both features corresponding to their total strengths σ_r in the inelastic channel $v=0\rightarrow1$ of the ground state. Since the dip in the profiles of these resonances is very weak when perceptible (see Fig. 10), σ_a is clearly very small and the total strength is essentially equivalent to $q^2\sigma_a$ (the maximum is then very close to E_r, *i.e.* to $\varepsilon=0$); given the attenuation effect of the instrumental energy resolution, one probably has $|q|\geq3$. Actually our determination[15] shows that the resonant cross section σ_r is at least 40 times larger than σ_{or}, the off-resonance cross section (which is of the order of 2×10^{-19} cm^2 sr^{-1}). This illustrates the fact that in this channel there is little amplitude with which the resonant amplitude can interfere.

Equation (1) can be rewritten

$$\frac{\partial\sigma}{\partial\Omega}=b^2+\frac{4(a^2-ab\Gamma\sin\phi)}{\Gamma^2(1+\varepsilon^2)}+\frac{4ab\varepsilon\cos\phi}{\Gamma(1+\varepsilon^2)}\quad, \tag{5}$$

where b^2 corresponds to the off-resonance cross section σ_{or}. The symmetric part of the resonance is given by the second term of the right-hand part of this equation, while the antisymmetric part is given by the third term and vanishes if the off-resonance amplitude (b) vanishes. In such a case, the cross section around the resonance would be given only by the second term, and be equal to $4a^2/\Gamma^2=\sigma_r$ at E_r. The fact that some interference is seen at 18°, even with a resonant cross section so much larger than the off-resonant cross section, implies that one has $|\cos\phi|\approx1$ for the two resonant peaks. The asymmetries which are on opposite sides also implies that one has $\phi\approx0$ for the first peak and $\phi\approx\pi$ for the second peak.

At large angles, for example 162°, no interference is clearly visible probably because the $|\cos\phi|$ term is less favorable. Then considering $|\cos\phi|\approx0$ with the same phase difference of π for the two peaks, one can easily demonstrate[15] that this can result in a large difference in intensity, due to the difference in the sign of $\sin\phi$ in the second term of Eq. 5, which is comparable to the difference observed in Fig. 10. It may also be shown that this difference of intensity is very sensitive to a small variation of b^2, the off-resonance cross section.

Thus it appears that our observations in N_2 can be seen as a consequence of special conditions of interference with the non-resonant scattering amplitude. However, given the narrow relation between these two resonant states belonging to the same electronic state, it remains to be explained why the value of the phase ϕ is not the same for the two peaks and why it may differ by as much as π. This is a problem which hopefully will interest theoreticians.

6. Conclusion

In this work, we tried to illustrate the fact that, in electron scattering spectroscopy as in many other fields, more efficient and more sensitive techniques and instruments can be helpful to reexamine unexplained phenomena and those formerly assumed as explained. This may lead to interesting questions and new challenges as in the case of the spectacular phenomenon of negative ion resonances in electron scattering which still involves many misunderstood aspects.

Acknowledgments

The authors are grateful to L. Dubé, A. Muscat and B. Tremblay for their collaboration. This work has been supported by NSERC-Canada and Fonds FCAR du Québec. D. R. is member of Centre de Recherches sur les Propriétés des Interfaces et la Catalyse (CERPIC) and the Network of Centers of Excellence on Molecular and Interfacial Dynamics (CEMAID/CEDMI).

References

1. D. Tremblay, D. Dubé, and D. Roy, *Rev. Sci. Instrum.* **60** (1989) 879.
2. D. Roy, D. Tremblay, B. Tremblay, and D. Dubé, *J. Electron Spectrosc.* **52** (1990) 787.
3. U. Fano and J. W. Cooper, *Phys. Rev.* **138** (1965) A400.
4. J. A. Simpson, M. G. Menendez, and S. R. Mielczarek, *Phys. Rev.* **150** (1966) 76.
5. J. Comer and F. H. Read, *J. Phys. B* **4** (1971) 1055.
6. D. Roy, *Phys. Rev. Lett.* **38** (1977) 1062.
7. D. Roy, A. Delâge, and J.-D. Carette, *J. Phys. B* **11** (1978) 4059.
8. D. Dubé, D. Tremblay, and D. Roy, in *Sixteenth International Conference on the Physics of Electronic and Atomic Collisions, (AIP Conference Proceedings 205)*, eds. A. Dalgarno, R.S. Freund, M. S. Lubell, and T. B. Lucatorto (American Institute of Physics, New York, 1989), p. 204.
9. J. A. Preston, M. A. Hender, and J. W. McConkey, *J. Phys. E* **6** (1973) 661.

10. P. J. Hicks, S. Cvejanovic, J. Comer, F. H. Read, and J. M. Sharp, *Vacuum* **24** (1974) 575.
11. R. Gosselin and P. Marmet, in *Sixteenth International Conference on the Physics of Electronic and Atomic Collisions, (AIP Conference Proceedings 205)*, eds. A. Dalgarno, R.S. Freund, M. S. Lubell, and T. B. Lucatorto (American Institute of Physics, New York, 1989), p. 206.
12. K. Smith, D. E. Golden, S. Ormonde, B. W. Torres, and A. R. Davies, *Phys. Rev. A* **8** (1973) 3001.
13. H. G. M. Heideman, C. E. Kuyatt, and G. E. Chamberlain, *J. Chem. Phys.* **44** (1966) 355.
14. D. S. Newman, M. Zubek, and G. C. King, *J. Phys. B* **16** (1983) 2247.
15. D. Tremblay and D. Roy, *J. Phys. B* **24** (1991) 1867.
16. E. S. Chang, *Phys. Rev. A* **16** (1977) 1841.
17. D. F. Register, S. Trajmar, and S. K. Srivastava, *Phys. Rev. A* **21** (1980) 1134.

THE MEASUREMENT OF IONIZATION CROSS SECTIONS: Ar AND CF_4 AS CASE STUDIES*

R. A. BONHAM§
Argonne National Laboratory, Argonne, IL 60439 U.S.A.

and

M. R. BRUCE and CE MA¶
Department of Chemistry, Indiana University, Bloomington, IN 47405 U.S.A.

ABSTRACT
Recent work carried out in the authors' laboratory on the measurement of absolute electron impact partial ionization cross sections is reviewed. Emphasis is placed on the techniques used. Efforts on Ar and CF_4 are used as illustrative examples of problems that arise in the measurement of cross sections for multicharged species production and dissociative ionization in cases where the fragment ions are produced with high translational energies.

1. Introduction

The measurement of absolute electron impact cross sections for the ionization of atoms and molecules has always been relevant to applications in astrophysics and radiation physics. Recently these cross sections have been in demand for the modeling of low-temperature or glow discharges of the type used in commercial plasma processing.[1-4] Despite the importance of cross section measurements and the fact that absolute cross sections have been measured since the early 1930s, surprisingly few accurate results are available.[5,6] Some of the best recent work can be found in papers by V. Schmidt and coworkers[7] and S. K. Srivastava and coworkers.[8] We will review work on Ar[9] and CF_4[10] carried out in our laboratory as illustrative examples of the techniques and problems involved in making absolute measurements. All major research groups[7-9] in this field currently employ pulsed electron beams with

* Work supported in part by the U.S. Department of Energy, Office of Energy Research, Office of Health and Environmental Research, under Contract W-31-109-Eng-38, and by NSF Grant Number PHY-8913096.

§ On leave from Department of Chemistry, Indiana University, Bloomington, IN 47405.

time-of-flight mass analysis. Interestingly, in the field of photodissociation, some of the leading research groups employ almost identical techniques.[11] Hence, the methods that will be discussed here can also be considered for use in the new Brazilian Synchrotron Light Source.

2. Experimental Techniques

Experimental Apparatus

The functions of the experimental apparatus fall into three areas. The controlled preparation of a source of ions is followed by extraction and transport of the ions to a detector, then detection of the ions and storage of the information.

Figure 1 shows a cutaway view of the apparatus currently in use in our laboratory. It has a very large entrance aperture (40 mm diameter) and a relatively short flight path (20 cm) to ensure maximum collection efficiency. Obviously, mass resolution had to be sacrificed to achieve this goal. The system operates by passing a 40 to 50-nsec-wide pulse of electrons (about 10^6 electrons per pulse) between two 91% transmission gold grids separated by 12 mm. The grids are shorted to ground just before the electron pulse arrives and are released 5-10 nsec after the last pulse electron has cleared the region. The DC potential on the extraction grids is maintained somewhere between -30 and -100 V and can be shorted to ground in less than 4 nsec.

After the ions clear the extraction tube lens of length 2.5 cm, they are accelerated over a distance of 6 mm into a constant-field time-of-flight (TOF) tube of length 15 cm held at a DC potential set at a value between -300 and -1,000 volts. The exact settings of the extraction and TOF tube potentials are determined by SIMION[12] ion trajectory calculations to guarantee the best transmission of ions to the detector depending on whether the ions are initially translationally hot. Figure 2 shows trajectories for electrons with identical initial translational energies of 5 eV but with different initial directions. After they leave the TOF cell, the ions travel 6 mm to a fifth gold grid and then another 6 mm to the front face of a pair of microchannel plates

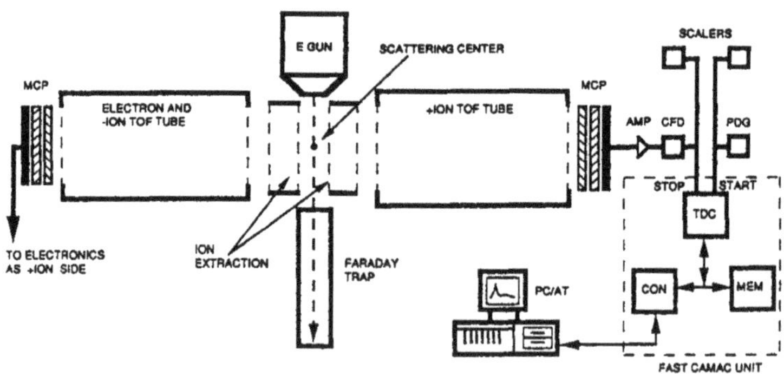

Fig. 1. Cutaway view of the time-of-flight apparatus. Abbreviations not explained in text are: E gun, electron gun; AMP, fast amplifier; CFD, constant-fraction timing discriminator; PDG, programmable delay gate generator; CON, CAMAC crate controller; MEM, homemade fast memory for quick storage of the TDC's output data.

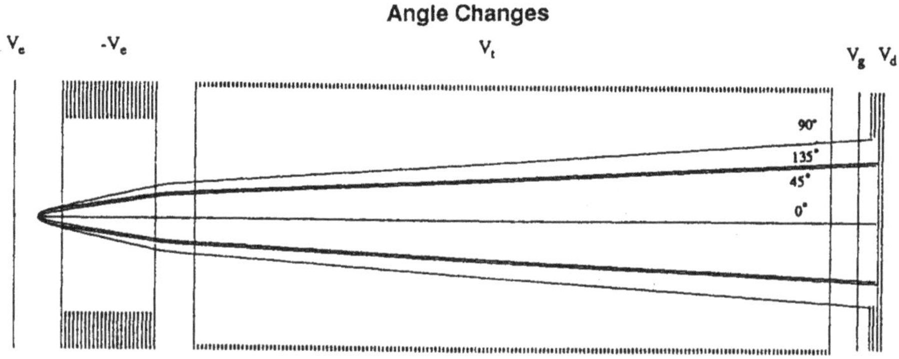

Fig. 2 Trajectories of ions with initial translational energy of 5 eV with starting directions indicated in figure for -100 V extraction ($-V_e$) and -1000 V TOF field ($V_t = V_g$). Trajectories with real grids will have somewhat less dispersion due to punch through fields.

(MCPs) held at a negative potential of 2.5-4 kV.

The fifth grid measures the collection efficiency of the MCPs as a function of the voltage applied to them. Applying a negative potential greater than that on the front face of the detector causes secondary electrons produced on the front face to return to it and increases the detection efficiency as shown in Fig. 3. Although a high absolute detector efficiency is thus attained, we normally choose not to take advantage of this feature to assure that operating conditions are stable. We have also achieved suitable results with the last three grids removed (those nearest the MCP). The MCPs are an ideal detector for pulse counting because flight timing resolutions of better than 250 psec can be routinely obtained. The MCPs can also be used as a current amplifier.[13] A schematic view of the homemade MCP detector mounts is shown in Fig. 4. The MCP detector is followed by commercial amplifiers and constant-fraction discriminators as described in detail in Ref. 9. The time-to-digital convertors (TDC) are also commercially available, but a specially designed bin with dedicated fast memory was constructed so that experimental repetition rates of 100 kHz could be obtained. Directions for obtaining the wiring diagrams for this circuitry are contained in Ref. 9. Figure 5 presents a schematic view of the data collection system.

Cross Section Measurements

Two approaches for making absolute cross section measurements are commonly used. In the first, a beam-beam experiment is performed, and the ratios of currents of all detected ionic species are measured relative to the ion current of the most intense mass peak as a function of the electron impact energy. The data are made absolute by recording the ion current of a standard gas such as argon under identical experimental conditions at an impact energy below the threshold for production of all ions except the parent ion (i.e., Ar^+ in the example cited) and then finding the normalizing factor that matches this result to the absolute measurements of Rapp and Englander-Golden[14] at the same impact energy. One, of course, must be certain that all ions are extracted from the ionizing region and that they are all

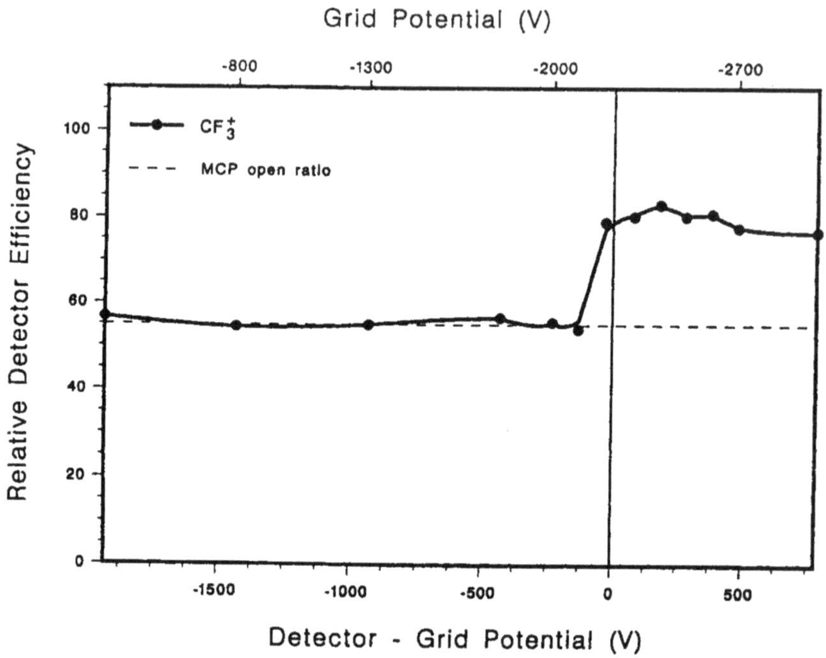

Fig. 3. CF_3^+ ion count rate as a function of the voltage applied to the grid nearest the MCP front surface. Points to the left of the solid vertical line are in a field that draws electrons away from the front face of the detector, and points to the right are in a field that attracts electrons back to the plate.

transported and detected with the same efficiency, or the efficiencies of all ions must be measured. Further details of this approach are given in Ref. 8.

The second method, employed in our laboratory, consists of the use of a beam-beam experiment as in the first method, to establish the ratios of all ion currents to the current observed for the most abundant ion. This task must be carried out separately for each ion energy as in the first method. The second step, also employing the beam-beam geometry, measures the ion current of the most abundant species at a preselected standard energy, usually around 80 eV, then at a particular energy, E, and finally at the standard energy again. If the results for the two standard energy measurements are the same, then experimental conditions are assumed to have remained constant for the duration of the three measurements, and the measurement at the energy E is added to the final data set. This procedure is repeated for every energy point needed with identical or measured data collection times. The shape of the most intense mass peak in the spectrum is thus established as a function of energy. Finally, the previously collected ratios are used to obtain the relative partial ionization cross sections for all the observed ion masses. Up to this point, the approach being described is essentially identical to the previously described method.

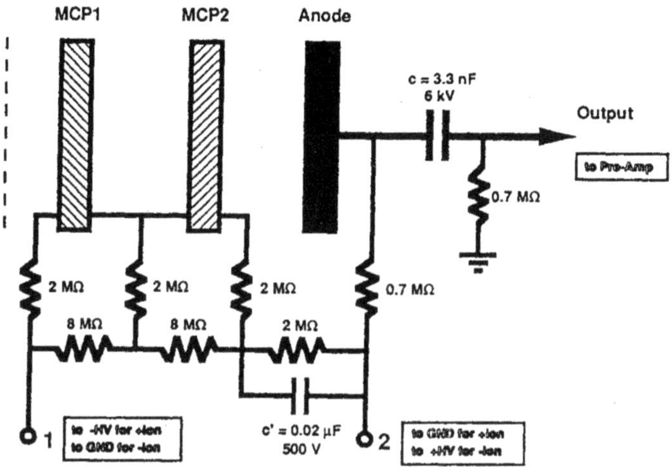

Fig. 4. Diagram of the MCP detector assembly with all circuit components specified.

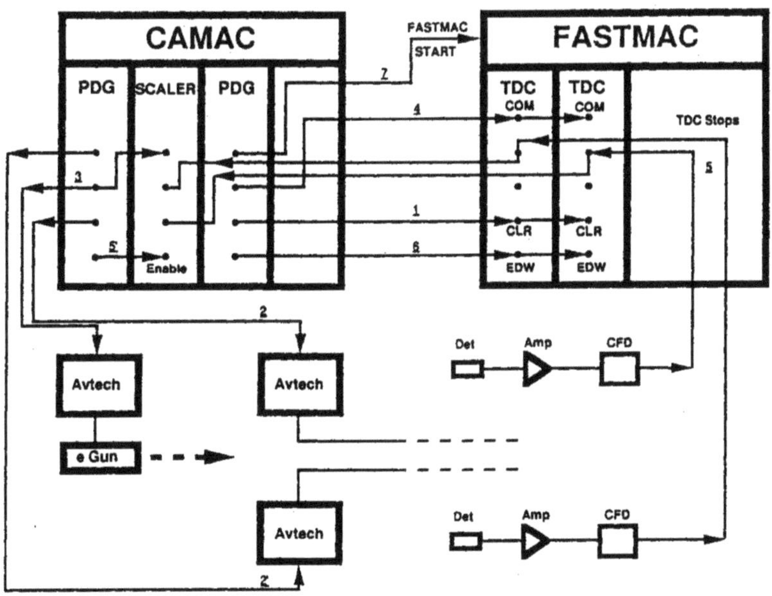

Fig. 5. Schematic view of the signal processing electronics. All symbols are defined in the legend to Fig. 1 except Avtech, which is the manufacturer's name for our commercial pulsers.

The major difference occurs in the normalization to an absolute scale. To establish the absolute scale, the mode of experimental data collection is changed as described in the next paragraph.

To set the absolute scale, the entire apparatus is filled with the gas to be studied to a pressure of 1-12 μ Torr. An electron pulse is passed through the apparatus as in the beam-beam experiment, but now a column of ions of length ℓ is extracted from the ionization region and transported to the detector. A small amount of attenuation to the ion beam occurrs during transport. If experimental conditions are such that the Beer-Lambert law holds, we may write an expression for the observed ion current for any particular ion as

$$I^{(j)} = I_F n \ell \varepsilon \sigma_{ion}^{(j)} \exp\{-n[<L\sigma_{im}> - L_o \sigma_{tot}(E)]\}, \quad (1)$$

where $I^{(j)}$ is the observed ion current for a particular ion, j, and $\sigma_{ion}^{(j)}$ is the corresponding absolute partial ionization cross section to be determined. The electron beam current measured in the Faraday trap simultaneously with the ion current measurement is denoted by I_F, the number density of the gas is n, the length of the extracted ion column is given by ℓ, the ion transmission and detection efficiencies are given by ε, the distance from the end of the ion column nearest the electron gun to the entrance to the Faraday trap is denoted by L_o, and $\sigma_{tot}(E)$ is the total electron impact cross section for the target gas at the electron impact energy E. The term involving this last quantity arises because the electron beam current incident on the front of the ion column is given by $I_F \exp\{+ nL_o\sigma_{tot}(E)\}$. The final term, $<L\sigma_{im}>$, accounts for the attenuation of the ion beam by the background gas. Since the ions travel at different velocities through the gas in the preextraction, extraction, and TOF regions, this final term can be written approximately as

$$<L\sigma_{im}> = L_1\, \sigma_{im}(E_1) + L_2\, \sigma_{im}(E_2) + L_3\, \sigma_{im}(E_3), \quad (2)$$

where 1, 2, and 3 refer to the preextraction, extraction, and drift regions, respectively.

For 40 to 50-nsec pulses with extraction taking place within 5-10 nsec of the end of the electron pulse, the term in the preextraction region can usually be neglected because the distance traveled by the ion, L_1, is on the order of a few microns unless the ion is extremely hot translationally. In the energy range used in the extraction and TOF regions (-80 and -900 volts, respectively), typical cross section values for ion-molecule collisions range from 50 to 100 Å^2 ($\times 10^{-16}$ cm^2),[15] which leads to an ion beam attenuation of less than 10% at pressures of 10 μ Torr and below. The testing of the assumption of the validity of the Beer-Lambert Law and the extraction of the desired cross section can be accomplished by plotting $\ln\{I^{(j)}/(I_F n)\}$ vs. n. The intercept of such a plot yields $\ln\{\ell\varepsilon\sigma_{ion}^{(j)}\}$, from which the cross section can be calculated if the transmission and detector efficiencies and the length of the extracted ion column are known. To the extent that the aforementioned plot is linear, the validity of the experiment can be claimed. A typical result is shown in Fig. 6. In addition, the slope of the plot yields a rough estimate of $<L\sigma_{im}>$, whose accuracy is limited by the low attenuation and the collection of forward ion scattering into the detector. In practice, the term $L_o\sigma_{tot}(E)$ is quite small compared to $<L\sigma_{im}>$ and can usually be neglected. Estimates of $\sigma_{im}(E)$ appear to be correct within an order of magnitude.[16]

In principle, the length of the ion column can be measured by taking two or more measurements, with each using an extraction grid of a different diameter, of a partial ionization cross section at a single electron impact energy. If we label the

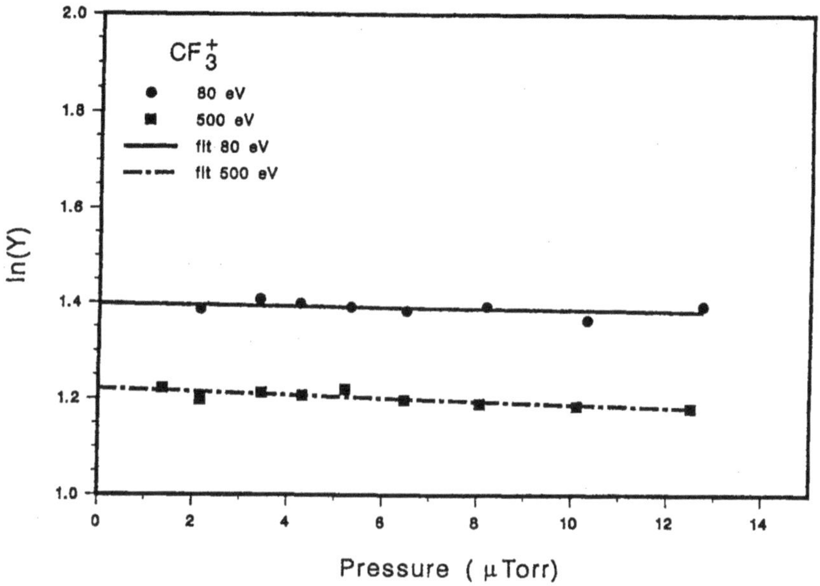

Fig. 6. A plot of $\ell n(I^{(j)}/I_{Fn}) = \ell nY$ vs. pressure, P, for 80-eV and 500-eV electron impact.

intercept value obtained from an experiment using an aperture opening of ℓ_1 as A and that for an aperture opening ℓ_2 as B and we assume that end effects in both cases (i.e., deviations from the geometric distances due to nonlinear extraction field effects at the ends of the ion column) are identical, we can calculate the corrections to ℓ_1 and ℓ_2, given by Δ, as

$$\Delta = [\ell_1 - \ell_2\, e^{(A-B)}]/[e^{(A-B)} - 1]. \tag{3}$$

In most cases so far we have relied on SIMION trajectory calculations to compute the correction Δ. A curve of the calculated value of ℓ for various ion initial energies is given in Fig. 7, which shows that for thermal translational energies, the value of Δ is close to zero and is less than 10% of the pathlength for translational energies below 5 eV. Two other corrections to the pathlength are worth mentioning. First, the electron beam is attenuated as it passes through the gas, so that less ionization may take place at the end of the ion column than at the beginning. In addition, the scattered electrons may produce ions outside the width of the ion column, some of which will also be collected. We can get an order-of-magnitude estimate of this effect by integrating over the electrons scattered out of the beam to obtain the approximate multiplicative correction factor $\{1 - \exp[-n\ell\sigma_{tot}(E_o)]\}$, which is less than 0.1% for a total electron impact cross section of 10 Å^2 if the number density of the gas does not exceed 3×10^{11} molecules/cm^3 (a pressure of about 10 μ Torr at room temperature). The second correction stems from the fact that the electrons in the incident beam have velocity components perpendicular to the beam propagation direction either because a collimating magnetic field is used or because of the momentum uncertainty in the electron beam itself. The perpendicular velocity

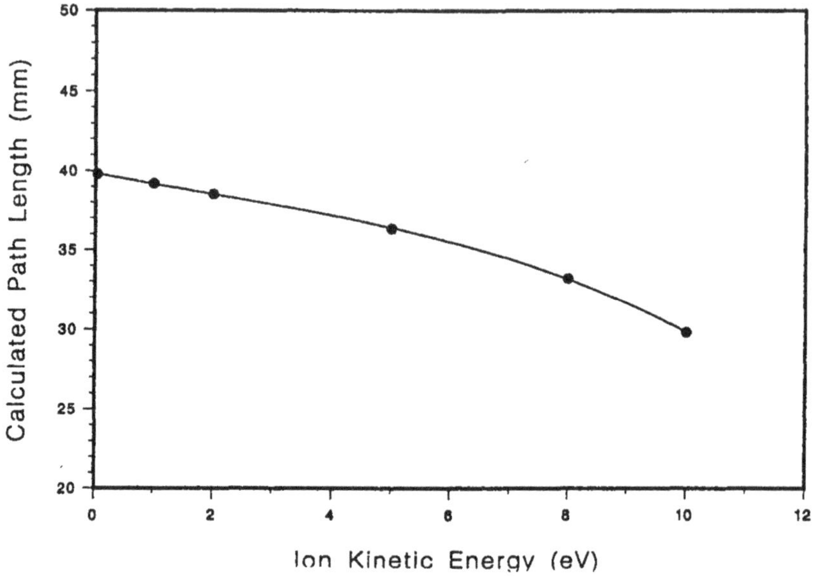

Fig. 7. SIMION calculations of the length, ℓ, of the extracted ion column as a function of the initial ion translational energy.

component of the electron beam, $v_\perp$, causes the electrons' flight path through the gas to be longer and hence increases the number of ions. The multiplicative correction factor for this effect can be written as $\{1 - (v_\perp/v)^2\}^{-1/2}$, where v is the electron velocity. In our experiment, the residual magnetic field has been reduced below 2 m Gauss so that no major correction results. Note, however, that if the energy spread in the incident beam is large (i.e., 0.5 eV) and the beam energy is small (i.e., 5 eV) serious corrections can occur.

The remaining factor, ε, is the product of the grid transmission efficiencies and the detector efficiency. The optical transmission of the five grids taken together was measured by using visible light, which yielded a value of 91.7% transmission per grid. SIMION simulations indicated that deviations from this value due to local field effects should be less than 5%. This was recently confirmed in our laboratory in experiments in which only the first two grids were installed, enabling the direct measurement of the actual grid transmission. The detector efficiency has been assumed from the work of others.[9,16] Measured absolute detector efficiencies for vertical ion impact on the front surface of an MCP detector show that the ion efficiency is simply the ratio of the open hole area to the active surface area of the plate, if ion impact energies are above 2 keV and count rates are less than 100 kHz/cm^2. Typical ratio values lie between 0.55 and 0.60.[17,18]

3. Experimental Results

In the case of Ar, the results of our measurements are in excellent agreement with those of Srivastava's group,[19] but they disagree with results reported by Märk's group[20] and derived by the fast neutral beam method.[21] If we probe more deeply into the nature of the disagreement, we find that the average level of agreement at electron impact energies of 50, 100, 150, and 200 eV is 4% for the partial ionization cross section of Ar^+ and 15% for Ar^{2+}. Since the claimed experimental accuracies are only about 15%, the results would appear to be in excellent agreement. The difficulty is that the ratio of the Ar^{2+} to Ar^+ cross section shows an average agreement of only 17%, when all experiments claim that their ratios are measured to an accuracy of 3-5%. We summarize these results in Table 1. While at present we do not completely understand the nature of this problem, we do know that it seems to apply to other cases where multiply charged ions are involved.

Table 1. Summary of results on the partial cross section measurements for Ar^+ and Ar^{2+} and the ratio Ar^{2+}/Ar^+.

Electron Energy (eV)	Reference	Ar^+ (10^{-16} cm^2)	Ar^{2+} (10^{-16} cm^2)	Ar^{2+}/Ar^+
50	20	2.49	0.016	0.0064
	21	2.62	0.018	0.0069
	19	2.497	0.0124	0.00497
	9	2.48	0.012	0.0048
100	20	2.46	0.168	0.0683
	21	2.68	0.177	0.0660
	19	2.522	0.228	0.0904
	9	2.47	0.211	0.0870
150	20	2.35	0.167	0.0711
	21	2.54	0.171	0.0673
	19	2.319	0.224	0.0966
	9	2.32	0.211	0.0909
200	21	2.32	0.153	0.0659
	19	2.09	0.187	0.0895
	9	2.10	0.177	0.0843

The second major problem area, about which we now have some understanding, is what happens when ions dissociate as the result of electron impact excitation. The molecule CF_4, which is an important etching gas, is a case in point. This molecule appears to have no stable excited states accessible to electron impact excitation. The evidence for this is the absence of a parent ion, CF_4^+, in the mass spectrum and the complete absence of any, as yet unassignable, fluorescence from the CF_4 molecule under electron bombardment. Hence, for this molecule we believe that the total dissociation and total inelastic cross sections are one and the same. It is also interesting that the lowest observed appearance potentials for fluorescence[22-26] and production of metastable F atoms[27] all lie above 29-30 eV. This observation implies that the molecular fragments produced by electron bombardment below about 30 eV must be mainly in their ground electronic or low-lying nonfluorescing states.

When we first measured the partial ionization cross sections for CF_4, we discovered good agreement with the earlier work of Märk's group[20] for the main ion fragment, CF_3^+, but progressively poorer agreement for the lighter fragments, with our values for F^+ lying higher by nearly a factor of six. Figure 8 reveals the reason for this problem. We carried out an experiment in which the extraction grid was kept at ground potential for the duration of the experiment so that only those ions originally ejected in a direction 90 ± 30° to the incident electron beam direction could be detected. In addition, the incident electron beam energy was set at 20 eV so that CF_3^+ was the only ion being produced. From the time-of-flight curve we were able to estimate the distribution of initial translational energy of CF_3^+ by correcting the TOF distribution for the differing pathlengths to the detector caused by the wide opening angle of the TOF tube. The results show that the average translational kinetic energy of CF_3^+ is between 1.5 and 2 eV. Because of conservation of energy and linear momentum, it can be argued that the lighter fragments can be translationally even hotter.[28] Since this observation was made, Märk and coworkers have used SIMION to model their Nier ion source and have discovered that the fraction of total ions produced that can be drawn from the source is a strong function of the initial kinetic

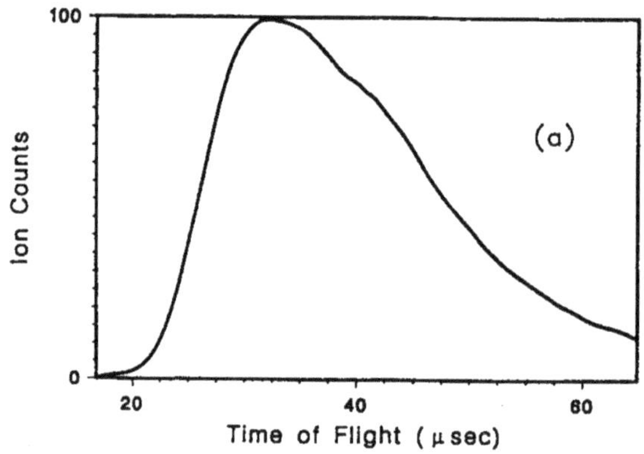

Fig. 8a. CF_3^+ counts as a function of flight time with zero extraction field and 800-V TOF field for 20-eV electron impact.

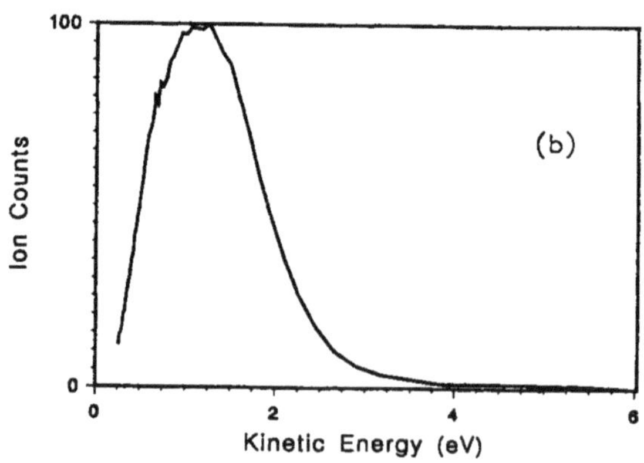

Fig. 8b. Figure 8a corrected for path flight differences and converted to a plot of the CF_3^+ count rate vs. CF_3^+ initial translational energy.

energy of the ions. They have estimated the initial kinetic energies of all their CF_4 fragment ions and have corrected their original measurements so that they are now in reasonable agreements with ours. However, we still observe serious and, as yet, unexplained discrepancies in the case of the multicharged ions CF_3^{2+} and CF_2^{2+}. As shown by Märk's laboratory, the extraction efficiency of a Nier type ion source varies from about 40% for ions with thermal translational energy to 4% for ions with 10 eV of translational energy. It seems reasonable to conclude that a Nier type source is unsuitable for quantitative work with molecules where translationally hot fragments are the norm. We believe that the current available evidence supports the use of open type ion sources with large extraction apertures such as the types employed in Refs. 8, 9, and 11 for quantitative work involving the possible existence of translationally hot molecules.

Recently, we have extended our experimental capability to the measurement of double ion formation and have measured the partial ionization cross sections for the following processes:

$$CF_4 + e \rightarrow CF_3^+ + F^+ + 3e$$

$$CF_4 + e \rightarrow CF_2^+ + F^+ + F + 3e$$

$$CF_4 + e \rightarrow CF^+ + F^+ + (F_2 \text{ or } 2F) + 3e$$

$$CF_4 + e \rightarrow C^+ + F^+ + (F + F_2 \text{ or } 3F) + 3e \qquad (4)$$

The limit of detection of these experiments was about 10^{-20} cm^2. Within this limit, no ionizing processes other than the four listed above were observed up to impact energies of 500 eV. Experiments of this type are quite easy to carry out.[29] In their simplest form, one can simply split the output from the positive ion detector into three parts and use one to record a normal mass spectrum for scale calibration purposes and use the second and third to stop and start a second TDC. The spectrum in the second TDC is then an autocorrelation spectrum that gives peaks at the differences between arrival times of all the fragments. In general, if the probability for forming an ion in each experiment is less than one, only the true coincidence peaks will have appreciable intensities. A slightly more refined version involves gating the start of the second TDC by use of an external veto pulse so that only the arrival of an ion with a preselected mass can start the experiment. An example of this procedure is shown in Fig. 9, where, because of the preset time window on the start of the TDC, only F^+ ions are allowed to start the experiment.

The time scale in Fig. 9 is a measure of the difference in arrival times of F^+ and the ion peak in question. Since C^+ is lighter than F^+, it gets to the detector before the start signal, and its time difference is negative, which is an interesting capability of the LeCroy model 4208 TDCs (dynamic range of -16 msec to +16 msec with 1-nsec resolution). Still more sophisticated experiments can be carried out if the arrival time of the start ion is measured relative to the electron pulse and plotted as a function of the stop ion flight times.[11] From a two-dimensional plot like the one in Fig. 9, information about the separation energies of the fragments and the number of fragments involved in the dissociation can be inferred.

If all the multiionization processes can be individually measured, one can define both a gross and a counting total ionization cross section. If $\sigma_{ion}^{(j)}$ is the partial ionization cross section for a particular process and if n_j is the number of charges formed in the ionization process, then the gross cross section is defined as

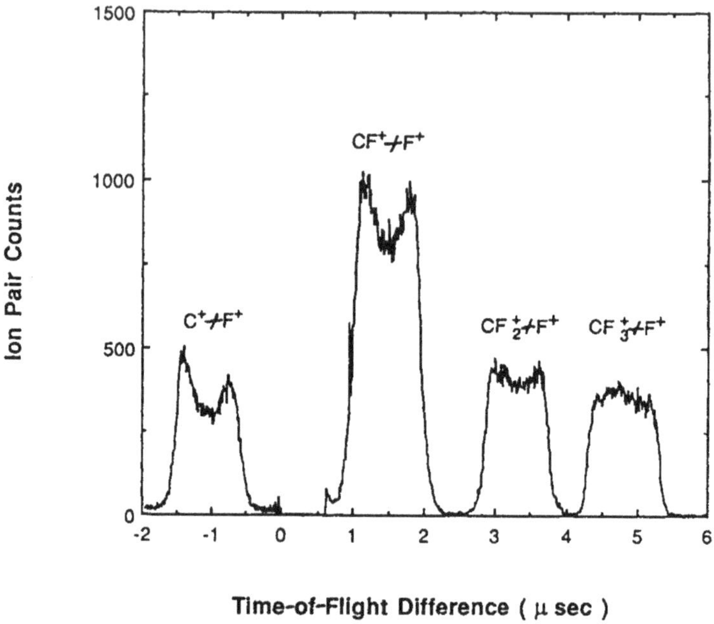

Fig. 9. Time-of-flight autocorrelation spectrum for coincidences between F^+ ions and other fragment ions following the electron impact dissociative ionization of CF_4 at 150 eV. The axes indicate the TOF difference with respect to F^+. Four coincidences are observed. (The F^+ selfcoincidence peak has been deleted.)

$\sum_j n_j \sigma_{ion}^{(j)}$ while the counting cross section is defined as $\sum_j \sigma_{ion}^{(j)}$, where the sums are carried out over all possible processes. Note that noncoincidence experiments, which count individual ions, count multiion formation in the gross cross section sense, while multicharged ions are counted in the counting cross section sense, and their cross sections must be multiplied by their charge before they are added to the gross cross section sum. To obtain the counting cross section, the multiion cross sections must be removed from the partial single ion cross sections, where they have been counted at least twice (note that F^+ has been overcounted by the sum of all four reactions shown above), and the multicharged single ion cross sections must be added in without the charge weighting. Recalling that the counting cross section got its name from the use of particle detectors that recorded the number of ionizing events as opposed to the number of ions formed will help to prevent confusion between these two cross section types. In other words, the counting cross section simply counts the total number of ion-producing collisions without regard to the charge state or the number of ions formed in the event. Counting the number of individual ions formed in the case of molecules does not in general lead to either the counting or gross cross section. Measuring the ion current by means of a Faraday trap does, however, lead to the gross cross section. As we have shown for the case of CF_4, it is possible to measure both total cross section types in an ion counting experiment only if coincidence studies are carried out.

In the case of CF_4, it is possible to obtain a rough estimate of the total neutral dissociation cross section by subtracting the total counting ionization cross section from the total dissociation cross section measured by Winters and Inokuti.[30] The difference between using the correct counting cross section for ionization (the total dissociation cross section is a counting cross section) and the gross ionization cross section is a factor of two in the estimate of the neutral dissociation cross section. Hence, such seemingly insignificant differences can lead to significant problems if the definitions are not clearly understood. Another benefit of carrying out coincidence measurements is that they can provide indirect proof about neutral particle production. For example, if we only observe the partial ionization cross section for CF_3^+ in a noncoincidence experiment, we do not know whether the other fragment involved was F or F^+. When the coincidence measurement is made, we can subtract the contribution from $CF_3^+ + F^+$ and determine the contribution from the CF_3^+ cross section to neutral F production, which turns out to be the leading source for energies above 35 eV. Our findings are summarized in Table 2, where the counting dissociative ionization cross section is compared to the neutral dissociation cross section and to upper and lower bounds for neutral F production. It is of interest to note that the neutral F production cross section is of the same order of magnitude as the total dissociation cross section. This observation explains why CF_4 is an excellent etching gas.

Table 2. The counting ionization cross section σ_{ion}^{count}, the neutral dissociation cross section σ_{neut}, and lower, σ_F^{min}, and upper, σ_F^{max}, bounds for the neutral F production cross section for CF_4 as a function of electron impact energy, E.

E (eV)	σ_{ion}^{count} (10^{-16} cm^2)	σ_{neut} (10^{-16} cm^2)	σ_F^{min} (10^{-16} cm^2)	σ_F^{max} (10^{-16} cm^2)
20	0.365	0.45 ± 0.13	0.37	0.81
25	0.916	0.77	0.88	1.69
30	1.58	0.92	1.47	2.50
35	2.15	0.93	1.99	3.07
40	2.59	0.91	2.39	3.97
45	3.02	0.90	2.72	4.66
50	3.34	0.96 ± 0.82	2.96	5.21
55	3.56	1.04	3.12	5.59
60	3.86	0.98	3.32	5.99
65	4.11	0.91	3.46	6.30
70	4.25	0.90	3.42	6.50
75	4.33	0.94	3.52	6.63
80	4.37	0.98 ± 1.04	3.50	6.73

4. Conclusions

We have attempted to outline some of the major problems in making measurements of absolute ionization cross sections. Also included are recent experimental results on Ar and CF_4, which illustrate some of these problems. We have also briefly described our most recent coincidence experiments, which have enabled us to obtain estimates of the total gross and counting ionization cross sections and, from these, estimates of total neutral dissociation and the identification of the major source for neutral F production for electron impact energies above 35 eV.

References

1. G. Ecker and K. U. Riemann, Exp. Tech. Phys. 35 (1987) 119.
2. H. F. Winters, R. P. H. Chang, C. J. Mogab, J. Evans, J. A. Thornton, and H. Yasuda, Mat. Sci. Eng 70 (1985) 53.
3. G. S. Oehrlein, Phys. Today 39 (1986) 26.
4. Y. Sakamoto, O. Matsumoto, S. Morita, S. Hattori, K. Fukimoto, Y. Ikada, E. Miyazaki, I. Kojima, and Y. Takahashi, *Japan Materials Report* (Japan Technical Information Service) (ASM International, Metals Park, OH, 1988).
5. L. J. Kieffer and G. H. Dunn, Rev. Mod. Phys. 38 (1966) 1.
6. T. D. Märk and G. H. Dunn, eds., *Electron Impact Ionization* (Springer-Verlag, Wien, 1985).
7. P. Nagy, A. Skutlartz, and V. Schmidt, J. Phys. B 13 (1980) 1249.
8. S. K. Srivastava (lecture notes in this volume).
9. C. Ma, C. R. Sporleder, and R. A. Bonham, Rev. Sci. Instrum. 62 (1991) 909.
10. C. Ma, M. R. Bruce, and R. A. Bonham, Phys. Rev. A XX (1991) XXX.
11. L. J. Frasinski, K. Codling, and P. A. Hatherly, Science 246 (1989) 1029.
12. D. A. Dahl and J. E. Delmore, *The SIMION PC/AT User's Manual, Vol. 3.0,* (1987) Idaho National Engineering Laboratory, EGG-CS-7233.
13. J. E. Rowe, S. B. Christman, and R. D. Plummer, Rev. Sci. Instr. 42, 1733 (1971).
14. D. Rapp and P. Englander-Golden, J. Chem. Phys. 43 (1965) 1464.
15. A. V. Phelps, J. Phys. Chem. and Ref. Data 20 (1991) 557.
16. R. A. Bonham and M. R. Bruce, Comm. At. Mol. Phys. XX (1991) XXX.
17. R. S. Gao, P. S. Gibner, J. H. Newman, K. A. Smith, and R. F. Stebbings, Rev. Sci. Instrum. 55 (1984) 1756.
18. K. Tobita, H. Takeuchi, H. Kimura, Y. Kusama, and M. Nemoto, Jap. J. Appl. Phys. 26 (1987) 509.
19. E. Krishnakumar and S. K. Srivastava, J. Phys. B 21 (1988) 1055.
20. K. Stephan, H. Helm,, and T. D. Märk, J. Chem. Phys. 73 (1980) 3763.
21. R. C. Wetzel, F. A. Baiocchi, T. R. Hayes, and R. S. Freund, Phys. Rev. A 35 (1987) 559.
22. S. Wang, J. L. Forand, and J. W. McConkey, Can. J. Phys. 67 (1989) 699.
23. H. A. Van Sprang, H. H. Brongersma, and F. J. de Heer, Chem. Phys. 35 (1978) 51.
24. J. F. M. Aarts, Chem. Phys. Lett. 114 (1985) 114.
25. K. A. Blanks and K. Becker, J. Phys. B 20 (1987) 6157.
26. K. A. Blanks, A. E. Tabor, and K. Becker, J. Chem. Phys. 86 (1987) 4871.
27. P. J. M. van der Burgt and J. W. McConkey, J. Phys. B XX (1991) XXX.
28. T. J. Sommerer and M. J. Kushner, J. Appl. Phys. XX (1991) XXX.
29. R. A. Bonham and M. R. Bruce, Joint Symposium on Electron and Ion Swarms and Low Energy Electron Scattering, Bond University, Gold Coast, Australia, 18-20 July 1991, Book of Abstracts, p. 5.
30. H. F. Winters and M. Inokuti, Phys. Rev. A25 (1982) 1420.

INVESTIGATION OF NO^+, SF^+ AND SF^{2+} BY MEANS OF TRANSLATIONAL ENERGY SPECTROSCOPY

M. Hamdan
GLAXO RESEARCH LABORATORIES
VIA FLEMING, 4
37100 VERONA, ITALY

ABSTRACT

The electronic states of the two singly charged diatomic ions, NO^+, SF^+ and the doubly charged SF^{2+} ion are investigated by means of relatively high resolution (0.1 eV FWHM) translational energy spectroscopy (TES). The spectra pertaining to the two singly charged ions display a number of excitation and de-excitation features which can be attributed to transitions between the ground and a number of low lying excited electronic states. The TES of SF^{2+} scattered off helium shows no such features, however the collision induced dissociation of the same ion clearly indicates that such dissociation involves more than one dissociative electronic state.

INTRODUCTION

Recent advances in quantum chemical methods[1-4] as well as in experimental techniques[5-8] have provided much needed information on the structural properties, energies and symmetries of the electronic states of both doubly and singly charged diatomic ions.

In translational energy spectroscopy (TES), a mass selected and nearly monoenergetic ion beam is focused into a collision cell containing a target gas, where collisions between the incident ion and the neutral target induce transitions between the various energy levels of the collision partners. These transitions are generally manifested as discrete changes in the translational energy of the ion beam and can be precisely measured by scanning a post-collision electrostatic analyser. In a number of recent publications the author and co-workers demonstrated that the TES technique is capable of yielding valuable information on the electronic states of atomic[9], diatomic[10,11] as well as triatomic[12] ions. In the present work the same technique is employed to investigate the energies as well as the lifetimes of the electronic states of two singly charged as well as a doubly charged diatomic ions.

EXPERIMENTAL

The present measurements were performed on a novel double focusing translational energy spectrometer which has been fully described in an earlier publication[13]. Briefly, the ions were formed by 100eV electron imapct, extracted from the source, accelerated to 3kV, mass selected by a sector magnet and allowed to pass through an electrostatic energy monochromator followed by a collision cell containing low pressure (about 10^{-4} Torr) research grade He which insured single-collision conditions. Post collision ionic products were monitored by scanning in tandem two electrostatic energy analysers, the second of which was used to eliminate low level interferences which may be caused by ion fragmentation and ion (or neutral) reflections from walls and slits. In the present investigation the following collision processes are investigated.

$$AB^{+} + He \longrightarrow AB^{+*} + He - \Delta E \quad (1)$$
$$AB^{+*} + He \longrightarrow AB^{+} + He + \Delta E \quad (2)$$
$$AB^{+(*)} + He \longrightarrow AB^{+(*)} + He \quad \Delta E \sim 0 \quad (3)$$
$$SF^{2+} + He \longrightarrow S^{+} + F^{+} + \Delta E \quad (4)$$

where AB^{+} and AB^{+*} are the incident ions in the ground and excited states and ΔE is the net energy loss (gain) of the scattering process. Equation (1) represents an inelastic process in which part of the translational energy, ΔE, is converted into internal excitation energy of either the incident ion and/or the target. Equation (2) on the other hand represents the opposite process, where internal excitation energy of the incident ion is converted into translational energy. Equation (3) refers to an elastic scattering process in which interchange between translational and internal excitation energy does not take place. The process expressed in (4) refers to the combined unimolecular and collision induced dissociation of the doubly charged SF^{2+} ion where ΔE, is the kinetic energy release associated with the coulombic repulsion between the two separating singly charged atomic fragments, S^{+} and F^{+}.

RESULTS AND DISCUSSION

The translational energy spectra of 3 keV NO^{+} and 3 keV SF^{+} scattered off He are given in Figs. 1(a) and (b) respectively.

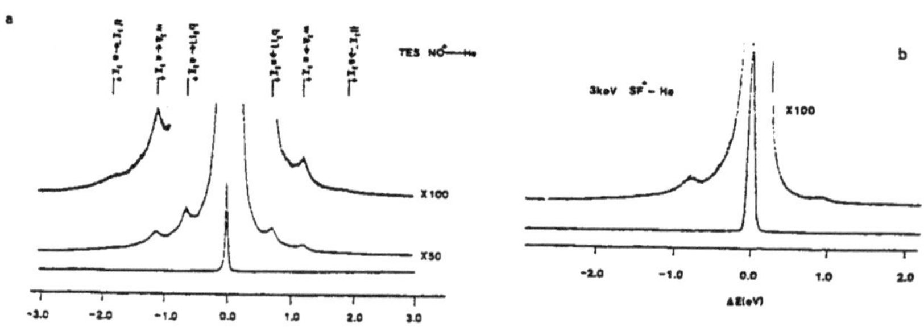

Fig.1.Translational energy spectra of:(a)NO^{+} and(b)SF^{+} scattered off He.

The first spectrum shows three superelastic peaks and three inelastic peaks centered at energy gains and energy losses of ± 0.7, ± 1.2 and ± 1.8 eV. The similarity in transition energies and peak widths infer that the two sets of peaks are due to de-excitation and excitation among the same electronic states of NO^+ incident ion. Table 1 shows very good agreement between these values and the values reported in an earlier TES measurements[14] and the values cited in the photoelectron spectrum of NO (see Fig. 1c)[15,16].

Table 1 Observed electronic transitions and corresponding energies of 3keV NO^+ scattered off He. The measured energies are compared with values reported in reference 14.

Energy (eV) Present	Previous$^{(a)}$	Transition$^{(b)}$
± 0.7 ± 0.1	± 0.6	$W^3\Delta$ <--> $b^3\Pi$
± 1.2 ± 0.1	± 1.2	$b'^3\Sigma^-$ <--> $b^3\Pi$
± 1.8 ± 0.2	± 1.9	[$b^3\Sigma^-$, $W^3\Delta$] <--> $a^2\Sigma^+$

(a) ref. 14
(b) ref. 15, 16

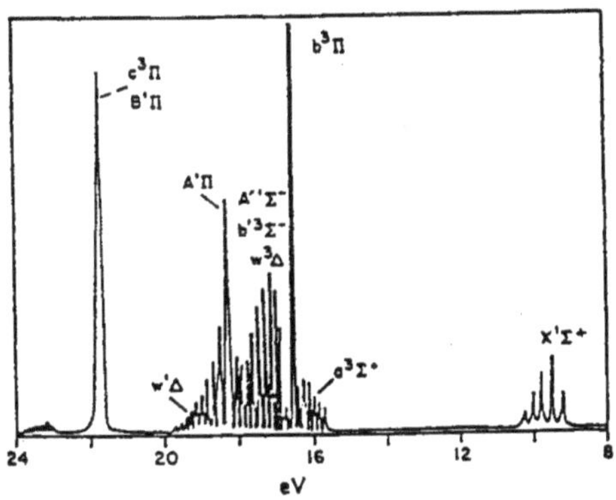

Fig.1(c).Photoelectron spectrum of NO^+ taken from refs.15,16.

Despite the good agreement between the three sets of data, it is pertinant to underline the following observations: first, it is well established that the first electronic excited state of NO^+ is ($a^3\Sigma^+$) and lies at 6.39 eV[17] above the NO^+ ($X^1\Sigma^+$) ground state. Excitation/de-excitation between these two electronic states is spin-forbidden which explains the absence of such transitions in the present as well as in the earlier TES measurements reported by O'Keefe et al.[14]. This particular observation is in full accord with the work of Moore[18] in which he demonstrated that transitions between the various energy levels observed in TES spectra are expected to be spin-conserving. Second, because of the relatively high density of bound electronic states (about 2 states/1eV) which characterises the region, 7-11 eV above the $NO^+(X^1\Sigma^+)$ ground state[15,16] the assignments of the observed electronic transitions given in table 1 have to be considered purely tentative. However, the potential energy curves of NO^+ [19] predicts vertical excitation energy for the transition, $NO^+(w\ ^3\Delta$ <--- $b^3\Pi$) of 0.6 eV and 1.2 eV for the transition $NO^+(b'\ ^3\Sigma^-$ <--- $b^3\Pi$). Both energies are in reasonable agreement with the present values of 0.7 and 1.2 eV. The lack of vibrational and rotational resolution precludes a precise and detailed analysis of the presented data. However, the present data may provide a useful guide for the spectrum regions where emission measurements can be attempted. Third, the observation of the three superelastic peaks associated with the collisional de-excitation of the incident, NO^+ ion infer that the ground state as well as a number of low-lying electronic states of NO^+ have a minimum lifetime of few microseconds (ion transit time between the ion source and the collision cell).

The TES of 3 keV SF^+ scattered off He is given in Fig. 1(b). As well as the elastic peak which is centered at E = 0, the spectrum shows a superelastic and an inelastic peaks which are centered at ΔE = 0.9 eV and - 0.9 eV respectively. The author is not aware of any spectroscopic data on the electronic states of SF^+ however this ion is isoelectronic with the SO radical which has ($X^3\Sigma^-$) ground state[20]. The excitation/de-excitation features observed in Fig. 1(b) are tentatively attributed to transitions between $SF^+(X^3\Sigma^-)$ ground and the first excited state.

The TES of 6keV SF^{2+} scattered off He and the collision induced dissociation of the same ion are given in Figs. 2(a) and 2(b). The latter spectrum pertaining to the S^+ fragment

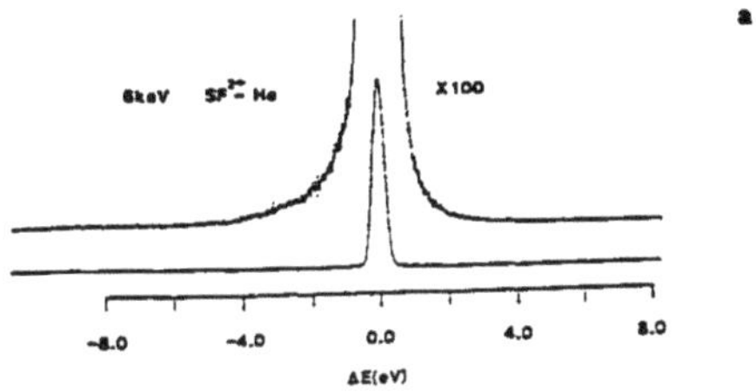

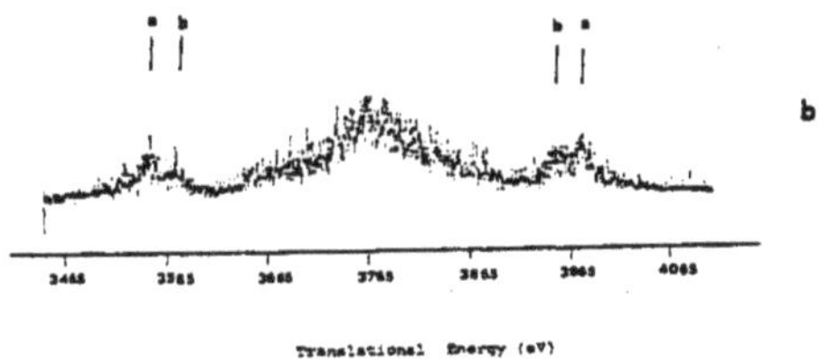

Fig.2(a).Translational energy spectrum of SF^{2+} scattered off He.

Fig.2(b).Collision induced dissociation of SF^{2+}

shows two partially resolved kinetic energy releases designated aa, bb and a single broad peak centered at the translational energy of 3765 eV (this value is in the laboratory-frame of reference). This particular peak is attributed to the process of dissociative single electron capture which is expressed in equation (5).

$$SF^{2+} + He \longrightarrow SF^{+} + He^{+}$$
$$\qquad\qquad\qquad\quad \llcorner\!\longrightarrow S^{+} + F \qquad (5)$$

By measuring the widths aa and bb and using a well known formula[21] the translational energy released in the center of mass frame of reference is calculated to be 10.2 ± 0.2 and 8.0 ± 0.2 eV. These values together with the spectrum in Fig. 2(b) infer that the observed S^{+} ion fragments are due to the dissociation of SF^{2+} ion from two discrete energy levels which are separated by about 2.2 ± 0.2 eV. This statement of course is based on the assumption that the measured energy releases are entirely due to the coulombic repulsion between the two charges. The translational energy spectrum of 6 keV SF^{2+} scattered off He is given in Fig. 2(b). Neither superelastic nor inelastic peaks are observed. The absence of the first is a strong indication that the SF^{2+}

incident ion beam which reaches the collision cell is predominantely in the ground electronic state. However, the absence of the inelastic peaks may infer that most of the electronic states which may be populated in collision with He are of a dissociative nature. This is of course not in conflict with the collision induced dissociation spectrum in Fig. 2(a) where two of these states are clearly evident. It has to be added that the complete absence of both theoretical calculations and experimental data regarding the dissociative and non dissociative electronic states of this ion make any detailed interpretation of the presented data rather difficult.

CONCLUSIONS

The two singly charged NO^+ and SF^+ ions as well as the doubly charged SF^{2+} ion have been investigated by means of high resolution translational energy spectroscopy and collision induced dissociation. The measured energies of transitions observed in the TES of NO^+ are in full agreement with existing TES and photoelectron results. A long lived electronic excited state lying at 0.9 eV above the ground state is observed in the TES of SF^+ ion. The collision induced dissociation of SF^{2+} display two dissociation channels which are separated by about 2.2 eV.

Acknowledgement

The author would like to thank Miss Elena Iamonte for her assistance in preparing this manuscript.

References

1. R.W. Wetmore, R.J. Leroy and R.K. Boyd, J. Chem. Phys. 88 (1984) 6318.

2. J.A. Pople, B. Tidor and P. Von R. Schleyer, Chem. Phys. Lett. 88 (1982) 533.

3. H. Agren, J. Chem. Phys. 75 (1981) 1267.

4. E.W. Thulstrup and A. Anderson, J. Phys. B. 6 (1975) 965.

5. D. Cossart and F. Launay, J. Mol. Spectros. 113 (1958) 159.

6. G. Dujardin, S. Leach, O. Dutuit, P.M. Guyon and M. Richard Viard, Chem. Phys. 88 (1984) 339.

7. J. Appell, J. Durup, F.D. Fehsenfeld and P. Fournier, J. Phys. B. 6 (1973) 197.

8. G. Dujardine, L. Hellner, M. Hamdan, A.G. Brenton, B.J. Olsson and M.J. Besnard-Ramage, J. Phys. B. 23 (1990) 1165.

9. M. Hamdan and A.R. Lee and A.G. Brenton, J. Phys. B. 21 (1989) L561.

10. M. Hamdan and A.G. Brenton, Chem. Phys. Lett. 155 (1989) 321.

11. M. Hamdan and A.G. Brenton, J. Phys. B. 22 (1989) L45.

12. M. Hamdan, D.P. Almeida and A.G. Brenton, J. Phys. B. 22 (1989) 1817.

13. P. Jonathan, M. Hamdan. A.G Brenton and G.D. Willett, Chem. Phys. 119 (1988) 159.

14. A. O'Keefe, A.J. Illies, J.R. Gilbert and M.T. Bowers, Chem. Phys. 82 (1983) 471.

15. J.E. Collin, J. Delwiche and P. Natalis, Int. J. Mass Spectrom. Ion Phys. 7 (1971) 19.

16. O. Edquist, L. Asbrink and E. Lindholm, Z. Naturforsch. 26 (1971) 1407.

17. D.W. Turner, C. Baker, A.D. Baker and C.R. Brundle, "Molecular photoelectron spectroscopy", Wiley Interscience, 1969 pp. 35.

18. J.H. Moore, Jr. Physical Rev. A 8 (1973) 2359.

19. D.L. Albritton, A.L. Schmeltekopf and R.N. Zare, J. Chem. Phys. 71 (1979) 3271.

20. G. Herzberg, "Spectra of Diatomic Molecules", Van Nostrad Reinhold, 1950 pp. 573.

21. R.G. Cooks, J.H. Beynon, R.M. Caprioli and G.R. Lester, "Metastable Ions", Elsevier, Amsterdam 1973.

EXPERIMENTAL WORKSTATIONS FOR THE BRAZILIAN NATIONAL SYNCHROTRON LIGHT LABORATORY

Aldo Craievich

Laboratório Nacional de Luz Síncrotron (LNLS)/CNPq, Campinas and Instituto de Física/USP, São Paulo, Brazil

ABSTRACT

The LNLS, a Brazilian National Laboratory located in Campinas, is building a synchrotron light facility which consists of a 1.15 GeV electron storage ring, its injection system and a set of light beam-lines. The light source will be used for applications in various fields of science and technology. The basic storage ring will provide users with high brilliance and tunable photon beams over a wide spectral range extending from infrared to soft X-Rays. The synchrotron source will be equipped with insertion devices (wigglers) to produce the emission of hard X-Rays (up to about 25 KeV). The design and construction of the instrumentation to use the synchrotron light, has begun at LNLS and at several laboratories interested in applications of the light source. We expect several of the designed workstations to be ready for operation by 1994, when the light source is scheduled to be available for users.

1. Introduction

The Laboratório Nacional de Luz Síncrotron (LNLS), a National Laboratory belonging to the Brazilian National Council for Scientific and Technological Development (CNPq), started its activities in 1987 in Campinas, São Paulo State[1]. This Laboratory is building a synchrotron light facility to be utilized for a wide range of applications in science and technology[2]. Applications of this facility cover basic research areas such as atomic, molecular and condensed matter physics, molecular and solid state chemistry, molecular biology and different aspects of materials science. Technological applications, namely the use of X-Ray microlithography[3] to produce electronic microcircuits and for mechanical micromachining, and that of crystallography for drug design, will also be possible.

Each one of the many and varied applications of synchrotron light requires specially designed experimental workstations. They should be equipped with a) optical systems to focus and monochromatize the incoming beam, b) special chambers for positioning, orienting and also for conditioning the samples under the required conditions of temperature, pressure, humidity, etc. and c) detection devices for counting, analyzing and recording photons, electrons and ions produced by the interaction between the incident (usually monochromatic) beam and the studied samples.

Two electron accelerators (a LINAC and a circular storage ring), the first part of the planned workstations (front ends, mirrors and monochromators) and a few basic workstations are being built at LNLS. The specific parts of most of the other planned workstations are being designed in external laboratories by future users.

2. Synchrotron Light Source

The synchrotron light source which is being built at LNLS consists of an electron storage ring of 1.15 GeV, named UVX2 ring. It will be injected with an electron LINAC of 100-200 MeV(4). The energy of the electrons will be ramped inside the ring up to the final energy of 1.15 GeV. The design of the UVX2 ring is based on a six-fold symmetric Chasman-Green achromat lattice(5).

The nominal electron current and the approximate electron life-time are 100mA and 10 hours, respectively. The nominal emittance of the source is 65.3 nm.rad. The flexible design of the ring allows for different modes of operation. In the low and high emittance modes the natural emittances are 33.9 and 117.0 nm.rad., respectively(5).

The electron energy E and the field of the bending magnets B define the critical energy E_c, and consequently the critical wavelength λ_c, of the photon emission spectrum. These parameters represent the energy, or wavelength, values which divide the white spectrum in two parts of equal emission power. Since E=1.15 GeV and B=1.4 Tesla, the critical parameters of the UVX2 ring are λ_c=10 Å and $E_c \cong 1.2$ KeV. The flux produced by storage rings is high enough to be useful, for most of the applications, up to $E_m \cong 5.E_c$. Therefore the beam-lines associated with the bending magnet ports of the UVX2 ring will provide useful photons with energy up to 6 KeV.

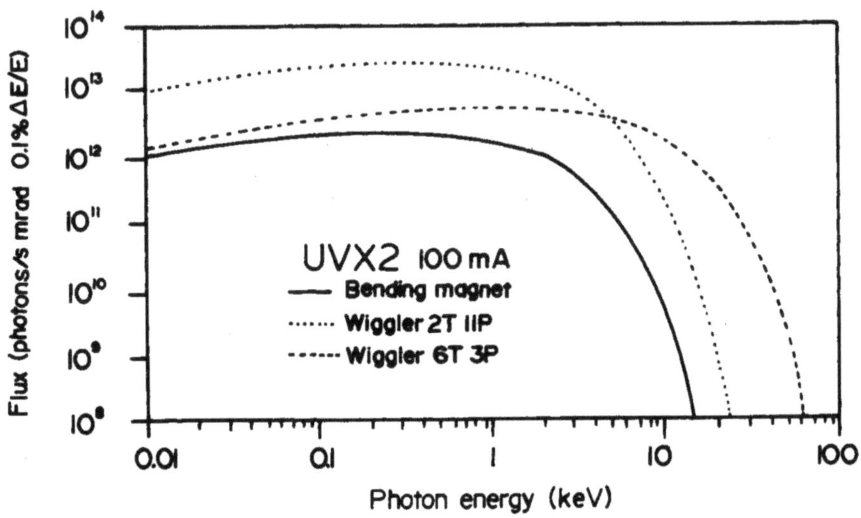

Fig. 1: Emission photon flux spectra associated with bending magnet and wiggler beam-lines(5).

The insertion of an 11-pole wiggler, with normal magnets of 2 Tesla, is planned. The beam-lines associated with this wiggler will provide useful photons up to about 10 KeV and, consequently, will allow for the installation of workstations for X-Ray diffraction and scattering applications. The insertion of a superconductor 3-pole 6 Tesla wiggler (wave-shifter) is at a discussion stage. This insertion is able to provide high-flux photon beams with an energy up to 30 KeV.

The straight sections of the ring, which have 2.8 m free for insertions, will also allow for the installation of undulators to provide photons for the workstations which need VUV beams with special features (very high brightness within narrow energy bands, particular polarization and collimation conditions).

The spectra of photon flux from bending magnet and wiggler ports are illustrated in Fig. 1[(5)].

3. Experimental Workstations

Up to now seventeen experimental stations are scheduled to be installed around the UVX2 ring. Four of them are designed or planned at LNLS, ten were proposed by future user groups from several Brazilian laboratories and three by future users from other countries. An schematic view of the planned locations of the workstations is shown in Fig. 2. We can see that nine beam-lines will be necessary to accommodate the seventeen workstations. This planning assumes that the two wiggler beam-lines have beam splitters and several of them have a time sharing mode of operation.

The workstations which were proposed by external user groups and those which are planned to be built at LNLS are listed below. This list includes the name of the stations, their spectral range, the type of monochromator (see the caption of Fig.2), the description of their applications[(6)] and the name and address of the scientist who may be contacted for further information.

1) Time Resolved Spectroscopy (4 - 6.5 eV, NIM):
Single bunch fluorescence spectroscopy. Measurements of life-time of excited states of macromolecules in solution, molecular dynamics, time-resolved microscopy, excited state reactions (photosynthesis), macromolecular rotations, heterogeneous emission (petrochemistry), double excitations (photobiology and photochemistry) - J. L. da Silva, Departamento de Bioquímica, Instituto de Ciências Biomédicas/UFRJ, Cidade Universitária, Ilha do Fundão, 21910 Rio de Janeiro, RJ, Brasil.

2) Molecular Spectroscopy (10 - 1000 eV, TGM and SEM):
Study of different processes associated with the interaction of monochromatic photons with atomic and molecular complexes, angle-resolved photoelectron spectroscopy of valence and core states, Auger spectroscopy (normal and resonant), mass spectroscopy, electron energy loss spectroscopy - G. G. B. de Souza, Laboratório de Impacto de Elétrons, Instituto de Química/UFRJ, Centro de Tecnologia, Bloco A, 6º andar, 21910 Rio de Janeiro, RJ, Brasil.

3) Photoelectron Spectroscopy (10 - 300 eV, TGM):
Angle-resolved photoelectron spectroscopy for solid state research, chemical analysis of semiconductor surfaces and interfaces, electronic structure of semiconductor surfaces and interfaces, photodiffraction studies of surface atomic structures - W. N. Rodrigues, Departamento de Física, ICEX/UFMG, Caixa Postal 702, 30161 Belo Horizonte, MG, Brasil.

4) Photodesorption (10 - 100 eV, TGM):
Mass spectroscopy of desorbed ions from surfaces illuminated with UV photons, surface analysis (electrons, molecules and aggregates), photodesorption and photoionization processes, instrumentation development - E. F. da Silveira, Laboratório Van der Graaf, Departamento de Física/PUC, Rua Marquês de São Vicente 225, 22453 Rio de Janeiro, RJ, Brasil.

5) Surface Physics (200 - 1000 eV, SEM):
Photoelectron spectroscopy studies of surfaces of solids, electronic structure of first atomic layers of solids and of adsorbed atoms, chemical analysis and atomic coordination in surfaces (PES, EXAFS, LEEDS, AES and XAES) - R. Landers, Instituto de Física/UNICAMP, Caixa Postal 6165, 13081 Campinas, SP, Brasil.

6) X-Ray Microlithography (around 1KeV, filters):
Microlithography for applications to microelectronics, studies of resists for microelectronic devices of very high (submicronic) spatial resolution - Scientific Division, LNLS, Caixa Postal 6192, 13081 Campinas, SP, Brasil.

7) Soft X-Ray Spectroscopy (800 - 4000 eV):
Core level spectroscopy. Electronic structure of light atoms (K edge), transition metals ($L_{2,3}$ edges) and rare earths ($M_{4,5}$ edges). Magnetic and superconducting materials - H. Tolentino, LNLS, Caixa Postal 6192, 13081 Campinas, SP, Brasil.

8) EXAFS, Extended X-Ray Absorption Fine Structure (1 - 15 KeV, 2C):
Local atomic structure of crystalline and amorphous solids and liquids - H. Tolentino, LNLS, Caixa Postal 6192, 13081 Campinas, SP, Brasil.

9) Powder Diffractometry (5 - 15 KeV):
Phase transformations and surface reactions in alloys, "in situ" studies of corrosion (oxidation, sulfatation and nitration) at high temperatures - F. Rizzo, Departamento de Ciência dos Materiais e Metalurgia/PUC, Rua Marquês de São Vicente 225, 22453 Rio de Janeiro, RJ, Brasil.

10) AWAXS, Anomalous Wide Angle X-Ray Scattering (4 - 15 KeV, 2C):
Local atomic structure of amorphous materials (oxide and metallic glasses, and others) - J. C. de Lima, Instituto Politécnico do Rio de Janeiro, Caixa Postal 97282, 28630 Nova Friburgo, RJ, Brasil.

11) X-Ray Fluorescence (1 - 20 KeV, 2C):
Quantitative X-Ray fluorescence analysis, space resolved analysis (mapping),total reflection analysis, slab analysis in bulk - M. Rubio, FaMAF/ Universidad Nacional de Córdoba, Laprida 854, 5000 Córdoba, Argentina.

12) SAXS, Small Angle X-Ray Scattering (6 - 12 KeV, 1C):
Low resolution structural studies of macromolecules in solution, porous materials, microphase separation in solids, aggregation in solids and liquids, fractal structures - I. Torriani, Instituto de Física/UNICAMP, Caixa Postal 6165, 13081 Campinas, SP, Brasil.

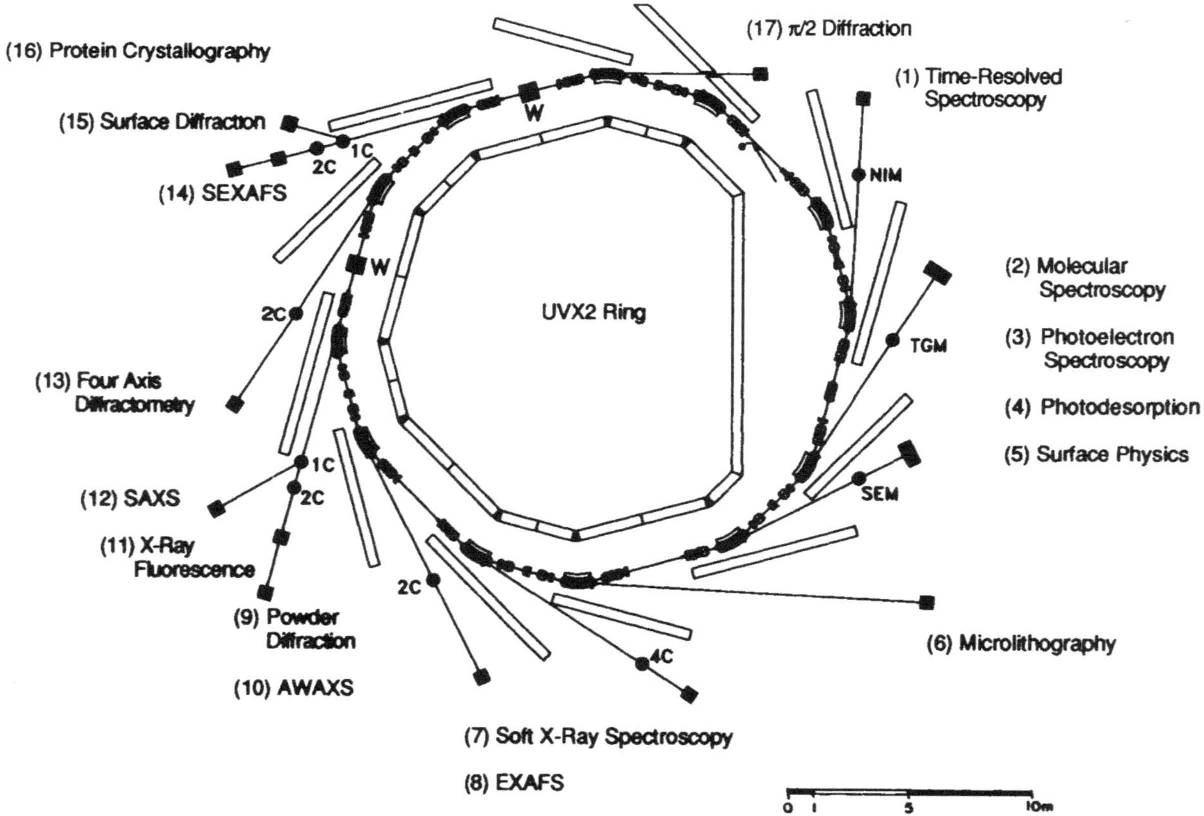

Fig. 2: Preliminary lay-out of the experimental workstation locations around the UVX2 storage ring. The type of monochromator for each workstation is indicated (NIM: Normal Incidence, TGM: Toroidal Grating, SEM: Spherical Element, 4C: Four Crystals, 2C: Two Crystals, 1C: One Crystal).

13) Four Axis Diffractometry (1 - 10 KeV, 2C):
Measurement of optical parameters (f' and f"), "perfect crystal" optics (Bragg line profile, X-Ray topography, grazing incidence diffraction, standing waves), interferometry, instrumentation development - C. Cusatis, Departamento de Física/UFPr, Caixa Postal 19091, 81504 Curitiba, Pr, Brasil.

14) SEXAFS, Surface Extended X-Ray Absorption Fine Structure (4 - 12 KeV, focussing 2C):
Surface atomic structure, surface reactions, solid-gas and solid-solid interfaces, catalyzers. (Photoelectron and Auger electron SEXAFS) - J. L. del Barco, INTEC, Guemes 3450, 3000 Santa Fé, Argentina.

15) Surface Diffraction (3 - 12 KeV, focussing 2C):
Grazing incidence X-Ray diffraction, atomic structure of semiconductor surfaces, studies of "in situ" prepared MBE multilayers - V. H. Etgens, LURE, Bat. 209, Université Paris-Sud, 91405 Orsay, França.

16) Protein Crystallography (6 - 12 KeV, 1C):
High resolution structure of crystallized proteins and other macromolecules - G. Oliva, Instituto de Física e Química de São Carlos/USP, Caixa Postal 369, 13560 São Carlos, SP, Brasil.

17) $\pi/2$ Diffraction (1 - 10 KeV, 1C $\pi/2$):
X-Ray optics, high resolution diffractometry - C. Cusatis, Departamento de Física/UFPr, Caixa Postal 19091, 81504 Curitiba, Pr, Brasil.

Several of the listed workstations, involving diffraction and scattering techniques, require rather hard X-Rays (about 8 KeV). This implies that at least two wigglers will be necessary in order to avoid the sharing of the wiggler beams between a too large number of stations. The wiggler locations are indicated in Fig. 2 (W).

LNLS intends to provide the beam-lines, including the front-ends, monochromators, and mirrors if necessary, for the proposed workstations within its budgetary possibilities. The user groups, which are in charge of workstation design and construction, are responsible for funding their instrumentation projects.

The details of the policies for sharing beam-lines and experimental stations, in accordance with the National Laboratory objectives, are being worked out. The Governing Board of LNLS has the ultimate responsibility for deciding on these policies. Input from future users, during the forthcoming annual Users' Meeting, will be an important element for policy making.

4. Beam Lines: Front ends, Monochromators and Detectors

The front-ends are the components used to extract the light beam from the bending magnets and from the insertion devices (wiggler and undulators). They include a primary shutter, a valve to isolate the storage ring from the workstations, and a safety system composed of a delay-line, a fast valve and a pressure monitor . The first front-end has been built and vacuum tested(7). The construction of five additional ones will be started soon.

The first monochromator system built at LNLS is a toroidal grating monochromator (TGM). It consists of three toroidal gratings, which can be used alternatively, and two toroidal mirrors. The spectral range covered by this system is 12 eV<E<310 eV[8]. This monochromator was recently assembled[9] and will provide photons for the Molecular Spectroscopy, Photoemission and Photodesorption workstations.

A second grating monochromator is being designed[10]. It is based on spherical elements (SEM) and will be tuned over a spectral domain ranging from 200 to 1000 eV. It will be used for the Surface Physics and Molecular Spectroscopy workstations.

Three crystal monochromators are being built at LNLS. They will be used to provide monochromatic soft and hard X-Ray photons to different workstations, depending on the flux and energy resolution requirements.

A bent one-crystal silicon, high vacuum compatible, monochromator was designed for workstations needing low energy resolution and high flux [11] such as Small Angle X-Ray Scattering (SAXS) and Protein Crystallography.

A two crystal monochromator is at an advanced stage of construction. A precision goniometer produces a wide range θ-rotation and two elastic translation stages are used for keeping a fixed beam offset[12]. This type of monochromator will be utilized to provide photons to the EXAFS and Four Axis Diffractometry workstations.

A four-crystal monochromator was built and tested[13]. It consists of two symmetrical pairs of silicon crystals. Two coupled and opposite rotations, driven by an external translation stage, assure the constancy of the Bragg reflection conditions during the tuning. This monochromator will be used for the Soft X-Ray Spectroscopy workstation.

Two electronic systems associated with a one dimensional position-sensitive Reticon photon detector, for visible light and X-Ray, respectively, were completed[14] and a project for a two-dimensional position sensitive detector, based on a microchannel plate and a resistive anode, is under way[15].

5. Present Status

The construction of the first 50 MeV sections of the electron LINAC was completed by December 1989. The prototype of 1/12 of the storage ring was assembled and submitted for evaluation to and approved by an International Review Panel, which met at LNLS last July. The production phase of the project will start in January 1992. The construction of the first beam-line including the mirror and monochromator system was completed and the production of five additional front-ends will start within the next months. The construction of the first three crystal monochromators is well advanced.

The LNLS is temporary installed in a 2000 m^2 building. The civil engineering works concerning the buildings for the permanent installation started in December 1990 at a 380,000 m^2 site donated by the São Paulo state government. The staff of LNLS is composed at present of 80 engineers and technicians in the Project Division and 15 scientists and technicians in the Scientific Division. The first group of the Project Division will move to the new buildings by September 1991.

The current schedule for the construction of the synchrotron source permits us to expect the first photons to be delivered to users by the beginning of 1994.This represents a delay of one year when compared with the initial time schedule.

The geometry of the UVX2 electron storage ring allows for the location of 24 front-ends (there are two beam ports per bending magnet). Consequently the development of additional workstation projects may be decided, this number depending on the interest of qualified users and funding possibilities. The LNLS is open to the participation of future users, Brazilian and from other Latin American countries, in activities involving planning, design, construction and funding of new experimental workstations and their related instrumentation.

References

1. C. Gonçalves da Silva, A. R. Rodrigues and A. Craievich, *The National Laboratory for Synchrotron Light of Brazil* , Synchr. Rad. News 1 (1988) 28.
2. A. F. Craievich, ed., Synchrotron Light: Applications and Related Instrumentation, World Scientific Publishers, Singapore (vol. I, 1989 and vol. II, 1990).
3. A. F. Craievich, G. G. Bezerra de Souza and V. Baranauskas, eds., Microlithography: High Integration in Microelectronics", World Scientific Publishers, Singapore (1990).
4. L. Lin, L. Jahnel and P. Tavares, *The injection system for the LNLS synchrotron light source*. Proceedings of the Particle Accelerators Conference, Chicago (1989) 993.
5. L. Lin, L. Jahnel and P. F. Tavares, *The LNLS UVX2 soft X-ray source*, LNLS MeT-09/91.
6. G. G. Bezerra de Souza, ed., *Aplicações da luz síncrotron em química, física e ciência dos materiais : desenvolvimento integrado de instrumentação,* LNLS Report (1991).
7. A. R. B. de Castro and P. T. Fonseca, *Primary shutter and gamma ray trap*, Synchrotron Light: Applications and Related Instrumentation II, World Scientific, Singapore (1990) 273.
8. A. R. B. de Castro, *Optics for a VUV beam line at LNLS*, Synchrotron Light: Applications and Related Instrumentation II, World Scientific, Singapore (1990) 20.
9. P. T. Fonseca, J. G. Pacheco, E. d'A. Samogin and A. R. B. de Castro, *VUV beam lines at LNLS, the Brazilian synchrotron source*, to be published Rev. Sci. Instrum. (1991).
10. A. R. B. de Castro, *Projeto conceitual de uma linha e monocromador de luz síncrotron para espectrocopia de alta resolução (200 - 1000 eV)*, Surface *Physics* Workstation, LNLS Report (1991).
11. L. Bernardes, H. Tolentino, A. R. Rodrigues, A. Craievich and I. Torriani, *A SAXS experimental station for LNLS*, to be published Rev. Sci. Instrum. (1991).
12. M. C. Corrêa, H. Tolentino, A. Craievich and C. Cusatis, *A two-crystal monochromator for the UVX ring of LNLS*, to be published Rev. Sci. Instrum. (1991).
13. H. Tolentino and A. R. Rodrigues, *High resolution monochromator for soft and hard X-rays*, to be published Rev. Sci. Instrum. (1991).
14. A. R. B. de Castro, *Solid state parallel detectors*, Synchrotron Light: Applications and Related Instrumentation I, World Scientific, Singapore (1989) 53.
15. J. E. Verdugo, F. Ferreira and A. R. B. de Castro, *Electronics for 2-Dim-sensitive photon counting*, Synchrotron Light: Applications and Related Instrumentation II, World Scientific, Singapore (1990) 278.

NEW MATERIALS PROCESSING WITH ION BEAMS*

I.J.R. Baumvol

Instituto de Física, Universidade Federal do Rio Grande do Sul, 91500 Porto Alegre, RS, Brasil

ABSTRACT

Ion beams of low and medium energies (100 eV - 200 keV) have been extensively used in the obtainance of new materials for specific applications.

In this paper we summarize our experience in this field describing different research efforts made using ion beams in shallow implants in silicon for VLSI doping technology; in metallurgical applications like surface alloying and surface protection of metals against corrosion, friction and wear; and in the use of ion beam mixing and ion beam assisted deposition of thin films to produce magnetic thin films with specific characteristics.

1. SHALLOW JUNCTIONS IN SILICON

A considerable effort is presently being made in order to obtain higher packing density in the production of integrated circuits. However, when the linear dimensions in the silicon wafer plane are reduced, the scaling rules require that the vertical dimensions must also be reduced by a similar factor, imposing the necessity of producing shallow doped regions with precise control of the dopant profile.

This control of the dopant profile can be attained by using ion implantation to introduce the dopants into the single cristal silicon substrate, but one has to minimize the profile redistribution due to channeling and to diffusion during the annealing of the damage.

More recently some ion beam assisted techniques have been considered to achieve shallow implantations compatible with the requirements of very large scale integration of silicon-based microelectronic devices, in between them the recoil implantation method and the preamorphization method.

1.1 Recoil Implantation[1]

Recoil implantation for doping purposes is performed by bombardment, usually with some inert gas ion beam, of a thin layer of the dopant deposited on the surface of the semiconductor silicon. This causes a fraction of the atoms in the film to cross the film-substrate interface, thus penetrating into the silicon surface layers. Backward sputtering also arises as a consequence of the bombardment, removing a considerable amount of the deposited film. The remaining film is usually removed by chemical etching.

*Work supported in part by CNPq - Brasil.

The case to be presented here is the recoil implantation of antimony into (100)-oriented silicon by means of Ar^+ bombardment, trying to correlate parameters like bombardment dose, energy, current density and substrate temperature with the recoil yield of Sb atoms into the Si substrate and also with the depth profile of Sb atoms.

Figure 1 represents the Rutherford Backscattering Spectrometry (RBS) results using a $^4He^{++}$ beam at incident energy of 760 keV, where only the region of the spectra corresponding to scattering from Sb atoms is shown. The RBS spectra were measured after removal of the remaining Sb film by chemical etching.

The increase of the depth profile of Sb for increasing Ar^+ bombardment dose (Fig. 1a) can be attributed to multiple recoils of Sb atoms taking place at the highly alloyed Sb-Si interface. A further evidence of this is given in Fig. 1b where we notice that the increase of the energy of the bombarding Ar^+ beam does not influence the depth profile of the recoiling Sb atoms, showing that the recoil energies of the Sb atoms do not depend on the energy share on the primary collisions but they would correspond much more to secondary collision energies.

The current density of the Ar^+ beam and the sample temperature up to 300 °C also do not influence significantly the Sb depth profile, indicating again that we are in the presence of cascade recoil mechanisms.

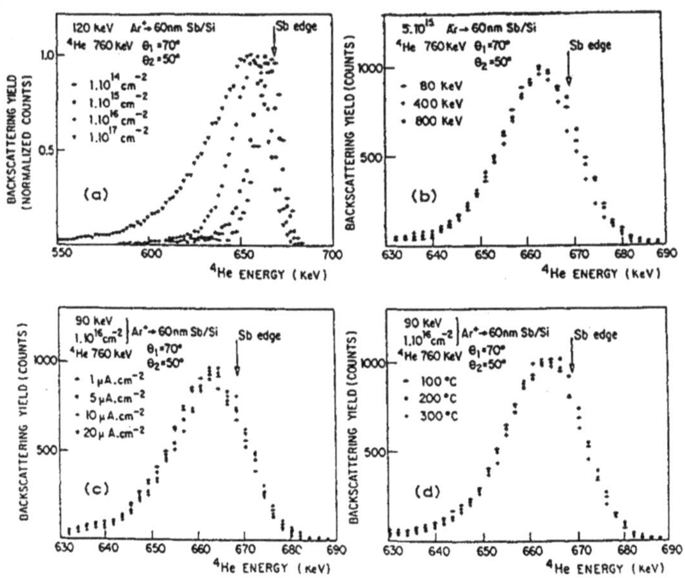

Figure 1. RBS spectra from a 60 nm Sb film deposited on (100) Si and bombarded with an Ar^+ beam at energies, total doses, current densities and substrate temperature as indicated. Only the Sb part of the RBS spectra are shown in each figure.

1.2 Preamorphization[2,3]

An alternative doping process to obtain shallow implants in silicon is to perform an amorphization step in the surface layers of the single cristal silicon wafer prior to the implantation of electrically active dopants. This preamorphization step is accomplished by the damage produced by a previous implantation of electrically inactive species like Si, Ge or Sn.

In this way the channeling during implantation of the electrically active species like B, P or As can be avoided and high electrical activation of the dopants can be obtained by solid phase epitaxial (SPE) regrowth of the amorphous layer, leading to shallower doping profiles.

The complete set of processing steps leading to a shallow substitutional dopant profile in silicon, free from major profile tails, can be seen in Fig. 2, where we show the preamorphization of Si by Ge^+ implantation followed by implantation of As^+ which is a most common dopant species of Si. The epitaxial regrowth was accomplished by RTA at 1050 °C during different times.

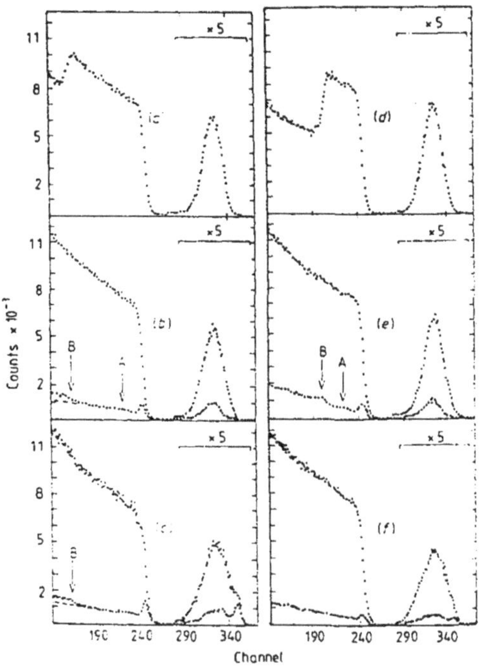

Figure 2. (a) Random and <100> - aligned RBS spectra of silicon samples pre-amorphised by means of Ge^+ implantation and followed by a As^+ (100 keV, 1×10^{16} cm^{-2}) implantation; (b) after RTA at 1050 °C, 20 s and (c) after RTA at 1050 °C, 60 s.

2. ION BEAM MIXING

The doping of semiconductors by ion implantation was the first practical use of ion beams to modify materials properties. The interest was the direct introduction of dopants by a ballistic process to concentrations capable of modifying the electrical properties, i.e., about 0.1 at % or less. The corresponding doses were less than 10^{15} ion.cm^{-2}. Whenever higher concentrations of an impurity are necessary for modifications of other material properties, as in the case of metallurgical applications, one needs higher doses. However, owing to sputtering there is a limit of composition achievable by high-dose implantation. To overcome this difficulty one can first deposit on the sample surface a thin film layer of the desired impurity by some conventional means, and then subject the system to ion beam bombardment, causing mixing. The primary aim is now the energy deposition at the thin film-substrate interface which allows atomic rearrangements to occur. This reorganization of constituent target atoms due to irradiation of solids with energetic particles is called ion beam mixing, which can be suplemented with heating of the target either during or after bombardment to promote even furtherly the formation of a mixed region.

We describe here a few examples of ion beam mixing and some of the new material properties that can be achieved by this process.

2.1 Radiation Enhanced Diffusion of Tin into Iron[4,5]

Fig. 3 gives an example of the mixing of a 800 Å Sn film into an iron substrate due to Ar^+ bombardment, keeping the sample temperature at 170 °C during bombardment.

Fig. 4 shows the consequences of the mixing on the high temperature oxidation behaviour of iron.

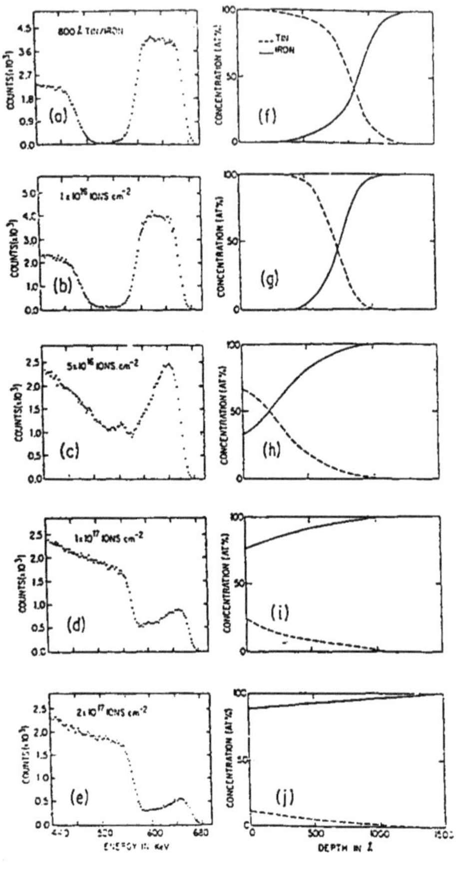

Figure 3. RBS spectra and concentration depth profiles for an 800 Å tin film deposited on an iron substrate and bombarded with 300 keV Ar^+ ions to the doses specified in the figure.

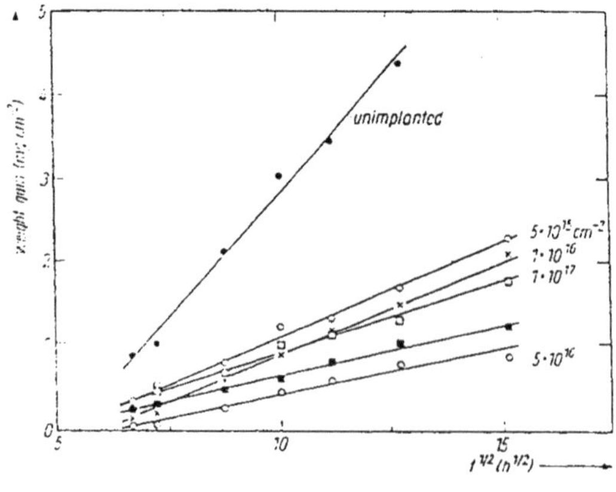

Figure 4. Oxidation kineties at 500 °C of iron samples uncoated and coated with a Sn film and bombarded with Ar^+ as shown in Fig. 3.

2.2 Ion Beam Mixing of Fe-Al Bilayers[6]

The Fe-Al thin film bilayer is described here as an illustrative example of crystalline intermetalic phases formation by heavy ion bombardment. In particular we compare the interdiffusion and reaction in Fe/Al bilayered samples submitted to thermal annealing, ion beam mixing and to the simultaneous combination of these two types of metallurgical treatments.

The Fe/Al bilayers were 100 nm/500 nm thick, deposited on SiO_2 substrate. Fig. 5 shows the RBS spectra of samples bombarded with Ar^+ ions at several different doses and substrates temperatures.

Fig. 6 shows the X-rays diffractograms of furnace annealed and ion bombarded samples, where the phases of the Fe-Al system corresponding to the reflection peaks are identified.

In can be verified that both the interdiffusion of the bilayer constituents and the phase formation observed in the ion bombarded samples replace those obtained in furnace annealing. The main difference is that under ion bombardment all these phanomena occur at appreciably lower substrate temperatures.

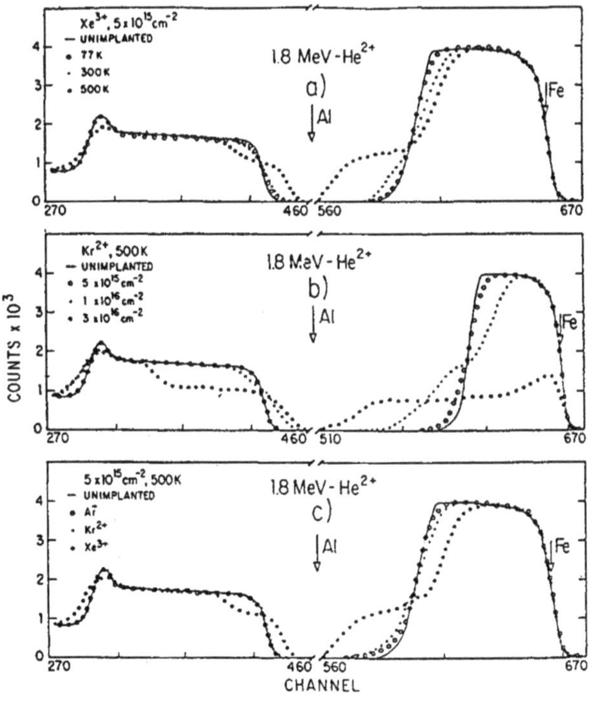

Figure 5. RBS spectra of 100 nm/500 nm Fe/Al bilayers evaporated on SiO_2 substrates and bombarded with different ionic species at doses and sample temperatures as indicated.

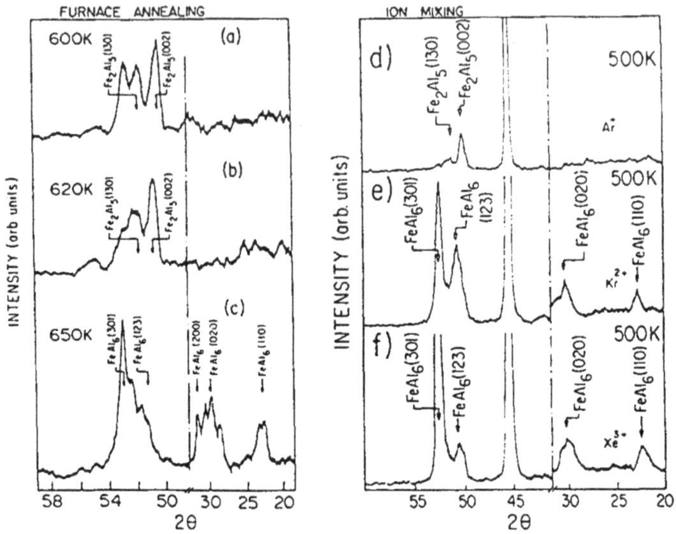

Figure 6. X-ray diffractograms for Fe-Al thin film bilayers submitted do high-vacuum furnace annealing and ion beam mixing with different ionic species.

2.3 Ion Beam Mixing of Magnetic Thin Film Multilayers[7]

There is a large interest in depositing transition metal/rare earth metal multilayers because these structures can act as an effective starting point to study magnetic thin film intermetallic phases. As a good example we describe here the case of the Fe-Nd and Fe-Nd-B systems. Nd rich, as well as Fe rich phases were found in some way to be responsible for the very high coercivities of the super-hard NdFeB sintered magnets and several authors have addressed the question of which phases are formed and what are the magnetic characteristics of FeNd and FeNdB thin films.

In Fig. 7 we show the RBS spectra for the as-deposited 40/160 and 80/80 Å Fe/Nd multilayered samples. In Fig. 8 we show the effect of ion bombardment with B^+ and Fe^+ and subsequent annealing at 600 °C during 150 min in a ultra-high vacuum furnace, where we notice the complete disappearing of the multilayer periodic structures and the outcome of flat signals for the first about 3000 Å beneath the surface which indicate rather homogeneous Fe-Nd alloys in this region.

The magnetic hysteresis curves for some of the Fe/Nd samples as measured by the vibrating sample magnetometer are shown in Fig. 9. The magnetic hysteresis curves submitted to ion bombardment and subsequent thermal annealing show two independent contributions due to different, non-interacting magnetic phases.

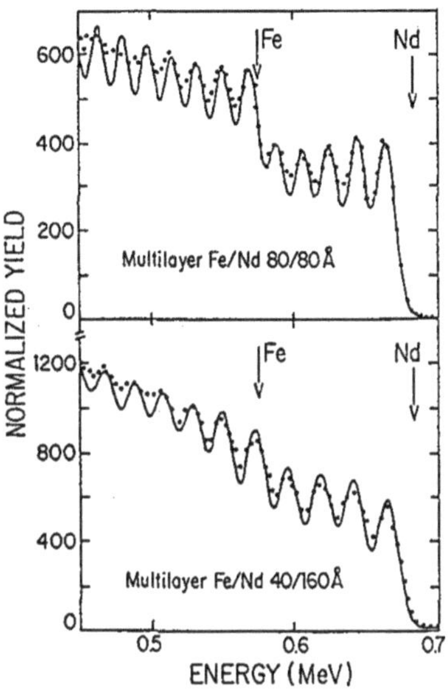

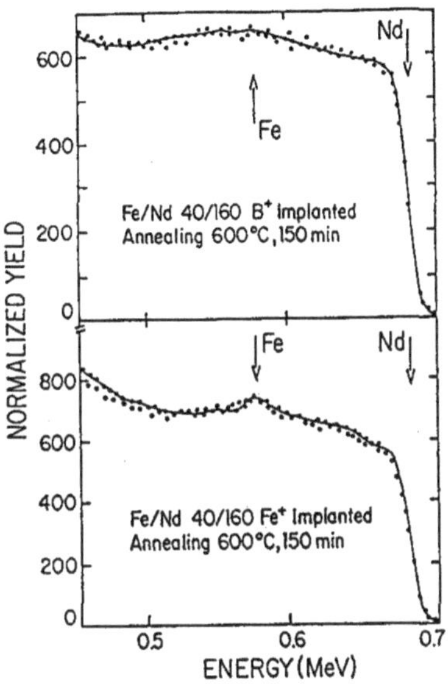

Figure 7. RBS spectra of as-deposited Fe/Nd multilayers. The projected wavelengths are indicated in the figure (left).

Figure 8. RBS spectra of the Fe/Nd 40 Å/160 Å multilayers implanted with 1×10^{16} cm^{-2} B^+ and Fe^+ and after annealed at 600 °C (right).

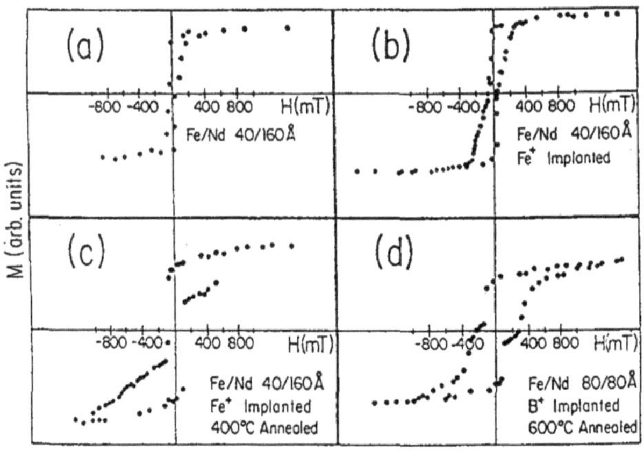

Figure 9. Hysteresis loops of the magnetization (in arbitrary units) of the 40 Å/160 Å and 80 Å/80 Å Fe/Nd multilayers implanted and annealed in different conditions.

3. ION BEAM ASSISTED DEPOSITION OF THIN FILMS[8]

The bombardment of thin films with energetic ion beams during the growing process (IBAD) may result in either physical, chemical or structural changes in the nature of the film. Many physical properties which are sensitive to these microstructural aspects can be expected to reflect the bombardment induced changes.

We describe here the modifications of the magnetic properties of iron films grown under Ar^+ bombardment in a vacuum chamber pressurized with nitrogen. It is verified that the magnetic coercivity and also the squareness of the magnetic hysteresis curve are strongly modified as the partial pressure of N_2 in the deposition chamber is varied, as well as if Ar^+ bombarding beam is switched on or off.

In Fig. 10 we show the hysteresis curves obtained for a high vacuum evaporated pure Fe film, and for Fe films evaporated in some specific conditions of Ar^+ bombardment and N_2 partial pressures.

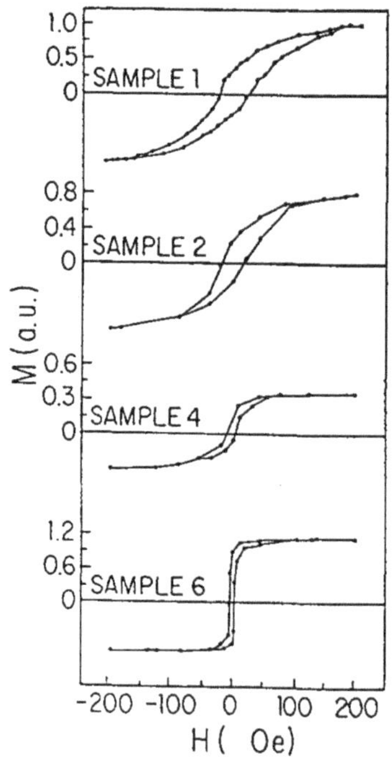

Figure 10. Hysteresis loops of the magnetization (in arbitrary units) for a Fe film evaporated in high vacuum (sample 1), Fe film evaporated at high deposition rate (15 Å.s^{-1}) under Ar^+ bombardment at 1.5 keV (samples 2 and 4) and Fe film evaporated at low deposition rate (1.5 Å.s^{-1}) under Ar^+ bombardment (sample 6). The deposition of samples 2, 4 and 6 were performed in the chamber pressurized with N_2 gas.

4. CONCLUSIONS

We have discussed several different uses of ion beams as a tool to assist the formation of new materials, many of them with unique properties. The examples given above included only research done in the author's Institute, but a more comprehensive overview should include also optical, superconducting and several other new materials that are currently being ontained with the use of ion beams.

Nevertheless the survey here presented was intended to give an idea of this well established and fastly expanding field of materials science.

5. REFERENCES

1] Erichsen Jr., R.; de Souza, J.P. and Baumvol, I.J.R., Nucl. Instr. Meth. B7/8, 316 (1985).
2] Scherer, E.M.; de Souza, J.P. and Baumvol, I.J.R., Semicond. Sci. Technol. 1, 241 (1986).
3] Hasenack, C.M.; de Souza, J.P. and Baumvol, I.J.R., Semicond. Sci. Technol. 2, 477 (1987).
4] Dionisio, P.H.; Teixeira, S.R. and Baumvol, I.J.R., Nucl. Instr. Meth. B16, 345 (1986).
5] Dionisio, P.H. and Baumvol, I.J.R., Phys. Stat. Solidi (a) 101, 63 (1987).
6] Vasconcellos, M.A.Z.; da Costa, J.A.T.B.; Schreiner, W.H. and Baumvol, I.J.R., Nucl. Instr. Meth. B37/38, 653 (1989).
7] Pereira, L.G.; Teixeira, S.R.; Schreiner, W.H. and Baumvol, I.J.R., Phys. Stat. Solidi (b) 166, 214 (1991).
8] Hübler, R.; Teixeira, S.R.; Schreiner, W.H. and Baumvol, I.J.R., to be published in Radiat. Effects and Defects in Solids (1991).

CONTRIBUTED PAPERS

ABSOLUTE ELASTIC DIFFERENTIAL CROSS SECTIONS FOR ELECTRON SCATTERING FROM SF_6 IN THE 75-1000 eV RANGE

J. C. NOGUEIRA and M. J. DALLAVALLI*

DEPARTAMENTO DE QUIMICA - UFSCAR
CAIXA POSTAL 676, 13560 SAO CARLOS, BRASIL

ABSTRACT

Absolute elastic differential cross sections have been measured for incident electron energies between 75 and 1000 eV and in the angular range between 10° to 120°. The relative flow technique was used and nitrogen was the secondary gas standard. Integral cross sections have also been determined from extrapolation of the differential cross sections. The data are compared with previous experimental data, showing good agreement.

1.INTRODUCTION

Sulphur hexafluoride (SF_6) is a large and very symmetrical assembly of atoms. SF_6 is a widely used gas as a dielectric in high-voltage equipment. The advantage of using this gas is a reduction of the size of the equipments. However, SF_6^- is formed from SF_6 at near zero electron energies; high voltage breakdown is also a problem. Accurate measurements of elastic differential cross sections (EDCS) for electrons scattered by SF_6 are necessary from the applied point of view and for the test of theoretical scattering models. However, few experimental and theoretical results have been reported in the literature for energies above 100 eV. Studies of this gas are required over a broad energy range. Sakae et al[1] measured elastic differential cross sections for electrons scattered by SF_6 in the 75 - 700 eV energy range and 5°-135° angular range. More recently Jonhstone and Newell[2] reported results for energies between 5

*COPEL/UFPR(Cia Paranaense de Energia) Curitiba, Paraná, Brasil).

and 75 eV, and 10° to 120°. In the present paper values of the absolute elastic differential cross section have been obtained for electron scattering by SF_6 in the energy range 75 to 1000 eV and in the angular range 10° to 120°.

2. EXPERIMENTAL

The equipment used in the present measurements was previously described by Iga et al[3] and will be only briefly discussed here. The scattering chamber is formed by an aluminium cylinder with 60 cm of diameter and 30 cm of height. The background pressure, of the order of 2x10^{-7} Torr, is provided by a six-inch oil diffusion pump equipped with a liquid nitrogen trap. When the gas is admitted to the system, the pressure of work is normally about 1.5x10^{-6} Torr for SF_6 and 8.0x10^{-7} Torr to N_2 , the secondary gas used as standard. The gas flows through a capillary array of 1 mm in diameter, with individual capillary of 0.5 mm diameter and 5 mm long (aspect ratio τ=0.01). An electron beam crosses perpendicularly the target beam and the scattered electrons are selected by a Mollenstedt velocity analyzer[4]. The energy resolution (FWHM) is around 0.6 eV. With this resolution, the inelastically scattered electrons arising from the electronic transitions are discriminated, but the vibrational and rotational transitions are not resolved. The detector view cone is limited by a set of apertures which allows an angular resolution of 0.2^{o}. In the scattering region the diameter of this view cone is about 2.0 mm. With this system, one expects the entire scattering volume to be visible to the detector. The electron beam has a diameter of the order of 1 mm. During the measurements, the primary beam current is set to be around 90 nA at small scattering angles (up to 50°) and 300 nA at large angles (above 40°). It is possible to rotate the electron gun around the scattering center in the angular range from -120° to +120° with an accuracy of 0.1^{o}. In addition, there is a set of six Helmholtz coils placed in the x, y and z direction which reduce the residual magnetic field to around 10 mG in the scattering region.

The relative flow technique[5,6] was used to obtain EDCS values in the absolute scale. The advantage of this technique is to avoid determination of difficult experimental parameters involved during the experiment. The EDCS can be obtained from the following expression:

$$(Ne)_1/(Ne)_2 = \sigma_1(\theta)/\sigma_2(\theta))(P_2/P_1) \qquad (1)$$

where P_1 and P_2 are the absolute gas pressures, $(Ne)_1$ and $(Ne)_2$ are the number of scattered electrons per second and $\sigma_1(\theta)$ and $\sigma_2(\theta)$ are the differential cross sections of gases 1 and 2. If the EDCS for the electron scattered by a certain specimen (the secondary standard) is known, one can obtain from equation (1) the absolute value of the EDCS for the sample gas under study. In the present study, nitrogen was chosen as the secondary standard because of the numerous experimental absolute cross sections reported in the literature for this gas in the energy range under study. Some precautions were taken during the measurements. The stability of the current of the electron beam was monitored by a Faraday Cup. For each energy, the measurements were subsequently repeated for another gas under the same experimental conditions. The pressure was continuously checked by a MKS Baratron. Measurements of the scattering intensity as a function of the gas pressure in the reservoir were performed for both gases and the results showed that this intensity is a linear function of the pressure. The background contribution was determined (measuring the scattering intensity in a sample gas introduced into the system) through a side inlet. At small scattering angles ($<50^{o}$) the contributions were less than 5% and never exceed 10% in the entire angular range. They are subtracted from the intensity of the beam scattering the target gas. For each incident energy, this procedure was repeated several times over different periods. The overall uncertainty in the cross sections is obtained from the equation

$$\Delta = (\Sigma d_i^2 + D^2) \qquad (2)$$

were d_i is the uncertainty of each experimental parameter controlled during the measurements and D is the quoted uncertainty in the absolute cross sections of the secondary standard. We used as the secondary standard the absolute EDCS values for N_2 reported by DuBois and Rudd[7] at impact energies up to 800 eV, with the quoted uncertainty of 12%. At 1000 eV, the absolute EDCS values for N_2 reported by Jansen et al[8] with quoted uncertainty of 6% were used. The various sources which contribute to the experimental uncertainties are: 5% in the ratio of the measured intensities, 1% in the measured pressures for each gas and 1% in the fluctuations of the primary beam current during the measurement. In this way, an overall error of 8% is attributed for the measurement of the EDCS at energy of 1000 eV and 13% for the other energies.

3.RESULTS AND DISCUSSION

The measured EDCS at 75 eV is showed in Figure 1, along with the results of Johnstone and Newell[2], Srivastava et al[9] and Sakae et al[1]. The results measured by Srivastava et al[9] are the results given by Trajmar et al[10]. At 75 eV our results are lower than those of Sakae et al[1] at 10° and higher than those of Johnstone and Newell[2], but within the experimental errors.

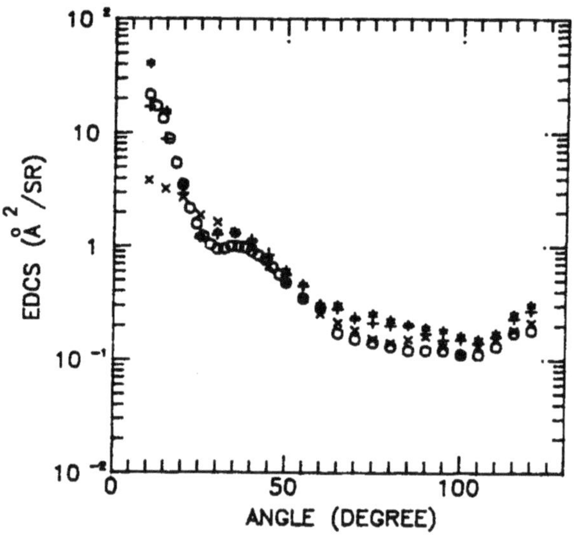

FIGURE 1. Results of differential cross sections for electron scattered by SF_6 at 75 eV. ° our results, + results of reference 1, ∗ results of reference 2 and × results of reference 5.

the maximum around 35° is clearly shown. The results of Srivastava et al[9] at

10, 15 and 20 degree are results extrapolated from their data and are lower than the other results. At high angles (higher than 60°) our results agree with the others in shape but are lower than the results of Johnstone and Newell[2] and Sakae et al[1].

In Figure 2 the results of the EDCS for electrons energies between 100 to 1000 eV are shown. At 100 eV our results agree with the values measured by Sakae et al[1] in the whole angular region except between 60° and 80° . Our results do not show a pronounced maximum. At 200 eV our results agree with the results of Sakae et al[1] except between 85° and 110°, where our results are lower than the results of Sakae et al[1]. At 500 eV the agreement is good in the whole angular region. At an energy of 1000 eV there are no other data to compare our results with.

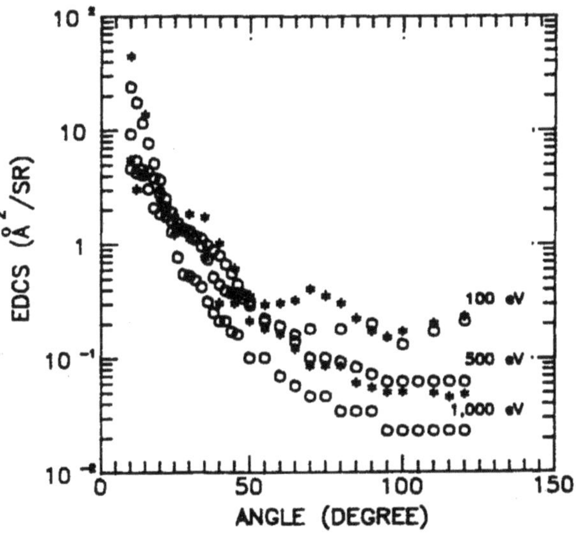

FIGURE 2. Results of elastic differential cross sections for electron incident energy between 100 and 1000 eV. ° our results, * results of Sakae et al[1].

In order to get the integrated total elastic and momentum transfer cross section we extrapolated our data to 0° at small angles and to 180° at angles above 120°.

The results for the integration of the EDCS over all scattering angles are shown in Table 1. While the values of total cross section of Sakae et al[1] decrease slowly between 75 and 700 eV, our results increase between 75 and 100 eV and then decrease up to 800 eV. The values at 800 and 1000 eV are almost the same (within experimental error). Even at 800 and 1000 eV the extrapolation to zero degree is difficult because the cross section measured does not start to increase at 10°. Hence, there is a large error in the value of the extrapolation to zero degree, but the error in the cross section is low because the $\sin(\theta)$ in this region is low to and the product becomes small.

4.CONCLUSIONS

The absolute elastic differential cross section was measured in the 75 to 1,000 eV energy range and 10° to 120° angular region. The overall agreement with previous measured results is good.

The integrated and momentum transfer cross sections were obtained by integration from the measured values and extrapolation was carried out to small and large angles. In this way, the data for SF_6 was extended in energy range.

TABLE 1. Integral cross section for elastic scattering of electrons by SF_6 (units of 10^{-16} cm^2)

ENERGY	75	100	200	300	400	500	800	1000
ICS	9.49	10.2	9.71	8.04	5.67	4.04	2.54	2.47

5.ACKNOWLEDGEMENTS

One of us, M. J. Dallavalli, would like to acknowledge COPEL for the absence. We are indebted with J. R. Vossenaar and Roque Rodrigues for the valuable mechanical workshop support. We acknowledge CNPq, RHAE, FINEP and PADCT for the financial support.

6.REFERENCES

1. T. Sakae, S. Sumiyoshi, E. Murakami, Y. Matsumoto, K. Ishibashi and A. Katase, J. Phys. B: At. Mol. Opt. Phys. 22 (1989) 1385.

2. W. M. Johnstone and W. R. Newell, J. Phys. B: At. Mol. Opt. Phys. 24 (1991) 473.

3. I. Iga, Lee Mu-Tao, J. C. Nogueira and R. S. Barbieri, J. Phys. B: At. Mol. Phys. 20 (1987) 1095.

4. E. M. A. Peixoto, G. G. B. de Souza and M. C. A. Santos, Optik 53 (1979) 405.

5. S. K. Srivastava, A. Chutjian and S. Trajamar, J. Chem. Phys. 63 (1975) 2659.

6. J. C. Nickel, P. W. Zetner, G. Shen and S. Trajmar, J. Phys. E: Sci. Instrum. 22 (1989) 730.

7. DuBois R. D. and Rudd M. E. J. Phys. B: At. Mol. Phys. 9 (1976) 2657.

8. R. H. Jansen, F. J. de Heer, H. J. Luyken, B. van Wingerden and H. J. Blaauw, J. Phys. B: At. Mol. Phys. 9 (1976) 185.

9. S. K. Srivastava, S. Trajmar, A. Chutjian and W. Williams, J. Chem. Phys. 64 (1976) 2767.

10. S. Trajmar, D. F. Register and A. Chutjian, Phys. Rep. 97 (1983) 219.

STUDY OF GAS PHASE PRODUCTS FORMED BY FLASH-VACUUM PYROLYSIS OF HALOMETHANES USING ELECTRON ENERGY-LOSS SPECTROSCOPY

I. M. Xavier Jr.[1], K. N. Walzl[2], M. R. Giorgi and A. Kuppermann
Arthur Amos Noyes Laboratory of Chemical Physics[3],
California Institute of Technology, Pasadena, CA 91125, U.S.A.

ABSTRACT

Thermal decomposition of chlorodifluoromethane, chloroform, dichloromethane and chloromethane under flash-vacuum pyrolysis conditions (900-1100°C) was investigated by the technique of electron energy-loss spectroscopy, using an impact energy of 50 eV and a scattering angle of 10^0. The pyrolytic reaction follows a hydrogen chloride α-elimination reaction pathway. The difluoromethylene radical was produced from the chlorodifluoromethane pyrolysis at 900°C and identified by its $\tilde{X}^1A_1 \rightarrow \tilde{A}^1B_1$ band at 5.04 eV.

1. Introduction

The thermal decomposition of four halomethanes molecules that follow similar pyrolysis pathways provides clues for generation of gas phase halocarbenes. By comparison between the electronic spectra of the pyrolytic precursor molecule at room temperature and at high temperature, one can determine the efficiency of decomposition and identify the possible pyrolytic products.

---

[1] Work performed in partial fulfillment of the requirements for the Ph.D. degree in Chemistry at the California Institute of Technology.

Present address : Departamento de Química Fundamental, Universidade Federal de Pernambuco, 50739 Recife, PE, Brazil.

[2] Present address : Institute for Defense Analyses, 1801 N. Beauregard Street, Alexandria, VA 22311-1772, U.S.A.

[3] Contribution No. 8509

This report presents an electron energy-loss study of the hydrogen chloride (HCl) α-elimination reaction from halomethanes under flash-vacuum pyrolysis (FVP) conditions.[1,2] The FVP conditions employed in this study were 1-10 $mTorr$ of the gas sample over a quartz pyrolyzer surface at 900-1100°C. Under these conditions, thermal activation takes place primarily by contact with the hot pyrolyzer walls, so that processes observed in this study result mainly from very fast heterogeneous reactions.[1]

Electron-impact spectroscopy has been previously used in our laboratory for investigation of the methyl (CH_3) radical and some of its pyrolytic precursors.[3] In that work, the number density of CH_3 radicals in the jet at the pyrolysis temperature was estimated to be 10^{13} molecules/cm^3 on the basis of the intensity of the 5.73 eV band.

2. Experimental

The spectrometer used in the present experiment has been described previously.[4] Briefly, electrons are emitted from a tungsten filament and focused into a hemispherical monochromator. The monoenergetic electrons are then focused into the scattering region and, after interaction with the target molecules, enter a hemispherical energy-loss analyzer prior to detection.

The spectrometer resolution (as measured by the full width at half-maximum of the elastically scattered feature) varied between 70 and 100 meV. For all reported spectra, the incident electron energy was 50 eV and the scattering angle was 10°.

In order to produce thermal decomposition of the halomethanes, an *in situ* pyrolysis technique was employed.[3] The pyrolysis takes place within a graphite-coated quartz tube with an inner diameter of 0.060 *in*. The pyrolyzer is heated by tantalum sheathed double stranded tungsten-rhenium alloy wires wound over a 0.35 *in* length of the quartz tube, and temperatures up to 1100°C (as measured by an optical pyrometer) can be reached. Since the pyrolysis is done at low pressures, the residence time in the pyrolyzer is kept in the few milliseconds range and thermal activation results mainly from collisions with the wall. Because the distance from the end of the pyrolyzer to the scattering center region is short, most of the reactive intermediates should be detected. The background pressures in the scattering chamber are quite low ($< 5 \times 10^{-5} Torr$), so recombination rates are slow.

The investigation of four halomethanes are reported: chlorodifluoromethane (Aldrich, 99.9%); chloroform (J.T.Baker, 99.5%); dichloromethane (J. T. Baker, 99.7%) and chloromethane (Aldrich, 99.5%). The chloroform and dichloromethane samples were subjected to three liquid-nitrogen, freeze-pump-thaw cycles. Hydrogen chloride (Matheson, 99.0%) was used as reference in the study of the high-temperature spectra.

3. Results and Discussion

Figure 1 shows the electron energy-loss spectrum of the HCl jet. The spectrum is featureless in the region 2.0-6.5 eV, which is followed by a broadband continuum with a maximum around 8 eV. The four peaks in the region above 9 eV consist mainly of the $C^1\Pi$ and $b^3\Pi$ states, which are Rydberg states corresponding to the $\pi \rightarrow 4s$ transition.[5] The strongest peak at 9.62 eV is assigned as $C^1\Pi(v' = 0)$, while the two peaks at 9.93 and 10.28 eV are the $C(v' = 1, 2)$ bands, respectively. The small peak at 9.31 eV corresponds to $b^3\Pi(v' = 0)$.[6]

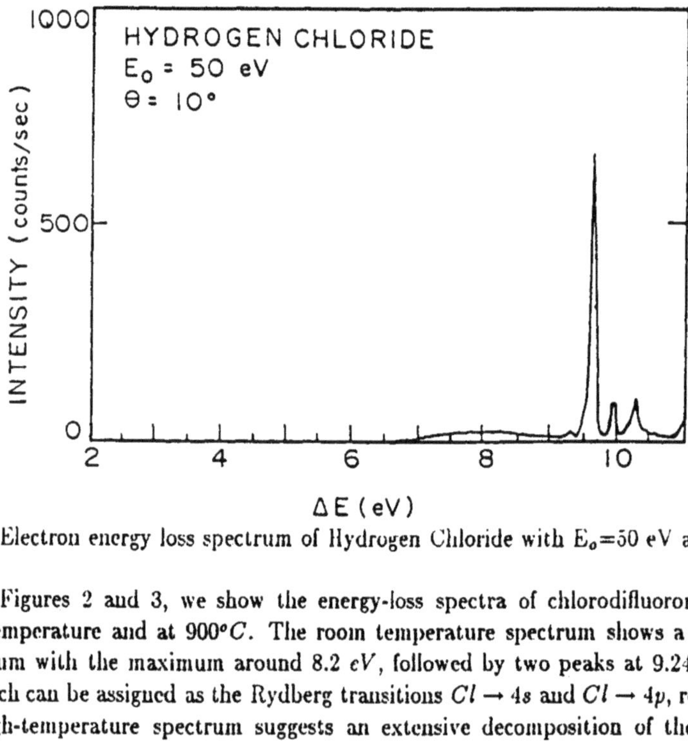

Fig. 1 Electron energy loss spectrum of Hydrogen Chloride with E_o=50 eV and θ=10°.

In Figures 2 and 3, we show the energy-loss spectra of chlorodifluoromethane at room temperature and at 900°C. The room temperature spectrum shows a broadband continuum with the maximum around 8.2 eV, followed by two peaks at 9.24 and 10.00 eV, which can be assigned as the Rydberg transitions $Cl \rightarrow 4s$ and $Cl \rightarrow 4p$, respectively. The high-temperature spectrum suggests an extensive decomposition of the precursor molecule and we can observe the familiar $b(v' = 0)$ and $C(v' = 0, 1, 2)$ bands of HCl at the high-energy-loss side. However, there is a new broadband feature starting at 4.2 eV and peaking at 5.04 eV. We attribute this band to the $\tilde{X}^1A_1 \rightarrow \tilde{A}^1B_1$ transition of the difluoromethylene (CF_2) radical. The spectrum of this radical was first observed and identified in the region 3300-2300 Å by Venkateswarlu.[8] This transition was assigned later

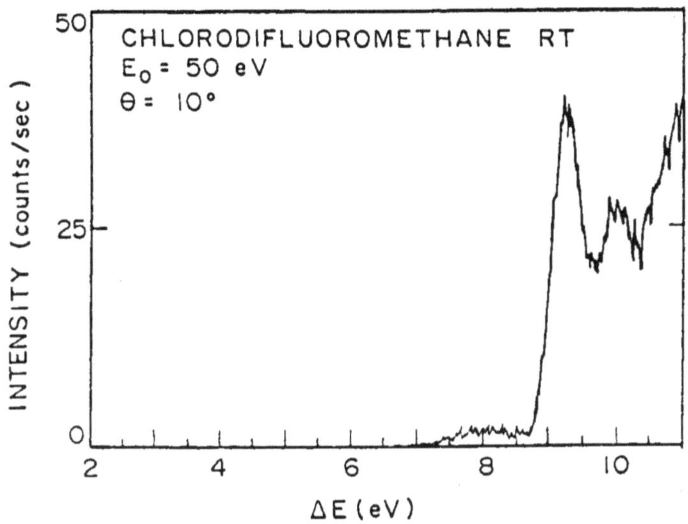

Fig. 2 Electron energy-loss spectrum at room temperature of Chlorodifluoromethane.

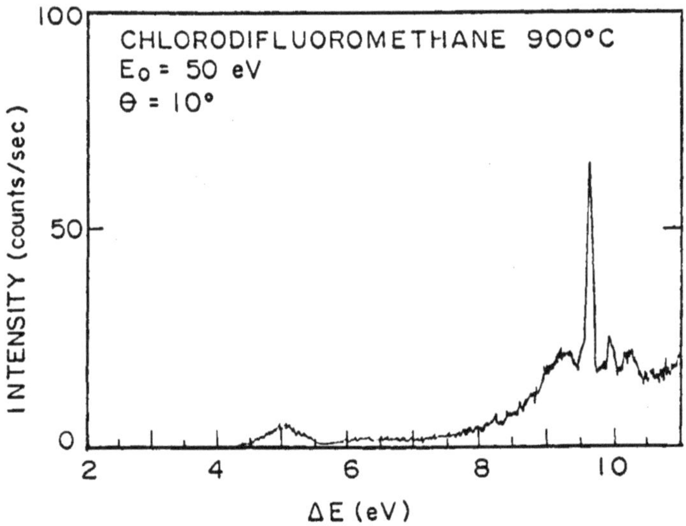

Fig. 3 Electron energy-loss spectrum at 900°C of Chlorodifluoromethane.

by Mathews[9] as $\tilde{X}^1A_1 \rightarrow \tilde{A}^1B_1$. In our case, the possibility of the combination of two CF_2 can be eliminated by comparison with the electron energy-loss spectrum of tetrafluoroethylene.[10] The CF_2 radical appears to be unreactive with itself and is thermochemically stable ($\Delta H_f^\circ = -49 \pm 3\ kcal/mol$).[11]

Figure 4 shows the chloroform spectrum at room temperature. The first band with the maximum at 7.10 eV can be assigned as an $n \rightarrow \sigma^*$ Rydberg transition.[12] The shoulder at 8.31 eV and the peak at 8.72 eV are both believed to stem from the $n \rightarrow 4s$ Rydberg manifold.[12] The higher energy transition at 9.26 eV can be attributed as an $n \rightarrow 4p$ Rydberg. The one dominant feature at still higher energy is a peak at 10.62 eV.

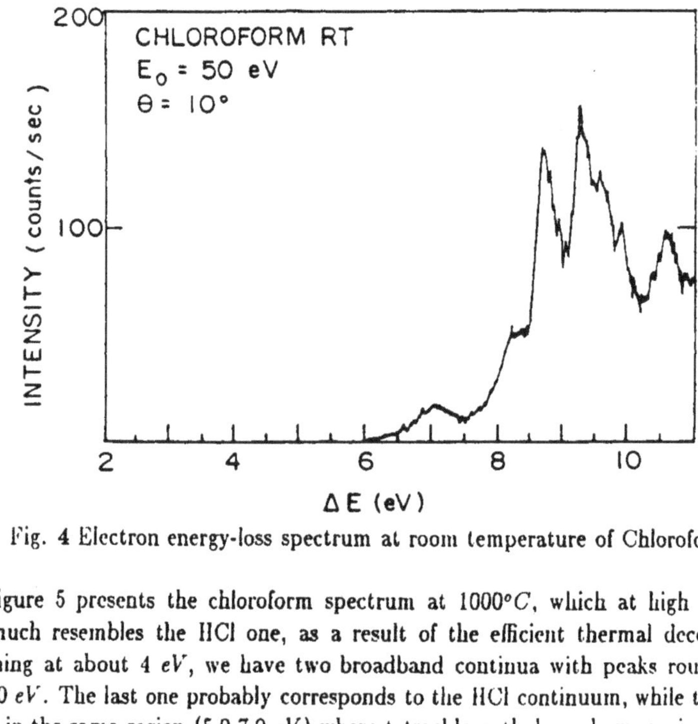

Fig. 4 Electron energy-loss spectrum at room temperature of Chloroform.

Figure 5 presents the chloroform spectrum at 1000°C, which at high energy loss very much resembles the HCl one, as a result of the efficient thermal decomposition. Beginning at about 4 eV, we have two broadband continua with peaks roughly at 6.3 and 8.0 eV. The last one probably corresponds to the HCl continuum, while the first one occurs in the same region (5.0-7.0 eV) where tetrachloroethylene shows an intense broad transition.[4] Tetrachloroethylene can be formed by bimolecular processes that usually do not occur under FVP conditions except in the case of radical-radical reactions.[2] These facts suggest that the first broadband is probably due mainly to the combination of two dichloromethylenes (CCl_2) to give tetrachloroethylene. It is known that the radical CCl_2 has a weak visible absorption band in the region 2.21-2.81 eV[13] and an ultraviolet absorption band centered at 3.76 eV.[14] The shoulderlike band in the region 4.0-5.0 eV could tentatively be attributed as the $^1A_1 \rightarrow {}^1A_2$ band of CCl_2, which *ab initio* and CI calculations predict to be at 4.5 eV.[15] However, there is no conclusive evidence for the presence of thermochemically unstable CCl_2 ($\Delta H_f^\circ = 39 \pm 3$ $kcal/mol$).[11]

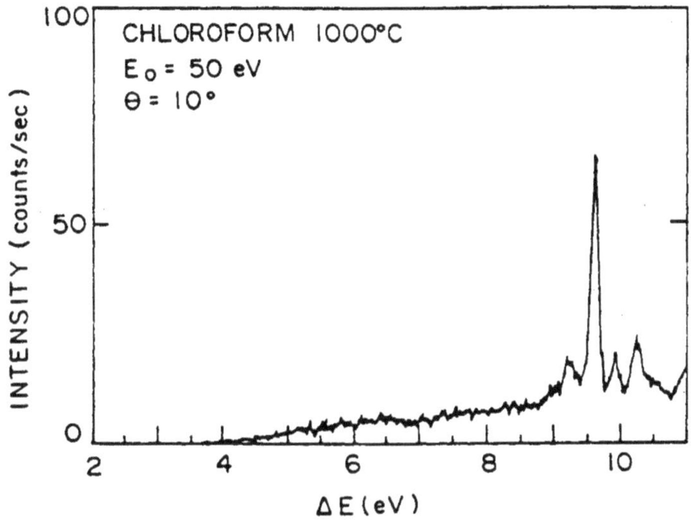

Fig. 5 Electron energy-loss spectrum at 1000°C of Chloroform.

In Figures 6 and 7, the dichloromethane spectra are displayed at room temperature and at 1100°C, respectively. The room temperature spectrum has a shoulder at 7.16 eV that can be assigned as an $n \rightarrow \sigma^*$ Rydberg transition.[12] The first peak with the maximum at 8.26 eV can be attributed to degenerate $n \rightarrow 4s$ Rydberg transitions, while the broadband at 9.12 eV is due to $n \rightarrow 4p$ Rydberg transitions.[12] At still higher energy is a sharp peak at 10.30 eV. The high-temperature spectrum suggests that the dichloromethane is only partially decomposed at 1100°C. The strong and sharp peak at 9.6 eV is the $C^1\Pi(v' = 0)$ transition of the HCl. Except for a shoulderlike feature in the region 5.0-6.0 eV that could not be assigned, the rest of the spectrum resembles a broad and noisy version of the room temperature one. No evidence for the presence of the thermochemically unstable CClH ($\Delta H_f^\circ = 71 \pm 5$ $kcal/mol$)[11] was found at this temperature.

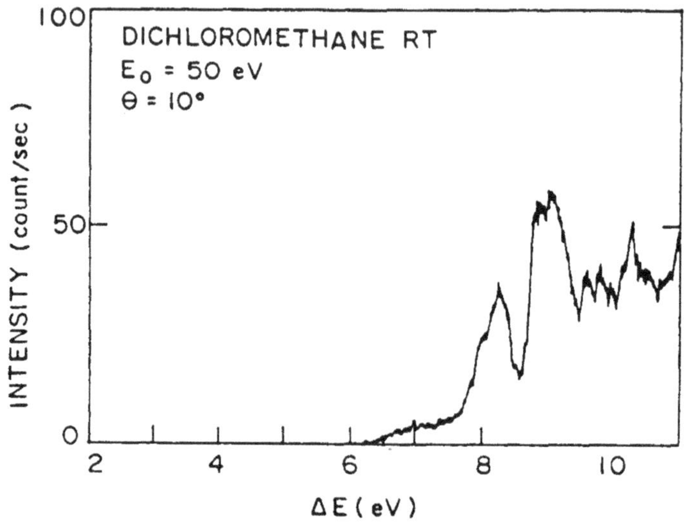

Fig. 6 Electron energy-loss spectrum at room temperature of Dichloromethane.

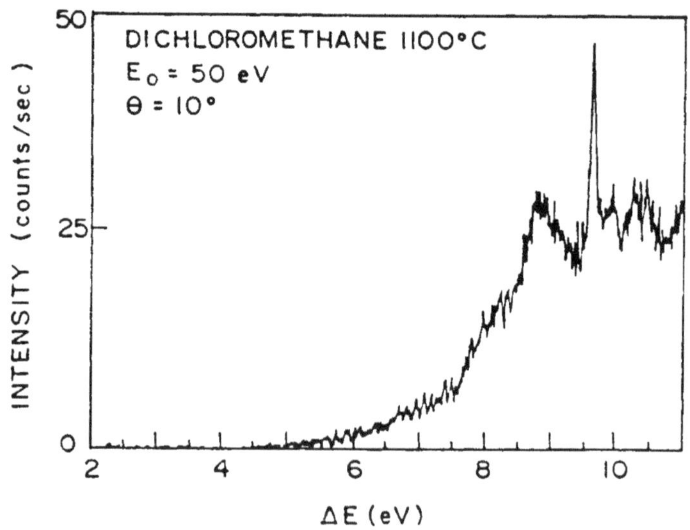

Fig. 7 Electron energy-loss spectrum at 1100°C of Dichloromethane.

Figure 8 shows the room temperature spectrum of chloromethane. The shoulder from 6.50 to 7.65 eV can be assigned as an $n \rightarrow \sigma^*$ Rydberg transition.[12] The next band centered at 7.91 eV is attributed to an $n \rightarrow 4s$ Rydberg transition.[12] Then, we have an intense feature at 8.91 eV assigned as an $n \rightarrow 4p$ Rydberg transition, followed by a sharp and intense peak at 9.22 eV, which has been tentatively assigned to the $n \rightarrow 3d$ Rydberg transition.[12] Another intense peak is at 10.20 eV and can be considered as the $n \rightarrow 5p$ Rydberg transition.[12]

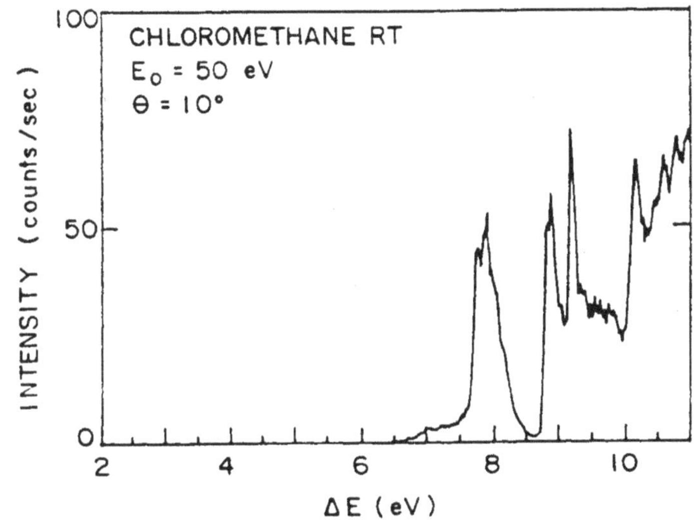

Fig. 8 Electron energy-loss spectrum at room temperature of Chloromethane.

Finally, Figure 9 presents the spectrum of chroromethane at 1100°C, which is noisy and the lacks fine structure of the one at room temperature. There is no HCl feature about 9.6 eV, and we can assume that there was no measurable thermal decomposition at this temperature.

4. Summary

In conclusion, we have used the method of electron energy-loss spectroscopy to investigate the pyrolysis of four halomethanes via HCl α-elimination reaction. At sufficiently high temperature we observed a complete and clean thermal unimolecular decomposition of halomethanes that contain at least one hydrogen and one chlorine atom. The difluoromethylene radical can be easily generated by the chlorodifluoromethane pyrolysis at 900°C. We also assigned the far-ultraviolet bands of the pyrolytic precursor molecules based only on the optical spectra. To our knowledge, no low-energy-loss study of these compounds has been published.[16,17]

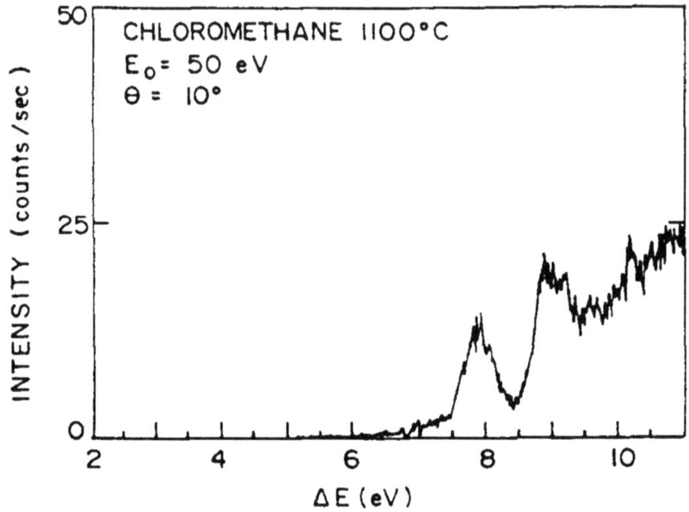

Fig. 9 Electron energy-loss spectrum at 1100°C of Chloromethane.

5. Acknowledgements

We thank the U.S. Departament of Energy (Contract No. DE-AM03-76F00767, Project Agreement No. DE-AT03-76ER72004) for financial support.

6. References

1. D. M. Golden, G. N. Spokes, and S. W. Benson, *Angew. Chem. Int. Ed. Engl.*, **12**, 534 (1973).
2. E. Hedaya, *Acc. Chem. Res.*, **2**, 367 (1969).
3. K. N. Walzl, C. F. Koerting, I. M. Xavier Jr., and A. Kuppermann, *J. Chem. Phys.*, **86**, 89 (1987).
4. C. F. Koerting, K. N. Walzl, and A. Kuppermann, *Chem. Phys. Lett.*, **109**, 140 (1984).
5. M. Bettendorff, S. D. Pluerimhoff, and R. Buenker, *Chem. Phys.*, **66**, 261 (1982).
6. J. B. Nee, M. Suto, and L.C. Lee, *J. Chem. Phys.*, **85**, 719 (1986).

7. R. Gilbert, P. Sauvageau, and C. Sandorfy, *J. Chem. Phys.*, **60**, 4820 (1974).
8. P. Venkateswarlu, *Phys. Rev.*, **77**, 676 (1959).
9. C. W. Mathews, *Can. J. Phys.*, **45**, 2355 (1967).
10. M. J. Coggiola, W. M. Flicker, O. A. Mosher, and A. Kuppermann, *J. Chem. Phys.*, **65**, 2655 (1976).
11. S. G. Lias, Z. Karpas, and J. F. Liebman, *J. Am. Chem. Soc.*, **107**, 6089 (1985).
12. B. R. Russell, L. O. Edwards, and J. W. Raymonda, *J. Am. Chem. Soc.*, **95**, 2129 (1973).
13. D.E.Milligan and M. E. Jacox, *J. Chem. Phys.*, **47**, 703 (1967).
14. T. Ha, H. U. Gremlich, and R. E. Bühler, *Chem. Phys. Lett.*, **65**, 16 (1979).
15. M. T. Nguyen, M. C. Kerins, A. F. Hegarty, and N. J. Fitzpatrick, *Chem. Phys. Lett.*, **117**, 295 (1985).
16. A. Kuppermann, W. M. Flicker, and O. A. Mosher, *Chem. Rev.*, **79**, 77 (1979).
17. S. Trajmar, D. F. Register, and A. Chutjian, *Phys. Rep.*, **97**, 219 (1983).

LASER-EXCITED FLUORESCENCE SPECTRUM OF PbI

P. PATIÑO, A. TOVAR and D. IACOCCA
Laboratorio de Espectroscopia Molecular y Plasmaquimica,
Escuela de Quimica, Facultad de Ciencias,
Universidad Central de Venezuela.
Apartado 47102, Caracas, Venezuela.

ABSTRACT

The PbI fluorescence spectrum has been obtained from excitation by the 4880 Å line of an argon ion laser. From a fit to the observed lines, preliminary rotational constants have been estimated and vibrational ones for the ground state have been improved.

1.- Introduction

It is of interest to study the PbI spectrum based on of the heavy character of case c for the A and X states, specially for the former which is expected to be a pure (c) 1/2 state[1]. Nevertheless, the fact that only one paper has been published on the analysis of the lead monoiodide band spectrum reflects some difficulty on achieving it. The relatively small values of the rotational constants and the resulting density of lines would make the analysis of classical absorption and emission spectra laborious; moreover, among the lead monohalides, PbI presents the weakest spectrum, the A - X visible system being the faintest one[2].

In the hope of making some progress, presented here are the preliminary results of a study of laser-excited fluorescence in PbI, recorded at modest resolution with conventional grating spectroscopy. This simple technique has prooved to be very reliable in yielding useful information for BiS[3] and BiBr[4,5]. A respectable number of extensive fluorescence series have been induced on these heavy molecules by several lines of both argon ion and kripton ion lasers.

2.- Experimental

The experimental conditions were similar to those employed for BiS and BiBr. The spectrum of PbI was obtained from a sample in a 2 cm^3 cylindrical quartz cell with flat window at one end containing about 0.2 mg of enriched ^{208}Pb plus 0.2 mg of I_2 and heated to about 750 °C. Lead monoiodide, which is readily produced at this temperature, was pumped by the argon ion laser which was tuned on the 4880 Å line at the power of 1.5 W. The fluorescence spectrum was photographed in the second order of a 3.4 m JARRELL-ASH Ebert spectrograph equipped with a plane 590 lines/mm grating on Kodak 103a-O and 103a-F plates and Ilford HP5 films. Exposures ranged from 9 to 12 hours. Fe and Ne hollow cathode lines were used as standards for calibration. The reciprocal dispersion in the region 4600-6070 Å was approximately 2.4 Å/mm. Fluorescence lines were measured with a precision of $\pm$ 0.06 cm^{-1}.

3.- Observation and analysis

Four fluorescence series, which consist of transitions in the A - X system, were observed. The pattern was that of the usual doublets R(J-1) and P(J+1). Due to the fact that no previous rotational structure had been observed in the PbI spectrum, the assignment of levels was tentatively made. Lumely and Barrow[1] have concluded that the A state of PbF is a pure case (c) 1/2 state with high spin splitting ($p/B \simeq$ 3.0, very close to the theoretical value obtained from the Kopp and Hougen equations[6]). Ω-type splitting in $X_1{}^2\Pi_{1/2}$, although smaller ($p/B \simeq -0.6$) is significant as well. For PbI higher splitting is expected in both states. Unfortunately, as no additional weaker lines arising from collision-induced changes $\Delta J'$ were observed, it was impossible to determine the parity of the lines, e or f. Hence the approach for this preliminary analysis was first to estimate a value for B_e''. From certain regularities in the B_e'' values of CF, CCl, SiF, SiCl, SiI, SnF and SnCl[7] it seemed reasonable to assume $B_e(PbF)/B_e(PbI)$ to be $\simeq$ 5, and with $B_e''(PbF) = 0.2288$ cm^{-1}, we could deduce $B_e''(PbI) \simeq 0.046$ cm^{-1}. From the Peckeris relation[8] a value for α_e'' was also estimated and an attempt to determine molecular constants was done by simultaneous fits to measured lines, in which upper and lower state term values:

$$T_{v,J} = T_e + \omega_e(v+1/2) - \omega_e x_e(v+1/2)^2 + \omega_e y_e(v+1/2)^3 + \ldots$$

$$+ (B_e - \alpha_e(v+1/2) + \ldots)J(J+1) + \ldots$$

were represented by an appropriate number of adjustable parameters.

With the vibrational and rotational numbering tentatively proposed in table I, constants shown in table II were derived. From the values obtained for $B_7' = 0.04458$ and $B_{10}' = 0.03435$ cm^{-1} compared to that of B_e'', the high splitting in the rotational levels of PbI was confirmed. Its effect (not taking into account as previously explained) is clearlyly observed through the constants and it seems that the level excited in the series labeled as R(44.5), $v'=7$ (excitation in 7-6 band) is f and that of the series R(85.5), $v'=10$ (excitation in 10-8 band) is e. Therefore, more lines

Table I. Tentative assignment of the fluorescence series excited by the 4880 A Ar II line in A(1/2) - $X^2\Pi_{1/2}$, ^{208}PbI.

v′	v″	Transition	Range of v″ observed
7	6	R(44.5)	5 - 32
10	8	R(88.5)	4 - 34
11	8	P(108.5)	21 - 35
15	11	P(161.5)	3 - 8

Table II. Constants for $X^2\Pi_{1/2}$, ^{208}PbI (cm^{-1}). Values in parentheses are of 2σ. The 96 lines were fitted and weighted standard deviation of the fit was 0.234 cm^{-1}. [ ] indicates a fixed value.

ω_e	160.631(108)	B_e	0.04073(112)
$\omega_e x_e$	0.3167(58)	α_ex10^6	0.78(360)
$\omega_e y_e$	0.363(98)	γ_ex10^9	[1.0]
		D_ex10^9	[14.1]

Table III. Vacuum wavenumbers of observed lines

SERIES 1

v″	R(44.5)	P(46.5)	v″	R(44.5)	P(46.5)
5	20644.17	-	20	18355.11	18348.83
6	LASER	20480.75	21	18207.16	18201.24
7	20331.07	20324.87		-	-
8	20175.43	20169.34	23	17913.21	17907.38
9	20020.45	20014.40	24	17767.23	17761.28
	-	-	25	17621.72	17615.72
12	19558.71	19552.46		-	-
13	19406.10	19400.04	27	17332.43	17326.80
	-	-	28	17188.66	17182.80
16	18952.13	18945.99	29	17045.35	17039.55
17	18801.80	18795.86		-	-
18	18652.29	18646.25	31	16760.84	16754.99
	-	-	32	16619.41	16613.62

SERIES 2

v"	R(88.5)	P(90.5)	v"	R(88.5)	P(90.5)
4	21112.59	21100.57	20	18666.11	18654.86
5	20995.39	20943.49		-	-
6	20798.67	20786.78	22	18370.95	18359.84
7	20642.71	20630.54	23	18224.65	18213.38
8	LASER	20475.33		-	-
	-	-	26	17788.19	17777.02
12	19869.96	19858.52	27	17643.95	17632.59
13	19717.26	19705.72		-	-
14	19565.34	19552.46	29	17356.93	17345.89
15	19415.32	19402.95	30	17214.40	17203.25
16	19263.60	19252.05	31	17072.24	17061.27
17	19113.04	19101.69		-	-
18	18963.27	18952.13	33	16790.09	16779.03
19	18814.60	18803.10	34	16649.81	16638.78

SERIES 3

v"	R(106.5)	P(108.5)
21	18517.48	18503.00
22	18370.42	18355.11
	-	-
25	17932.04	17917.75
26	17786.68	17772.59
	-	-
28	17498.64	17484.57
	-	-
32	16929.09	16915.02
33	16788.29	16774.11
	-	-
35	16508.14	16494.19

SERIES 4

v"	R(159.5)	P(161.5)
3	21719.80	21699.55
4	21563.98	21543.35
	-	-
7	21126.04	21105.31
8	20970.98	20950.12

would be necessary in order to include the effect of parameter p in the fits and hence to get reliable constants. Nevertheless, it is considered appropriate to present in table III the frequencies of the observed lines as a contribution for future calculations. Apart from the two series previously mentioned, shown as series 1 and 2, excitation in series 3 is in P(108.5), 11-8 band, and in series 4 is in P(161.5), 15-11 band. For the moment, from the v" levels observed, vibrational constants of the ground state have been improved, $\omega_e''y_e''$ having been calculated for the first time.

4.- Acknowledgements

This investigation has been supported by CONICIT (Grant S1-1530) and CDCH-UCV (Grant C-03.19/83). Thanks to them.

5. References

1. D. J. W. Lumley and R. F. Barrow, *J. Phys. B: Atom. Molec. Phys.* **10** (1977) 1537.
2. K. Wieland and R. Newburgh, *Helv. Phys. Acta* **25** (1952) 87.
3. P. Patiño, J. H. D. Eland and R. F. Barrow, *J. Phys. B: At. Mol. Phys.* **17** (1984) 1009.
4. P. Patiño, M. Caetano, D. Iacocca and M. Marcano, *Spectrochim. Acta* **42A** (1986) 35.
5. M. Caetano, P. Patiño, D. Iacocca and M. Marcano, *Acta Cient. Ven.* **37** (1986) 151.
6. I. Kopp and J. T. Hougen, *Can. J. Phys.* **45** (1967) 2581.
7. K. P. Huber and G. Herzberg, *Molecular Spectra and Molecular Structure. Vol. IV, Constants of Diatomic Molecules*, (Van Nostrand Reinhold, New York, 1979).
8. G. Herzberg, *Molecular Spectra and Molecular Structure, Vol. I. Spectra of Diatomic Molecules*, (Van Nostrand Reinhold, New York, 1950).

A DISTORTED WAVE MODEL FOR ELECTRON CAPTURE IN LARGE ANGLE COLLISIONS WITH CORRECT ASYMPTOTIC CONDITIONS

O.A. FOJON and J.M. MAIDAGAN

Instituto de Fisica Rosario, Facultad de Ciencias Exactas e Ingenieria, Consejo Nacional de Investigaciones Cientificas y Tecnicas and Universidad Nacional de Rosario, Av. Pellegrini 250, 2000 Rosario, Argentina.

ABSTRACT

We have calculated the K-K electron capture probability at large scattering angles by using a symmetric second order model which is closely related with the Continuum Distorted Wave approximation. The influence of the Coulombic distortion of the nuclei is investigated by comparison with OBK and CIS-type calculations. Moreover, the influence of a nuclear resonance on the capture probabilities is analyzed. Numerical results for proton impact on N at intermediate energies are compared with experimental data.

1. Introduction

Electron capture in close encounter collisions has been subject of active experimental works in the last years[1-7]. In these works the study was focused on the charge exchange at large scattering angles by protons on multielectronic targets. The related problem of the influence of a nuclear resonance on the electron capture process has been recently investigated by experimentalists[8-10]. The interplay between the atomic and nuclear phenomena becomes evident only at large angles where the Rutherford scattering is not dominant. Therefore, the theoretical description of large angle-close encounter collisions seems to be essential in order to investigate this interplay. However, only limited theoretical work has been made on the subject[11-13]. In previous works, the Shakeshaft and Spruch (SS)[14] and Continuum Intermediate State Type (CIST)[15] approximations, which are closely related to the standard OBK and CIS models, respectively, have been presented. In the OBK (Oppenheimer-Brinkman-Kramers) approximation, the initial and final wavefunctions of the active electron are taken as non perturbed wavefunctions bounded to the target and to the projectile, respectively. This is equivalent to omit the distortion produced by the projectile or the target charge on the initial or final evolutions, respectively. In this sense, this is a first order approximation. The absence of any distortion leads to a general overestimation of the cross sections at high and intermediate energies. In the CIS (Continuum Intermediate State) approximation, the non perturbed final wavefunction is distorted by a Coulomb factor associated with the electron-target nucleus interaction. Implicit electron continuum states are thus introduced. So that, two step collision mechanisms are considered in the description. In this way, this is a second order approximation. The main conclusion of these previous works was that the second-order effects included in the CIST model are of great importance to describe adequately the experimental data. In asymmetric collisions the CIST calculation takes into account the Coulombic distortion in the final wave function due to the presence of the strong charge of the target nucleus. This distortion changes drastically the angular dependence of the capture probabilities with respect to the OBK results. In spite of the good agreement among the CIST probabilities and the experimental data, the CIST model is not quite satisfactory from the theoretical point of view. As has been pointed out by Belkić et al[16], the asymptotic boundary condition in the entrance channel is not preserved. The aim of the present work is to take into account the weak interaction between the active electron and the projectile in the entry channel in order to preserve the correct asymptotic behaviour of the initial wave function. To this end, we propose the symmetrization of the CIST approximation which leads to a CDW-type model (hereafter CDWT). From a physical point of view is to be expected that the effect of this symmetrization on the capture probabilities should not be too large for charge-asymmetric systems. But a quantitative study of this effect appears to be useful in order to test the validity of the different approximations. It must be remarked that the present model is not *the large angle version* of the standard CDW approximation[17]. Our aim is less ambitious: we are interested here in the study of the influence of the projectile-active electron interaction in the entrance channel on the capture probabilities and we have selected one among various simples choices in order to include the projectile distortion on the initial wave function. This permits us to investigate the influence of the Coulombic distortion of both nuclei by comparison with the SS and CIST models. Atomic units will be used throughout.

2. Theory

We consider a K-K electron capture process by proton impact on a multielectronic atom. We adopt here the active electron model as described in section 12 of reference 16 in which the presence of the passive electrons is taken into account by means of an effective one-electron potential V_T. We indicate with $\vec{r}_T$ and $\vec{r}_P$ the position vector of the active electron with respect to the target nucleus and projectile, respectively. The position vectors of the projectile relative to the target nucleus will be denoted by $\vec{R}$. We will adopt the semi classical picture in which the heavy particles follow a classical path. This classical path will be assumed to be a broken straight line as described in references 13 and 14. The initial and final relative velocity will be denoted by $\vec{v}^{(-)}$ and $\vec{v}^{(+)}$, respectively. In the following, we will consider that the target remains at rest in the laboratory reference frame. Farther on, we will include the effect of the target recoil on the capture amplitude.

The final wave function is defined as in reference 15 and reads:

$$\chi_f^{CDWT} = \chi_f^{CIST} = \mathcal{L}_f \; \Phi_f \tag{1}$$

$$\Phi_f = \phi_f(\vec{r}_T - \vec{v}^{(+)} t) \;\; \exp(-i\varepsilon_f t) \;\; \exp\left[i(\vec{v}^{(+)} . \vec{r}_T - \frac{1}{2} v^2 t) \right] \tag{2}$$

$$\mathcal{L}_f = \exp(\frac{1}{2} \nu_T \pi) \;\; \Gamma(1+i\nu_T) \;\; {}_1F_1(-i\nu_T, 1, -i(v\, r_T + \vec{v}^{(+)} . \vec{r}_T)) \tag{3}$$

where $\nu_T = z_T^*/v$, v is the magnitude of the relative velocity and ε_f the energy of the final orbital ϕ_f. The effective target distortion z_T^* has been selected following the criterion of Belkić et al[16]. This choice of the final distorted wave χ_f^{CDWT} takes into account partially inertial effects related to the abrupt change in the projectile velocity: the final electronic momentum distribution cannot follow (in a time reversed picture) this change. This is the motivation for the inclusion of $\vec{v}^{(+)}$ for all time in Eq. 2 and 3. The orbital Φ_f moves always on a straight line $\vec{v}^{(+)}t$ even if the actual projectile moves along the broken straight line $\vec{v}^{(+)}t$ for t>0 and $\vec{v}^{(-)}t$ for t<0. In this model the relative velocity between the electronic orbital Φ_f and the distorting target nucleus is $v^{(+)}$. Therefore, there is no ambiguity in the choice of the distorting wave $\mathcal{L}_f$.

To include in a simple manner the Coulombic distortion of the projectile on the initial wave function, we define:

$$\chi_i^{CDWT} = \mathcal{L}_i \; \Phi_i \tag{4}$$

$$\Phi_i = \phi_i(\vec{r}_T) \;\; \exp(-i\varepsilon_i t) \tag{5}$$

$$\mathcal{L}_i = \exp(\frac{1}{2} \nu_P \pi) \;\; \Gamma(1-i\nu_P) \;\; {}_1F_1(-i\nu_P, 1, i(v\, r_P + \vec{v}^{(-)} . \vec{r}_P)) \tag{6}$$

where $\nu_P = Z_P/v$ and ε_i is the energy of the initial orbital ϕ_i. With this choice, the capture amplitude reads:

$$a^{CDWT} = a_{(-)}^{CDWT} + a_{(+)}^{CDWT} \tag{7}$$

$$a^{CDWT}_{(\mp)} = \pm i \int_0^{\mp\infty} dt < \chi_f^{CDWT} \mid A_i \chi_i^{CDWT} > \tag{8}$$

$$A_i \chi_i^{CDWT} = \Phi_i \left[- \frac{\nabla \phi_i}{\phi_i} + i \left(\vec{v}^{(\mp)} - \vec{v}^{(-)} \right) \right] . \nabla_{\vec{r}_T} \mathcal{L}_i \tag{9}$$

The second term in the right side of Eq.9 is the result of our particular choice of the distorting wave in Eq. 6. In order to take into account even approximately the electronic inertial effects (as we have made in the final wave function), we have kept the velocity $\vec{v}^{(-)}$ unchanged in the out way even if the actual distorting charge moves with velocity $\vec{v}^{(+)}$. This procedure preserves the continuity of χ_i for t=0. However, we have included implicitly the actual relative velocity through $\vec{r}_P$ in the argument of the distorting wave $\mathcal{L}_i$. This is the reason for the appearance of this non standard perturbation in Eq. 9. Due to the presence of the second term in Eq. 9, it is clear that this choice will become inadequate when the collision energy increases. However, for the intermediate impact energies which are of interest in the present calculation, the two terms in Eq.9 are of the same order of magnitude. For these energies, the initial orbital velocity, which is implicit in the term containing $\nabla\phi_i$, is comparable to the impact velocity v. Therefore, the choice of the distortion in Eq. 6 is consistent in this sense. In writing $\vec{r}_P$ in equation Eq. 6 we have partially broken the symmetry of our procedure in what respect to the inertial effects related to the abrupt change in the projectile velocity. An alternative and, without any doubt, more adequate and symmetric choice for high impact energies, is the replacement of $\vec{r}_P$ for $\vec{r}_T - \vec{v}^{(-)}t$ for all time in Eq. 6. This leads to a perturbing potential $V_P(\vec{r}_T-\vec{v}^{(-)}t) - V_P(\vec{r}_T-\vec{v}^{(+)}t)$ instead of that which is described by the second term of Eq. 9. However, its numerical manipulation is more difficult than the one involved in our actual choice.

In writing Eqs.1-9 we have assumed that the target remains at rest in the laboratory reference frame. The corrections to the capture amplitude due to the target recoil can be evaluated perturbatively[13-15] by adding to a^{CDWT} the following recoil amplitude:

$$a^{CDWT}_{Rec} = < \chi_f^{CDWT} \mid \exp(-i\vec{v}_{Rec} . \vec{r}_T) - 1 \mid \chi_i^{CDWT} > \tag{10}$$

where $\vec{v}_{Rec}$ is the target recoil velocity and the bracket must be calculated at t=0.

3. Numerical Results

We shall not detail here the numerical calculation of the capture amplitude because the scheme developed in references 14 and 15 can be modified easily to take into account the distortion in the initial channel. As in the referred papers, we have restricted our calculation to the cases where the impact velocity is much larger than the final orbital velocity. Although the calculation is straightforward, the final expression of the capture amplitude is somewhat cumbersome and we shall not give it here. It involves a three-dimensional numerical integration of elementary functions.

In order to apply the present model to the K-shell electron capture from multielectronic targets, we have used hydrogenic orbitals with effective nuclear charge $Z_{Teff}=Z_T-5/16$ and orbital energies from reference 18. In Fig. 1 we show the theoretical electron capture probabilities for 1.058 MeV proton impact on Nitrogen. The experimental data from reference 7 include electron capture from K and L shell of Nitrogen and , therefore, the comparison with theoretical curves should be made with caution. However, it is expected that contributions coming from other shells than K shell be lesser than 20 % of the total capture probabilities. In order to show the relative importance of the Coulombic distortion of the nuclei, we present SS, CIST and CDWT probabilities. It is apparent from examination of this figure that the inclusion of the projectile Coulombic distortion modifies slightly the CIST predictions. The main effect of this distortion is to increase the capture probabilities in the forward direction ($\theta < 90$) and to reduce them in the backward angles. On contrary, the target distortion appears to be crucial in order to describe the trend of the experimental data. It must be noted that some of the present impact energies are below the ones which follow from the empirical rule for the validity of the CDW model[16]. In our case, this rule reads $E(\mathrm{KeV}) > 80|\varepsilon_1|$.

In order to complete this study of charge transfer on Nitrogen, we calculate capture probabilities in the presence of a nuclear resonance by using the following simple two-amplitude formula[20]:

$$P^{cap}(E,\theta) = \frac{d\sigma}{d\Omega} \Big/ |f(E,\theta)|^2 = \left| a_{(-)} \frac{f(E-\Delta E,\theta)}{f(E,\theta)} + a_{(+)} \right|^2 \tag{11}$$

where: $\frac{d\sigma}{d\Omega}$ is the differential cross section for charge exchange; $f(E,\theta)$ is the nuclear amplitude; ΔE is the atomic energy transfer and $a_{(-),(+)}$ the atomic amplitudes. In writing (7) it is assumed that the elastic scattering is the dominant process so that $|f|^2$ can be regarded as the differential cross section.

In Figure 2 we show the theoretical capture probabilities for protons on Nitrogen at θ_{lab} =150 deg around the 1.058 MeV nuclear elastic resonance. We present CIST and CDWT probabilities. In the case of CDWT probabilities we have not included the recoil amplitude in the evaluation of the total amplitude; however, we have taken into account its effect through a multiplicative factor in the capture probabilities. The nuclear amplitudes were calculated in the same way as in reference 20. The experimental results are from reference 6. In spite of the fact that the CDWT underestimates some experimental values it is able to describe the gross features of the capture probabilities.

It is worth to mention here that the comparison of atomic amplitudes

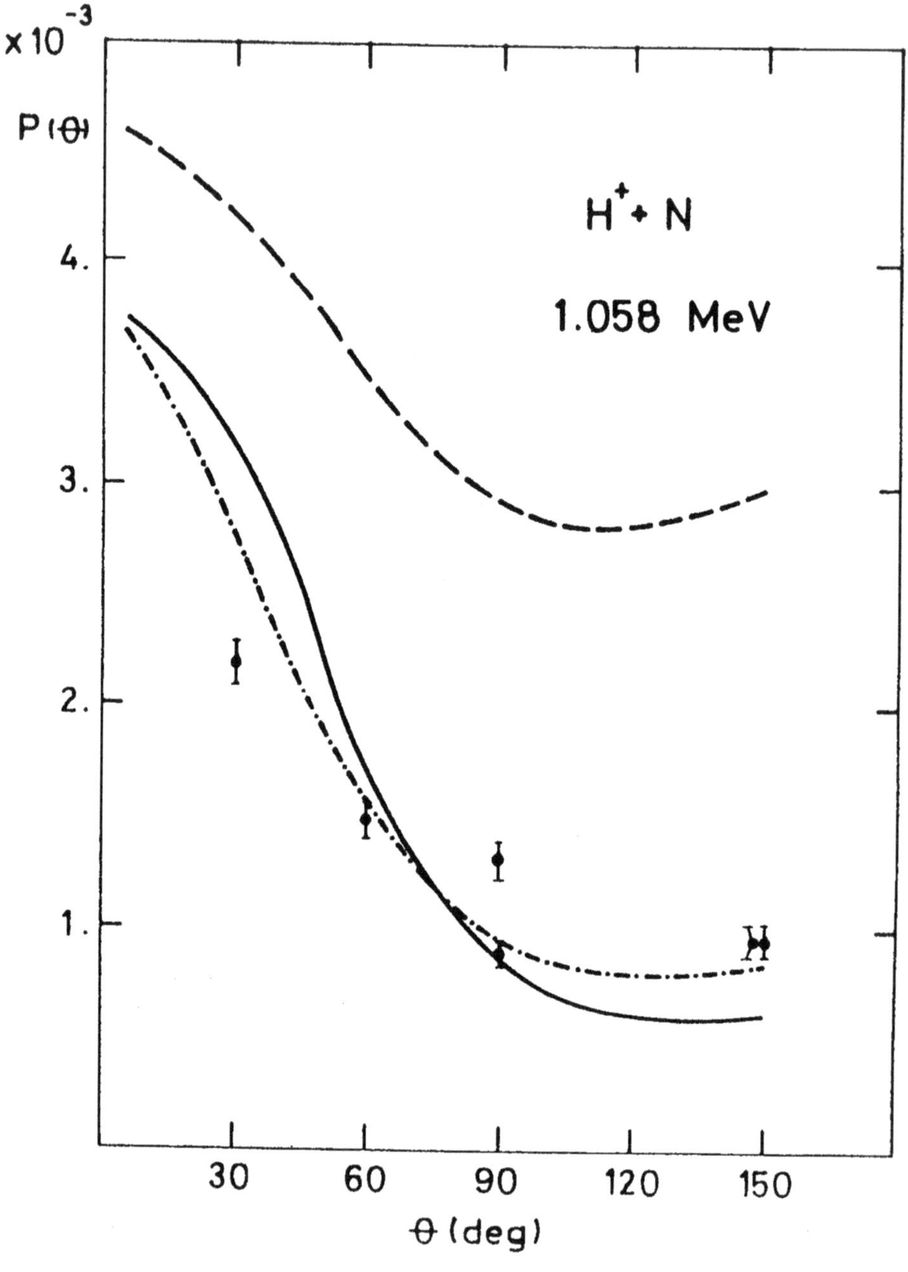

Figure 1. Probabilities for electron capture from N by impact of 1.058 Mev protons : ———— present CDWT ;-.-.- present CIST; - - - - present SS; experimental data (●) from Ref. 7.

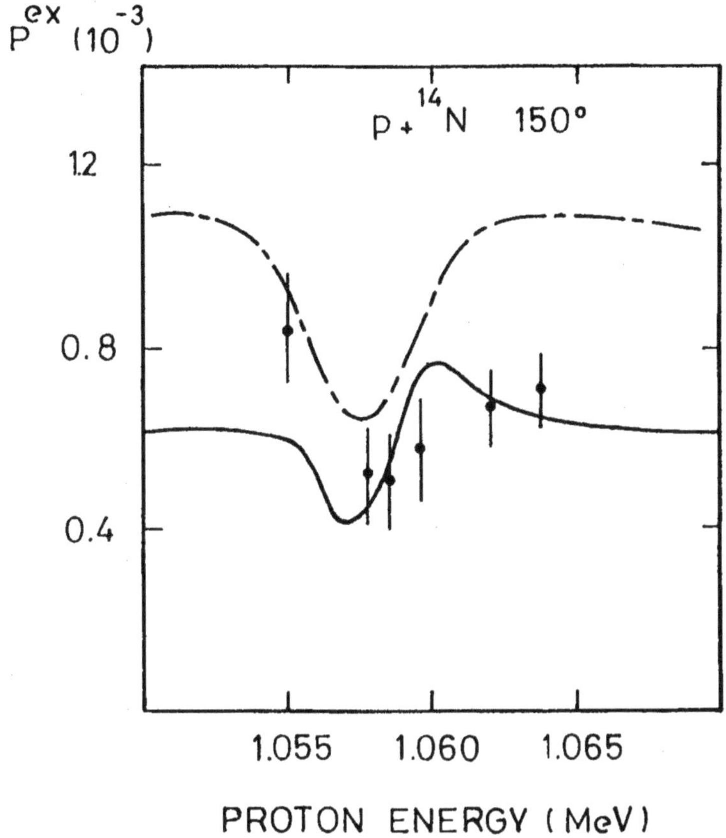

Figure 2. Probabilities for electron capture from N by impact of 1.058 Mev protons at θlab= 150 deg.: ——— present CDWT ; -.-.- present CIST; experimental data (●) from Ref. 6.

in the presence of a nuclear resonance is a more severe test than the one done by comparing only probabilities: the relative magnitude of modules and phase shift between in and out amplitudes are also tested in the first case.

4. Conclusions

We have developed the calculation of a perturbative model which preserves Coulomb conditions and is closely related to the standard CDW approximation. We have found by comparison with the CIST model that the inclusion of the weak projectile distortion does not modify essentially the capture probabilities in asymmetric collisions. However, the present model is the only among the ones considered here that is theoretically well-founded.

References

1. E Horsdal-Pedersen, P Loftager and J L Rasmussen, J. Phys B 15 (1982) 2461.
2. E Horsdal-Pedersen, P Loftager and J L Rasmussen, J. Phys. B At. Mol. Phys. 15 (1982) 4423.
3. E Horsdal-Pedersen and J L Rasmussen, IEEE US-30 (1983) 891.
4. E Horsdal-Pedersen, J. Phys. B 20 (1987) 785.
5. E Horsdal-Pedersen, Nucl.Instr.Meth. B24/25 (1987) 130.
6. J. N. Scheurer, O. K. Baker and W.E. Meyerhof, J. Phys. B 18 (1985) L85.
7. J. N. Scheurer, S. Andriamonje, F. Azaiez, D. Bertault, J. F. Chemin, H. Oubaha "XIIeme. Colloque National sur la Physique des Collisions Atomiques et Electroniques" Caen, 21-24 Juin 1988.
8. E Horsdal, B Jensen and K O Nielsen, Phys. Rev. Lett. 57 (1986) 675.
9. O K Baker, C Stoller and W E Meyerhof, Nucl.Instr.Meth. B24/25 (1987) 89.
10. O K Baker, W E Meyerhof, D W Spooner , C H Stoller and J N Scheurer, Phys. Rev. Lett. 60 (1988) 913.
11. R Shakeshaft and Spruch, Phys. Rev. A 29 (1984) 605.
12. P A Amundsen and D H Jakubassa-Amundsen, J.Phys.B At. Mol.Phys. 17 (1984) 2671.
13. L Kocbach and J S Briggs, J.Phys.B At.Mol.Phys. 17 (1984) 3255.
14. J M Maidagan, J.Phys.B 21 (1988) 1395.
15. J M Maidagan, J.Phys.B At. Mol. Opt. Phys. 21 (1988) 4177.
16. Dz Belkić, R Gayet and A Salin , Phys.Rep. 56 (1979) 279.
17. I M Chesire, Proc. Phys. Soc. 84 (1964) 89.
18. E Clementi and C Roetti, At. Data Nucl.Data tables 14 (1974) 177.
19. E Horsdal-Pedersen, extracted from ref. 12 (1982).
20. O Fojón and J M Maidagan, J Phys B At. Mol. Opt. 24 (1991) 2529-2542

SADDLE ELECTRON EMISSION FROM IONIZING H^+ AND $^3He^{2+}$ COLLISIONS ON Ne

S. Suárez*, G. Bernardi, C.R. Garibotti* and W. Meckbach*

División Colisiones Atómicas, Centro Atómico Bariloche
8400 S.C. de Bariloche, Argentina

ABSTRACT

We measured doubly differential ionization cross sections for 100 keV/u H^+ and $^3He^{2+}$ on a Ne target. We studied systematically the effect of a distributed target along the beam trajectory on the electron emission. Low energy electron emission is increased when the gas target is extended. We attribute this effect to the scattering of electrons with distributed target atoms. This result has a direct consequence on the interpretation about how the saddle emission mechanism manifests itself on observed electron distributions.

1. Introduction

When a nude projectile with velocity v_p interacts in an ionizing collision with an atomic gas target under single collision conditions, electrons are emitted from the target and their double differential distribution, in angle and energy, presents some structures with distinguished features:

- The binary peak, produced by a direct knock-on collision between the projectile and an electron of the target and localized on a sphere of radius v_p centered at the projectile velocity. The shape of this peak reflects the momentum distribution of the electron in its initial bound state.

- The peak of electron capture to the continuum (ECC) of the projectile, produced essentially by pure coulombic interaction between the projectile and electrons emitted with velocities $\vec{v}_e \sim \vec{v}_p$.

- The peak of soft electrons, due to collisions of large impact parameters and centered around the residual ionized target at $\vec{v}_e = 0$.

The cusp shape of these two latter electron distributions puts into evidence the expected divergence of electron states around an

* Consejo Nacional de Investigaciones Científicas y Técnicas

isolated charge, represented by the projectile and the residual ion respectively.

Each one of the mentioned structures has been identified with a particular mechanism, associated to the main interaction of the ejected electron with the ionized target or the projectile, respectively. However much experimental and theoretical effort has been devoted to the understanding of ionization processes in which an emitted electron is subject to the simultaneous and comparable interaction with both ions resulting from the collisions.

In this connection, the first experimental observation, at intermediate energies, that showed a structure in the double differential electron distribution in the way of a ridge stretched between the ECC and the soft electron peak, was reported by Meckbach et al.[1]. This structure, interpreted as due to the evolution of emitted electrons on the saddle of the potential of the ion target and projectile, was subsequently shown to be a broad hump centered at the beam direction (θ= 0 deg)[2,3].

On the other hand, from classical trajectory Monte Carlo (CTMC) calculations, Olson[4,5] pointed out that a large fraction of electron emission could be found close to half the relative velocity of the two coulombic centers. This concentration of electrons was later understood as a manifestation of a new mechanism for electron ionization, in which an electron standing on the equiforce point of the "two ion" potential, has a great probability to evolve in the continuum of two centers. This mechanism then was called saddle point electron emission[6] and several experimental results were reported[6-11]. However, there subsist differences between authors in in both measurements and interpretation. As was reported by Olson et al.[6], Irby et al.[7] and Gay et al.[8,9] the doubly differential cross section $d\sigma/dv_e d\Omega$ as a function of angle and electron velocity of emission for 60 and 100 keV protons in collision with He, Ne and Ar targets, presents a maximum near $\vec{v}_p/2$. This feature was understood as a predominant electron emission in this velocity region. In particular, for $^3He^{2+}$ on the same targets, they found a shift of that maximum to lower electron velocities according to the displacement of the saddle potential when the projectile charge is increased.

We discussed recently the interpretation of such a maximum in the cross section $d\sigma/dv_e d\Omega$ and showed that it is originated by a kinematic artifact[10-12], so it can not be associated with a predominant electron emission close to the saddle point velocity. For $^3He^{2+}$ on He and Ne, our experimental results[10-11] do not show any shift in the electron emission towards lower electron velocities.

In this paper, we present doubly differential electron distributions obtained from H^+ and $^3He^{2+}$ in collision with a Ne target. We find that an extended gas target produces an increase of low energy electron emission that affects the interpretation of measurements in relation with the mentioned saddle point mechanism.

2. Experimental Setup and Procedure

Doubly differential electron distributions $Q(E_e,\theta)$ as a function of electron energy E_e at an emission angle of θ = 10 deg have been measured with our coaxial cylindrical mirror spectrometer shown in Fig. 1. A detailed description of this equipment and its performance has been given elsewhere[3,13].

The ion beam delivered from the Bariloche Cockcroft-Walton accelerator facility is collimated to 0.5 mm diameter and 0.1 deg of angular spread. The target, provided by an atomic gas stream emerging from an hypodermic needle of 0.25 mm bore, is localized at the intersection of the ion beam, the spectrometer axis and a third vertical axis around which the spectrometer can be rotated for measurements at different angles θ covering the full angular range from θ= 0 to±180 deg.

Electrons emitted from the target at the entrance focus of the spectrometer are deflected and focused back on the spectrometer axis where an orifice O_1 (0.5 mm diameter) defines an energy resolution of 1%. The half angle θ_o of the accepted electron cone, is defined by O_1 and a second orifice O_2 that can be choosen to obtain six angular resolutions from θ_o= 0.25 to 2.5 deg. For the present measurements we used θ_o= 2 deg.

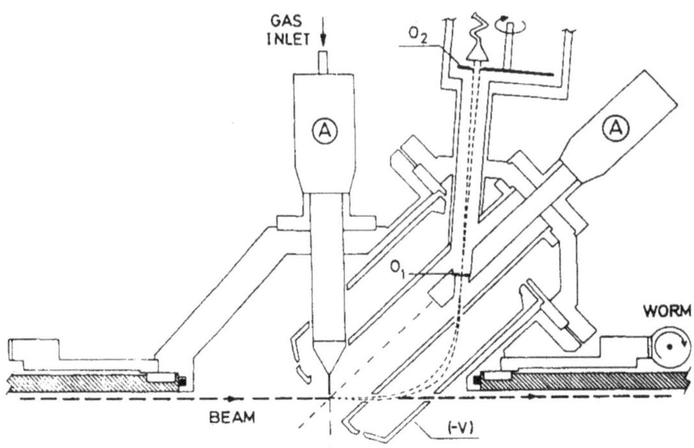

Fig. 1: Schematic diagram of the coaxial cylindric spectrometer.

The collision chamber was pumped by a 1500 l/s diffusion pump down to a base pressure of 2×10^{-7} torr. The working pressure was of 2×10^{-5} torr for the Ne target. Stray magnetic fields were reduced to less than 7 mG by a single conveniently oriented coil. Counts due to residual gas at base pressure were found to be less than 1% of the target-in count rate.

The highly localized gas target at the entrance focus of the spectrometer precludes distortions of measured spectra due to extended gas target effects. Two distinct causes contributed to an extended target: a) an overall, approximately uniform, gas pressure in the chamber and b) the shape of the effusing gas stream from the needle. They were minimized using a high speed pumping system and placing the needle tip close enough to the ion beam path. This results in a local density of gas target over two orders of magnitude larger than that corresponding to the pressure measured in the chamber, far from the target region. We have performed a quantitative study of these effects for small angles of electron emission[13,14].

We reproduce here in Fig. 2 part of such an investigation in which an extended gas target was artificially generated by separating gradually the needle tip (source of gas) from the beam line.

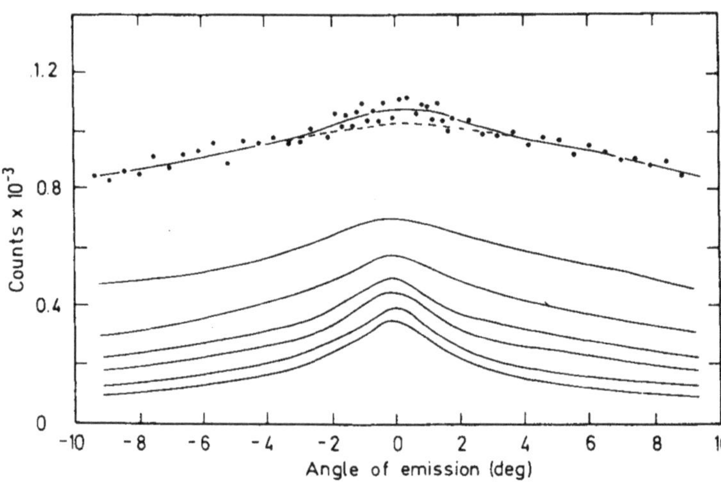

Fig. 2: Angular spectra at E_e = 74.8 eV (v_e = 1.73 au) for H^+ on He collisions at 170 keV. The distances between needle tip and ion beam from above are 0.1, 0.5, 1.0, 1.5, 2.0, 3.0 and 4.0 mm.

We observe in this figure that, as a function of the distance from the needle tip to the beam line, not only the overall count rates decrease, but they are relatively enhanced at small angles. This latter effect was explained as due to electrons produced from ionizing collisions along the beam line, which could be still collected by the spectrometer acceptance θ_o. For larger angles θ, an electron produced out of the target region cannot be detected unless it had a second collision.

3. Results and Discussion

In Fig. 3, doubly differential cross sections $d\sigma/dE_e d\Omega$ obtained for 100 keV H^+ on Ne are shown, but now measured as a function of the electron energy E_e at an emission angle of θ = 10 deg. Again these spectra were taken as a function of the distance from the needle tip to the beam trajectory. "0.1 mm" means the minimum distance to which the needle tip could be approached to the beam without having background due to secondary electron emission. A more extended gas target is obtained for increasing distances while a uniform gas

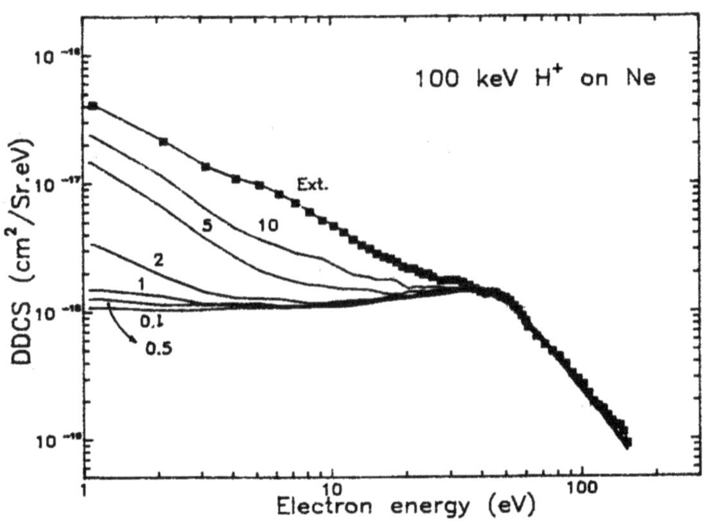

Fig. 3: Doubly differential cross sections measured at θ = 10 deg for 100 keV H^+ on Ne, taken for distances from the needle tip to the beam path of 0.1, 0.5, 1.0, 2.0, 5.0, 10.0 mm and when the target is uniformly distributed (Ext).

distribution ("Ext.") is achieved introducing the target into the spectrometer chamber by another lateral source, far from the collision region. Fig. 4 shows the same as Fig. 3 but for 100 keV/u $^3He^{2+}$ on Ne.

We observe from these two figures how low energy electron emission is increased by a factor of up to 60 when the target region is enlarged. The spectra obtained for each distance to the needle tip have been normalized at higher energies where no dependence of the shape of electron distributions on extension of gas target was observed. In addition, for a distance smaller than 0.5 mm we do not observed changes in the shape of the electron spectra. This shows that, for normal conditions of measurements, that means with the needle tip close enough to the beam, our spectra are independent of the gas target distribution, as it is expected for a well localized target.

We attribute the enhancement of the emission observed for an extended gas target at lower electron energies to a scattering of electrons originally emitted along the beam path, with target atoms distributed along the visual line of the electron spectrometer close to its entrance focus. In this way, electrons originated away from this focus can be collected into the acceptance cone of the spectrometer, even for large angles θ of detection. On the other hand, for small angles, electrons coming from large distances from the focus do not need an eventual scattering with an atom to be detected, giving rise to an artificially enhancement centered at θ= 0 deg, as shown in Fig. 2.

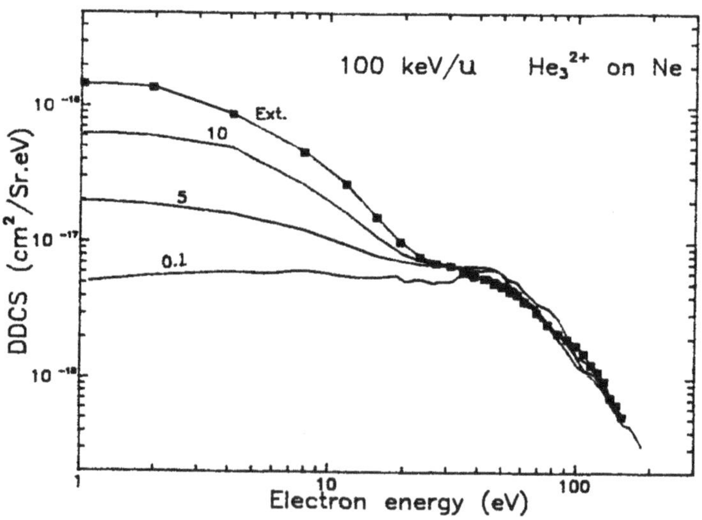

Fig. 4: Same as in Fig. 3 but for 100 keV/u $^3He^{2+}$ on Ne.

For a distributed gas target, a large fraction of electron emission is produced along the beam trajectory within the collision chamber compared to that originated at the spectrometer focus. Then, even for detection at angles which are not small, the contribution from electrons born at the focus and that from those produced off the focus and subsequently scattered by target atoms could be comparable.

A quantitative evaluation of the observed enhancement at lower electron energies must account for the number of collisions along the beam trajectory, the energy and angular distributions of emitted electrons and the elastic and inelastic cross sections for electron-atom collisions. We suggest that the enhancement of the low energy electron detection is due to elastic scattering of low energy electrons as well as inelastic scattering, for which scattering into larger deflection angles is more favorable.

For a sake of evaluation of this low energy effect as a function of the projectile charge, we show in Fig. 5 the ratio between doubly differential cross sections resulting from $^3He^{2+}$ and H^+, extracted from Fig. 3 and 4, for gas targets distributed artificially under the same conditions. For the well localized target, that results for 0.1 mm distance from the needle tip to the beam path, we obtain, as shown previously[3,11], a weak dependence on electron energy up to the energy of ECC peak, where the sudden step reported for θ= 0 deg can be still clearly distinguished for θ= 10 deg. In this connection much endeavor, both experimental and theoretical, has been devoted to the investigation of the so called "two center electron emission"[3,15-17]. Much attention was fixed on the departures from the perturbative Z_p^2 scaling as a function of the projectile charge Z_p.

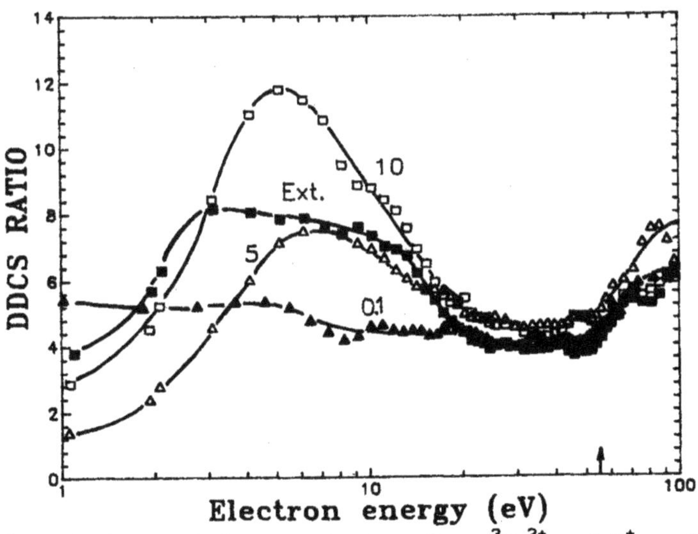

Fig. 5: Ratios between emissions from $^3He^{2+}$ and H^+ on Ne for different target distributions. The arrow indicates the ECC peak energy.

The ratios obtained for extended gas targets produced when the distance to the needle tip was 5 and 10 mm are also shown. Here we found a surprising growth of these ratios in the form of maxima centered between 5 and 10 eV. A similar feature is observed for the ratio "Ext." corresponding to a completely extended gas target.

The behavior shown by these ratios calls the attention about the interpretation of the projectile charge dependence on saddle electron emission. Experimental results reported by Gay et al.[8] in the way of doubly differential cross sections $d\sigma/tv_e d\Omega$, for the same projectiles and target used in this work, show a shift to lower energies of the electron emission for the larger projectile charge that gives rise to a quotient very similar, in both shape and magnitude, to those obtained from our extended gas target spectra. The projectile charge dependence that they found gave support to an interpretation based on the displacement of the saddle point of the two center potential for an increasing projectile charge that, in view of the present results, must be attributed to an extended gas target effect.

We have presented our results of projectile charge dependence of electron emission by means of ratios to obtain a picture that is independent of a particular choice of cross sections.

At this point it is convenient to state that the absence of a maximum in the ratios of cross sections for different projectile charges, does not speak against the existence of a saddle electron emission mechanism. The distribution of electrons subjected to a comparable interaction of both Coulomb centers, the projectile and the residual target ion, was shown to appear as a broad ridge stretching between the ECC and the soft electron peaks[2,3]. This was interpreted[1] as a result of a post-collisional evolution of electrons standing on the saddle potential. This statement is supported by theclassical CTMC calculations of Olson[4] in the sense that "a considerable number of small angle scattering events for ejected electron velocities approximately equal to one half the projectile velocity" was found.

Finally, we mention that recently several theoretical approaches have been used to evaluate the saddle electron emission. They include studies from collisions with zero range potentials[18], classical trajectory Monte Carlo calculations[19] and a model based on electron molecular states promotion[20]. These studies show that a large contribution to the total electron emission could be attributed to the saddle mechanism, but that the final electron distribution spreads in velocity space, such that a correlation between this distribution and the velocity of the saddle point cannot be observed.

References

1. W. Meckbach, P.R. Focke, A.R. Goñi, S. Suárez, J. Macek and M.G. Menendez, *Phys. Rev. Lett.* **57**, (1986) 1587
2. G. Bernardi, S. Suárez, P.R. Focke and W. Meckbach, in *High-Energy Ion-Atom Collisions*, Vol. **294** of *Lecture Notes in Physics* (Springer-Verlag, Berlin, 1987), p. 286.
3. G. Bernardi, S. Suárez, P.D. Fainstein, C.R. Garibotti, W. Meckbach and P. Focke, *Phys. Rev.* **A40**, (1989) 6863
4. R.E. Olson, *Phys. Rev.* **A16**, (1983) 1871
5. R.E. Olson, *Phys. Rev.* **A33**, (1986) 4397
6. R.E. Olson, T.J. Gay, H.G. Berry, E.B. Hale and V.D. Irby, *Phys. Rev. Lett.* **59**, (1987) 36
7. V.D. Irby, T.J. Gay, J.Wm. Edwards, E.B. Hale, M.L. McKenzie and R.E. Olson, *Phys. Rev.* **A37**, (1988) 3612
8. T.J. Gay, H.G. Berry, E.B. Hale, V.D. Irby and R.E. Olson, *Nucl. Instr. and Meth.* **B31**, (1988) 336;
9. T.J. Gay, M.W. Gealy and M.E.Rudd, *J. Phys.* **B23**, (1990) L823
10. G. Bernardi, P. Fainstein, C.R. Garibotti and S. Suárez, *J. Phys.* **B23**, (1990) L139; G. Bernardi, S. Suárez, P. Fainstein, C.R. Garibotti, W. Meckbach and P. Focke, *J. Phys.* **B23**, (1990) L829
11. S. Suárez, C.R. Garibotti, W. Meckbach, G. Bernardi, P. Focke, R. Pregliasco and G. Sigaud, in *High-Energy Ion-Atom Collisions*, Vol. **376** of *Lectures Notes in Physics* (Springer-Verlag, Berlin, 1990), p. 88
12. W. Meckbach, S. Suárez, P. Focke and G. Bernardi, *J. phys.* **B**, in press.
13. G. Bernardi, S. Suárez, P. Focke and W. Meckbach, *Nucl. Instr. and Meth.* **B33**, (1988) 321
14. G. Bernardi, P.R. Focke, S. Suárez, and W. Meckbach, in *High-Energy Ion-Atom Collisions*, Vol. **294** of *Lecture Notes in Physics* (Springer-Verlag, Berlin, 1987), p. 295.
15. N. Stolterfoht, D. Schneider, J. Tanis, H. Altevogt, A. Salin, P.D. Fainstein, R. Rivarola, J.P. Grandin, J.N. Scheurer, S. Andriamonje, D. Bertault and J.F. Chemin, *Europhys. Lett.* **4**, (1987) 899
16. C.O. Reinhold and R.E. Olson, *Phys. Rev.* **A39**, (1989) 3861
17. J.E. Miraglia and J. Macek, *Phys. Rev.* **A43**, (1991) 4321
18. J. Wang, J. Burgdörfer and A. Bárány, *Phys. Rev.* **A43**, (1991) 4036
19. G. Bandarage and R. Parson, *Phys. Rev.* **A41**, (1990) 5878
20. M. Pieksma and S. Yu Ovchinnikov, *J. Phys.* **B24**, (1991) 2699

Neutralization of slow ions before collision with surfaces

F.P. Santos, M. Matos e E.F. da Silveira
Departamento de Física, PUC-Rio - Cx. Postal 38071,
Rio de Janeiro, 22453 RJ, Brazil

Abstract

A model to describe the neutralization process of a slow ion approaching a surface is analysed. In the formalism it is considered explicitly the over-barrier and the tunneling contributions. For the cases treated, it is found that the contribution of the former is one order of magnitude higher than that of the latter.

I. Introduction

Despite of the fact that electron emission by surfaces under bombardment of ions has been studied for a long time [1,3], the subject is still under development [4-6]. The many-body interaction of an ionic projectile with solids is complex, giving rise to several processes of electron emission. Secondary electrons, delta electrons, projectile Auger electrons, target atom Auger electrons, convoi electrons, shock wave electrons are the most important examples of electron emission. In order to characterize the mechanism of each process, quantities such as total yield (number of electrons emitted per projectile), energy distribution and angular distribution of electrons are considered. Radiative spectroscopy and projectile charge exchange measurements are also useful to understand electronic excitation/de-excitation processes.

In the last years, highly charged ion beams have become available and experiments were set up to analyze the dependence of the electron emission on the charge state of the projectile. Interactions of slow [4-6] and fast [7] ions with surfaces have been treated. One major point in the understanding of these processes is the separation of the description of the electron emission before ion-surface collision (above-surface process) from that of the electron emission after the ion has crossed the first layer of the target (bellow-surface process). In particular, electron emission before ion-surface collision followed by electron capture may neutralize totally or partially the incoming ion, leading to neutral or low charge projectile collision with the surface. It is believed that this process is very important for slow ions (projectiles with velocities comparable with valence electron velocities).

Electron emission induced by an approaching ionic projectile is described by the potential ejection model [2]. Depending on the ion-surface distance and on the positive ion charge state, one or more electrons, attracted by the coulombian potential of the projectile, may jump over the surface barrier or even "tunnel" through it. The relative importance of these two processes has been addressed recently by Burgdöorfer et al [6]. Using a classical model for the over-the-barrier process and a rough approximation to describe the quantum mechanical tunnelling process, they have found that the latter is negligible with respect to the former. In their tunneling calculation it was considered that electrons could escape with a perpendicular transmission probability given by the one dimensional Nordheim Fowler formula for field ionization.

The goal of the present work is to analyze in more detail the tunneling process, comparing the neutralization cross section expected in this case with that predicted by the over-the-barrier model. For both processes, we took into account the contribution of one single energy band level. It is considered the three-dimensionallity of the problem by including electrons emitted in directions other than that perpendicular to the surface. Transmission probabilities of electrons emitted from different points of the surface are calculated according to the corresponding geometry. Calculations were carried out for an insulating target, in which case image potentials are not present. Effects of image charges, which are important for metallic surfaces, are also discussed. Finally these results are compared with those obtained by Burgdörfer et al [6]. It is shown that the contribution of the over-the-barrier process is much higher than the tunneling contribution in accordance with ref [6]. However, three-dimensional effects lead to tunnelling emission simultaneously with the emission over the barrier and, in this respect, our results differ substantially from theirs.

II.1 – The model

It is assumed in this model that the projectile captures every electron leaving the surface, either by tunneling or jumping over the barrier. It it known [8] that for ion-gas collision in which the relative velocities are below the velocities of the electrons in the atoms, the electron capture cross section generally dominates over the impact ionization cross section.

In addition, the adiabatic approximation is considered: during the electron movement from the surface towards the incoming projectile the latter is assumed to be fixed. This occurs when the projectile velocity is very low as compared with the velocity of the tunneling-electron. In particular, this condition is satisfied when the electron has an energy close to the Fermi level of most materials and the projectile velocity is smaller than the Bohr velocity.

The incoming ion will induce the electron emission and capture them with the rate

$$\frac{dN}{dt} = \phi\sigma(Z) \tag{1}$$

where ϕ is the flux of electrons which are available to leave the surface and $\sigma(Z)$ is the electron capture cross section for a projectile at a distance Z from the surface. In general ϕ is time dependent; in the case of metal surfaces, it may be considered constant and estimated, for instance, from a free electron model.

Considering now the electron leaving the surface at a distance between ρ and $\rho + d\rho$ from the point of impact, $\sigma(Z)$ writes:

$$\sigma(Z) = \int_o^\infty T(\rho, Z)2\pi\rho d\rho \tag{2}$$

where the transmission coefficient T is equal to 1 for the classical over-the-barrier process (classical contribution) or is determined by the WKB approximation whenever the tunneling effect occurs (quantum mechanical contribution). Fig. 1 shows the potential energy diagram of the interaction for two cases: a) the projectile is close to the surface (Z_1); b) the projectile is far away (Z_2). Note that the barrier becomes lower and thiner as the projectile approaches the solid. There is a critical distance Z_{cr} below which only classical contributions are present.

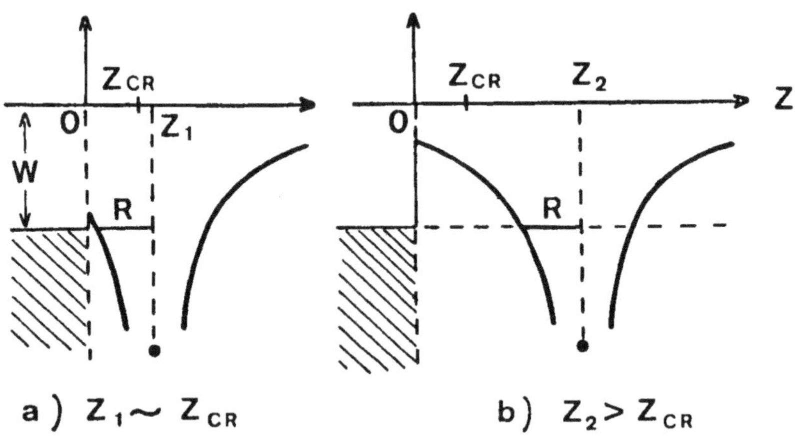

Figure 1

Given the surface work function W, a particular spherical coulomb equipotential of radius R is defined by $-W = -\frac{e^2 q}{4\pi\epsilon_0 R}$

The surface potential is supposed to be abrupt (step function) when no image charge effect is considered. $Z_{cr} = R$ in this case. When $Z \leq R$ electron emision over the barrier takes place (Fig. 2a), in the surface region of area $\pi\rho_0^2 = \pi(R^2 - Z^2)$. When $Z > R$, electrons cannot be ejected except by tunneling (Fig. 2b).

The WKB transmission coefficient is given by

$$T(Z,\rho,\Theta) = \exp\Big[-2\int_X \sqrt{\frac{2m}{\hbar^2}(W + V(x))dx}\Big] \tag{3}$$

where $V(x) = -\frac{eq^2}{4\pi\epsilon_0\ r(x)}$, W is defined positive and x is the coordinate along the line X corresponding to the "classical trajectory" inside the barrier, i.e., the region in which the total energy of the electron is smaller than the potential energy.

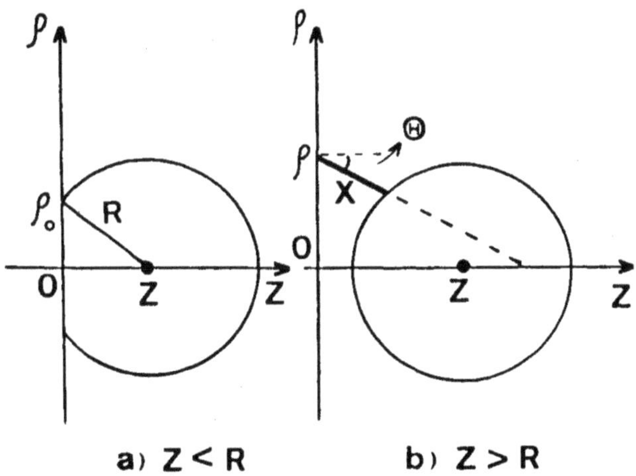

Figure 2

In order to consider different ways to tunnel, an average transmission coefficient is calculated:

$$T_{WKB}(\rho, Z) = \frac{1}{2\pi}\int_{2\pi} T(\rho, Z, \Theta)\ d\Omega \tag{4}$$

where $d\Omega$ is the emission solid angle centered at the point where the electron leaves the surface. It is supposed that electrons leave isotropically the surface.

Using eqs. (3) and (4), and $T(Z < R) = 1$ for $\rho < \rho_0$, in (2), one gets

$$Z \leq R: \quad \sigma(Z) = \pi(R^2 - Z^2) + \int_{\rho_0}^{\infty} T_{WKB}\ \ 2\pi\rho d\rho \tag{5a}$$

$$Z > R: \quad \sigma(Z) = \int_0^{\infty}\ T_{WKB}\ \ 2\pi\rho d\rho \tag{5b}$$

It is interesting to point out that from eq.5 $\sigma(Z)$ can be expressed as

$$\sigma(Z) = \sigma_c(Z) + \sigma_q(Z)$$

i.e., $\sigma(Z)$ is the sum of classical and quantum cross sections. Of course, when $Z > R$, $\sigma_c(Z) = 0$.

II.2 – Electron emission from metals

When the ionic projectile approaches a metal surface and an electron is emitted, it is necessary to include the contributions of the projectile and electron image charges.

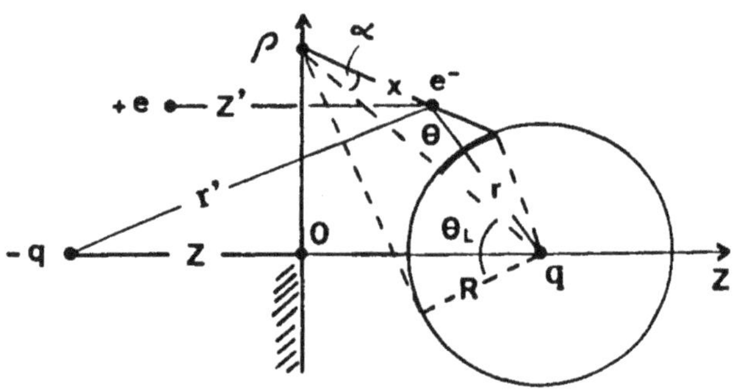

Figure 3

Using the distances defined in Fig. 3, the total potential energy seen by the electron writes:

$$\begin{aligned} V_t(r, r', Z') &= V_{\text{solid}} + V_{\text{ion}}(r) + V^I_{\text{ion}}(r') + V^I_e(Z') \\ &= U(Z') + \frac{e}{4\pi\epsilon_0}\Big[-\frac{q}{r} + \frac{q}{r'} - (1/2)\,\frac{e}{2Z'}\Big] \end{aligned} \tag{6}$$

where $U(Z')$ is, in a first approximation, a step function. The image charge of the projectile is $-q$. Note that the potential energy of a dipole in which the charges $+e$ and $-e$ are separated by the distance $2Z'$ is $V = -\frac{1}{4\pi\epsilon_0}\frac{e^2}{2Z'}$. The 1/2 factor in the image potential energy of the electron, $V^I_e(Z')$, takes into account that any displacement of the electron implies an equivalent movement of the image charge. Due to the dependence of V_t on r' and Z', it is clear that the equipotentials around the projectile will lose the spherical symmetry, which they have in the absence of image charges.

Fig. 4 illustrates the total potential energy $V_t(Z')$ for the particular case where the electron moves along the same line as the projectile (perpendicular to the surface), so that $r = Z - Z'$ and $r' = Z + Z'$. We have used in this example $W = 5\ eV$, $q = 6$, $Z = 1.5R$, where $R = e^2q/4\pi\epsilon_o W$. The points Z'_1 and Z'_2 define the limits of the WKB integral. The critical distance Z_{cr} is found by solving $V_t(Z_{cr}, Z'_{sad}) = -W$, where the saddle point Z'_{sad} corresponds to the maximum of V_t in a plane perpendicular to the solid surface and to the minimum in a plane parallel to the same surface. From this figure, the effects of the image potentials may be visualized: the narrower and lower barrier will increase σ_q. As $Z_{cr} > R$ now, emission over the barrier will be enhanced as well: electrons will start to jump earlier and $\sigma_c(Z)$ will be larger as compared to the value obtained for the insulator.

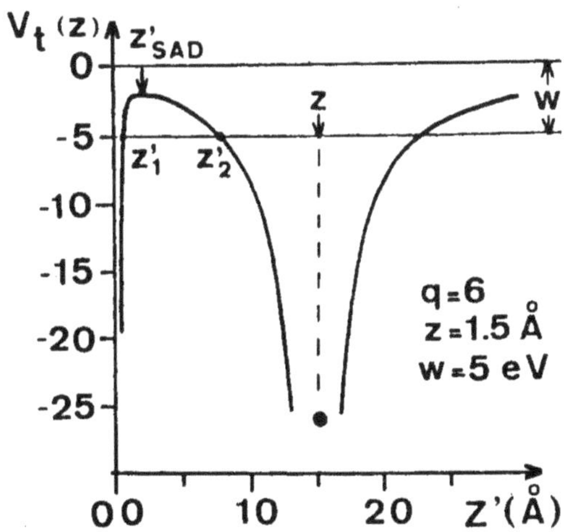

Figure 4

III - Results

In this section results obtained for $\sigma(Z)$, the cross section for electron emission given by eq.(5), are analysed. In the absence of image charges, the solid angle integration, given by eq.(4), is performed very simply by taking advantage of the azimutal symmetry with regard to the line drawn from the projectile to the point where the electron leaves the target. Eq.2, then, becomes

$$\sigma(Z) = \int_0^\infty \rho d\rho \int_0^{\theta_L} \exp\left[-2\int_x \sqrt{2m/h^2(W + V(x)}\, dx\right] \frac{R^2}{X^2} \cos(\theta+\alpha)\, \varphi \sin\theta d\theta \quad (7)$$

where the notation of Fig.3 is used. For $Z > R$, φ equals 2π. In order to eliminate electron emission from the surface towards the bulk, $\varphi = \varphi(Z, \rho, \theta)$ varies between π and 2π for $Z < R$.

As mentioned before, for a metallic surface, image charges effects must be considered. The image potential, given by eq.(6), requires a more time consuming mathematical treatment. For the scope of the present work, we will limit the study to the simpler case of a spherical $V(r)$, which corresponds to the insulating target.

Numerically, results depend on the number of steps used in the simple quadrature formula and ρ_{max}, the upper integration limit. These quantities were chosen from a compromise between accuracy and computational convenience. For the surface work function and the charge state, typical values of $W = 5$ eV and $q = 6$ were used.

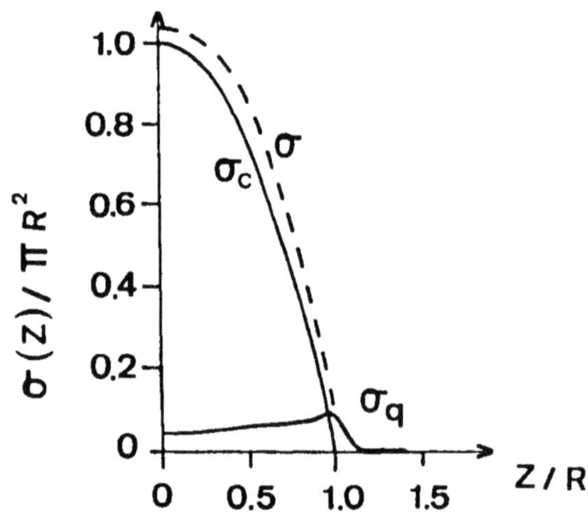

Figure 5

In Fig.5 the dashed line represents the calculated values for the total cross section, $\sigma(Z)$, as a function of the distance, Z, between the ion and the surface. It is also shown the specific contributions of the classical $\sigma_c(Z)$ and quantum, $\sigma_q(Z)$, cross sections, defined in eq. (6). Far from the surface $\sigma(Z > R)$ is uniquely due to quantum contribution and is negligible until $Z/R \approx 1.2$. From this point σ_q increases and is the sole contribution until σ_c starts to contribute at $Z/R = 1$. From $Z/R = 1$ to $Z/R = 0$, σ_q decreases slowly due to the balance of factors: a) the decreasing of the solid angle inside of which the electrons may be emitted; b) the increasing of the area $2\pi\rho\Delta\rho$ from which the electrons are preferably emitted. That is, for an electron emitted from the surface in a point located at $\rho > \rho_0$, the solid angle, through which it "sees" the sphere corresponding to the equipotential W, decreases for $Z < R$ (see Fig. 2). On the other hand, as far as the "sphere W enters into the solid" the peripherical area $2\pi\rho_0\Delta\rho$, corresponding to a region with the thinest potential barier, incrases as ρ_0 increases.

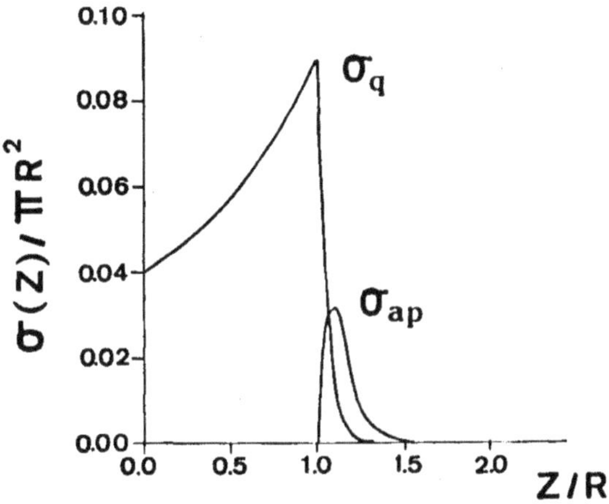

Figure 6

For the sake of comparison, Fig. 6 shows $\sigma_q(Z)$, once again, as well as $\sigma_{ap}(Z)$, the approximate formula for the electron capture cross section suggested in ref. 6. In that paper $\sigma_{ap}(Z)$ was estimated as $\sigma_{ap}(Z) = T(Z,0,0)z^2$, where z is the shortest distance between the equipotential and the surface. In our notation $z = Z - R$.

The approximate formula was obtained by assuming that the electrons are emitted perpendicularly to the surface, while the model used in this paper allows for the multidirectional electron emission, as discussed above. These different assumptions must be taken into account when comparing $\sigma_q(Z)$ and $\sigma_{ap}(Z)$. By considering that electrons behave as molecules in a gas we have divided by 3 the calculated values obtained from the approximate formula. The assumption of a free electron system is reasonable for simple metals as, for example, the alkaline metals. For other materials, such as insulators, semiconductors, transition metals, organic substrates, the proportionality factor may no longer be 3. Nevertheless, with the aim of taking into partial account the perpendicular versus multidirectional emission, we consider this factor satisfactory for the present discussion. As will become clear, its particular value does not affect the main conclusions obtained as a results of comparing $\sigma_q(Z)$ and $\sigma_{ap}(Z)$.

From Fig.6 it can be noted that $\sigma_{ap}(Z)$ and $\sigma_q(Z)$ have the same asymptotic behavior as Z tends to infinity; σ_{ap} slightly overestimates $\sigma_q(Z)$ for large distances. When the projectile approaches the surface, $\sigma_{ap}(Z)$ increases, as expected. It presents a maximum value close to $Z = R$ due to the competition between the transmission factor $T(z, \rho, \theta)$ and the effective area πz^2, which decreases with Z. Moreover, as Z approches the equipotential radius $\sigma_{ap}(Z)$ drops to zero. In this region the $3-d$ model gives for $\sigma_q(Z)$ a non zero value quantitatively significant, as seen from Fig.6. The strikingly distinct behavior of $\sigma_{ap}(Z)$ and $\sigma_q(Z)$ can be explained by the different approaches taken to treat the infinite surface. It shows that, when the ion is close enough to the surface, so that the classical process is about to start, or has already started, contributions due to tunnelling are still present.

As the ion approaches the surface two regions can be identified from the above discussion. The first one, defined by $Z > R$, corresponds to distances where all electrons are emitted by tunneling. For large distances all points of the surface are approximately equivalent and have small transmission probability. As the ion-surface distance decreases an effective area for tunnelling begins to appear, which corresponds to the portion of the surface, close to the the point of impact, where there is a significant transmission probability. This effective, circular area increases as Z decreases, up to the onset of classical emission. The second region, defined by $Z < R$, corresponds to the ion-surface distances where both classical and quantum processes contribute to $\sigma(Z)$. In this case the effective area for tunnelling will be formed by the region ouside the classical area and inside the effective tunnelling circle.

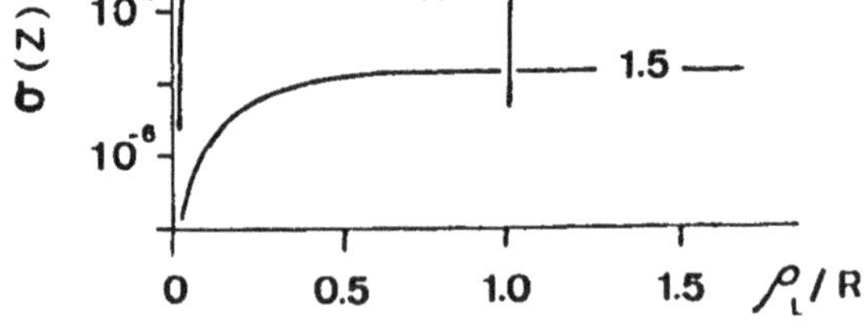

Figure 7

These effects can be observed in Fig.7. In this figure the quantum cross section $\sigma_q(Z)$, integrated in a finite range, to $\rho = \rho_L$, is plotted as a function of ρ_L, for three different values of Z. It can be noted that $\sigma_q(Z)$ increases steadly with ρ_L; the form of the curves shows the convergence behavior of the integral given by Eq.(7) as a function of ρ_L. A slow increases of $\sigma(Z)$ indicates that the tunnellling probability does not vary significantly with ρ. From Fig. 7 it is seen that this occurs for large Z, as expected, since in these cases all surface points are roughly equivalent.

IV - Conclusions

In this paper we have analysed classical and quantum cross sections for electron emission from surfaces, induced by slow ion impact, before collision. The quantum tunnelling cross section was calculated by considering with detail the infinite target, and surface electrons were allowed to escape in any direction. The tunnelling probability was given by the WKB approximation.

It was shown that significant contributions from the quantum tunnelling process exist for ion-surface distances such that the classical process is about to start or has just started. At the beggining of the classical emission the quantum process dominates the total cross section. When the classical emission area increases, what happens when the ion approaches the surface, the quantum contribution to the total cross section decreases, being about one order of magnitude smaller than the classical cross section for $Z \approx 0$.

Our main conclusion is that, in the study of ion neutralization processes by solid surfaces, tunnelling effects may be significant and should not be neglected. The analysis of image charge effects must be done in order to establish the relative importance of quantum and classical processes.

References

[1] P.L. Kapitza, Phi. Mag. **45** (1923) 987.

[2] H.D. Hagstrum, Phys. Rev. **96** (1954) 336.

[3] E.J. Sternglass, Phys. Rev. **108** (1957) 1.

[4] P. Varga, Comments At. Mol. Phys. **23** (1989) 111.

[5] J.P. Briand, XVth Conference on X-Ray and Inner-Shell Processes. Knoxville, July 1990.

[6] J. Burgdörfer, P. Lerner and F. Meyer, Phys. Rev. (1991). Also F. Meyer et al., Phys. Rev. Lett. **67** (1991) 723.

[7] S. Della-Negra, Y. Le Beyec, B. Monart, K. Standing and K. Wien, Nucl. Instr. and Meth. **B32** (1988) 360.

[8] See for instance D.H. Crandall, M.L. Mallory and D.C. Kocher, Phys. Rev. **A15** (1977) 61.

IMPROVEMENTS IN THE DESCRIPTION OF THE BACKGROUND SPECTRUM IN EPMA

Jorge Rubio - Gustavo Castellano - José Riveros

FaMAF - Universidad Nacional de Córdoba
Laprida 854 - 5000 Córdoba - Argentina

Abstract

With the purpose of advancing in the Peak-To-Background technique in EPMA, Lifshin's and Reed's corrections of Kramers' model for the background spectrum are studied. Special attention is paid to the low energy region (below 3 keV), in which there is no proper description of experimental data. Bremsstrahlung radiation from the beryllium window is also taken into account, considering the energy distribution of backscattered electrons.

1. Introduction

When an electron beam enters a bulk sample, the resulting X-rays conform characteristic peaks which are superposed to a continuous spectrum. Kramers[1] developed an analytical expression which quite accurately describes this continuous spectrum above 5 keV. A better agreement in the whole energy range of interest in EPMA is achieved by means of Lifshin's[2] and Reed's[3] modifications to it.

Several analytical expressions have also been given for characteristic intensities; lately most of them have been based on the ionization distribution function $\phi(\rho z)$[4–12], developed from experimental and Monte Carlo data.

Both characteristic and continuum intensities strongly depend on the target composition, which allows for quantitative analysis. At present, most analytical techniques are not absolute, since standard samples are required as references. An absolute technique is being developed, which is based on the peak-to-background ratio, taking the continuum intensity as a reference standard. Thus, geometrical factors and absorption corrections are avoided, and less strict conditions are required for the sample surface.

With the purpose of advancing in the development of this technique, Lifshin's and Reed's modifications to Kramers' model are studied in this paper and a generalization of these expressions is discussed.

Characteristic intensities are not dealt with in this paper, since a great number of models have been repeatedly tested[4–12].

2. Theoretical basis

If adequate descriptions for characteristic and continuum intensities are available, experimental data must properly be predicted by the relation

$$\left(\frac{P}{B}\right)^{exp} = \left(\frac{P}{B}\right)^{theo} \quad (1)$$

where P and B point to peak and background intensities.

In order to describe the characteristic radiation, Castaing and Descamps[13] developed the tracer method to experimentally determine the depth distribution of characteristic X-ray production $\phi(\rho z)$. Using one of the analytical models developed for $\phi(\rho z)$, the characteristic radiation measured for element i can be expressed as

$$C_i \frac{Q(E_o)\, N_o}{A}\, \varepsilon(E_\nu)\, f(\chi)\, \frac{\Delta\Omega}{4\pi} \int \phi(\rho z)\, d(\rho z) \quad (2)$$

where C_i is the concentration of element i, $Q(E_o)$ is the ionization cross-section of the sample for the incident electron energy E_o, N_o is Avogadro's number, A is the atomic weight of element i, $\varepsilon(E_\nu)$ the detector efficiency for energy E_ν, $f(\chi)$ the absorption correction where χ is a parameter related to the mass absorption coefficient of the sample and to the take-off angle, and $\Delta\Omega$ is the solid angle subtended by the detector. A similar expression can be given for the background, canceling common factors when the intensity ratio is taken. Therefore, we may write

$$C_i = \frac{\left(\frac{P}{B}\right)^{theo}}{\frac{\int \phi(\rho z)\, d(\rho z)}{B^{theo}}} \quad (3)$$

B^{theo} being related to a theoretical model for the background radiation generated in the sample.

This paper will not deal with analytical expressions based on the depth distribution of continuum x-ray production, since this has not been properly developed due to the lack of experimental data. This is due to the fact that it is not possible to determine at which depth the Bremsstrahlung radiation is emitted (a few papers on Monte Carlo simulations only show qualitative behaviour of the distribution function[5,14]).

Kramers[1] developed the first satisfactory description of the continuous spectrum, in which the intensity I_ν (number of photons) with energy E_ν per incident electron, is given by

$$I_\nu = \alpha \frac{I\, \Delta t}{e}\, \varepsilon(E_\nu)\, \frac{\Delta\Omega}{4\pi}\, f_s(\chi)\, R_s\, Z_s\, \frac{E_o - E_\nu}{E_\nu} \quad (4)$$

where α is a constant, independent of Z, E_ν and E_o, I is the current intensity, Δt the live time and subindex s refers to the specimen. For simplicity, Eq. (4) will henceforth be called $I_\nu(K)$ —Kramers' model.

This model is not suitable for EPMA, since it was originally proposed for describing the spectrum emerging from an x-ray tube, the most important energy range being above 5keV (low energies are strongly absorbed in the tube window).

For this reason Lifshin[2] suggested an energy dependent correction factor,

$$I_\nu(L) = I_\nu(K)\,[1 + \beta(E_o - E_\nu)] \tag{5}$$

and a similar modification was proposed by Reed[3],

$$I_\nu(R) = I_\nu(K)\,E_\nu^{-\beta} \tag{6}$$

where β is constant in both expressions.

Fiori et al.[15] suggested that α and β could be determined by considering the intensities at two separate photon energies, free of peak interference, and solving the resulting two equations in two unknowns.

3. Experimental and data analysis

A complex mineral sample has been taken for the present paper, its composition being (wt. %): Al 11.67, Si 17.91, Fe 23.62, Mn 0.81, Mg 4.2, Ca 0.97, O 40.95. All assessments will be performed considering the Gaussian absorption correction[16] and the backscattering factor of Love et al[17].

On the other hand, we must bear in mind that the detector is isolated by means of a beryllium window. This material is chosen because it is almost transparent to radiation in the range of interest. However, since it is opaque for electrons, the models for the background spectrum considered in the previous section will be complemented with the contribution due to the slowing down of backscattered electrons in the beryllium window, as suggested by Statham[14], i. e.,

$$I_\nu = \alpha\,\frac{I\,\Delta t}{e}\,\varepsilon(E_\nu)\,\frac{\Delta\Omega}{4\pi}\,\frac{E_o - E_\nu}{E_\nu}\,[f_s(\chi)\,R_s\,Z_s + \overline{\eta}\,f_{Be}(\chi)\,R_{Be}\,4] \tag{7}$$

where indices Be refer to the beryllium window, and $\overline{\eta}$ is the backscattering coefficient of the sample.

In order to determine if Fiori's suggestion is adequate for the purposes of this work, two different pairs of energies were chosen, 3 and 10.6 keV, 5.15 and 10.6 keV. Fig. 1 shows that background predicted by Lifshin's model changes when the energies chosen are changed. Since this arbitrariness must be overcome, most data excepting characteristic peaks were taken into account.

Figs. 2 and 3 show fittings (dotted line) for both models when E_o=20 keV and the energy intervals chosen were [2.62,3.42] keV, [4.86,5.61] keV, [7.76,9.29] keV and [10.19,11.03] keV. It can be seen that the prediction achieved is apparently good.

If the accuracy of the models at low energies must be tested, a spectrum free of peaks in this region is needed. For this purpose, a copper specimen was measured, considering the energy intervals [4.00,6.00] keV, [6.64,7.20] keV and [9.60,10.96] keV. Fittings for this spectrum are shown in figs. 4 and 5 (dotted line); it can be seen that a good agreement is achieved only above 3 keV.

It should be pointed out that Eq. (7) assumes that the energy of backscattered electrons (E_r) is E_o. However, these electrons reach the beryllium window with an energy spectrum. For this reason it can be assumed that they arrive at the detector with a mean energy value E_m, which has been assessed using the distribution proposed

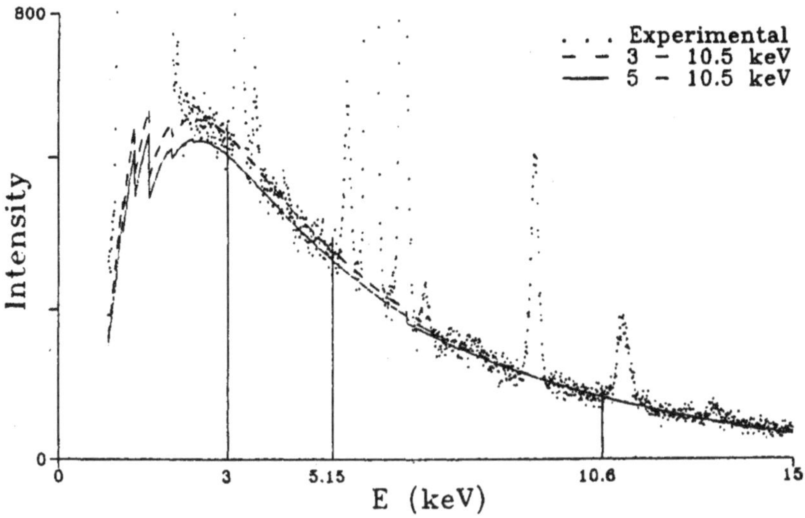

Figure 1: Background predicted by Lifshin's model for two different pairs of energies.

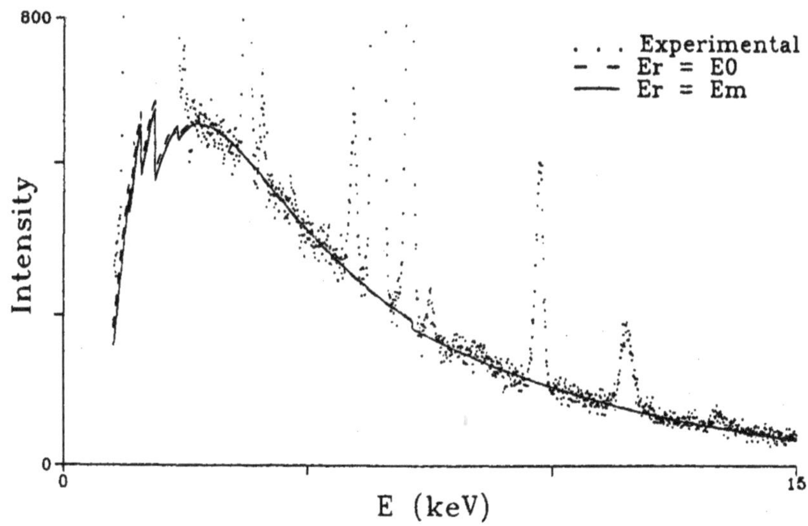

Figure 2: Lifshin's model for a mineral sample, fitted using two different energies for backscattered electrons.

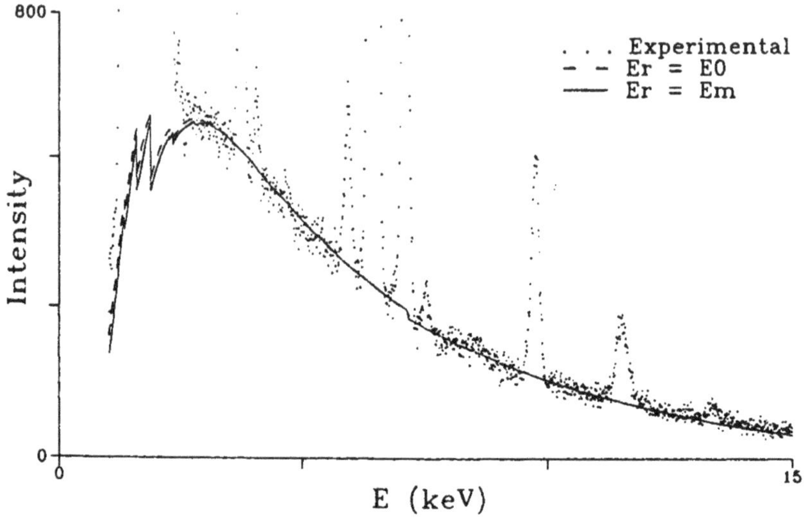

Figure 3: Reed's model for a mineral sample, fitted using two different energies for backscattered electrons.

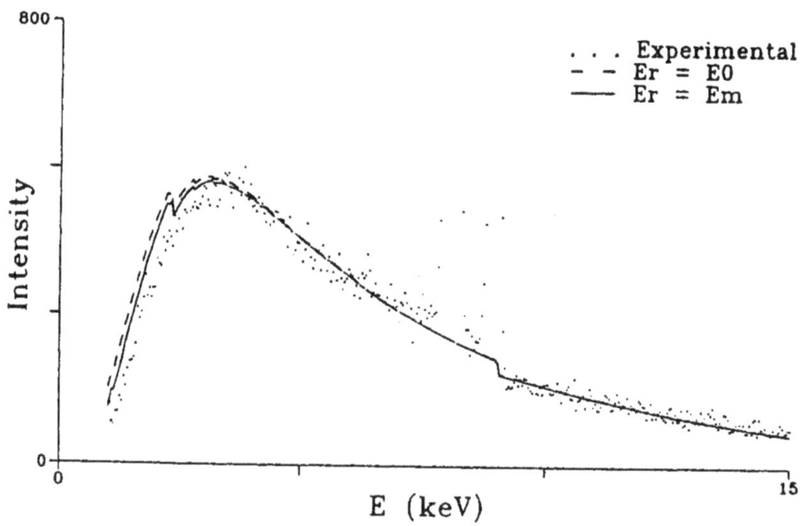

Figure 4: Lifshin's model for a copper sample, fitted using two different energies for backscattered electrons.

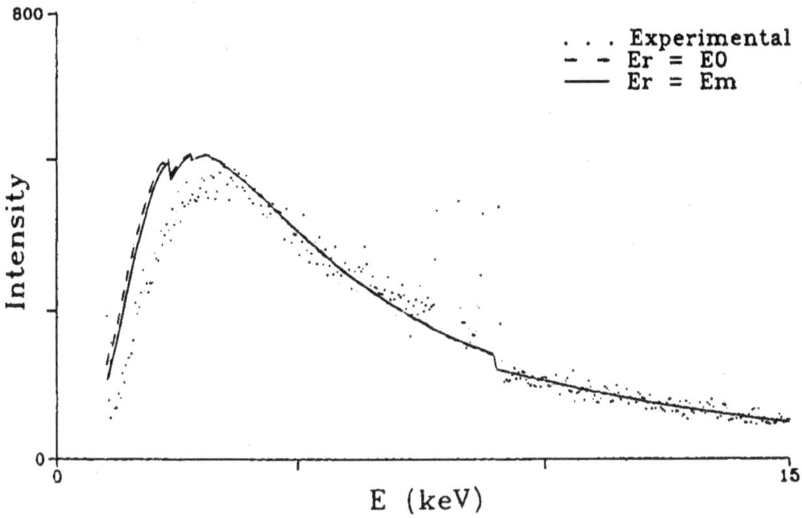

Figure 5: Reed's model for a copper sample, fitted using two different energies for backscattered electrons.

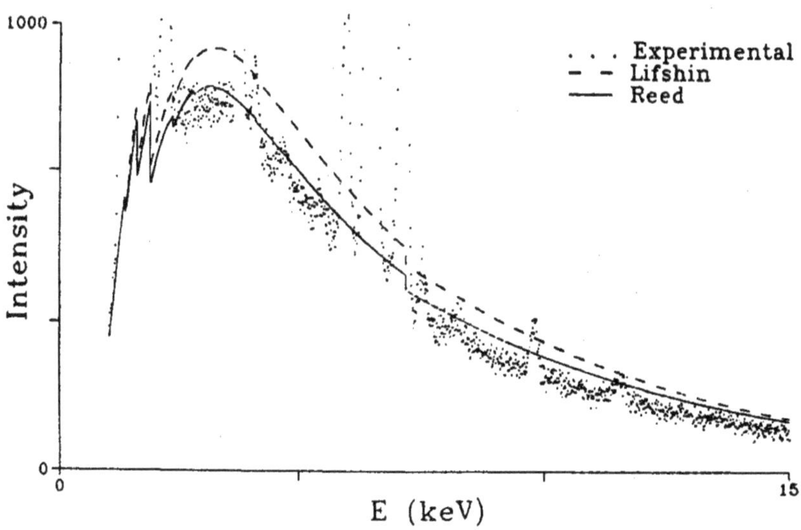

Figure 6: Lifshin's model with parameters adjusted for E_o =20keV used in a sample with E_o =25keV.

by del Giorgio et al.[18]. Therefore Lifshin's correction factor must be separately applied to the intensities generated within the sample and within the detector respectively:

$$\begin{aligned} I_\nu(L) &= \frac{I\,\Delta t\,\varepsilon(E_\nu)}{e\,E_\nu}\frac{\Delta\Omega}{4\pi}\,\{\,f_s(\chi)\,R_s\,Z_s\,(\,E_o-E_\nu\,)\,[\,\alpha+\beta\,(\,E_o-E_\nu\,)\,] \\ &+\ \bar{\eta}\,f_{Be}(\chi_{Be})\,R_{Be}\,4\,(\,E_m-E_\nu\,)\,[\,\alpha+\beta\,(\,E_m-E_\nu\,)\,]\,\} \end{aligned} \qquad (8)$$

A similar procedure must be performed with Reed's model.

These considerations lead to the fittings shown in figs. 2 – 5 (solid line). Both models produce a slight improvement between 2 and 3 keV when the previously mentioned intervals are used —all data correspond to energies above that of the maximum of the spectrum. However, if data with lower energies are included in the fitting, predictions are not so good.

On the other hand, since both models give an explicit dependence on Z and E_o, α and β are expected to be constant parameters. However, fig. 6 shows that descriptions are not appropriate when the parameters fitted for $E_o = 20$ keV are used for 25 keV.

4. Conclusions

Parameters α and β in both models analyzed have shown certain dependence on the incident electron energy, as previously pointed by Statham[19], although they are expected to be independent of Z as well as E_o.

When a good description is achieved in the energy range above that of the maximum of the spectrum, both models fail in the low energy region. This is a serious inconvenience, since the spectrum varies rapidly in this region. Therefore, if we need to work with low energy peaks, the background subtraction process will be strongly affected, even if conventional analysis is carried out.

These considerations show that although the modifications given by Lifshin and Reed improve Kramers' description of the continuous spectrum, they are still not appropriate for the low energy region. Therefore, a new model for the background spectrum must be developed for the whole energy range of interest (0 – 15 keV). In the meanwhile, the peak-to-background method will be restricted to specimens without light elements.

5. References

1. H. Kramers, *Phil. Mag.* 46, 836 (1923).

2. E. Lifshin, *Proc. Ninth Annual Conference of Microbeam Analysis Society*, Ottawa, Canada, # 53 (1974).

3. S. Reed, *X-Ray Spectrom.* 2, 16 (1975).

4. J. Philibert, in S. Reed, *Electron Microprobe Analysis, Cambridge University Press*, 244 (1975).

5. W. Reuter, in S. Reed, *Electron Microprobe Analysis, Cambridge University Press*, 247 (1975).

6. H. Bishop, in S. Reed, *Electron Microprobe Analysis, Cambridge University Press*, 260 (1975).

7. D. Sewell, G. Love, V. Scott, *J. Phys. D: Appl. Phys.* **18**, 1245 (1985).

8. R. Packwood, J. Brown, *X-Ray Spectrom.* **10**, 138 (1981).

9. J. Tirira, J. Riveros, *X-Ray Spectrom.* **16**, 27 (1987).

10. J. Tirira, M. del Giorgio, J. Riveros, *X-Ray Spectrom.* **16**, 243 (1987).

11. J. Pouchou, F. Pichoir, in *X-Ray Optics and Microanalysis*, eds. J. Brown and R. Packwood (Ontario: Univ. of Western Ontario), 249 (1987).

12. G. Bastin, H. Heijligers, *Scanning* **12**, 234 (1990).

13. R. Castaing, J. Descamps, *J. Phys. Rad.* **16**, 304 (1955).

14. P. Statham, *X-Ray Spectrom.* **5**, 163 (1976).

15. C. Fiori, R. Myklebust, K. Heinrich, *Analit. Chemistry* **48**, 172 (1976).

16. D. Ugarte, G. Castellano, J. Trincavelli, M. del Giorgio, J. Riveros, *X-Ray Spectrom.* **16**, 249 (1987).

17. G. Love, M. Cox, V. Scott, *J. Phys. D: Appl. Phys.* **11**, 7 (1978).

18. M. del Giorgio, J. Trincavelli, J. Riveros, *X-Ray Spectrom.* **18**, 229 (1989).

19. P. Statham, *Microchim. Acta Suppl.* **8**, 229 (1979).

CRITICAL BEHAVIOR OF THE WEAKLY DILUTED ISING ANTIFERROMAGNETS AND FERROMAGNETS.

Gloria M. Buendía
Departamento de Física. Universidad Simón Bolivar.
Apartado 89.000 Caracas 1080. Venezuela.
and
Julio F. Fernandez
Centro de Física. Instituto Venezolano de Investigaciones Cientificas
Apartado 21827. Caracas 1020. Venezuela.

ABSTRACT

In this work we present a Monte Carlo study of the effects of small site or bond dilution in a nn/nnn Ising model in zero field. For the two dimensional model we show that the impurities generate random field type effects, we verified that the Harris criterion for the cross-over exponent, $\phi=\alpha$, is satisfied for the ferromagnetic system, but it is not for the antiferromagnetic, for which we found that ϕ, is consistent with the relation, $\phi=\gamma-2\beta$. We expect similar results for the three dimensional model on a bcc lattice, we advance some results of this case.

1. Introduction

The properties of the phase transition of disordered systems have been the subject of a great deal of interest. In systems with random exchange constants, Harris[1] predicted that the critical behavior of the random system would be similar to that of the pure system if the specific heat exponent, α, is less than zero. If α is greater than zero a different behavior is expected, such that a cross over from the regime free of impurities is given by,

$$t_C \sim (\delta J)^{\frac{1}{\phi}}, \tag{1}$$

where t_C is the cross-over from the reduced temperature and δJ is the random variation in the exchange constant. For ferromagnetic systems, the Harris criterion implied, $\phi=\alpha$.

It has been suggested[2] that, in absence of an external field, for antiferromagnetic Ising models with ground states $n \geq 2$ fold degenerate (in two or more spatial dimensions), small amounts of

random site dilution (substitution of magnetic atoms, chosen at random, by non-magnetic ones) as well as small amounts of random bond dilution (removal of some exchange bonds chosen at random) in zero external field, generate spatially random anisotropy, and in these cases the expected cross-over exponent must be, $\phi=\gamma-2\beta$.
In this work we are going to present a Monte Carlo study of an Ising model with nearest (nn) and next-nearest (nnn) neighbors[3] that exhibits the different cross-over behavior of the antiferromagnetic from the ferromagnetic system.
The Hamiltonian of the pure system is,

$$H_0 = -J_1 \sum_{nn} S_i S_j - J_2 \sum_{nnn} S_i S_j \tag{2}$$

If a small perturbation is introduced such that,

$$H = H_0 + \sum_{nn} \delta J_{ij} S_i S_j \tag{3}$$

where δJ_{ij} is an independent random variable, we can perform a cumulant expansion of the free energy, F, in powers of $\delta J^2=(\delta J_{ij})_{av}{}^2$ such that ,

$$-\beta F = -\beta F_0 + \frac{\beta^2 \delta J^2}{2} \sum_{<nn>} [<(S_i S_j)^2> - <S_i S_j>^2] \tag{4}$$

note that since $(S_iS_j)^2=1$ the term that gives de singular contribution to F of order $(\delta J)^2$ is $< S_iS_j >^2$.
From scaling

$$F \sim t^{2-\alpha} f\left(\frac{\delta J^2}{t^\phi}\right) \approx t^{2-\alpha} \left[f_0 + f_0' \frac{\delta J^2}{t^\phi}\right] \tag{5}$$

comparing (4) and (5) we get,

$$t^{2-\alpha-\phi} \sim < S_i S_{i+\delta} >^2, \tag{6}$$

where i and i+δ are nearest neighbors. For a ferromagnet $<S_i S_{i+\delta}> \sim t^{1-\alpha}$ and the Harris criteron is satisfied. It has been argued (see reference 2) that for antiferromagnets of type AF_2, $<S_i S_{i+\delta}> \sim t^\beta$, that with the scaling relation, $\alpha + 2\beta + \gamma = 2$, gives $\phi = \gamma - 2\beta$.
In the following we are going to describe how the value of the cross-over exponent, ϕ, is obtained from the behavior of′ the pure system.

Consider a system with Hamiltonian[4]

$$H \rightarrow H_0 - g \sum_i S_i \, S_{i+\delta} \tag{7}$$

where H_0 is given in (2) and $\sum_i$ stands for a sum over all the sites in the lattice, and $i+\delta$ is a nearest neighbor to i (always in the same direction), no sum over δ is performed. By translational invariance

$$\frac{1}{N}\left(\frac{\partial F}{\partial g}\right)_{g\rightarrow 0} = -< S_i S_{i+\delta} > \tag{8}$$

where F is the free energy. Also, from finite size scaling

$$F \sim t^{2-\alpha} f_g\left(\frac{g}{t^{\omega}}, \frac{\zeta}{L}\right) \tag{9}$$

From (8) and (9) we get,

$$< S_i S_{i+\delta} > \sim t^{2-\alpha-\omega} f_g\left(\frac{\zeta}{L}\right) \tag{10}$$

If we define a quantity R such that,

$$R = \frac{1}{N}\left(\frac{\partial^2 F}{\partial g^2}\right)_{g\rightarrow 0} = \sum_i \frac{<(S_i S_{i+\delta})^2> - <S_i S_{i+\delta}>^2}{N} \tag{11}$$

where i and $i+\delta$ are nearest neighbors.
From (9) it is clear that,

$$R \sim t^{2-\alpha-2\omega} f_R\left(\frac{\zeta}{L}\right) \tag{12}$$

Comparing (6) and (10) we get,

$$\phi = -2 + \delta + 2\omega \tag{13}$$

Then,

$$R \sim t^{-\phi} f_R\left(\frac{\zeta}{L}\right) \tag{14}$$

and, since $t^{-\phi} = L^{\frac{\phi}{\nu}} \left(\frac{\zeta}{L}\right)^{\frac{\phi}{\nu}}$ then

$$R \sim L^{\frac{\phi}{\nu}} \qquad (15)$$

2. The 2-D Ising model[5]

We will consider two cases: the case where the ordered state is ferromagnetic, $J_1=J_2 > 0$, and the case where the ordered state is antiferromagnetic, $J_1= J_2 < 0$. The ground state of the antiferromagnetic model is n=2 fold degenerate.

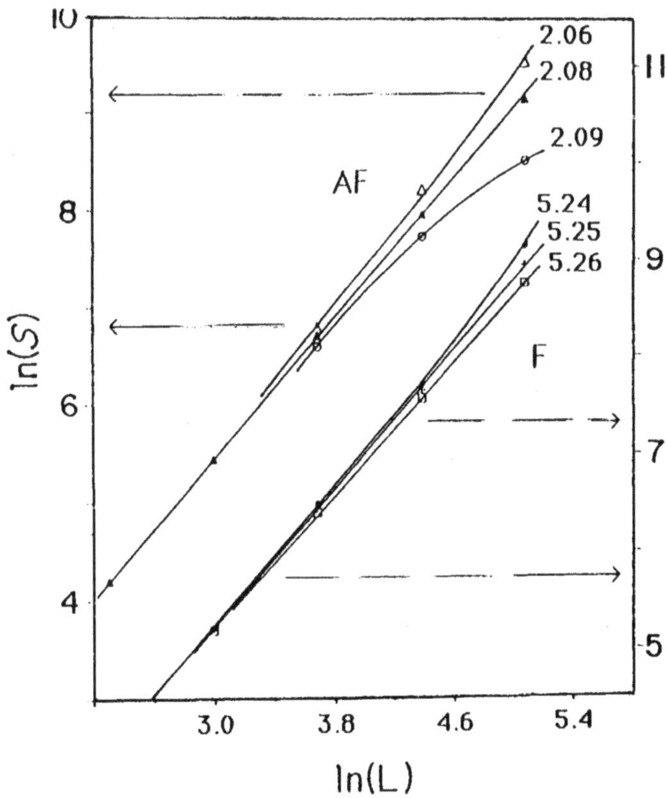

Fig 1. ln(S) vs ln(L), at different temperatures, for 2-d Ising F and AF.

The values of T_C, γ/ν and α/ν were calculated for the pure system in the following way. Defining a quantity S such that,

$$S = \frac{1}{N}\left[< \left(\sum_1 n_i S_i\right)^2 > + < \left(\sum_2 n_i S_i\right)^2 > \right] \qquad (16)$$

where Σ_1 (Σ_2) is a sum over all i and j on the one (the other)

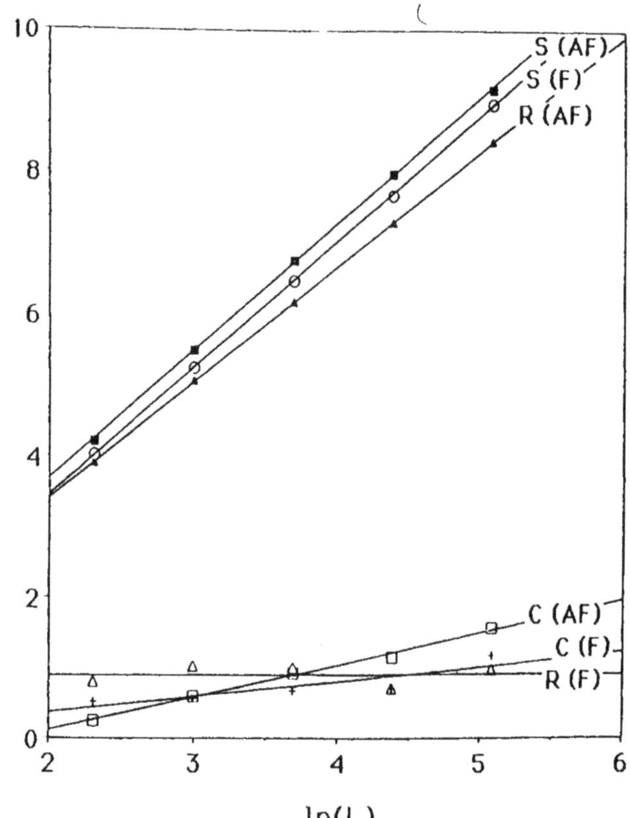

Fig 2. ln(S), ln(R) and ln(C) vs ln(L) at kT_C=2.08 (5.25) for the 2-d Ising F (AF).

sublattice. For the ferromagnetic case $n_i=1$, and for the antiferromagnetic case n_i takes alternatively the values of ± 1 when summing over the lattice sites.

From finite size scaling we know that as $L \to \infty$, for $T< T_c$, $S \to L^d$, (d is the dimensionality of the system, and L its linear size) whereas for $T>T_c$, as $L\to\infty$, $S \to$ constant. At $T=T_c$, $S \sim L^{\alpha/\nu}$.

The Metropolis algorithm was applied to systems with periodic boundary conditions, at least 10^4 sweeps were made in equilibrium and 1500 sweeps to equilibrate.

In figure 1 we show ln(S) vs ln (L) , for different values of T. From these results we get, $kT_c=(2.08 \pm 0.01)J_2$, and $\gamma/\nu=1.79 \pm 0.05$ for the antiferromagnetic case. For the ferromagnetic case we get $kT_c=(5.25 \pm 0.01)J_2$. The errors are obtained from a statistical analysis of the Monte Carlo results.

From finite size scaling we know that the specific heat behaves as $C\sim L^{\alpha/\nu}$, at $T=T_c$. In figure 1 we also show a plot of ln(C) vs ln(L) at $T=T_c$. We get $\alpha/\nu =0.46 \pm 0.02$ for the antiferromagnetic case, and $\alpha/\nu=0.2 \pm 0.1$ for the ferromagnetic one.

In figure 2 we show a plot of ln (R) vs ln (L) at $T=T_c$. From this figure we get the value of $\phi/\nu = 1.63 \pm 0.05$ for the antiferromagnetic case, and $\phi/\nu = [-0.1, 0.1]$ for the ferromagnet.

The predicted value for the antiferromagnetic case, $\phi=\gamma-2\beta$, and the relations $\alpha + 2\beta + \gamma = 2$ and $d\nu = \alpha - 2$, yield $\phi/\nu = 2\gamma/\nu - d$, which upon substitution of the value $\gamma/\nu =1.78$ obtained from figure 1, leads to good agreement with the computed value for ϕ/γ for the antiferromagnet.

On the other hand, the slope of ln (R) vs ln (L) for the ferromagnet is close to the slope of ln (C) vs ln(L), as predicted by Harris.

2. The 3-D system on a bcc lattice

In this section we are going to perform a Monte Carlo simulation of the nn/nnn Ising model on a bcc lattice. For: a) $J_1>0$ and $J_2>-2/3\, J_1$, the ground state is one of the two ferromagnetic states with all the spins aligned, b) $J_1<0$ and $J_2 >2/3\, J_1$, the system is a type I antiferromagnet, AF_1, (can be transformed to a ferromagnet by reversing and redefining all spins in the body centered sites), c) $J_2< -2/3\, |J_1|$ the ground state is one of the type-II antiferromagnetic (AF_2) states in which each one of the two interpenetrating simple cubic sublattices orders independently as a simple two sublattice antiferromagnets, the ground state is invariant under reversal of all the spins on one sublattice, removal of a J_1 bond lifts the ground state degeneracy and produces a random anisotropy[6]. We expect that the Harris criterion for ferromagnets ($\phi=\alpha$) does not hold for the AF_2 system, instead we expect $\phi=\gamma-2\beta$.

We apply the Metropolis algorithm to perform simulations on lattices of $N=L^3/4$ spins with L=20,32,40 and 48, with periodic boundary conditions.

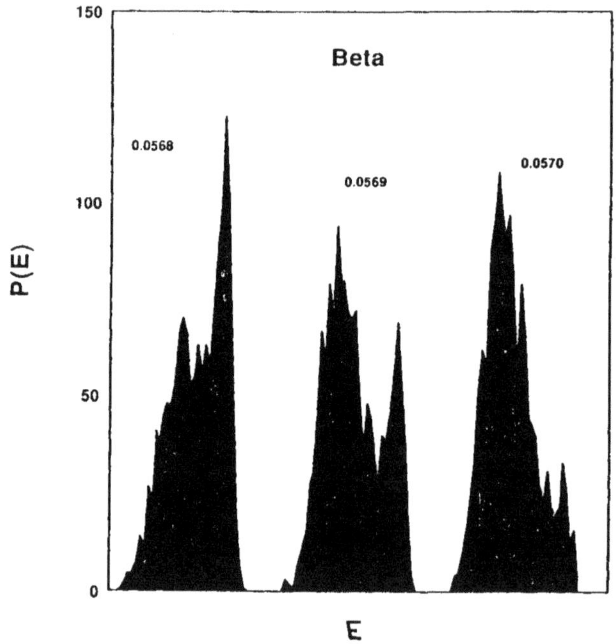

Fig 3. Energy distributions for a bcc Ising AF, for $J_1=7$, $J_2=5$, L=32, at different values of the inverse temperature,$\beta=(kT_c)^{-1}$.

Individual runs are typically of length 3×10^4 sweeps with 5×10^3 warming steps. We study the AF_2 región by selecting $J_1=7$, $J_2=-5$. In the vicinity of the expected bicritical point, $J1>0$; $J1/J2=-3/2$, Landau theory predicts a continuous transition between the paramagnetic phase and the AF_2 region[7].

Figure 3 shows energy distributions for a system of size $L=32^3$, that present a double peak around the expected critical point at this size, a strong suggestion of a first order phase transition. Similar results were found for lattices of different sizes. Our results seems to indicate that for the infinite system the transition happens at at β_c=0.05673, for J_1=7 and J_2=-5.

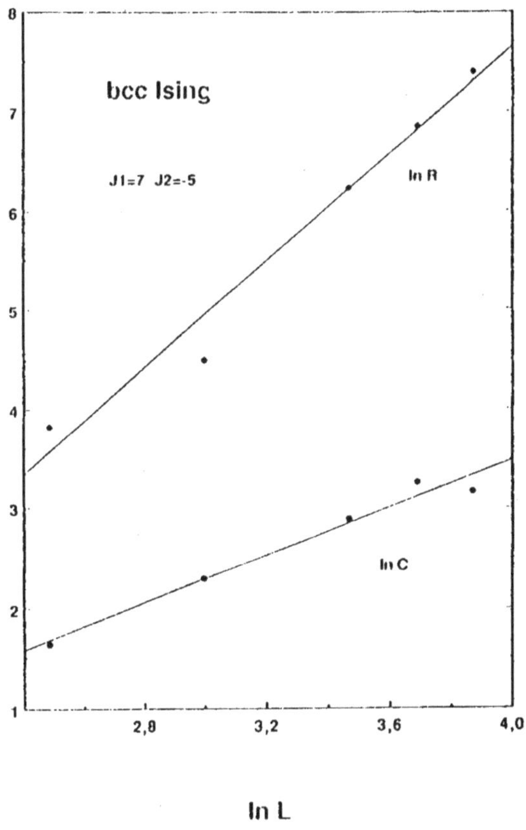

Fig 4. ln (R) and ln(C) vs ln (L) for the bcc Ising AF at $\beta=(kT_c)^{-1}$ and J_1=7, J_2=5.

In Figure 4 we show the results we got for the specific heat and the quantity R defined previously. The results are not conclusive enough to calculate the value of ϕ, however from the figure it is clear the inconsistency with the Harris criterion. Further studies with larger systems are been pursued in order to clarify the results.

3. Conclusions

We verified that the Harris criterion for the cross-over exponent, ϕ, is satisfied for the ferromagnetic Ising model with nearest and next-nearest neighbors, but not for the antiferromagnetic model for which we found that, at least for the 2-dimensional model, it is consistent with the relation, $\phi=\gamma-2\beta$. It is argue that this behavior is associated to the removal of the degeneration in the ground state by small amounts of non magnetic impurities.

1. A. B. Harris, J. Phys. C **7**, 1971 (1974); M. E. Fisher, Phys. Rev **176**, 257 (1968).
2. J. F. Fernandez, Europhys. Lett. 5, 129 (1988); Fernandez, J. F, Phys. Rev B **38**, 6901 (1988).
3. M. N. Barber, Phys. Reports **59**, 376 (1980).
4. A. Aharony, Europhys.Lett.**1**, 617 (1986).
5. G. M. Buendía and J. F. Fernandez, J. Physique C**8**, 1223 (1988).
6. J. Banavar, D. Jasnow, and D. P. Landau, Phys. Rev B **20**, 3820 (1979); M. Plischke and J. Oitmaa, Phys. Rev B **19**, 487 (1979).
7. D. Mukamel and S. Krinsky, Phys. Rev. B **13**, 5065 (1976).

THEORETICAL DOUBLE PHOTOIONIZATION STUDY OF HELIUM

V. D. Rodríguez, M. A. Kornberg and C. A. Falcón
Instituto de Astronomía y Física del Espacio
C.C. 67, Suc. 28 - 1428 Buenos Aires - Argentina

ABSTRACT

Double ionization cross sections of helium by photon impact are calculated within the wave-function approach. Electronic correlation is included in both the initial and final state wave-functions.The initial state is described by using a MCSCF wave-function that accounts for 94% of the correlation energy. The final state include some amounts of angular correlation through the variable effective charges method. Fithfold differential cross sections, electron kinetic energy spectra, distributions concerning the relative angle of the photoelectrons as well as total cross sections are shown and discussed.

1. Introduction

The formation of doubly charged ions by one-photon impact has deserved a growing attention during the last decade, mainly since it is known that this process is forbidden in the independent-electron model. Therefore, its study may shed light on the problem of electron correlation in the continuum.

The double photoionization (DPI) process of helium in the ground state, given by:

$$He(^1S_0) + h\nu \longrightarrow He^{2+} + e^-(\mathbf{k}) + e^-(\mathbf{k}') \tag{1}$$

seems to be the most simple case of a DPI process, even though, a theoretical description for this reaction has been obtained only in a partial form. The analysis of the total cross section, the main goal till recently, started with the pionnering work of Byron and Joachain[1] (BJ), and has continued with works by various authors[2,3,4]. Recently, Le Rouzo and Dal Capello[5] (LRDC) have obtained new results on the DPI process of Helium, specially concerning the fithfold differential cross sections and the distribution of electron kinetic energy. The analysis of LRDC is a reformulation of the partial wave analysis of BJ. These authors have further improved the calculation by using a ground state wave function of the multiconfiguration self-consistent-field (MCSCF) type[6], accounting for 98% of the correlation energy. They have also improved the final state description by the use of variable effective charges. However, their calculations when correlation is included in both the initial and final states are restricted to the fithfold differential cross sections, from where conclusions are difficult to extract. In this paper, we present a direct calculation of total and single differential cross sections for the DPI process. The symmetry properties of the transition matrix elements allow us to define a double differential cross section in the energy of one of the electrons and in the relative angle of both electrons. From this we calculate the relative angular distribution of the ejected electrons, which is perhaps the most sensitive magnitud for correlation studies.

The paper is organized as follows: section 2 presents the theory of the DPI process, the calculation of the double differential cross section and the wave functions employ in the calculus. Section 3 is a brief account of the mathematical details of the calculation, while in section 4 we present and discuss the results. Atomic units will be employed along this work except where otherwise explicitly stated.

2. Theory

The process that we consider is the impact of one linearly polarized photon on helium atoms in their ground state. For DPI to occur the photon energy (E_γ) must be greater than $-E_0$, where E_0 is the ground state energy. The total nonrelativistic energy for the ground state is[7] -2.903724 hartrees, or -79.014384 eV. Above the double ionization threshold, the ejection of both electrons may take place, with energies ϵ and ϵ' related to the photon and ground state energies through the conservation law $E_\gamma + E_0 = \epsilon + \epsilon'$.

The most detailed observable of the DPI process is the fithfold differential cross section (FDCS) $d^5\sigma^{2+}/d\epsilon d\Omega d\Omega'$ where ϵ is the energy of one of the electrons whose momentum $\mathbf{k}$ subtends an element of solid angle $d\Omega$. The prime quantities refer to the second electron. However measurements of the FDCS are very difficult to perform, because they require to measure in coincidence the dispersion angles of both electrons and the energy of one of them. Fortunately, more easily measurable observables can be defined, by proper integrations of the FDCS, namely the kinetic electron spectrum implying integrations over both angular directions $d\Omega$ and $d\Omega'$, and the total cross section (TCS) that require a further integration on energy $d\epsilon$. From the theoretical point of view, the total cross section require five integrations. In the present work a method is presented to reduce the number of integrations to only two. At the same time, this enables to introduce the double differential cross section (DDCS) $d^2\sigma^{2+}/d\epsilon \sin\theta_r d\theta_r$, describing the electron energy spectrum and the relative angle distribution.

In most calculations, the electric dipole approximation is employed[8]. This requires the evaluation of the dipole matrix element,

$$D^L_{fi}(\mathbf{k},\mathbf{k}') = < \psi_f | z_1 + z_2 | \psi_i > \qquad \text{(lenght form)} \tag{2}$$

$$D^V_{fi}(\mathbf{k},\mathbf{k}') = < \psi_f | \frac{\partial}{\partial z_1} + \frac{\partial}{\partial z_2} | \psi_i > \qquad \text{(velocity form)} \tag{3}$$

where $\psi_i(^1S_0|\mathbf{r}_1,\mathbf{r}_2)$ is the initial ground state wave function, $\psi_f(\mathbf{k},\mathbf{k}'|\mathbf{r}_1,\mathbf{r}_2)$ is the final state wave function. The incident photon is considered to be polarized along the z direction.

The FDCS is given by:

$$d^5\sigma^{2+}/d\epsilon d\Omega d\Omega' = (4\pi^2\alpha a_0^2)E_\gamma kk'|D_{fi}^L(\mathbf{k},\mathbf{k}')|^2 \quad \text{(lenght form)} \tag{4}$$

$$= (4\pi^2\alpha a_0^2)\frac{1}{E_\gamma}kk'|D_{fi}^V(\mathbf{k},\mathbf{k}')|^2 \quad \text{(velocity form)} \tag{5}$$

being α the fine-structure constant and a_0 the Bohr radius. Lenght and velocity forms of the FDCS give the same results if both the initial and final wave functions are exact.

We now turn to describe our way of integrating the FDCS. For this purpose we denote as $\mathbf{T}_{fi}^L$ and $\mathbf{T}_{fi}^V$ the vector matrix elements of lenght and velocity, respectively given by,

$$\mathbf{T}_{fi}^L = <\psi_f|\mathbf{r}_1 + \mathbf{r}_2|\psi_i> \tag{6}$$

$$\mathbf{T}_{fi}^V = <\psi_f|\nabla_1 + \nabla_2|\psi_i> \tag{7}$$

In what follows we denote by $\mathbf{T}_{fi}$ either the lenght or velocity matrix elements. The FDCS is proportional to $|\hat{\mathbf{e}}\cdot\mathbf{T}_{fi}|^2$, where $\hat{\mathbf{e}}$ denotes the photon polarization direction (usually the z axis). If we exploit the fact that the inner product is rotationally invariant, angular integrations may be easily performed. In the original coordinate system xyz the vectors $\mathbf{k}$ and $\mathbf{k}'$ have polar angles θ, ϕ and θ', ϕ', respectively. We can perform a rotation defined by the Euler angles[9] $\alpha = \phi$, $\beta = \theta$ and $\gamma = 0$, so that in the rotated system $x'y'z'$, $\mathbf{k}$ is along the z'-direction and $\mathbf{k}'$ have polar angles θ'', ϕ''. A subsecuent rotation of the coordinate system with Euler angles $\alpha' = \phi''$, $\beta' = 0$ and $\gamma' = 0$, leads the new system to $x''y''z''$ where $\mathbf{k}$ is in the z'' axis direction and $\mathbf{k}'$ is in the $x''z''$ plane. Taking into account that the solid angle element is rotationally invariant, it may be easily shown that

$$\int\int |\mathbf{T}_{fi}'' \cdot \hat{\mathbf{e}}''|^2 d\Omega d\phi'' = \frac{8\pi^2}{3}|\mathbf{T}_{fi}|^2 \tag{8}$$

In order to define the DDCS another consideration is required; the FDCS give the same result for both combinations of energy and angular variables ($\epsilon d\Omega$, $\epsilon' d\Omega'$) and ($\epsilon' d\Omega$, $\epsilon d\Omega'$), so when angular integration is performed a factor 1/2 must be introduced. LRDC[5] also introduce this factor in their evaluations of the TCS. The DDCS reads:

$$d^2\sigma^{2+}/d\epsilon \sin\theta_r d\theta_r = \frac{1}{2}(4\pi^2\alpha a_0^2)E_\gamma kk'(\frac{8\pi^2}{3})|\mathbf{T}_{fi}^L|^2 \quad \text{(lenght form)} \tag{9}$$

$$\mathrm{d}^2\sigma^{2+}/\mathrm{d}\epsilon\sin\theta_r\mathrm{d}\theta_r = \frac{1}{2}(4\pi^2\alpha a_0^2)\frac{1}{E_\gamma}kk'(\frac{8\pi^2}{3})|\mathbf{T}^V_{fi}|^2 \qquad \text{(velocity form)} \tag{10}$$

The single differential cross sections and the total cross section are obtain numerically by integration of the DDCS :

$$\mathrm{d}\sigma^{2+}/\mathrm{d}\epsilon = \int(\mathrm{d}^2\sigma^{2+}/\mathrm{d}\epsilon\sin\theta_r\mathrm{d}\theta_r)\sin\theta_r\mathrm{d}\theta_r \tag{11}$$

$$\mathrm{d}\sigma^{2+}/\sin\theta_r\mathrm{d}\theta_r = \int(\mathrm{d}^2\sigma^{2+}/\mathrm{d}\epsilon\sin\theta_r\mathrm{d}\theta_r)\mathrm{d}\epsilon \tag{12}$$

$$\sigma^{2+} = \int\int(\mathrm{d}^2\sigma^{2+}/\mathrm{d}\epsilon\sin\theta_r\mathrm{d}\theta_r)\sin\theta_r\mathrm{d}\theta_r\mathrm{d}\epsilon = \int\int(\mathrm{d}^2\sigma^{2+}/\mathrm{d}\epsilon\sin\theta_r\mathrm{d}\theta_r)\mathrm{d}\epsilon\sin\theta_r\mathrm{d}\theta_r \tag{13}$$

We finally describe the wave functions used for the calculations. Although there exists a lot of wave functions to describe the ground state of helium the most desirables ones are those which do not include the relative coordinate $\mathbf{r_{12}}$. For the present work we have chosen two wave functions which fit rather well to our purposes. One of them is the Hartree-Fock wave function of Clementi and Roetti[10] which is the basis of a non-correlated state. It takes the form

$$\psi(\mathbf{r_1},\mathbf{r_2}) = \phi(\mathbf{r_1})\phi(\mathbf{r_2}) \tag{14}$$

$$\phi(\mathbf{r}) = \sum_{j=1}^{5} C_j\varphi_j(\mathbf{r}), \qquad \varphi_j(\mathbf{r}) = \sqrt{\frac{\xi_j}{\pi}}\mathrm{e}^{-\xi_j r} \tag{15}$$

The ground state energy of the Hartree-Fock wave function is $E_{HF} = -2.86168$ hartrees, and the total correlation energy is defined as $E_{HF} - E_0$. A more adequate wave function to describe the ground state is the Tweed and Langlois[11] wave function, which is of the MCSCF type. It is of the form:

$$\psi(\mathbf{r_1},\mathbf{r_2}) = \sum_{l=0}^{2}\sum_{m=-l}^{l} C^{ll0}_{m\,-m\,0} Y_l^m(\hat{r_1})Y_l^{-m}(\hat{r_2}) \times \{\sum_{i,j} A^{(l)}_{ij}[R(n_i,\alpha_i;r_1)R(n_j,\alpha_j;r_2) + (r_1 \leftrightarrow r_2)]\} \tag{16}$$

where the normalized radial Slater wave function is

$$R(n,\alpha;r) = [(2n)!]^{-\frac{1}{2}}(2\alpha)^{n+\frac{1}{2}}r^{n-1}e^{-\alpha r} \tag{17}$$

The energy of the ground state using this wave function is $E_{TL} = -2.90125$ hartrees, which accounts for 94% of the total correlation energy.

As we have done for the initial state, we are interested in testing the influence of correlation in the final state. If the electrons have momenta $\mathbf{k}$ and $\mathbf{k}'$ the usual choice for a non-correlated final wave function is the symmetrized product of two individual Coulomb wave functions:

$$\psi_f(\mathbf{k},\mathbf{k}'|\mathbf{r}_1,\mathbf{r}_2) = \frac{1}{\sqrt{2}}\{\psi_{\mathbf{k}}(\mathbf{r}_1)\psi_{\mathbf{k}'}(\mathbf{r}_2) + \psi_{\mathbf{k}'}(\mathbf{r}_1)\psi_{\mathbf{k}}(\mathbf{r}_2)\} \tag{18}$$

The individual Coulomb wave functions are given by :

$$\psi_{\mathbf{k}}(\mathbf{r}) = \frac{1}{(2\pi)^{3/2}}e^{-\pi\xi/2}\Gamma(1-\mathrm{i}\xi)e^{i\mathbf{kr}}{}_1F_1(\mathrm{i}\xi,1,-ikr-\mathrm{i}\mathbf{k}\cdot\mathbf{r}) \tag{19}$$

where $\xi = -Z_T/k$, Z_T being the target charge, 2 in the case of helium. Because the final state of the DPI process of helium is a genuine three-body problem,the introduction of the correlation in this state is rather cumbersome. The method of variable effective charges, first introduced by Peterkov [12], could be exploit as was done by LRDC[5]. In order to satisfy the BJ condition on the matrix element[13] they orthogonalize the one electron Coulomb wave functions to a 1s ($Z_T = 2$) hydrogenic orbital. In the present work we employ the same variable charges used by LRDC[14] in their FDCS calculations, to obtain other more relevant cross sections.

3. Computation of the matrix element

The FDCS requires the computation of the z-component of the vectorial matrix element, which is needed to obtain the DDCS. The evaluation of the vectorial matrix element may be accounted for by employing the integral

$$B(nlm,\alpha,\mathbf{k}) = \int d\mathbf{r}\psi^*_{\mathbf{k}}(\mathbf{r})R(n,\alpha;r)Y_l^m(\hat{r}) \tag{20}$$

and in addition

$$\mathbf{B}^L(nlm,\alpha,\mathbf{k}) = \int d\mathbf{r}\psi^*_{\mathbf{k}}(\mathbf{r})\mathbf{r}R(n,\alpha;r)Y_l^m(\hat{r}) \tag{21}$$

$$\mathbf{B}^V(nlm,\alpha,\mathbf{k}) = \int d\mathbf{r}\psi^*_{\mathbf{k}}(\mathbf{r})\nabla R(n,\alpha;r)Y_l^m(\hat{r}) \tag{22}$$

for the lenght and velocity forms, respectively.

When Coulomb wave functions are chosen to be orthogonal to the $1s$ $(Z_T = 2)$ hydrogenic orbital, the following integral is required

$$J(nlm, \alpha, Z_T) = \int \mathrm{d}\mathbf{r} R(1, Z_T; r) Y_0^0(\hat{r}) R(n, \alpha; r) Y_l^m(\hat{r}) \tag{23}$$

and the following integrals

$$\mathbf{J}^L(nlm, \alpha, Z_T) = \int \mathrm{d}\mathbf{r} R(1, Z_T; r) Y_0^0(\hat{r}) \mathbf{r} R(n, \alpha; r) Y_l^m(\hat{r}) \tag{24}$$

$$\mathbf{J}^V(nlm, \alpha, Z_T) = \int \mathrm{d}\mathbf{r} R(1, Z_T; r) Y_0^0(\hat{r}) \nabla R(n, \alpha; r) Y_l^m(\hat{r}) \tag{25}$$

for the lenght and velocity forms, respectively. All these integrals may be obtained using the result of a basic integral calculated by S. Datta[15].

4. Results

In the present work, calculations are made with the velocity formulation. In a forthcoming paper the computation with the lenght form will be presented.

The FDCS are presented in the polar form adopted by LRDC[5]. We remind that in this representation a coplanar geometry is adopted for the momenta $\mathbf{k}$, $\mathbf{k}'$ and the direction of the electric field $\hat{\mathbf{e}}$. The photon energy is $E_\gamma \simeq 109$ eV, *i.e.*, it lays in the region where the total cross section (TCS) have its maximum. Two pairs of electron energies are considered. A symmetric one, where $(\varepsilon, \varepsilon') = (15, 15)$ eV, as shown in figure 1a, and an asymmetric case with $(\varepsilon, \varepsilon') = (5, 25)$ eV, shown in figure 1b. The initial wave function employed in both cases is the correlated one. For the final state either the non-correlated or the variable charge wave functions are used.

In figure 1a, we show the FDCS for three directions of $\mathbf{k}$: with respect to $\hat{\mathbf{e}}$ (0, 45 and 90 degrees). At 0 degrees the non-correlated cross section have a maximun along the direction of the electric field that corresponds to both electrons being ejected with the same momenta. On the other hand, the inclusion of correlation in the final state forbids the ejection at zero relative angles, appearing the two principal symmetric lobes and two secondary ones. At 45 degrees, the shape of the lobes has evolved to a highly asymmetric one. Finally at 90 degrees we found a typical dipole configuration for both cases. For the asymmetric case, figure 1b, a similar situation is found, but both electrons can be ejected in the same direction in the correlated case, and the secondary lobes have grown at the expense of those that where the principal ones, in the previous case.

The electron kinetic energy distributions are presented in figure 2 for $E_\gamma \simeq 109$ eV (intermediate photon energy) and $E_\gamma = 1000$ eV (high photon energy). The results reveals the symmetry derived from the identity of the electrons. When non-correlated final wave function (fixed charges) are used, the cross section for ejection of both electrons with the same energy is a minimum as also found in reference 5.

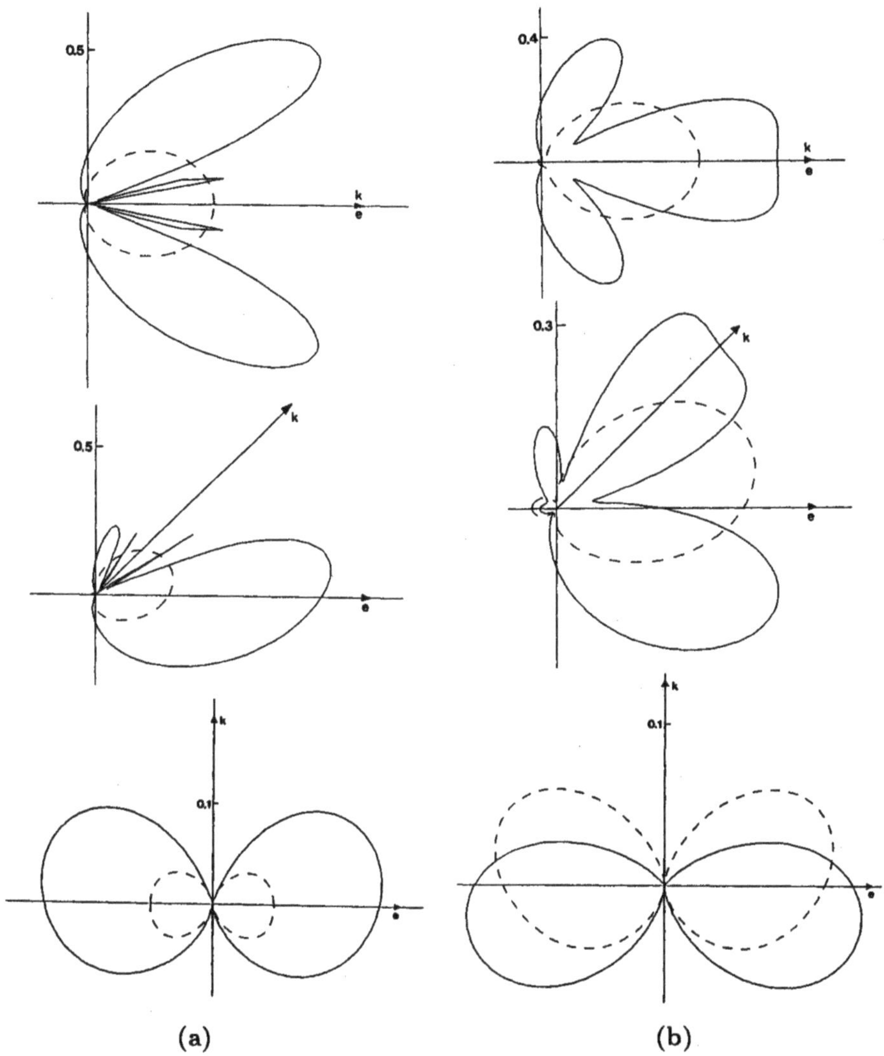

Fig. 1: The fithfold differential cross section (FDCS) (in kbarn/hartree) for the DPI of helium with coplanar geometry. For three fixed directions of **k** (0, 45 and 90 degrees) the FDCS are functions of the direction **k**′ of the other electron. The polar axis is the polarization direction (**e**). The initial state is describe by the wave function of Tweed and Langlois[11]. Final state wave function: ortogonalized Coulomb waves with variable charges (solid line); Coulomb waves with fixed charges ($Z = Z' = 2$) (dashed line). The electron energies are: (a) $\epsilon = \epsilon' = 15$ eV. (b) $\epsilon = 5$ eV, $\epsilon' = 25$ eV.

When variables charges are employed to described the final state, important changes in the shape of the curves are noted, namely a peak appears for $\varepsilon = \varepsilon'$, even when a non-correlated initial state is used. We may attempt a qualitative explanation as follows. The DPI process may proceed via two mechanisms, an uncorrelated one, known as shake-off implying the ejection of one electron by absortion of most of the photon energy and the subsecuent ejection of the other by sudden change in the effective nuclear charge seen by it. The second mechanism is due to correlation. One electron absorbs the photon energy then it interact with the other electron that then may pass to the contimuum, decreasing the final energy of the first electron. This last mechanism enhance the cross section in the region where the energy is equally distributed between both electrons. For $\epsilon = \epsilon'$ a relative maximum is obtained conversely to the situation described by the shake-off mechanism. At higher photon energies ($E_\gamma = 1000$ eV), this maximum becomes narrower and decreases in magnitude, what may indicate that for this case correlation is less important, as expected, since when the relative energy between the electrons is larger their 'interaction time' is shorter.

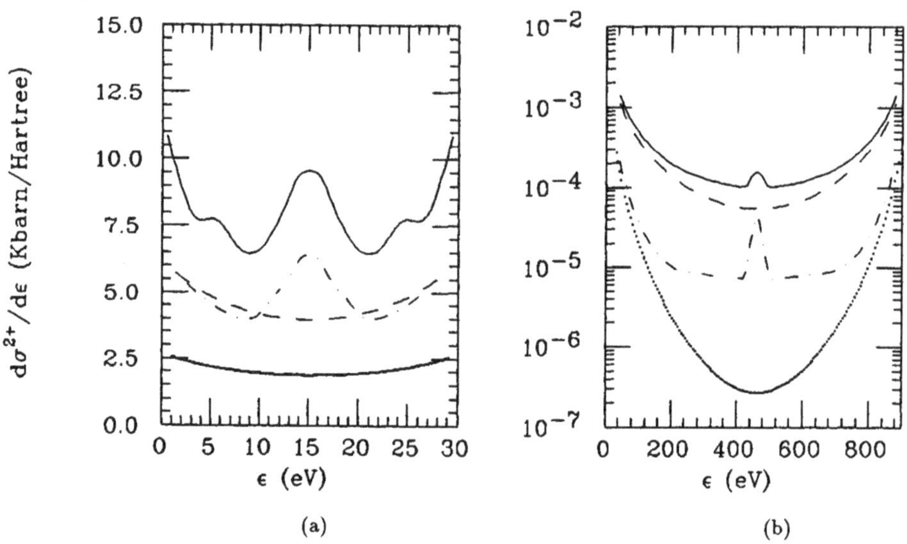

Fig. 2: Electron kinetic energy distributions. (a) $E_\gamma \simeq 109$ eV and (b) $E_\gamma = 1000$ eV. The wave functions employed are: correlated initial state wave function and variable effective charges final state (solid line); correlated initial state wave function and non-correlated final state (dashed line); Hartree-Fock initial state wave function and variable effective charges final state (dotdashed line) and Hartree-Fock initial state wave function and non-correlated final state (dotted line).

Relative angular distribution of the electrons are displayed in figure 3. For $E_\gamma \simeq 109$ eV (figure 3a) the ejection at zero relative angle is favored when correlation is not considered. The situation is similar when a correlated initial state is used. The situation is dramatically changed when variable charges are employed due to the appearence of two peaks that favor the ejection at relative angles of 25 and 55 degrees. For high photon energy (figure 3b) the non-correlated wave functions lead to a constant distribution. When correlation is included only in the initial state the maximun of the cross section is shifted to the larger relative angular region. Finally when correlation is included also in the final state two peaks appear again in the cross section but now at angles 15 and 150 degrees. The existence of these critical angles should be explained in terms of electron-electron scattering.

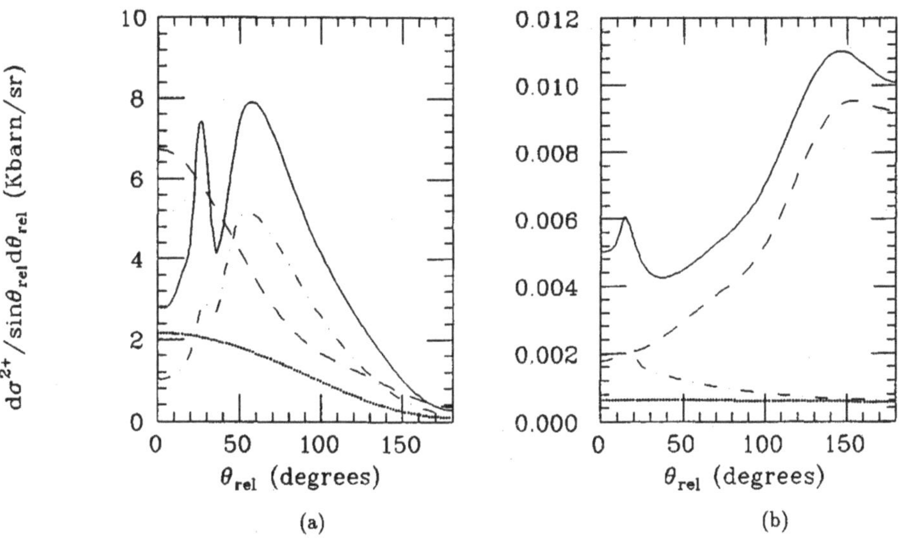

Fig. 3: Relative angle distributions for photon energies (a) $E_\gamma \simeq 109$ eV and (b) $E_\gamma = 1000$ eV. The curves are label as in figure 2.

We finally turn to the analysis of the total cross sections (figure 4). Our calculation with the Tweed and Langlois wave function and fixed charges give very similar results to those of BJ and LRDC, as expected; but the interest of these curves relies especially when correlation in the final state is included. The result obtained with a Hartree-Fock wave function in the initial state and fixed charges in the final one is too low when compared with the experimental data[16], but when variable charges are employed in the final state is very closed to the result obtained with the correlated initial wave function. The TCS obtained with the Tweed and Langlois initial state wave function (94% of the total correlation) and variable effective charges is the *best* result derived till now in the wave function approach (velocity formulation), though calculations employing the many body perturbation theory by Carter and Kelly[4] fits well to the data.

Acknowledgments One of the authors, V. D. Rodríguez, would like to thank Dr. J. Hanssen for providing part of the literature on this theme.

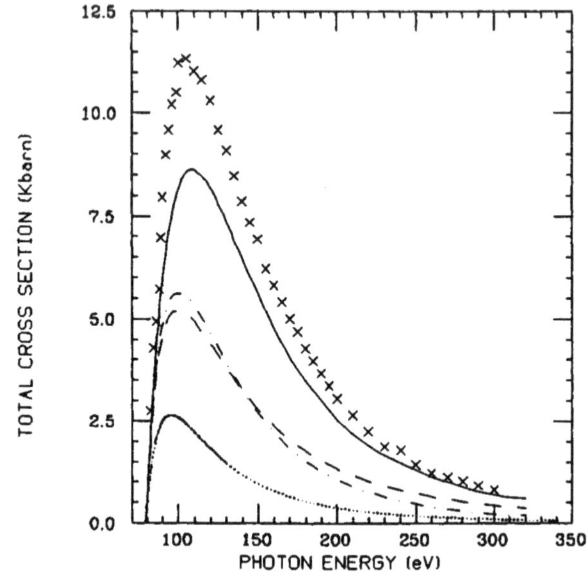

Fig. 4: Total cross sections. The crosses are the experimental data points from Bizeau *et al*[16]. T theoretical curves are label as in figure 2.

References

1. F.W.Byron and C.J.Joachain, Phys. Rev. A **164**, (1967) 1.
2. R.L.Brown, Phys. Rev. A **1**, (1970) 586.
3. S.N.Tiwary, J. Phys. B **15**, (1982) L323.
4. S.L.Carter and H.P.Kelly, Phys. Rev. A **24**, (1981) 170.
5. H.Le Rouzo and C. Dal Cappello, Phys. Rev. A **43**, (1991) 318.
6. N.Sabelli and J.Hinze, *J. Chem. Phys.*, **50** (1969) 684..
7. C.L.Pekeris, Phys. Rev. **126**, (1962) 143.
8. H.S.Bethe and E.E.Salpeter, *Quantum Mechanics of One- and two-electron aton* (Springer-Verlag, 1957), p.279.
9. M.E.Rose, *Elementary Theory of Angular Momentum*, (Wiley, 1957), p. 50.
10. E.Clementi and C.Roetti, *At. Data Nucl. Data Tables*, **14**, (1974), p. 177.
11. R.J.Tweed and J.Langlois, J. Phys. B **20**, (1987) 5213.
12. R.Peterkov, Sov. Phys.-*JETP* **16**, (1963) 442.
13. F.W.Byron and C.J.Joachain, Phys. Rev. Lett. **16**, (1966) 1139.
14. H.Le Rouzo and C. Dal Cappello, Phys. Rev. A **43**, (1991) 1395.
15. S.Datta, J. Phys. B **18**, (1985) 853.
16. J.M.Bizeau,F.Wuillemeunier,D.Ederer,P.Koch,P.Dhez,S.Krummacher and V.Schmi *unpublished.*

DISTORTED WAVE CALCULATIONS FOR DOUBLE ELECTRON TRANSFER

*ALEJANDRA E. MARTINEZ**
Division Colisiones Atomicas, Centro Atomico Bariloche
(8400) San Carlos de Bariloche, Republica Argentina

*ROBERTO D. RIVAROLA**
Instituto de Fisica Rosario (CONICET and UNR) and
Departamento de Fisica (ECEN - FCEIA - UNR), Avenida Pellegrini 250
(2000) Rosario, Republica Argentina

ROBERT GAYET and JOCELYN HANSSEN
Laboratoire des Collisions Atomiques#, Universite de Bordeaux I
351 Cours de la Liberation, 33405 TALENCE Cedex, France

ABSTRACT

The resonant double electron capture by alpha particles in helium targets is studied, at intermediate and high collision energies, using the Continuum Distorted Wave - Eikonal Initial State (CDW-EIS) model.

Differential and total cross sections for capture into the $He(1s^2)$ final state are calculated in the framework of an Independent Electron Approximation (IEA).

Theoretical results are compared with the experimental data available at present for capture into any final state of helium.

1. Introduction

The double electron capture process has been studied[1] in a very recent work, in the framework of an Independent Electron Approximation (IEA), by means of using the Continuum Distorted Wave (CDW) and Continuum Distorted Wave - Eikonal Initial State (CDW-EIS) approximations.

Calculations have shown that both theoretical models are suitable in order to describe the total cross sections for the mentioned process, in the energy range where they are applicable to the single

* Consejo Nacional de Investigaciones Cientificas y Tecnicas (CONICET)
# Unite Propre de Recherches 260 du CNRS

electron capture. The introduction of "angular" or "dynamical" electron correlations appears as not necessary in order to reproduce the general trend of the experimental data available at that time.

New experimental data for both, differential and total cross sections, were recently reported[2] for higher collision energies than those previously considered.

In the present work we extend our total cross section-calculations to this energetic range and also we present preliminary results for the differential cross sections in order to give a new test of our IEA at these high velocities.

Atomic units will be used, except otherwise stated.

2. Theory

We consider the double electron capture by a bare nucleus of charge Z_P in a helium-like target of nuclear charge Z_T and denote by $\bar{x}_j$, $\bar{s}_j$ and $\bar{r}_j$ the position of electron j referred to the target, the projectile and midpoint of the internuclear distance respectively.

In the framework of the impact parameter treatment the internuclear coordinate is given by $\bar{R} = \bar{\rho} + \bar{v}t$, where $\bar{\rho}$ is the impact parameter.

The prior form of the transition amplitude reads:

$$\mathcal{A}^{-}_{if} = \lim_{t \to -\infty} \langle \Psi^{-}_{f} \mid \Phi^{+}_{i} \rangle \tag{1}$$

where the wavefunction Ψ^{-}_{f} is the solution of

$$(H - i\, \partial/\partial t)\, \Psi^{-}_{f}(\bar{r}_1, \bar{r}_2, t) = 0 \tag{2}$$

with the boundary condition

$$\Psi^{-}_{f} \underset{t \to -\infty}{\to} \Phi^{-}_{f} \tag{3}$$

In equation 2, H is the total Hamiltonian

$$H = \sum_{k=1}^{2} \left(- \frac{\nabla^2_{r_k}}{2} - \frac{Z_T}{x_k} - \frac{Z_P}{s_k} \right) + \frac{1}{r_{12}} + \frac{Z_T Z_P}{R} \tag{4}$$

Furthermore, the asymptotic initial and final wavefunctions $\Phi^{\pm}_{j}$ satisfy

$$(H_j - i\, \partial/\partial t)\, \Phi^{\pm}_{j}(\bar{r}_1, \bar{r}_2, t) = 0 \qquad j = i, f \tag{5}$$

where (+) goes for $j = i$ and (-) for $j = f$ and

$$H_i = h_i + \frac{Z_P(Z_T - 2)}{R} =$$

$$= \sum_{k=1}^{2} \left(- \frac{\nabla^2_{r_k}}{2} - \frac{Z_T}{x_k} \right) + \frac{1}{r_{12}} + \frac{Z_P(Z_T - 2)}{R} \tag{6a}$$

$$H_f = h_f + \frac{Z_T(Z_P - 2)}{R} =$$

$$= \sum_{k=1}^{2} \left(- \frac{\nabla^2_{r_k}}{2} - \frac{Z_P}{s_k} \right) + \frac{1}{r_{12}} + \frac{Z_T(Z_P - 2)}{R} \tag{6b}$$

are the initial and final electronic Hamiltonians, corresponding to the entry and exit channel respectively.

Thus, the wavefunctions $\Phi^{\pm}_j$ result

$$\Phi^{+}_i = \varphi_i(\bar{x}_1, \bar{x}_2) \exp\{ - i\frac{v^2}{4}t - i\varepsilon_i t - i\frac{v}{2} (\bar{r}_1 + \bar{r}_2)\} \times$$

$$\times \exp\{ i \frac{Z_P(Z_T - 2)}{v} \mathrm{Ln}(vR - \bar{v}\cdot\bar{R})\} \tag{7a}$$

$$\Phi^{-}_f = \varphi_f(\bar{s}_1, \bar{s}_2) \exp\{ - i\frac{v^2}{4}t - i\varepsilon_f t + i\frac{v}{2} (\bar{r}_1 + \bar{r}_2)\} \times$$

$$\times \exp\{-i \frac{Z_T(Z_P - 2)}{v} \mathrm{Ln}(vR + \bar{v}\cdot\bar{R})\} \tag{7b}$$

where φ_i and φ_f are the initial and final bound wavefunctions that satisfy

$$(h_j - \varepsilon_j)\, \varphi_j = 0 \qquad\qquad j = i, f \tag{8}$$

with ε_j the corresponding atomic energy.

We introduce the CDW-EIS distorted wave model choosing distorted wavefunctions as:

$$\chi_i^+ = \mathscr{L}_i(\bar{r}_1,\bar{r}_2)\ \Phi_i^+ \exp\{\ i\ 2\ \frac{Z_P}{v} \operatorname{Ln}(vR-\bar{v}\cdot\bar{R})\} \tag{9a}$$

$$\chi_f^- = \mathscr{L}_f(\bar{r}_1,\bar{r}_2)\ \Phi_f^- \exp\{-i\ 2\ \frac{Z_T}{v} \operatorname{Ln}(vR+\bar{v}\cdot\bar{R})\} \tag{9b}$$

where $\mathscr{L}_j$ (j =i,f) is a product[3] of two functions $\mathscr{L}_j^k$ corresponding to each electron given by

$$\mathscr{L}_i^{CDW-EIS,k}(\bar{x}_k) = \exp(\pi Z_T/\ 2v)\ \Gamma(1-i\ \frac{Z_T}{v})\ \times$$

$$\times\ {}_1F_1(i\ \frac{Z_T}{v}\ ;1;\ i(vx_k+\ \bar{v}\cdot\bar{x}_k)) \tag{10a}$$

$$\mathscr{L}_f^{CDW-EIS,k}(\bar{s}_k) = \exp\{-i\ \frac{Z_P}{v} \operatorname{Ln}(vs_k+\ \bar{v}\cdot\bar{s}_k)\} \tag{10b}$$

The initial and final distorted wavefunctions defined in this way verify the corresponding correct boundary conditions.

If we call with $\hat{\Psi}_j^{\pm}$ and $\hat{\chi}_j^{\pm}$ (j= i,f) the reduced wavefunctions that do not contain the internuclear interaction, the transition amplitude $\mathcal{A}_{if}^-$ can be written as

$$\mathcal{A}_{if}^- = (\rho v)^{2iZ_PZ_T/v} \lim_{t\to-\infty} <\hat{\Psi}_f^-|\ \hat{\chi}_i^+> = (\rho v)^{2iZ_PZ_T/v}\ a_{if}^- \tag{11}$$

provided that

$$\lim_{t\to+\infty} <\hat{\chi}_f^-\ |\ \hat{\chi}_i^+> = 0 \tag{12}$$

in order to avoid contributions to the transition amplitude coming from the distorting potentials alone.

The bound wavefunctions φ_j are approximated by a product of K-shell electronic hydrogenic-like orbitals $\tilde{\varphi}_j^k$ with charges determined with the Raleigh-Ritz variational method ($Z^*_{j'} = Z_j - 5/16$; $j' = i,f$; $j' \neq j$) that are solutions of the Schrödinger equation

$$(\tilde{h}^k_j - \tilde{\varepsilon}^k_j)\,\tilde{\varphi}^k_j = 0\ , \quad \tilde{\varepsilon}^k_j = -\frac{(Z^*_{j'})^2}{2} \qquad j = i,f \tag{13}$$

with

$$\tilde{h}^k_j = -\frac{\nabla^2_u}{2} - \frac{Z^*_{j'}}{u_k} \tag{14}$$

where $\bar{u}_k = \bar{x}_k$ if $j = i$ or $\bar{u}_k = \bar{s}_k$ if $j = f$. In this way the reduced distorted wavefunctions $\hat{\chi}^{\pm}_j$ result as a product of individual distorted wavefunctions $\hat{\chi}^{k\pm}_j$ for each electron

$$\hat{\chi}^{k,+}_i = \mathcal{L}^{CDW-EIS,k}_i\ \tilde{\varphi}_i(x_k)\ \exp\{-i\frac{v^2}{8}t - i\,\varepsilon_i t - i\,\frac{\bar{v}}{2}\cdot\bar{r}_k\} \tag{15a}$$

$$\hat{\chi}^{k,-}_f = \mathcal{L}^{CDW-EIS,k}_f\ \tilde{\varphi}_f(s_k)\ \exp\{-i\frac{v^2}{8}t - i\,\varepsilon_f t + i\,\frac{\bar{v}}{2}\cdot\bar{r}_k\} \tag{15b}$$

Assuming that the electronic evolution of the system is made up by the independent evolution of each electron

$$\hat{\Psi}^-_f = \prod_{k=1}^{2} \hat{\Psi}^{k-}_f \tag{16}$$

the electronic transition amplitude can be written as

$$a^-_{if} = \prod_{k=1}^{2} \lim_{t\to-\infty} \langle\, \hat{\Psi}^{k-}_f \mid \hat{\chi}^{k+}_i \,\rangle = \prod_{k=1}^{2} a^{k-}_{if} \tag{17}$$

The one-electron transition amplitude given by equation 17 can be evaluated in first order resulting in

$$a^{k-}_{if} = -i \int_{-\infty}^{+\infty} dt\ \langle\, \hat{\Psi}^{k-}_f \mid \frac{(Z_P - Z^*_P)}{s_k} - \frac{(Z_T - Z^*_T)}{x_k} + W^k_i \mid \hat{\chi}^{k+}_i \rangle \tag{18}$$

where W^k_i is the perturbative potential resulting from the application of the operator $(H - i\partial/\partial t)$ on the distorted wavefunction χ^+_i and taking into account equation 15. In reference 1, calculations have been performed neglecting the two first addends.

Total cross sections can be calculated as:

$$\sigma_2 = 2\pi \int_0^\infty d\rho\ \rho\ |a^-_{if}(\rho)|^2 \tag{19}$$

and differential cross sections as:

$$\frac{d\sigma_2}{d\Omega} = \left| \int_0^\infty d\rho\ \rho\ J_0(\eta\rho)\ a^-_{if}(\rho)\ \rho^{2iZ_T Z_P/v} \right|^2 \tag{20}$$

The integrand in equation 20 is extremely oscillating and so its calculation can be carried out, in a simple even when slowly convergent way, by means of evaluating

$$\frac{d\sigma_2}{d\Omega} = \left| \sum_{\ell=1}^{\infty} \frac{(2\ell + 1)}{2\mu v}\ e^{\{2iZ_T Z_P/v\ \ln[(2\ell + 1)/2\mu v]\}}\ a^-_{if}\left(\frac{2\ell + 1}{2\mu v}\right) \times \right.$$

$$\left. \times\ P_\ell(\cos\theta_{CM}) \right|^2 \tag{21}$$

where we have considered that $\mu v \rho \cong \ell + 1/2$ (where μ is the reduced mass). P_ℓ indicates Legendre's polynomials.

3. Results and Discussions

In figure 1 total cross sections for the $He^{2+} + He \rightarrow He + He^{2+}$ reaction are shown for collision laboratory energies going from 100 keV up to 3000 keV. Theoretical CDW-EIS cross sections calculated as indicated in equation 19 and taking into account equations 17 and 18, for the resonant $1s^2 \rightarrow 1s^2$ process, are compared with the theoretical CDW-EIS cross section extracted from reference 1 and with experimental data for capture into all the states[4-6]. Contributions coming from capture into excited states is expected to be less than 35% [1,7], a value that is not likely to affect the behaviour of the total cross section as a function of the impact energy. In fact this contribution decreases when the collision velocity increases[8,9]. Good agreement with the experimental data is kept up in all the energy range. Discrepancies between both theoretical calculations are of the order of 35% for the lower energies and decrease with increasing energies reaching a value of the order of 10% for higher energies.

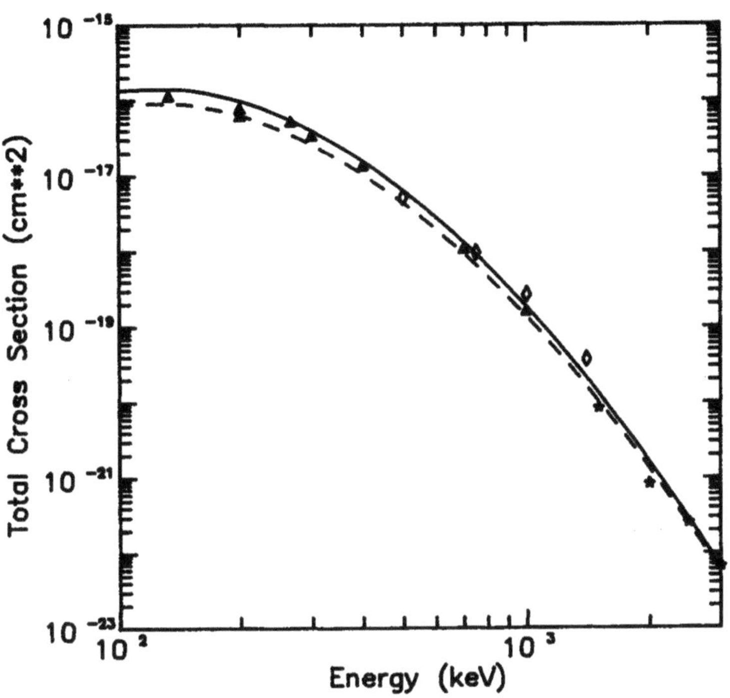

Figure 1: Total cross sections for the He^{2+} + He double electron capture reaction. Theories are for the resonant He^{2+} + $He(1s^2)$ collision; full curve: CDW-EIS this work, dashed curve: CDW-EIS from ref. 1. Experiments (for capture into all final states); triangles: ref. 4, lozenges: ref. 5 and asterisks: ref. 6.

In figure 2 total cross sections for the same reaction are present for collision laboratory energies going from 1000 keV up to 6000 keV. Theoretical CDW-EIS total cross sections calculated as in reference 1 are compared with CDW-$1s^2$ results evaluated in the same way, CDW-CI results evaluated with a static configuration interaction calculation for the bound He ground state[2] and with experimental data[2,4-6]. CDW-EIS theoretical predictions show a good agreement with experimental data for the lower energies but overestimate the new experimental results, while CDW-$1s^2$ calculations show the better agreement with the latter ones. However, it must be noted that experiments correspond to a partial cross section due to the finite size of the detector solid angle. As it has been estimated in reference 2, 30% of the experimental total cross section is missed and thus the measured data must be considered as a lower bound.

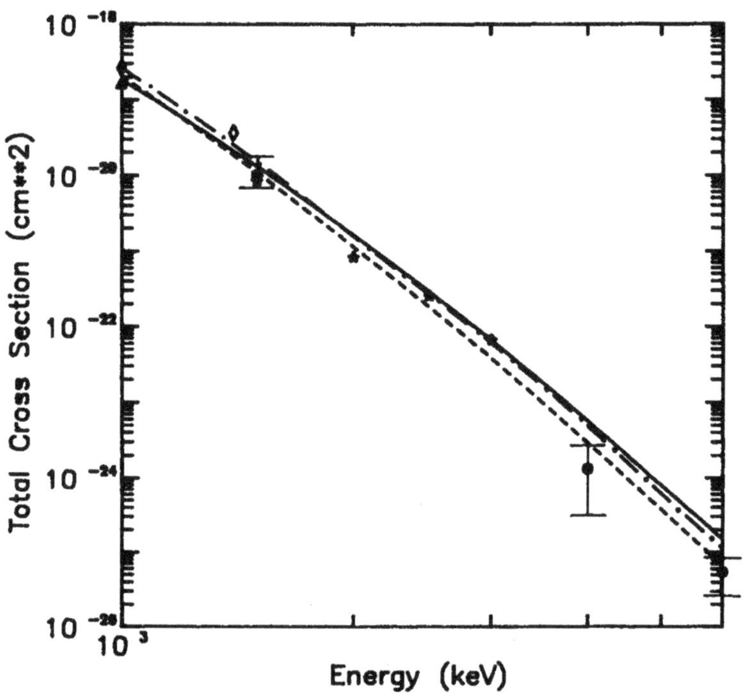

Figure 2: Total cross sections for the He^{2+} + He double electron capture reaction. Theories are for the resonant He^{2+} + He($1s^2$) collision; full curve: CDW-EIS this work, dashed curve: CDW-$1s^2$ this work and dash-and-dot curve: CDW-CI from ref. 2. Experiments (for capture into all final states); triangles: ref. 4, lozenges: ref. 5, asterisks: ref. 6 and full circles: ref. 2.

Differential cross sections for the same reaction are presented in figure 3 for a collision laboratory energy of 1500 keV. Equation 21 has been evaluated taking ℓ up to 30000 to obtain a convergent result. Good qualitative agreement is found with the experimental data. The inclusion of capture into excited states in the theoretical calculations could improve the agreement with the measured differential cross sections. It must also be noted that from indications given in reference 2 we can calculate that statistical errors are of the order of 10%.

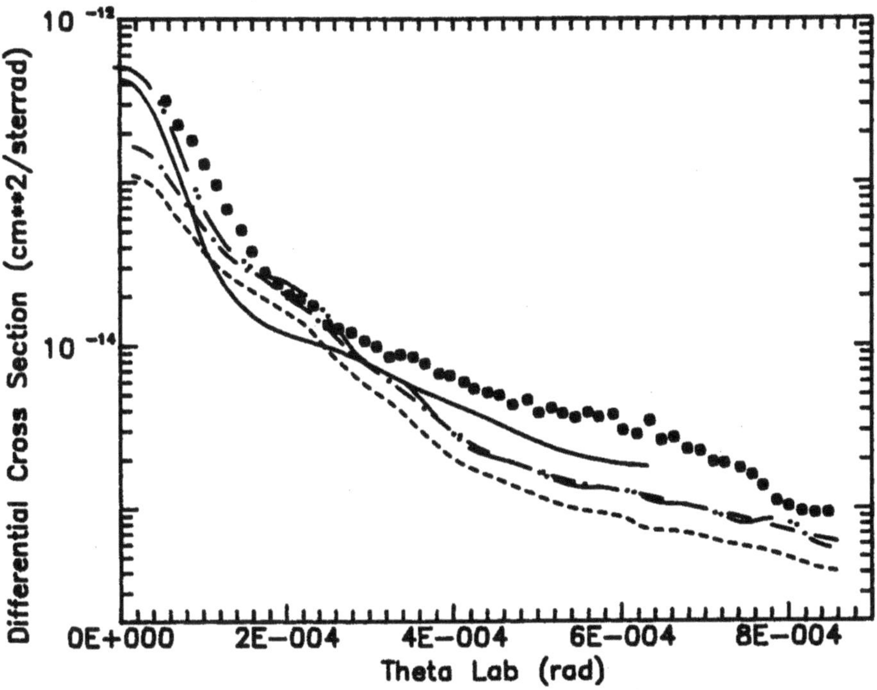

Figure 3: Differential laboratory cross section for the He^{2+} + He double electron capture reaction as a function of the laboratory scattering angle and for a laboratory collision energy of 1500 keV. Theories are for the resonant He^{2+} + He($1s^2$) collision; full curve: CDW-EIS (unfolded) this work, dashed curve: CDW-$1s^2$ (folded), dash-and-dot curve: CDW-CI (unfolded) and dash-and-two dots: CDW-CI (folded) from ref. 2. Experiments; full circles: from ref. 2.

Systematic errors coming from sequential capture events have been corrected in the data. The experiments do not account for the "head on" double electron capture collision (in which the target nucleus is scattered at zero degrees). However a simple calculation shows that this process is negligible in the determination of the total cross section at the impact energies here considered[10]. An elastic process in which the projectile hits on the target nucleus and as a consequence the complete target atom is scattered at zero degrees, is even less important

4. Conclusions

Theoretical total and differential cross sections for the double electron capture process are presented in the framework of the IEA for the He^{2+}+ He ($1s^2$) resonant reaction.

Very good agreement with the experimental data is found for a wide energetic range when total cross sections are analysed.

Good qualitative agreement with the experimental data is obtained in the case of differential cross sections.

We can conclude that in the framework of the IEA the CDW-EIS approximation may be used in order to make theoretical predictions for double electron transfer total and differential cross sections, for intermediate and high collision energies.

5. References

1. R. Gayet, J. Hanssen, A.E. Martinez and R.D. Rivarola, *Z. Phys. D* **18** (1991) 345.
2. R. Schuch, E. Justiniano, H. Vogt, G. Deco and N. Gruen, *J. Phys. B* **24** (1991) L133.
3. R. Gayet, R.D. Rivarola and A. Salin, *J. Phys. B* **14** (1981) 2421.
4. R.D. DuBois, *Phys. Rev. A* **36** (1987) 2585.
5. E. McDaniel, M.R. Flannery, H.W. Ellis, F.L. Eisele and W. Pope, *US Army Missile Research and Development Command Technical Report* (1977) 78.
6. N.V. de Castro Faria, F.L. Freire Jr. and A.G. de Pinho, *Phys. Rev. A* **37** (1988) 280.
7. C. Harel and A. Salin, *J. Phys. B* **13** (1980) 785.
8. Dz. Belkic and R. Gayet, *J. Phys. B* **10** (1977) 1911.
9. Dz. Belkic and R. Gayet, *J. Phys. B* **10** (1977) 1923.
10. M.C.R. McDowell and J.P. Coleman, *The introduction to the theory of ion-atom collisions* (North Holland Publishing Company, 1970), p. 392.

ELECTRON CAPTURE FROM MOLECULAR TARGETS BY ION IMPACT

SILVIA CORCHS, ROBERTO RIVAROLA
Instituto de Física Rosario, CONICET - UNR and Departamento de Física, ECEN, FCEIA, UNR; Avenida Pellegrini 250,2000 Rosario, Argentina

and

J.H. MCGUIRE
Department of Physics, Kansas State University, Manhattan Kansas 66506, USA

ABSTRACT

Electron capture from diatomic molecular targets by impact of bare ions is studied at high collision velocities. Interference patterns due to scatter from the two atomic centers of the target are determined by using different theoretical approximations. The importance of the inclusion of long range Coulomb interactions is investigated.

1. Introduction

Single electron capture from molecular hydrogen by high velocity protons has been studied by many authors. Tuan and Gerjuoy[1], Band[2] and Ray and Saha[3-4] have evaluated differential and total cross sections for this process averaging over the orientation of the molecular axis. Later, Deb et al[5] have calculated differential cross sections for fixed orientation obtaining interference patterns typical of scattering from two centers. Wang et al[6] formulated the impact parameter treatment of this capture process and obtained probabilities and total cross sections as a function of the molecular orientation. The results reported by these authors were obtained within the Oppenheimer - Brinkman - Kramers or the Jackson - Schiff approximations.

The aim of this paper is to analyze the influence of the inclusion of *appropriate Coulomb boundary conditions* in the description of the entry and exit channels for the molecular target case.

In section 2 we formulate the theory for scattering from diatomic molecules both in the wave quantum mechanical and in the impact parameter treatments. In section 3 we present differential cross sections and probabilities for single charge exchange from a molecular ion H_2^+ target.

Atomic units are used unless otherwise stated.

2. Theory

2.1 Quantum - mechanics formulation

Following Wang et al[6] we will treat the diatomic molecule as a system containing two effective atomic scatterers centered on each nuclei. The operator $\hat{T}^M$ corresponding to the molecular scattering can be expressed as:

$$\hat{T}^M \cong \hat{T}^1 + \hat{T}^2 \tag{1}$$

where $\hat{T}^1$ and $\hat{T}^2$ are the operators corresponding to the atomic scattering with the effective centers $\underline{1}$ and $\underline{2}$ respectively. In formula **1 we** have ignored rescattering between these centers.

We choose the center of mass of the molecule as the origin of the reference system. The atomic centers $\underline{1}$ and $\underline{2}$ are displaced from the center of mass by $(1-\mu)\bar{\rho}$ and $-\mu\bar{\rho}$ respectively; where $\bar{\rho}$ is the molecular axis vector, $\mu = M_1 / (M_1 + M_2)$ and M_j (j=1,2) is the nuclear mass of center j

Let t^j_{if} (j=1,2) denote the atomic T-matrix for center j when the process is described from a reference frame fixed on the center j and let T^M_{if} denote the molecular T-matrix. So, the following relation holds:

$$T^M_{if} = t^1_{if} \exp\{-i(1-\mu)\bar{\alpha}.\bar{\rho}\} + t^2_{if} \exp\{i\mu\bar{\alpha}.\bar{\rho}\} \tag{2}$$

where $\bar{\alpha} = \bar{K}_f - \bar{K}_i$ is the total momentum transfer given by the difference between the final and initial momenta $\bar{K}_f$ and $\bar{K}_i$.

For homonuclear molecules $t^1_{if} = t^2_{if} = t_{if}$ for gerade state, $t^1_{if} = -t^2_{if} = t_{if}$ for ungerade state and $\mu = 1/2$.

Equation 2 results

$$T^M_{if} = t_{if} \left(\exp\{-i\bar{\alpha}.\bar{\rho}/2\} \pm \exp\{i\bar{\alpha}.\bar{\rho}/2\} \right) \tag{3}$$

where + (-) stands for gerade (ungerade).

At fixed orientation of the molecular axis corresponding to a solid angle Ω_ρ , the double differential - cross section is given by

$$\frac{d\sigma_M}{d\Omega_\rho \; d\Omega} = \frac{\mu_i \; \mu_f}{4\pi^2} \; \frac{K_f}{K_i} \; |T^M_{if}|^2 \tag{4}$$

In equation 4, $\mu_i (\mu_f)$ is the initial (final) reduced mass of the system and Ω is the projectile solid angle.
Substituting equation 3 in equation 4 we obtain

$$\frac{d\sigma_M}{d\Omega_\rho \; d\Omega} = \frac{\mu_i \; \mu_f}{2\pi^2} \; \frac{K_f}{K_i} \; (1 \pm \cos(\bar{\alpha}.\bar{\rho})) \; |t_{if}|^2 \tag{5}$$

2.2 Impact parameter treatment

In the impact parameter treatment the position $\bar{R}$ of the projectile is given by

$$\bar{R} = \bar{B} + \bar{v} \; t \tag{6}$$

where $\bar{B}$ is the impact parameter with respect to the center of mass of the molecule and $\bar{v}$ is the initial velocity of the projectile which defines the z axis of the system. So, we can write: $\bar{\alpha} = \bar{\alpha}_\perp + \alpha_z \hat{z}$; $\bar{\rho} = \bar{\rho}_\perp + \rho_z \hat{z}$, where with the sub-index $\perp$ we indicate the component perpendicular to the z axis.

The probability amplitude $a_M(\bar{B})$ is obtained through the Fourier transform of T^M

$$a_M(\bar{B}) = \frac{1}{v} \int \exp\{i\bar{\alpha}_\perp.\bar{B}\} \; T^M_{if} \; d\bar{\alpha}_\perp \tag{7}$$

Using equation 2 in equation 7 the following expresion can be derived

$$a_M(\bar{B}) = [\; a_1(\bar{B}_1) + a_2(\bar{B}_2) \exp\{i\alpha_z\rho_z\} \;] \exp\{-i(1-\mu)\alpha_z\rho_z\} \tag{8}$$

In equation 8 $a_j(\bar{B}_j)$ (j=1,2) is the <u>atomic</u> transition amplitude corresponding to the center j and

$$\bar{B}_1 = \bar{B} - (1-\mu) \; \bar{\rho}_\perp \tag{9.a}$$

$$\bar{B}_2 = \bar{B} + \mu \; \bar{\rho}_\perp \tag{9.b}$$

are the impact parameters referred to the centers 1 and 2 respectively.

Then, the probability for scattering from a diatomic molecule results

$$P_M(\bar{B}) = |a_1(\bar{B}_1)|^2 + |a_2(\bar{B}_2)|^2 + 2\,\mathrm{Re}(a_1^* a_2 \exp\{i\alpha_z \rho_z\}) \tag{10}$$

Let us recall that equations 5 and 10 have been obtained using equation 1. The magnitudes t_{if} and a_j (j=1,2) that appear in these equations represent the "exact atomic transition amplitudes" in the wave quantum mechanical and impact parameter treatment respectively.

In the present work three different atomic models are used to calculate the transition amplitudes: i) the Oppenheimer - Brinkman - Kramers (OBK) , ii) the boundary corrected first Born approximation[7] (B1B) and iii) the first order approximation of the Bates series[8] (BA1). Both, the B1B and BA1 approximations, include the long range Coulombic electron-projectile and electron-target nucleus interactions in the description of the atomic entry and exit channels respectively.

Next we explicit the atomic T-matrix, t_A, for the models above mentioned

$$t_A = \langle\, \varphi_f \exp\{i\bar{K}_f.\bar{R}_f\}\, D_f \,|\, H - E \,|\, \varphi_i \exp\{i\bar{K}_i.\bar{R}_i\}\, D_i \,\rangle \tag{11}$$

where φ_i (φ_f) is the initial (final) atomic wavefunction, $\bar{R}_i$ ($\bar{R}_f$) is the position of the projectile (the center of mass of the electron-projectile system) with respect to the center of mass of the atomic scatterer (to the atomic nucleus scatterer). Also D_i and D_f are the distortion factors taken as

i) $$D_i^{OBK} = D_f^{OBK} = 1 \tag{12}$$

ii) $$D_i^{B1B} = \exp\{i/v \int_{-\infty}^{z} dz'\; Z_P(Z_T - 1)/R\,\} \tag{13.a}$$

$$D_f^{B1B} = \exp\{-i/v \int_{z}^{+\infty} dz'\; Z_T(Z_P - 1)/R\} \tag{13.b}$$

where Z_T (Z_P) is the nuclear charge of the target (projectile) and $\bar{R}$ is the internuclear coordinate.

iii) $$D_i^{BA1} = \exp\{\, i/v \int_{-\infty}^{z} dz'\; [Z_T Z_P/R + \langle\, \varphi_i \,|\, -Z_P/r_P \,|\, \varphi_i \rangle]\} \tag{14.a}$$

$$D_f^{BA1} = \exp\{-i/v \int_{z}^{+\infty} dz'\; [Z_T Z_P/R + \langle\, \varphi_f \,|\, -Z_T/r_T \,|\, \varphi_f \rangle]\} \tag{14.b}$$

where $\bar{r}_T(\bar{r}_P)$ is the electronic coordinate with respect to the nuclear target (projectile).

As shown by Wang et al[6] t_{if} and t_A are related by

$$t_{if} = N_i . t_A \tag{15}$$

where N_i is the normalization factor of the initial molecular wavefunction.

3. Results and Discussions

In figures 1a-c we have represented impact parameter probabilities by using formula 10 and in figures 2a-c the double differential cross sections are calculated employing expression 5. The reaction $H^+ + H_2^+(1s) \longrightarrow H(1s) + 2H^+$ is analyzed at a 1 Mev collision energy. The theoretical models used are the above presented OBK, B1B and BA1.

In the calculations we have considered $\rho = 2$ a.u. and the initial molecular wavefunction is taken to have the form

$$\Psi_i^M = N_i (\varphi_i^1 + \varphi_i^2) \exp\{-i\varepsilon_i . t\} \tag{16}$$

where N_i is the normalization factor given by

$$N_i = [2 (1 + (1 + Z_T^* . \rho + (Z_T^* . \rho)^2/3) \exp\{-Z_T^* . \rho\})]^{-1/2} \tag{17}$$

and φ_j (j=1,2) are the atomic hydrogenic functions with effective nuclear charge $Z_T^* = 1.228$. The electron orbital energy is taken as $\varepsilon_i = -1.083$ a.u.

In figure 1a we can observe that the B1B approximation presents a divergence at $B \cong 0$ when the polar and azimuthal angles ϑ_ρ and ϕ_ρ are equal to zero degrees. It is in correspondence with a similar behaviour previously obtained in the atomic case [9]. On the contrary, OBK and BA1 approximations show a smooth monotonic decreasing as B increases.

In figure 2a the typical dip structure of the B1B atomic differential cross section appears also for molecular targets. This is not the case for the OBK and BA1 calculations. The comparison between OBK, B1B and BA1 results permit us to determine the importance of the internuclear interaction, because, as indicated in equations 12-14.b the B1B and BA1 approximations include the nucleus-nucleus potential and OBK does not. The effect is to raise the differential cross sections in some orders of magnitude. As expected, for $\vartheta_\rho = \phi_\rho = 0$ no interference patterns appear.

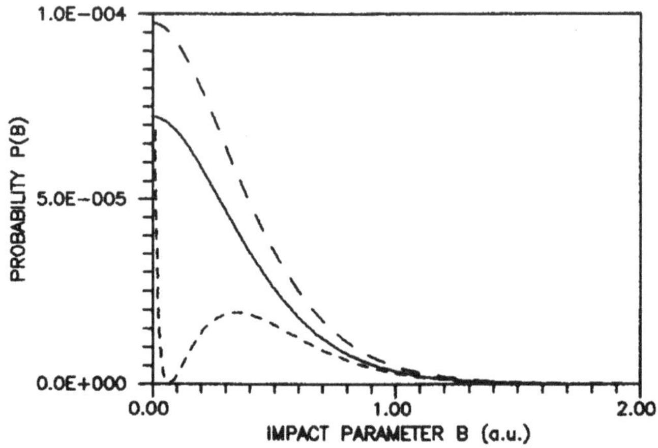

Figure 1.a: Impact parameter probabilities for electron capture in $H^+ + H_2^+$ collisions at 1 Mev for fixed orientation of the molecular axis: $\vartheta_\rho = \phi_\rho = 0°$. Theoretical results: — — — — OBK, ———————— BA1, - - - - - - - B1B.

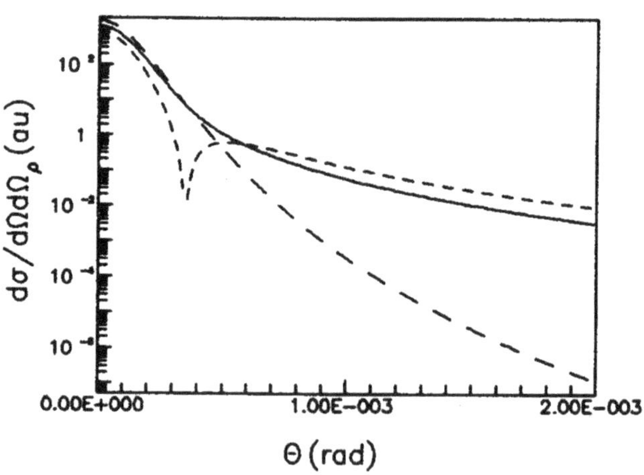

Figure 2.a: Double differential cross sections for electron capture in $H^+ + H_2^+$ collisions at 1 Mev for fixed orientation of the molecular axis: $\vartheta_\rho = \phi_\rho = 0°$. Theoretical results: idem fig. 1.a

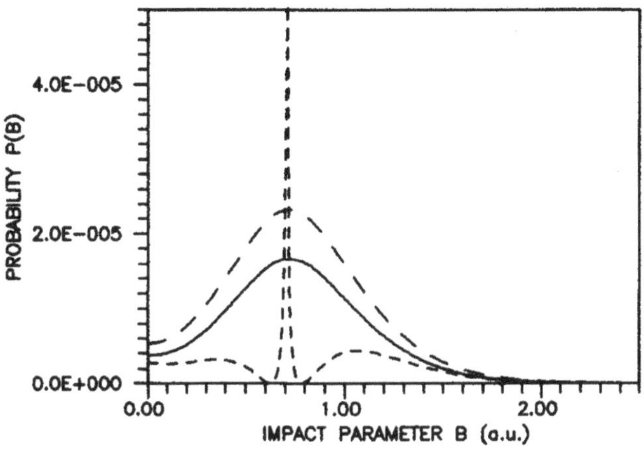

Figure 1.b: Impact parameter probabilities for electron capture in $H^+ + H_2^+$ collisions at 1 Mev for fixed orientation of the molecular axis: $\vartheta_\rho=45°;\phi_\rho=0°$. Theoretical results: idem fig. 1.a

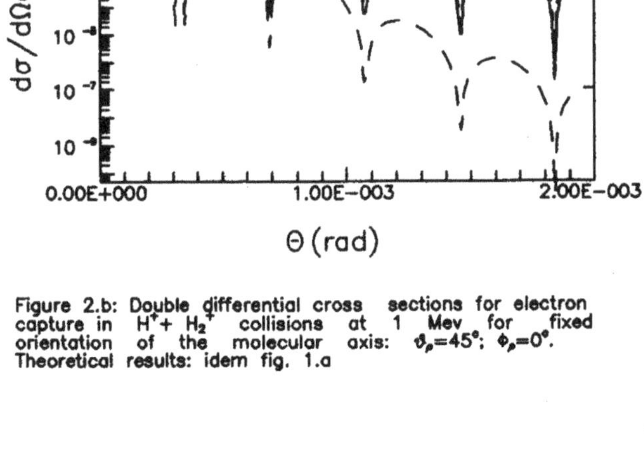

Figure 2.b: Double differential cross sections for electron capture in $H^+ + H_2^+$ collisions at 1 Mev for fixed orientation of the molecular axis: $\vartheta_\rho=45°$; $\phi_\rho=0°$. Theoretical results: idem fig. 1.a

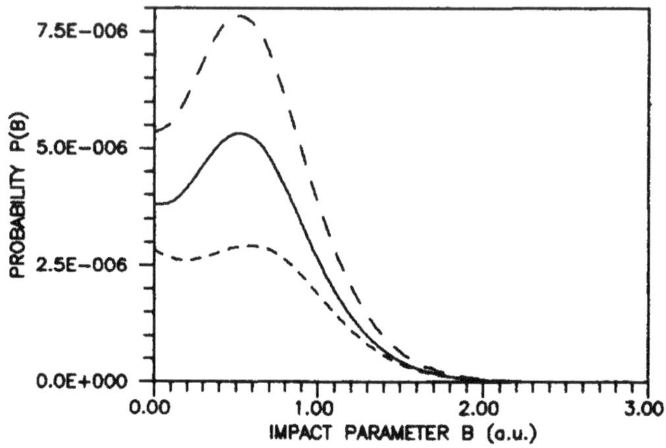

Figure 1.c: Impact parameter probabilities for electron capture in $H^+ + H_2^+$ collisions at 1 Mev for fixed orientation of the molecular axis: $\vartheta_\rho = 45°; \phi_\rho = 45°$. Theoretical results: idem fig. 1.a

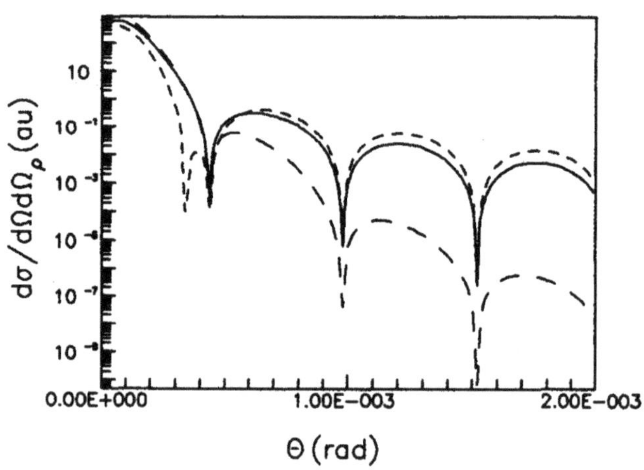

Figure 2.c: Double differential cross sections for electron capture in $H^+ + H_2^+$ collisions at 1 Mev for fixed orientation of the molecular axis: $\vartheta_\rho = \phi_\rho = 45°$. Theoretical results: idem fig. 1.a

In figure 1.b and 2.b the probabilities and cross sections are presented for ϑ_ρ = 45 deg. and ϕ_ρ = 0 deg. In this case, a zero atomic impact parameter corresponds to a non-zero value of B. So, the divergence of the B1B approximation appears at B ≅ 0.707 a.u. As a consequence, the general trend of the B1B probability is very different to the profiles shown by the OBK and BA1 calculations. In figure 2.b interference patterns are observed and again the influence of the internuclear potential can be seen. The dip in the B1B differential cross sections appears at the same angular position as for the previous studied molecular orientation.

In figures 1.c and 2.c probabilities and cross sections are shown when the molecular axis is placed out of the collision plane. In such a way we have set $\vartheta_\rho = \phi_\rho$ = 45 deg. No spurious divergences are observed for the B1B probabilities because the atomic impact parameters are always different from zero. Atomic interference patterns are again observed. Periodic interference structures are wider in the projectile angular coordinate than those of case of figure 2.b.

Finally, it must be noted that in all presented cases the OBK and BA1 probabilities show similar qualitative behaviours, being the OBK calculations larger than the BA1 ones.

Conclusions

Interference effects due to two atomic scatterer centers in ion-diatomic molecule collisions are determined by using different theoretical representations.

In particular, special attention is given to the inclusion of correct boundary conditions, associated with perturbative long range Coulomb interactions.

Spurious divergencies are found in electron capture probabilities when the B1B approximation is used to represent the case of molecular targets. These divergencies are corrected if the representation of the electron interactions are improved at finite distances.

References

1. T.F. Tuan and E. Gerjuoy, *Phys. Rev.* **117** (1960) 756
2. Y.B. Band, *J. Phys.* **B 7** (1974) 2055
3. P.P. Ray and B.C. Saha, *Phys. Lett.* **A 71** (1979) 415
4. P.P. Ray and B.C. Saha, *Phys. Rev.* **A 23** (1981) 1807
5. N.C. Deb, A. Jain and J.H. McGuire, *Phys. Rev.* **A 38** (1988) 3769
6. Y.D. Wang, J.H. McGuire and R.D. Rivarola, *Phys. Rev.* **A 40** (1989) 3673
7. Dz. Belkic, R. Gayet and A. Salin, *Phys. Rep.* **56** (1979) 279
8. R.D. Rivarola, J.M. Maidagan and J. Hanssen, *Nucl. Instrum. Methods* **B 27** (1987) 565
9. T.P. Grozdanov and P.S. Krstic, *Physica Scripta* **38** (1988) 32
10. Dz. Belkic, S. Saini and H.S. Taylor, *Z. Phys.* **D** 3 (1986) 59
11. Dz. Belkic and H. Taylor, *Phys. Rev.* **A 39** (1989) 6134

ION BEAM ANALYSIS OF ALUMINIUM NITRIDE THIN FILMS*

F.C. Stedile, I.J.R. Baumvol[1], W.H. Schreiner[1]

Instituto de Química, [1]Instituto de Física, Universidade Federal do Rio Grande do Sul, 91500 Porto Alegre, RS, Brasil

F.L. Freire Jr.

Departamento de Física, Pontifícia Universidade Católica do Rio de Janeiro, 22452 Rio de Janeiro, RJ, Brasil

ABSTRACT

Aluminium nitride thin films were deposited by D.C. magnetron reactive sputtering under various conditions of nitrogen partial pressure and incident power. Studying the reactive sputtering process some correlations among the deposition parameters were established. Based on this knowledge, aluminium nitride films were deposited and analysed by Rutherford Backscattering Spectrometry and Nuclear Reaction Analysis ($^{14}N(d,p)$, $^{14}N(p,\gamma)$, $^{27}Al(p,\gamma)$). Typical results from these techniques are shown as well as relationships between film characteristics (stoichiometry and thickness) and a deposition parameter (nitrogen partial pressure).

1. INTRODUCTION

Aluminium nitride thin films have been deposited by various methods, such as D.C. and R.F. reactive sputtering, CVD, ion implantation, ... Because of their piezoelectric, insulation and passivation characteristics they are used in surface acoustic wave devices[1] as well as in GaAs technology[2].

In the present work we deposited thin films of aluminium nitride by D.C. magnetron reactive sputtering. The plasma gas consisted in mixtures of Ar and N_2 in different proportions. From one deposition set to the other the incident power was also varied. Controling and varying the deposition parameters we established some relationship among them. The thickness and stoichiometry of the films were determined by Rutherford Backscattering Spectrometry (RBS) and by some Nuclear Reaction Analysis (NRA). Besides of the film characterization a dependence of the deposition rate and of the stoichiometric ratio N/Al on the nitrogen partial pressure was found.

*Work supported in part by CNPq-Brasil.

2. EXPERIMENTAL DETAILS

The aluminium nitride films were deposited in a Balzers BAS-450 magnetron sputtering system, where we used the D.C. source at constant power condition. The equipment allows to control the partial pressures of the gases (Ar and N_2 in our case) using a feedback mechanism that maintains the pressures at a constant level. The partial pressures of Ar and N_2, as well as other gases, can be monitored by a residual gas analyser based on a mass spectrometer. The distance between the Al 99.999% target (7cm x 17cm) and the substrates is 10cm. The vacuum chamber is pumped by a turbomolecular pump of 1500 l/sec down to a base pressure lower than 5×10^{-6} mbar and then throttled to start the gas admission. The substrates used to support the aluminium nitride films were Si(100) wafers. After a chemical cleaning of the substrates the Si wafers were submitted to a RF plasma etching (inside the vacuum chamber) during 7 minutes at an Ar pressure of 3×10^{-3} mbar. The aluminium target had to be presputtered in an Ar ambient ($P_{Ar} = 3\times10^{-3}$ mbar, Power = 200W) during 50 minutes to remove the Al_2O_3 layer prior to aluminium nitride deposition. During the film deposition we followed the process monitoring the partial and total pressures, the voltage, current and power.

The aluminium nitride thin films were analysed by RBS, with incident α of 2 MeV, and NRA. The reactions we used were the $^{14}N(d,p)^{15}N$ at 610 keV which has a plateau in the excitation curve allowing us to extract the total amount of ^{14}N in the thin film and the $^{14}N(p,\gamma)^{15}O$ and the $^{27}Al(p,\gamma)^{28}Si$, occuring respectively at 278 keV and 326 keV, both of which have a resonant peak in the excitation curve allowing us to profile the depth concentration of the elements in the film. The stoichiometric ratios N/Al were obtained from RBS and NRA and the thicknesses of the films from RBS and via profilometer measurements. All data agreed within an error of 5%.

3. RESULTS AND DISCUSSION

In the study of the D.C. reactive sputtering process we kept constant one deposition parameter and observed the dependence between two others. In Fig. 1a the incident D.C. power was kept at 240W and we follow the modification on the cathode voltage as we increased (●) and decreased (O) the nitrogen partial pressure inside the chamber. At low nitrogen partial pressures the two curves coincide but as the nitrogen concentration was increased the first curve always developed a lower cathode voltage forming a hysteresis loop. This fact was usually attributed to the formation of aluminium nitride at the surface of the target[3]. However, when we plotted the cathode voltage versus incident D.C. power keeping constant the nitrogen and argon partial pressures we obtained no hysteresis at all (Fig. 1b). Straight lines when plotting the deposition parameters were just observed in this specific combination of parameters (P_{Ar}, P_{N2}) and variables (Voltage, Power).

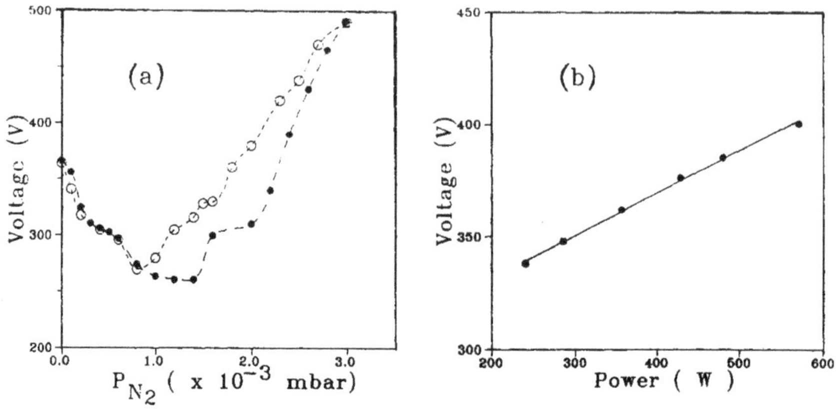

Figure 1a. Cathode voltage versus nitrogen partial pressure curves obtained as we increased (●) and decreased (O) the nitrogen pressure. The incident D.C. power was kept constant at 240W.

Figure 1b. Cathode voltage versus incident D.C. power. The argon and nitrogen partial pressures were kept constant at 2.8 and 0.2×10^{-3} mbar, respectively.

Based on the relationships extracted from this study we deposited aluminium nitride thin films under various conditions. The films were analysed by RBS and NRA. In Fig. 2 there is a typical RBS spectrum, where we pointed out the outermost edges on the Al and N signals and the Si-film interface, evidenced by the superposition of the Si and Al signals. To fit the experimental curve we assumed a density of 3.26 g/cm^3 for the deposited compound and then obtained a stoichiometric ratio N/Al of 0.9, i.e., the compound stoichiometry is $AlN_{0.9}$, and a thickness of 2300 Å for the film.

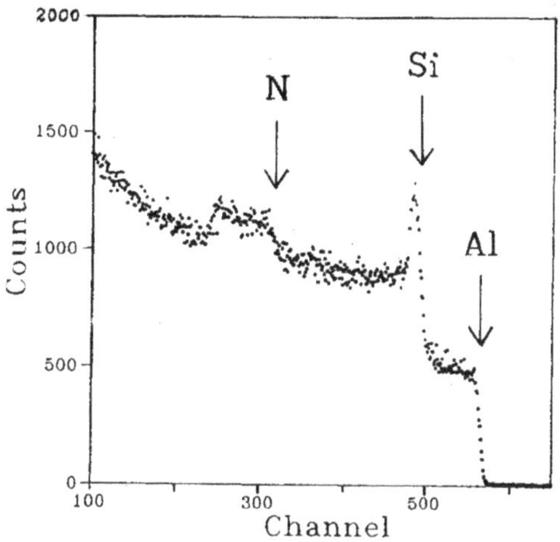

Figure 2. RBS spectrum obtained with 2.0 MeV incident α-particles backscattered by an aluminium nitride film deposited on Si(100). The incidence and detection angles were 0° and 160° to the film normal, respectively. The arrows in the Al and N signals indicate the outermost edge of the film and the other one, the Si-aluminium nitride film interface.

The nuclear reactions we used to characterize the films were the $^{14}N(d,p)^{15}N$, the $^{14}N(p,\gamma)^{15}O$ and the $^{27}Al(p,\gamma)^{28}Si$. Since the excitation curve of the $^{14}N(d,p)^{15}N$ reaction presents a plateau it was possible to obtain the total amount of nitrogen in the thin film. However, the $^{16}O(d,p)^{17}O$ and the $^{12}C(d,p)^{13}C$ reactions also have remarkable cross sections at this incident energy: E_d=610 keV, which allows us to determine the concentration of these two elements in the film. In Fig. 3 we have the spectrum of the detected protons. The area under the $^{14}N_{p5}$ signal was the one used to evaluate the amount of nitrogen in the film. The $^{16}O_{p1}$ + $^{14}N_{p6}$ and $^{16}O_{p0}$ peaks are exhibiting the effect of presputtering time of the target with respect to the oxygen content of the films. The solid curve was obtained from an aluminium nitride film deposited after a presputtering of the Al target of 30 minutes (P_{Ar} = 3x10^{-3} mbar, Power = 200W), while the dotted curve was obtained from a film deposited after 50 minutes of presputtering. We can conclude from this that a presputtering of at least 50 minutes is necessary to remove all aluminium oxide formed in the Al target surface.

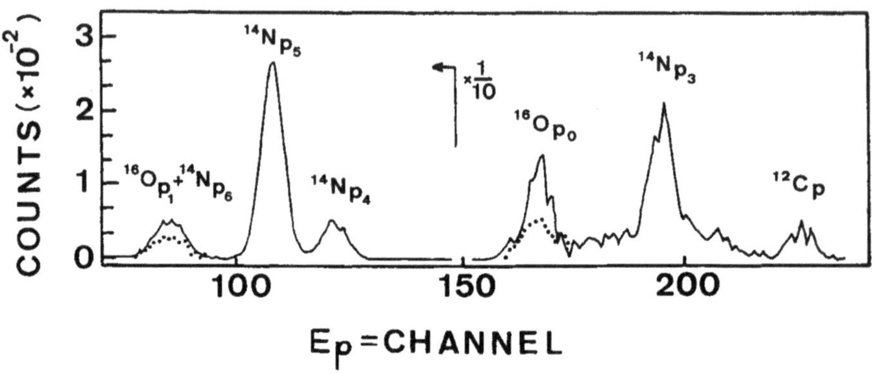

Figure 3. Spectra of the detected protons resulting from the incidence of 610 keV deuterons on an aluminium nitride thin film. The protons coming from different excited states are indexed in the figure. The solid line spectrum came from a film deposited after 30 minutes of presputtering of the aluminium target, while the dotted one, from a film deposited after 50 minutes presputtering.

The (p,γ) resonances of ^{14}N and ^{27}Al allowed us to obtain concentration depth profiles of these two elements. Since the resonance energies of these reactions are 278 keV and 326 keV for ^{14}N and ^{27}Al, respectively, we could make these profile determinations sequentially just increasing the incident energy of the protons. From the area under the γ-ray peaks we extracted the areal density of the isotopes, using Si_3N_4 and pure Al as standards for the correspondent reactions. In Fig. 4 we have the ^{14}N and ^{27}Al depth profiles in a film with stoichiometry $AlN_{0.9}$ and 2000 Å .

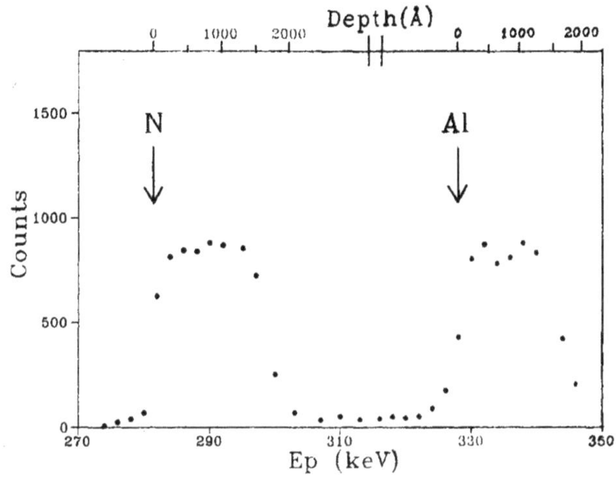

Figure 4. Nitrogen and aluminium depth profiles obtained respectively from the $^{14}N(p,\gamma)^{15}O$ and $^{27}Al(p,\gamma)^{28}Si$ resonances of an aluminium nitride thin film. The arrows indicate the outermost edges of the film in the nitrogen and aluminium depth profiles. The depth resolution in this case is about 200 Å.

With the thicknesses and stoichiometric ratios N/Al of the aluminium nitride films obtained from the above mentioned analytical techniques we plotted, in Fig. 5, the deposition rate and the N/Al ratio as functions of the percentage of nitrogen in the plasma mixture during deposition. In Fig. 5a we observe high deposition rates at low percentages of N_2, behaviour that is characteristic of metallic materials, and very low deposition rates at percentages of N_2 greater than 30, indicating a nitrided target. In between, the sudden drop in the deposition rate is attributed to the runaway situation that occurs in the transition of this two regimes[3]. From Fig. 5b we conclude that at low N_2 concentrations we have films poor in nitrogen and above 30% of N_2 in the plasma we always obtain films with stoichiometries around $AlN_{0.9}$ no mather what concentration of N_2 we had at the deposition time.

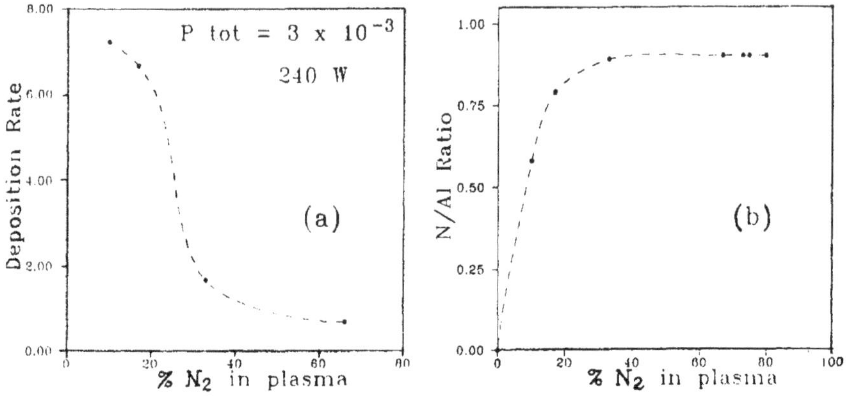

Figure 5. Dependence of a) the deposition rate and b) the stoichiometric ratio on the percentage of nitrogen in plasma during deposition.

4. CONCLUSIONS

Aluminium nitride thin films deposited by D.C. magnetron reactive sputtering under various conditions were analysed by RBS and nuclear reactions of ^{14}N and ^{27}Al.

In the present working conditions a hysteresis loop was observed when plotting the cathode voltage versus nitrogen partial pressure and a straight line when plotting the voltage versus incident power. The last behaviour is not usually found in this kind of process.

From the RBS spectra we extracted the stoichiometry and thickness of the films, while from the (d,p) reactions we found out that a long time (50 minutes) presputtering of the aluminium target is necessary to remove the aluminium oxide layer that covers it. This reactions should be used to monitore the relative concentrations of N, O and C in nitride, oxynitride and carbonitride films.

With the (p,γ) resonances in the ^{14}N and ^{27}Al we obtained the sequential depth profiles of this two elements with good depth resolutions (200 Å). Besides of the profiles these reactions allowed us to obtain the stoichiometric ratio N/Al of the films provided that we had standards with known concentrations.

In an attempt to correlate the deposition parameters with the film characteristics we sought for the dependence of the deposition rate and of the N/Al ratio on the percentage of nitride in plasma during deposition. The knowledge of these curves allows us to deposit films with predetermined characteristics.

5. REFERENCES

1] Shiosaki, T.; Yamamoto, T.; Oda, T. and Kawabata, A., Appl. Phys. Lett. 36(8), 643 (1980).
2] Hasegawa, F.; Takahashi, T.; Kubo, K.; Ohnari, S.; Nannichi, Y. and Arai, T., Jpn. J. Appl. Phys. 26, L1448 (1987).
3] McMahon, R.; Affinito, J. and Parsons, R.R., J. Vac. Sci. Technol. 20(3), 376 (1982).

The Different Steps in the Ion Beam Modification of Poly(ethylene terephthalate) (PET)*

R.M. Papaléo[1], M.A. de Araújo[2] and R.P. Livi[1]
[1]Institute of Physics, [2]Institute of Chemistry
Federal University of Rio Grande do Sul
91500 Porto Alegre, RS, Brazil

Abstract

The crystalline and chemical structures changes produced by ion beam bombardment of PET (mylar) were studied. The structural, thermal, optical, and electrical properties of the modified samples were followed by Fourier transform infrared spectroscopy(FTIR), ultraviolet - visible spectroscopy (UV-VIS), X-ray diffractometry, differential scanning calorimetry (DSC), thermogravimetrical analysis (TGA) and electrical resistance measurements. The melting temperature and the enthalpy of fusion, that are related to the crystalline structure, started to decrease at fluences where the chemical degradation was not significant. The hydrogen and oxygen losses led preferentially to new carbon-carbon bonds within the polymer chains and gradually increased the aliphatic and aromatic conjugation, also producing an optical gap decrease as function of the ion fluence. For $^{40}Ar^{++}$-bombarded samples the optical gap saturates around 0.7 eV for fluences of the order of 10^{15} cm^{-2}. At those fluences the electrical resistivity is relatively high ($\rho >> 10^6$ Ωcm), but for higher fluences it decreases by several orders of magnitude before saturation. The cross sections for the amorphisation, for the optical gap change and for the ester group bond reorganization processes were extracted.

1 Introduction

High energy ion beam implantation or irradiation of polymers yield remarkable changes in their original physical and chemical properties. For example, dramatic modifications in the molecular weigth distribution and solubility[1], as well as in the electrical, optical[2], mechanical[3] and thermal[4] properties have been reported in the last years. Electronic excitation, ionization, chemical bonds disruption and rearrangement are accepted as the fundamental events in the high energy irradiation of polymers[2,5,6]. One of the open and crucial points in this field is to stablish a correlation between microscopic changes, mainly at chemical level, and the modified samples macroscopic properties[6]. In this sense, we investigated the ion beam bombardment induced effects on the crystalline and chemical structure of PET and

*Work partially supported by CNPq, FINEP and FAPERGS, Brazil.

correlated them to the modifications in its thermal, optical and electrical properties. We also discuss the efficiency of the various projectiles used for the generation of irreversible damage in this material.

2 Experimental

The polymer foils used in this study were made of commercial grade poly(ethylene terephthalate) (mylar), 2 μm thick, biaxially oriented and semicrystalline. The bombardments were made in the HVEE 400 kV ion implanter at Porto Alegre, at room temperature and with vacuum between $3-6\times10^{-6}$ torr. $^{1}H^{+}$ (300 keV), $^{4}He^{+}$ (380 keV), $^{20}Ne^{++}$ (600 keV) and $^{40}Ar^{++}$ (760 keV) ions were used with fluences between 5×10^{11} to 2×10^{16} ions/cm^2 depending on the ion. The light ions ($^{1}H^{+}$ and $^{4}He^{+}$) ran through all the sample thickness, while $^{20}Ne^{++}$ and $^{40}Ar^{++}$ bombardments were performed on the two sides of the foils in order to modify them completely. The beam current density was kept always at 25 nA/cm^2, but samples irradiated with different currents in the range between (25-400 nA/cm^2) at a fixed fluence were also prepared in order to investigate the beam current effect on the polymer modification process. Thermal analysis was carried out with a Perkin-Elmer DSC-4 system at a heating rate of 20°C/min and a Perkin-Elmer TGS-2 thermogravimetrical system at a heating rate of 10°C/min, both under nitrogen atmosphere. X-ray diffractometry was taken in a Siemens D-500 diffractometer, using CuK_α radiation and a graphite monochromator in the step scanning mode (10s each 0.1°) through an angular range of 10° to 40°. The chemical degradation process was monitored by IR transmission measurements, performed in a Perkin-Elmer FTIR 1750 apparatus, and UV-visible spectroscopy, performed in a Intralab DMS 80 equipment. Electrical resistance measurements were made using the four point method for the less resistive samples. For low ion fluences effects two point probe was required.

3 Results and Discussion

The PET thermal properties are easily altered by the ion bombardment. Even at the lowest fluences the material already presents a decrease in the melting temperature and a broadening of the melting peak as indicated by the DSC thermograms (see fig. 1 and 2). Such behavior could be attributed to the ion beam induced formation of smaller and less perfect crystallites. It can be noticed also that the polymer is gradually undergoing an amorphisation process with increasing fluence as indicated by the decrease of the main x-ray diffraction peak area in fig. 3 and the enthalpy of fusion in fig. 4. The enthalpy of fusion ΔH is proportional to the polymer crystalline fraction represented here by the index of crystallinity X_c . Its decrease can be fitted nicely (solid lines in fig. 4) using a simple model based in a damage build up mechanism[4,7]. The ion beam induced amorphisation cross-sections

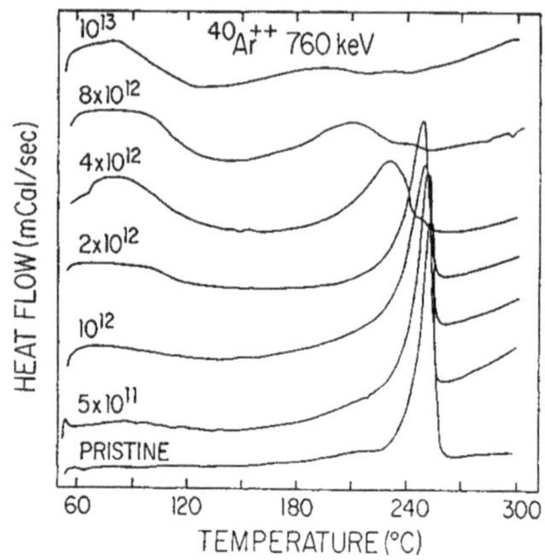

Figure 1: DSC curves for samples bombarded with different fluences of $^{40}Ar^{++}$ ions.

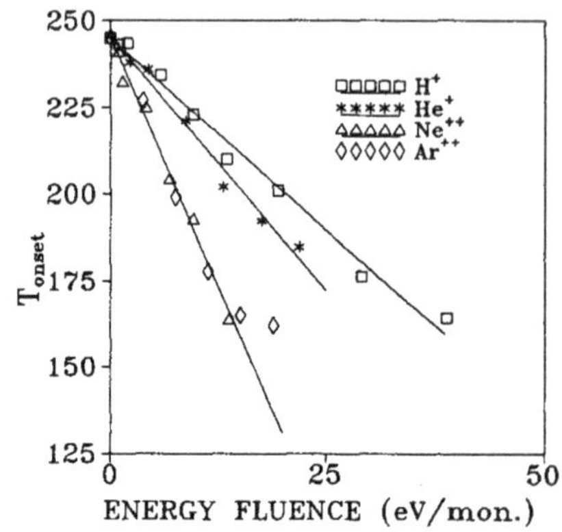

Figure 2: Melting onset temperatures as function of the ion energy fluence for different ions

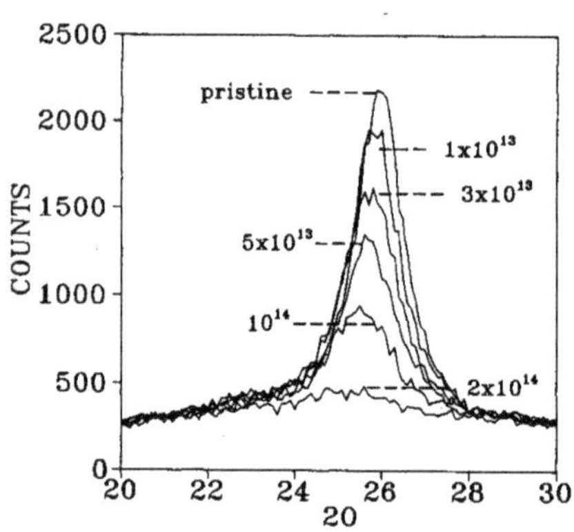

Figure 3: Main x-ray diffraction peak intensities for $^4He^+$ 380 keV bombardment for different fluences.

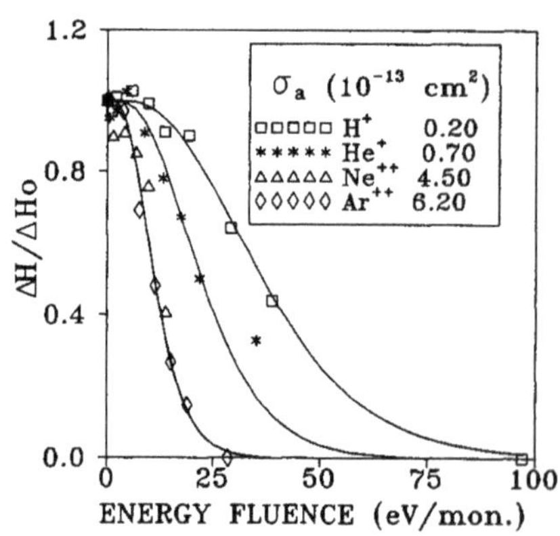

Figure 4: Enthalpies of fusion normalized by the pristine sample values plotted as function of the ion bombardment energy fluence

σ_a, obtained according to this model, are also shown in fig. 4. These results are plotted as function of the energy fluence ϕ^* which is an indicative of the mean deposited energy per monomeric unit at the implanted area. It does not take into account how this energy is deposited (for example in electronic or in nuclear processes) and it is given by the following equation, as proposed by Rössler[8]:

$$\phi^* = \frac{E\phi}{n} \quad (eV/mon.) \tag{1}$$

where E is the total energy deposited by the incident ion in eV, ϕ is the fluence in ions/cm^2 and n is the number of monomers per square centimeter at the implanted area.

The chemical bond reorganization process was studied by FTIR observing the changes of the different IR absorption band intensities as the energy fluence increases. Band assignments were done according to the literature[9–11] (see Table 1). We observed similar trends in the evolution of the IR spectra corresponding to samples bombarded with different ions, what indicates that there is no chemical effect involving these ions, even in the implantation cases (argon and neon). The results for Ne^{++} 600 keV are shown in fig.5. All kinds of chemical bonds are affected probably due to the high deposited energy (typically more than 10 eV/Å) by the incoming ions. However the main disruption route is through the ester group and C-H bonds localized at the molecule aliphatic segment, as indicated by the pronounced intensity decrease in the corresponding IR absorption bands. In fig. 6, the normalized absorbance A/A_o (modified sample absorbance/pristine sample absorbance) for the $\bar{\nu} = 1255\ cm^{-1}$ IR absorption peak (assigned to the ester C-O bond) is plotted as function of the energy fluence. The cross sections for the ion induced C-O bond reorganization process, responsible for the PET main chain scission, were extracted by an exponential fit and can also be seen in fig. 6. Diverging from this behavior, the C-C and C=C corresponding IR band intensities remain

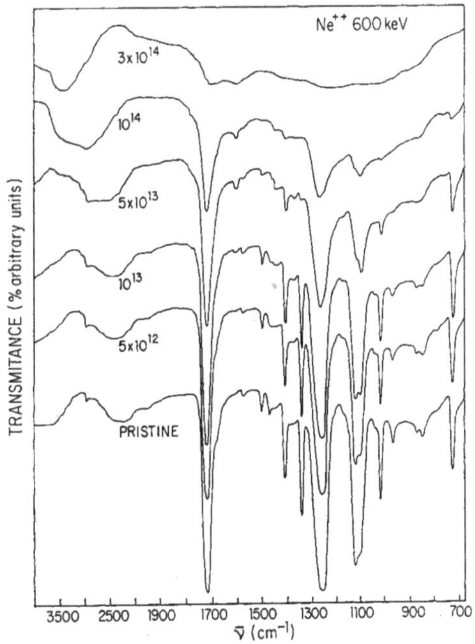

Figure 5: FTIR curves for samples bombarded with several Ne^{++} fluences.

Table 1: PET infrared band assignments for some C $\leftrightarrow$ C, C-H and ester group vibrations [9-11]

$C-C$ and $C=C$		$-COO$		$C-H$	
$\bar{\nu}$ (cm^{-1})	assignment	$\bar{\nu}$ (cm^{-1})	assignment	$\bar{\nu}$ (cm^{-1})	assignment
972	$^{*}\delta_{CCC}$	730	$\delta_{C-(C=O)}$	1022	δ_{C-H}
1506	$^{\dagger}\nu_{CC}$	1255	$\nu_{=C-O}$	1342	$^{\ddagger}w_{CH}$
1579	ν_{CC}	1719	$\nu_{C=O}$		

* bond rocking, † bond stretching, ‡ bond wagging

aproximately constant or increase strongly with energy fluence (as the 1611 cm^{-1} band shown in fig. 7). As the carbon atoms tend to rearrange and to be retained in the material, many of the carbon-carbon vibrations will still be active even after extensive chemical degradation. In this sense, there is no strict physical meaning in deriving disruption cross-sections for these particular bonds. Assuming a naive cilindrical track geometry, we can estimate the bond reorganization effective radius R_b corresponding to the cross-section. For the C-H and C-O bonds, values between 10 to 15 $\mathring{A}$ were obtained (argon implantation case).

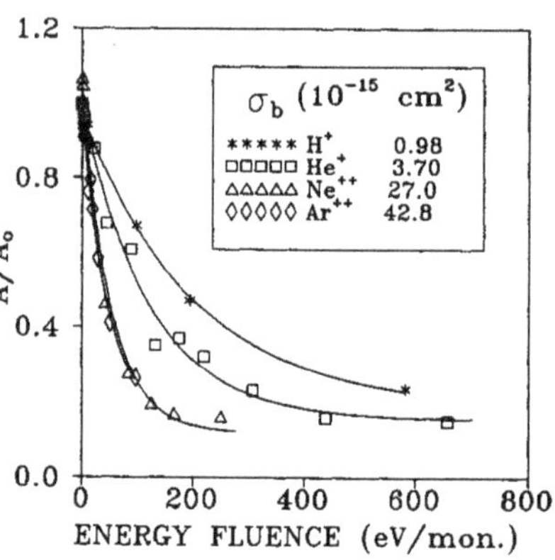

Figure 6: C-O (ester group) normalized IR absorbances as function of the ion fluence. The bond breaking process cross-sections shown were extracted from the exponential fittings (solid lines)

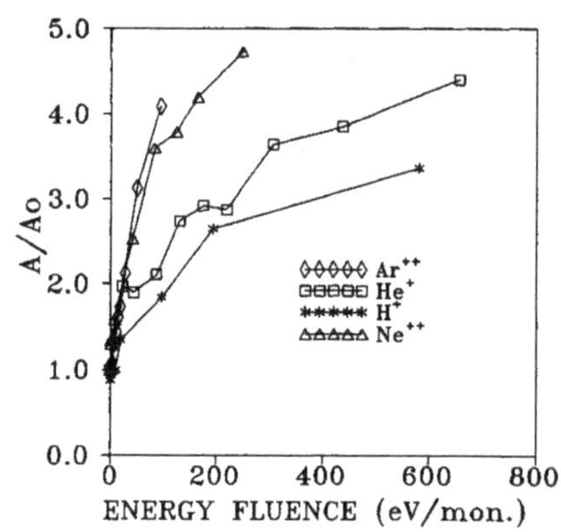

Figure 7: C=C bonds normalized IR absorbance as function of the ion energy fluence

The absorption bands broad patterns, noticeable since intermediate fluences, showed that there is a multiplicity of secondary products and hence of molecular decomposition paths in the ion beam induced chemical degradation of PET. A detailed study of this highly complex chemical process could not be made with the present analytical instrumentation. At high fluences the PET basic chemical units were completely destroied, characterizing a material without a well defined molecular structure.

PET is optically transparent in the visible region of the electromagnetic spectrum. It starts to absorb strongly at $\lambda \approx 310$ nm due to the $\pi \rightarrow \pi^*$ transitions of the $-CO - C_6H_4 - CO-$ cromophore. Under irradiation it was observed a red-shift of the optical absorption threshold with increasing fluence. This phenomenon is usually attributed to the conjugation of insaturated carbon bonds wich give rise to non localized π-electrons. This result is confirmed by the IR absorption intensity increase at the 1580-1615 cm^{-1} region. Absorption bands in this region are normally weak in the majority of the organic compounds, however in aromatic or aliphatic conjugated systems there is a significant absorption enhancement[12]. We can estimate from the UV-VIS spectra the optical gap, E_{op}, using the following expression[13]:

$$(\alpha E)^{1/2} = G(E - E_{op}), \tag{2}$$

where α is the linear absorption coeficient, E is the photon energy and G is a constant. E_{op} decreases from the pristine sample value (> 2.75 eV) to the saturation value (≈ 0.7 eV) reached for $\phi^* \approx 2000$ eV/mon. for $^{40}Ar^{++}$ bombardments. In fig. 8 the experimental points are fitted by an exponential curve $E_{op} = Aexp(-\sigma_c\phi) + B$, which gives a cross-section $\sigma_c = 3.3\times10^{-15} cm^2$. As observed for many polymers[14–17], high fluence IR and UV-visible spectra patterns indicate that the C-O and C-H bond breaking are resulting in the gradual formation of hydrogenated amorphous carbon clusters and probably small graphite crystallites[18]. We can estimate the mean size of these clusters from[13]:

$$E_{op} \cong 6/\sqrt{M} \tag{3}$$

where M is the number of carbon rings inside a cluster. We obtained for the Ar^{++} bombardment M $\approx$ 73, which represents cluster dimensions of $\approx$ 15-20 Å.

The bombarded PET samples thermal stability was followed by thermogravimetrical analysis (TGA). The weight loss onset temperature, that is 405 °C for pristine PET, deacreases with increasing fluence, as can be seen in fig.9. This fact is compatible with low molecular weight fragments formation, as degradation products, within the polymeric matrix. Nevertheless it was observed at high temperatures (T $> 450°C$) an increase of the degradation thermogram slope and of the residual mass as the fluence increases. The residue at 600 $°C$ for the $^4He^+$ 380 keV bombardment, which is 14% for fluences up to 4×10^{13} ions/cm^2 reach 47% at $\phi = 10^{15}$ ions/cm^2. From this we can conclude that at sufficiently high fluences the material is highly crosslinked exhibiting few low molecular weight fragments. TGA data confirms the PET ion-induced carbonization indicated in the FTIR and UV-VIS measurements.

Although exhibiting a relatively low E_{op} at the optical gap saturation fluence, the

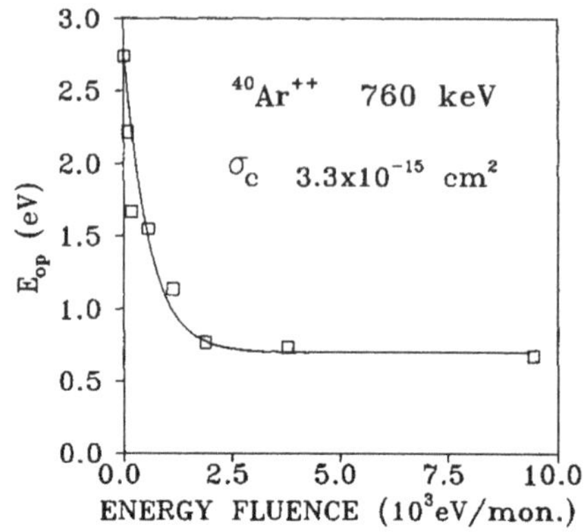

Figure 8: Optical gap E_{op} as function of Ar^{++} bombardment energy fluence. The carbonization process cross-section shown is obtained by an exponential fitting (solid line)

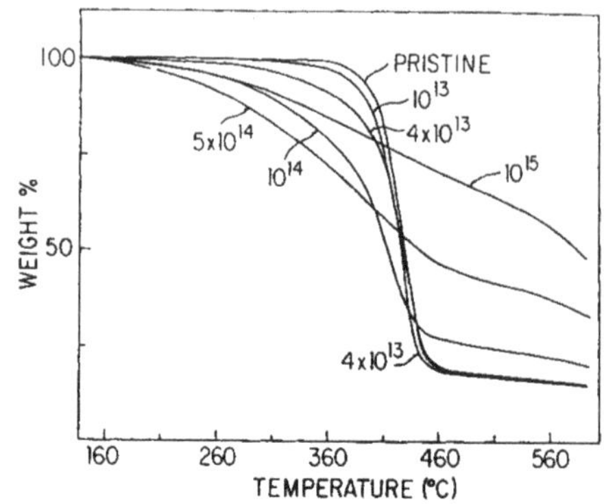

Figure 9: TGA thermograms of the He^+ bombarded samples

material still has insulating characteristics ($\rho >> 10^6\ \Omega cm$) (see fig 10). Only for fluences in excess of $5 \times 10^{15} cm^{-2}$, at which the carbon clusters should be sufficiently close or interconected, semiconducting behavior was found. This result is in accordance with the well stablished model of electrical conduction in high fluence irradiated polymers[2].

In fig. 10 we can see the ratio R/R_{min} plotted against $^{40}Ar^{++}$ energy fluence, where R is the electrical resistance of a sample bombarded with a certain fluence and R_{min} is the minimum sample resistance found for the $2 \times 10^{16} cm^{-2}$ fluence bombardment. The electrical resistivity corresponding to R_{min} for $^{40}Ar^{++}$ 760 keV ion bombardment is around 10 Ωcm.

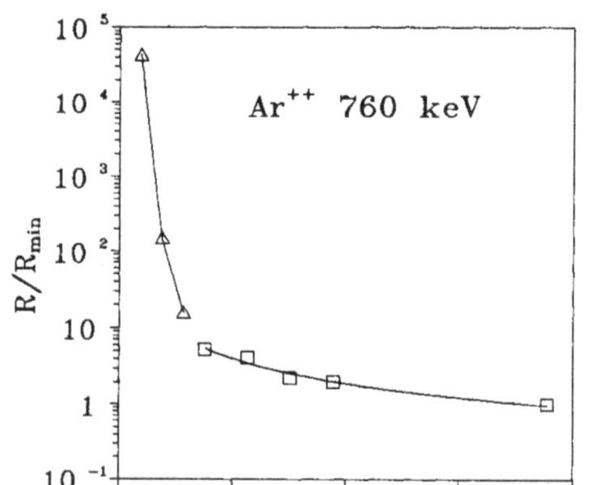

Figure 10: Electrical resistance change as function of Ar^{++} energy fluence

It is important to emphasize that all observed effects and numbers presented here depend strongly on the beam current. For example, variations of more than 40% in the IR band intensities were detected when the beam current density was changed from 25 nA/cm^2 to 400 nA/cm^2 at a fixed fluence. Similar current effects were also obtained in the DSC thermograms. The ion beam induced degradation efficiency seems to increase as the beam current gets higher.

We can make a correlation between the different effective modification radii or cross-sections for each process and the ion track radial characteristics. In the following lines, we will discuss the case of $^{40}Ar^{++}$ bombardment. The lowest effective radius obtained ($\approx 3\AA$) is related to the polymer carbonization and to processes which need direct ion impact or large amount of deposited energy (track core effects). Bond breaking occurs over larger distances ($\approx 10 - 15\AA$) around the ion path and involve events which need intermediate energies (of the order of 1 eV). Secondary electrons can play an important role on it. Polymer crystal order is maintained mainly by weak Van der Waals forces. The high effective radius ($\approx 44\AA$) obtained for PET amorphisation is compatible with its crystallites fragile stability (cohesive energies of few cents of eV[19]). Low energy interaction events, that occur far from the track core, involving for example low energy electrons and phonons are sufficient to damage the crystallites.

It can be noticed that the melting temperature, the enthalpy of fusion, as well as UV-VIS and IR absorption curves corresponding to bombardments with different ions do not collapse in a single curve in their respective plots as function of energy fluence. Such collapse would be expected if the mean energy deposited per repeating unit in the polymer was the unique relevant parameter for the physical and chemical modification processes induced by the bombardment. The efficiency for both, the amorphisation and bond disruption processes is progressively higher for the heavier projectiles. H^+ and He^+ ions must deposit larger amounts of energy per monomer than Ne^{++} and Ar^{++} in order to produce the same damage in the material. So, besides the mean deposited energy, the manner how this energy is deposited is crucial in the ion-macromolecules interaction effects. At least three factors should be taken into account: a) the stopping power, b) the ionization spatial distribution and c) the stopping mechanisms. A rigorous discussion of all these factors can not be done within the present experimental conditions, but some qualitative considerations could be raised.

In processes that depend on the local radiation induced defect concentration as the chemical bond breaking and recombination, the heavier ions (Ne^{++} and Ar^{++}) are more efficient than the light ones (H^+ and He^+). This is probably due to the fact that: *i*) the heavier ions mean stopping power is larger, generating a higher density of defects that will have a larger chance to interact producing irreversible recombinations; *ii*) besides the fact that the faster ions have smaller stopping power the produced δ rays have a larger range and the resulting radial ionization distribution is much wider[20]. In this case the spatial energy density will be much lower than for the heavier ions and hence part of the deposited energy could be dissipated without generating irreversible damage in the material.

The available data do not permit a conclusive discussion about the importance of nuclear and electronic stopping mechanisms in the modification processes. We can say that if the nuclear stopping was the more important mechanism, the H^+ bombarded samples for example which have a very low nuclear stopping ($Sn \approx 0,007 eV/\mathring{A}$) would show less pronounced modification of their physical and chemical properties with the energy fluence, than what was observed in this work. Actually, all the ions used deposited their energy almost only in electronic processes, at least $\approx$ 80 % of the total deposited energy. In this sense, particularly in the ion beam induced amorphisation process, the different ways of convertion of electronic excitation and ionization energy to atomic motion[21] is a very important theoretical problem to be considered.

4 Conclusions

The ion bombardment induced modifications in the physical and chemical properties of the PET foils were followed with several different techniques. The more sensible ones were DSC and X-ray Diffractometry that showed the amorphisation process that occurs for fluences where IR absorption bands intensities have less than 15% modification. For higher fluences the material undergoes a chemical degradation and carbonization that was followed by the FTIR and UV-VIS spectroscopies, TGA and electrical resistance measurements.

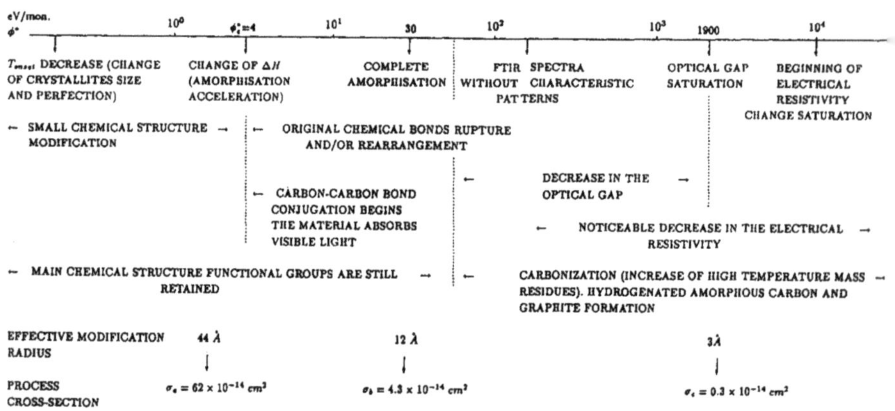

Figure 11: Summary of $^{40}Ar^{++}$ bombardment induced effects as function of the energy fluence. Effective modification radii and processes cross-sections are also shown.

A summary of all the different steps in the ion beam modification of PET is shown in fig. 11, that also presents the effective modification radii and processes cross-sections The results presented in fig. 11 correspond to $^{40}Ar^{++}$ bombarded samples. For different ion bombardments and beam currents the general behaviour is the same but large variations in ϕ^* values and cross-sections for each process are found.

The fact that the dependence of several properties with the energy fluence is not the same for the different ion bombardments, showed that the deposited energy per repeating unit in the polymer is not the only relevant factor in the modification processes. From the chemical point of view the local concentration of reactive defects seems to be the key point in the ion beam modification of this polymer. In this sense the stopping power, the stopping mechanisms and the ionization spatial distribution are the most relevant parameters that determine the ion induced modifications efficiency .

Acknowledgements

The authors would like to thanks Tintas Renner S.A. for providing FTIR analysis time and Icotron S.A. for providing thin mylar foils. We are also in debt with Benhur Felisberti and Alexandre Pedott for their collaboration at our laboratory.

References

1. O. Puglisi, A. Licciardello, L. Calcagno and G. Foti, *Nucl. Instr. and Meth.* **B19/20** (1987) 865
2. T. Venkatesan, L. Calcagno, B.S. Elman and G. Foti, in: *Ion Beam Modification of Insulators*, eds. P. Mazzoldi and G. Arnold (Elsevier, Amsterdam, 1986) p. 301 -379.
3. T. Sasuga, S. Kawanishi, T. Seguchi and I. Kohno, *Polymer* **30** (1989) 2054.
4. R.M. Papaléo, M.A. Araújo and R.P. Livi, *Mat. Res. Soc. Symp. Proc.* **157** (1990) 617.
5. C. David, in: *Comprehensive Chemical Kinetics*, eds. C.H. Banford and F.H. Tipper (Elsevier, Amsterdam, 1975) v.14, p.175-320
6. G. Marletta, *Nucl. Instrum. and Meth.* **B46** (1990) 295.
7. J.F. Gibbons, *Proceedings of the IEEE* **60** (1972) 1062.
8. K. Rössler, *Rad. Eff.* **99** (1986) 505
9. L. D'Esposito and J.L. Koenig, *J. Polymer Sci.: Pol. Phys. Ed.* **14** (1976), 1731
10. T.R. Manley and D.A. Williams, *Polymer* **10** (1969) 339

11. J. Stôrk, B. Schneider, D. Dostoĉilová, J. Lovy and P. Sedláĉek, *Polymer* **23** (1982) 714

12. L.J. Bellamy, *The Infrared Spectra of Complex Molecules*, Chapman and Hall, London, 1975

13. J. Robertson, *Advances in Physics* **35** (1986) 317.

14. T. Venkatesan, S.R. Forrest, M.L. Kaplan, P.H. Schmidt, C.A. Murray, W.L. Brown, B.J. Wilkens, R.F. Roberts, L. Rupp Jr. and H. Schonhorn, *J. Appl. Phys.* **56** (1984) 2778

15. C.L. Renschler, A.P. Sylwester and L.V. Salgado *J. Mater. Res.* **4** (1989) 432

16. J. Davenas, P. Thevenard, G. Boitaux, M. Fallavier, X.L. Xu,*Nucl. Instr. and Meth.* **B46** (1990) 317

17. L. Calcagno Reitano, G. Foti, *Nucl. Instr. and Meth.* **B59/60** (1991) 1153

18. T. Venkatesan, R.P. Livi, T.C. Banwell, T.A. Tombrello M.A. Nicolet, R. Hamm and A.E Meixner, *Mat. Res. Soc. Symp. Proc.*, **45** (1985) 189

19. R. P. Daubeny and C.W.Bunn, *Proc. Roy. Soc.* **A226** (1954) 1167

20. E.J. Kobetich and R. Katz, *Phys. Rev.* **170** (1968) 391.

21. R.E. Johnson and W.L. Brown, *Nucl. Instr. and Meth.* **198** (1982) 103

SIMPLE MODEL FOR CONVOY ELECTRONS TRANSPORT

W.R. CRAVERO,* M.L. MARTIARENA* AND C.R. GARIBOTTI*

División Colisiones Atómicas, Centro Atómico Bariloche
8400 San Carlos de Bariloche, Río Negro
Argentina

ABSTRACT

We study the energy and angular distribution of electrons emitted when fast heavy particles transverse a medium. We present a simple transport model, based on the Boltzmann Equation, which results in a partial differential equation for the electron distribution. This model requires the assumption of simple expressions for the cross sections. The transport equation can be analytically solved and provides the electron spectra near the Convoy Electron Peak, which shows a qualitative agreement with experimental observations in the forward direction and predicts the angular dependence.

1. Introduction

When fast ions or atoms penetrate a dense medium, electron emission take place[1]. The resulting electron velocity distribution presents a sharp cusp for velocities (v) very close to that of the projectile (v1). This cusp is known as the convoy electron peak[2] (CE).

The processes leading to this feature are a rather controversial matter[3]. Several mechanisms for the production of convoy electrons have been proposed such as capture of target electrons into continuum states of the projectile[4] (ECC), projectile electron loss[5] (ELC), Wake-riding electrons[6] and free electrons capture to the projectile continuum (FETC)[7]. The dominance of these processes depends on whether the convoy electrons come from inside the target or from the outgoing surface.

* Consejo Nacional de Investigaciones Científicas y Técnicas.

We will assume that the CE are produced inside the material. As the beam pass through the medium the particles capture and lose electrons, at a rate determined by the corresponding cross sections.

Now, if the medium density is high enough, multiple collision processes will modify the emitted e^- velocity distribution as a function of the medium thickness. This modification will depend on the elastic and inelastic electron- target atom cross sections.

Here we assume that the CE peak is a consequence of the ECC and ELC processes. Therefore the electron production distribution is determined by the corresponding cross sections, which near the projectile velocity $\mathbf{v}_1$ have the form[8]

$$\frac{d\sigma}{d\mathbf{v}} \propto 1/|\mathbf{v}-\mathbf{v}_1| \tag{1}$$

We write for the CE system an extended transport equation[9], where electrons, after being emitted, are considered as particles passing trough a rarefied medium. We consider the target atoms at rest, i.e. in the zero temperature limit, neglect recoil effects and assume that the initial distribution of electrons in the medium is zero.

2. The Transport Equation

Electrons produced by the beam particles, will play the role of a source term in our transport equation. As results from Eq. 1, emitted e^- will initially have velocities close to the projectile's, so we shall consider a source term $H(x,E,\vartheta)$ with the following structure:

$$H(x,E,\vartheta) = S(x)\ \delta(E-E_1)\ \delta(\cos\vartheta - 1) \tag{2}$$

where $S(x)$ is a factor weighing e^- production in the medium. It is proportional to the beam density which varies with deepness. Here E and E_1 are the energies of electrons and beam respectively.

In this framework we write a transport equation[10,11] for the electronic distribution $f(x,E,\vartheta)$ with cylindrical symmetry about the axis of the incident beam,

$$\partial_x f(x,E,\vartheta) = H(x,E,\vartheta) + \int d\varepsilon \int d\Omega\ \sigma'(\varepsilon,\vartheta') \times$$

$$\times \Big(f(x,E+\varepsilon,\vartheta+\Delta\vartheta) - f(x,E,\vartheta)\Big) \tag{3}$$

where x is the distance from the entrance surface and ϑ is measured from the beam direction. $H(x,E,\vartheta)$. As usual, the second (third) term on the right-hand side describes the source (sink) term of the collision integral. The transport kernel $\sigma'(\varepsilon,\vartheta')$ describes the rate per unit path length in which electrons transfer an energy ε and are scattered by an angle ϑ', and it is assumed to be independent of the path length x.

The expression

$$\int d\varepsilon \int d\Omega = \int d\varepsilon \int_0^{2\pi} d\phi' \int_0^{\pi} d\vartheta' \operatorname{sen}(\vartheta') \qquad (4)$$

is an integral carried on a spherical region centered on the distribution function coordinates, as shown in fig. 1.

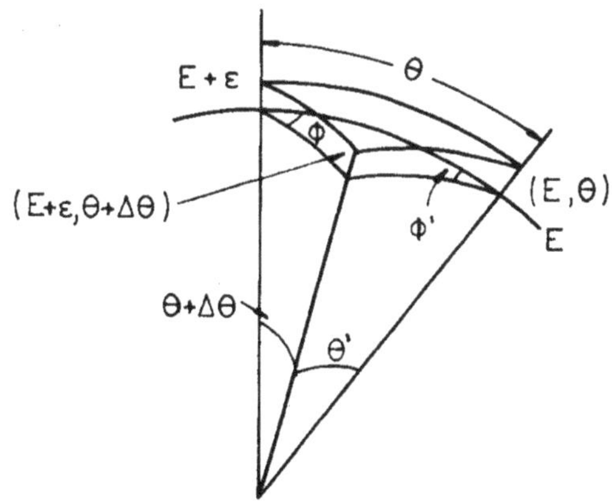

figure 1
Model Description Coordinates

Introducing adequate cross sections, it can be obtained from Eq. 3 the corresponding to Landau Theory for inelastic Multiple Scattering[12], and the transport equation for Goundsmidt-Saunderson Theory[13].

We propose the following form for the distribution function $f(x,E,\vartheta)$:

$$f(x,E,\vartheta) = N(x)\ \delta(E-E_1)\ \delta(\cos\vartheta-1) + g(x,E,\vartheta) \qquad (5)$$

where the first term represents the emitted electrons that suffer no collision. Replacing Eq. 2 and Eq. 5 in Eq. 3, we get

$$\partial_x N(x) = S(x) - \sigma^t N(x) \tag{6}$$

where

$$\sigma^t = \int d\varepsilon \int_0^{2\pi} d\phi' \int_0^{\pi} d\vartheta' \ \text{sen}\vartheta' \ \sigma'(\varepsilon,\vartheta') \tag{7}$$

is the total cross section. We assume that the beam density is exponentially decreasing with x. Therefore we choose $S(x)=S_0\exp(-\beta x)$. Solving Eq. 6 we obtain,

$$N(x) = S_0 \, e^{-\sigma^t x}\left(\frac{e^{(\sigma^t-\beta)x} - 1}{(\sigma^t - \beta)}\right) \tag{8}$$

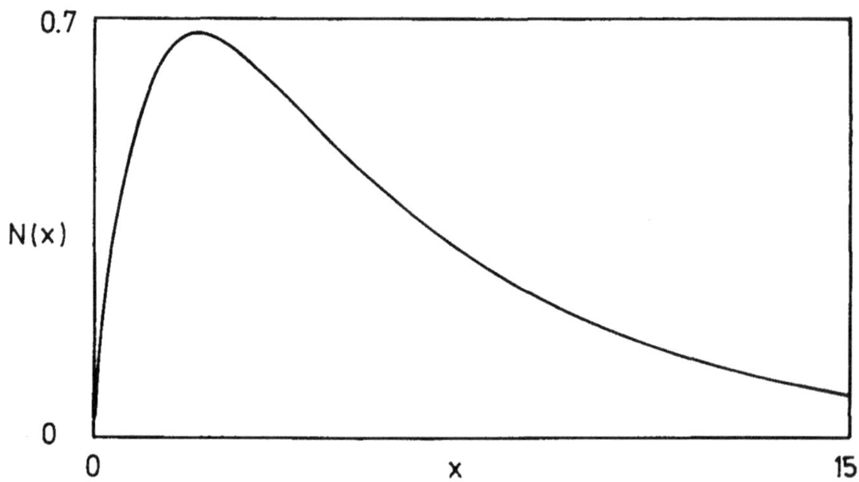

figure 2
CE density evolution with thickness

The remaining equation for $g(x,E,\vartheta)$ reads

$$\partial_x g(x,E,\vartheta) = N(x) \int d\varepsilon \int_0^{2\pi} d\phi' \int_0^{\pi} d\vartheta' \ \mathrm{sen}\vartheta' \ \sigma'(\varepsilon,\vartheta') \times$$

$$\times \ \delta(E+\varepsilon-E_1) \ \delta(\cos(\vartheta+\vartheta')-1) \ +$$

$$+\int d\varepsilon \int_0^{2\pi} d\phi' \int_0^{\pi} d\vartheta' \mathrm{sen}\vartheta' \sigma'(\varepsilon,\vartheta') \Big(g(x,E+\varepsilon,\vartheta+\Delta\vartheta) - g(x,E,\vartheta)\Big) \qquad (9)$$

This equation is still too complex, so we make some drastic approximations in order to handle it analytically. We take for the cross section the following simple form:

$$\sigma'(\varepsilon,\vartheta') = \begin{cases} \sigma' & 0 < \varepsilon < \varepsilon_0 \ \ \& \ \ \vartheta' < \vartheta'_0 \\ 0 & \varepsilon > \varepsilon_0 \ \ \text{ó} \ \ \vartheta' > \vartheta'_0 \end{cases} \qquad (10)$$

This assumption is reasonable since real cross sections for inelastic scattering fall quickly for increasing energy transfer or angular dispersion. Replacing Eq. 10 in Eq. 9, it reads

$$\partial_x g(x,E,\vartheta) = N(x) \ \sigma' \int_0^{\varepsilon_0} d\varepsilon \int_0^{2\pi} d\phi' \int_0^{\vartheta'_0} d\vartheta' \ \mathrm{sen}\vartheta' \times$$

$$\times\delta(E+\varepsilon-E_1) \ \delta(\cos(\vartheta+\Delta\vartheta)-1) \ +$$

$$+ \ \sigma' \int_0^{\varepsilon_0} d\varepsilon \int_0^{2\pi} d\phi' \int_0^{\vartheta'_0} d\vartheta' \ \mathrm{sen}\vartheta' \Big(g(x,E+\varepsilon,\vartheta+\Delta\vartheta)-g(x,E,\vartheta)\Big) \qquad (11)$$

We note that the first term on the right hand side is zero except for an interval around E_1. Elsewhere Eq. 11 reduces to

$$\partial_x g(x,E,\vartheta) = \sigma' \int_0^{\varepsilon_0} d\varepsilon \int_0^{2\pi} d\phi' \int_0^{\vartheta'_0} d\vartheta' \ \mathrm{sen}\vartheta' \Big(g(x,E+\varepsilon,\vartheta+\Delta\vartheta)-g(x,E,\vartheta)\Big) \qquad (12)$$

Assuming that both ε_0 and ϑ'_0ere small, we can expand[14]:

$$g(x,E+\varepsilon,\vartheta+\Delta\vartheta)-g(x,E,\vartheta) = \partial_E g(x,E,\vartheta)\ \varepsilon + \partial_{\vartheta'} g(x,E,\vartheta)\ \vartheta' +.. \quad (13)$$

replacing Eq. 13 in Eq. 12 , the integration can be carried on explicitly, leading to a partial differential equation for $g(x,E,\vartheta)$:

$$\partial_x g(x,E,\vartheta) = A\ \partial_E g(x,E,\vartheta) + B\ \partial^2_\vartheta\ g(x,E,\vartheta) \quad (14)$$

where

$$A = \sigma'\varepsilon_0^2\ \frac{\pi}{2}\ \vartheta'_0 \quad (15a)$$

and

$$B = \sigma'\varepsilon_0\ \frac{\pi}{2}\ \vartheta'^2_0 \quad (15b)$$

The general solution for Eq. 14 is found to be:

$$g(x,E,\vartheta) = 2\vartheta'_0 \left(\frac{A}{4\pi B\ (E_1-E)}\right)^{-1/2} g(x-(E_1-E)/A,E_1,0)\ \times$$

$$\exp\left(\frac{-A\ \vartheta^2}{4\pi B(E_1-E)}\right) \quad (16)$$

The equation for $g(x,E_1,0)$ can be obtained from Eq. 11:

$$\partial_x g(x,E_1,0) = N(x)\sigma' - \sigma^t g(x,E_1,0) \quad (17)$$

Using Eq. 8, Eq. 17 is immediately integrated,

$$g(x,E_1,0) = \frac{\sigma' S_0}{(\sigma^t-\beta)} \left(\frac{e^{(\sigma^t-\beta)x}-1}{(\sigma^t-\beta)} - x\right)\exp(-\sigma^t x) \quad (18)$$

With this boundary condition, Eq. 16 can be written

$$g(x,E,\vartheta) = \frac{\sigma' S_0}{(\sigma^t-\beta)}\left(\frac{\exp\left((\sigma^t-\beta)[x-(E_i-E)/A]\right) - 1}{(\sigma^t-\beta)} - [x-(E_i-E)/A]\right)\times$$

$$\times \exp\left(-\sigma^t[x-(E_i-E)/A]\right)2\vartheta'_0\left(\frac{A}{4\pi B\ (E_i-E)}\right)^{1/2}\times$$

$$\exp\ \left(\frac{-\ A\ \vartheta^2}{4\pi B(E_i-E)}\right)\Theta(x-(E_i-E)/A) \qquad (19)$$

The thickness-energy dependence of this solution can be decomposed into a sum of exponential terms. The first takes into account the decay of the ion beam, while the others stand for the dispersion of electrons themselves. It represents an electronic flux displacing to lower energies as x increases (see figure 3). This effect is observed in experimental spectra in the forward direction[15] and is due to inelastic collisions suffered by convoy electrons. However due to the drastic approximations made, we can only expect a qualitative description. We are not aware of experimental data about the angular distribution for which our model predicts a gaussian angular dependence.

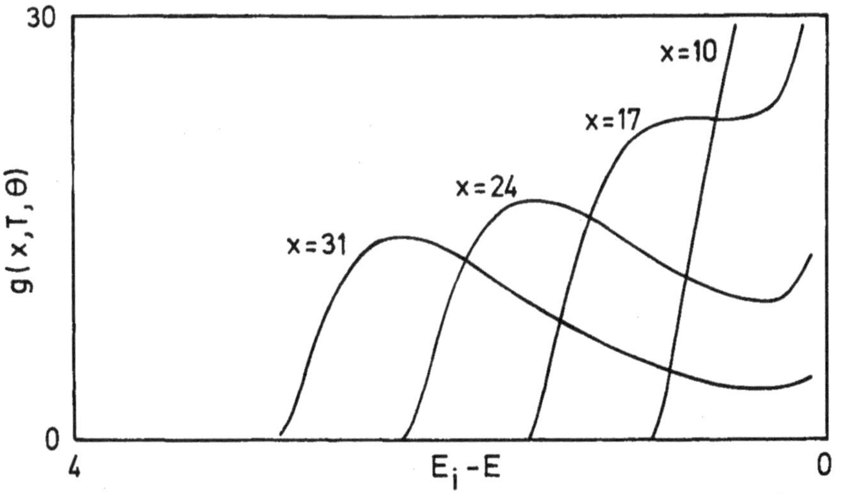

figure 3
Forward Electronic Distribution given by Eq. 19 for several values of x

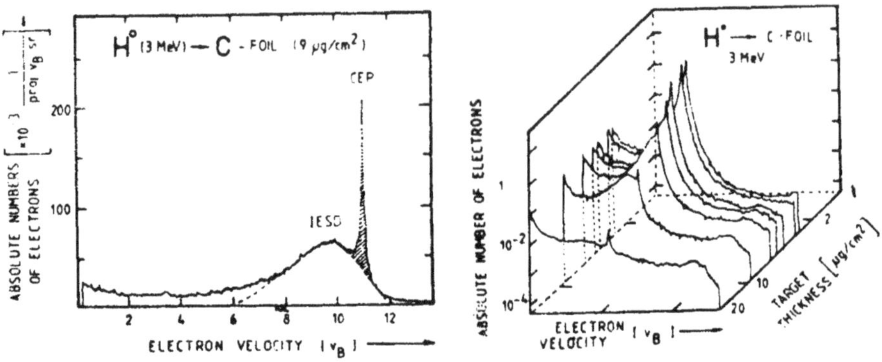

figure 4
Experimental electronic spectra[15]

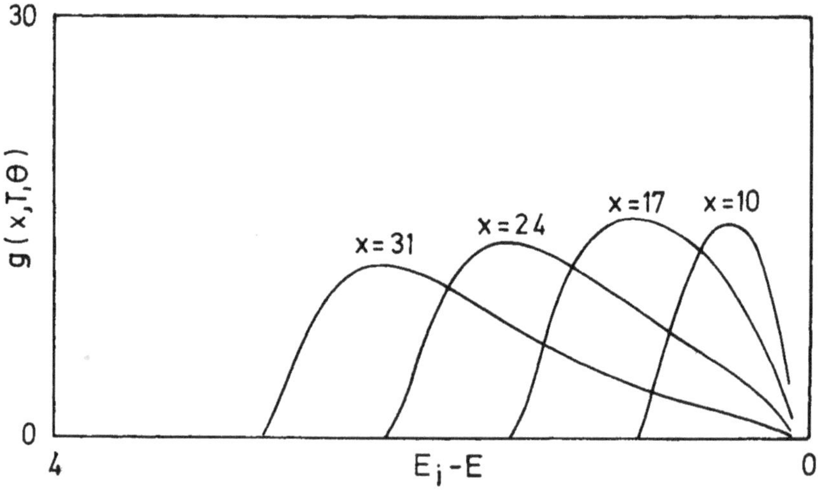

figure 5
Idem fig. 3 for a direction 5° apart from the beam path.

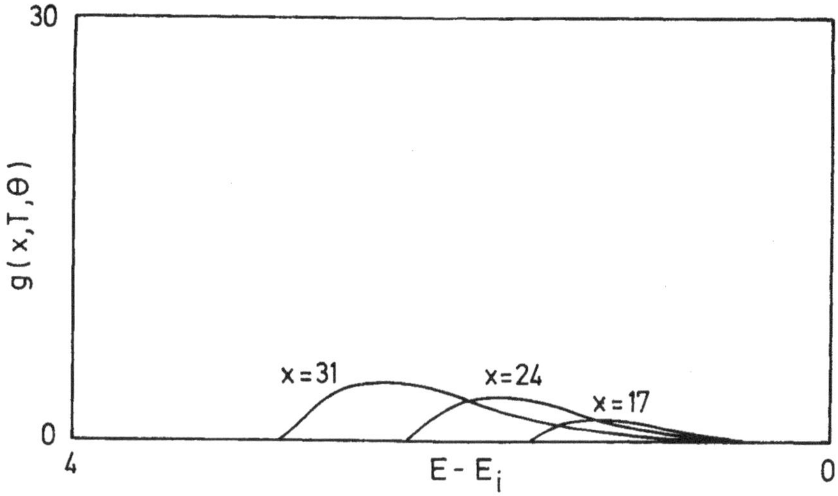

figure 6
Idem fig 3 for a direction 17° apart from forward direction.

3. Conclusions

We have presented a simple model for describing the angle and energy distribution function of electrons produced when an energetic ion beam passes through a medium.

This model takes up the physical hypothesis we used in a previous paper[16] to obtain an electron transport equation. In that case we studied the energy distribution function for forward emission, performing a numerical calculation of the transport equation, and using realistic elastic and inelastic cross sections for the collisions of electrons with target atoms.

Due to the mathematical complexity of the transport equation and the number of reaction channels, numerical calculations required hard computer work, even for the case of electron emission in the forward direction.

For these reasons, in the present work we have introduced plain mathematical approximations for the cross sections and for the decay of the ionic beam in the medium. This has made possible an analytical solution of the transport equation which shows explicitly the angular dependence. The electron energy distribution exhibits a bump at retarded velocities relative to the CE peak, which agrees qualitatively both with experimental observations and with numerical calculations for forward emission.

Besides, our model gives a first known hint about the angular distribution. It seems to point out that the structure generated at 0° is softly propagated with increasing angle, as a consequence of elastic collisions.

Quantitative comparisons between our model and experimental data are not possible at this level, because this requires a physical connection between real and model cross sections of the different collision processes.

Finally, we conclude that a model based in a kinetic transport equation is able to describe the main features of electron emission spectra.

4. References

1. E.Bonderup, *Penetration of Charged Particles trough Matter*, (University of Aarhus, Denmark, 1978).
2. W.Meckbach, K.Chiu, H.Brongresman and J.McGowan, *J.Phys.* **B10** (1977)3255.
3. W.Meckbach and P. Focke, *Nucl. Inst. Meth.* **B33** (1988) 255.
4. K. Dettmann, K.G. Harrison and M.W. Lucas, *J. Phys.***B7** (1974) 269.
5. Breimig et al. *Phys.Rev.* **A25** (1982) 3015.
6. V.Neslavathi, R. Ritchie and W. Brandt, *Phys.Rev.Lett.* **33**, (1974) 302.
7. Y. Yamazaki and N. Oda, *Phys.Rev.Lett.***52**, (1984) 29.
8. R.O. Barrachina, A. Goñi, P. Focke and W. Meckbach, *Nucl.Inst.Meth.***B 33**, (1988) 330.
9. M.L. Martiarena and C.R. Garibotti, *Physica* **A 166** (1990) 115.
10. M.L. Martiarena and C.R.Garibotti, *Actas III Encuentro Latinoamericano de Colisiones Atomicas, Moleculares y Electronicas,* (Cuernavaca, Mexico, 1989).
11. S. Lencinas and J. Burgdörfer, *Phys.Rev.***A 41**, (1990) 1435.
12. L.Landau, *J. Phys.* (Moscow) **8** (1944) 1.
13. S.Gaundsmidt and J.Saunderson, *Phys. Rev* **57** (1940) 24.
14. H.Bethe and M.Rose, *Proc. Am. Phil. Soc.* **78** (1938) 573.
15. P.Koshar et al., *Nucl. Inst. Meth.* **B 24/25** (1987) 153.
16 M.L. Martiarena and C.R. Garibotti, *Phys.Rev.* (1991) to be published.

APPENDIX

APPENDIX

SUGGESTIONS FOR NEXT DESORPTION STUDIES

Having Prof. Schweikert as Chairman, the lecturers participating in the round table on *"Future Trends of Desorption Methods of Mass Spectrometry"* have analyzed PDMS instrumentation and discussed about experimental and theoretical aspects of the subject. Very appropriated and certainly quite useful were their suggestions for next desorption experiments and studies.

The understanding of the desorption mechanisms is today an attractive problem of fundamental physics linked to energy deposit and transfer in solids. The great potential of the desorption processes, to investigate interactions between fast ions and materials and to develop analytical tools, explains the increasing number of scientists and laboratories devoting their interest to ion desorption. All domains of applications will gain from fundamental studies.

It is presented bellow the reproduction of the contents of the transparencies used during their oral expositions.

DEVELOPING A PROGRAM BASED ON PDMS

Ronald Macfarlane

I- Why PDMS ?

a. Possibility for important new studies in:

1) Applications - Analysis of complex molecules
2) Fundamental Studies in Ion-Solid Interactions
3) Study of physical/chemical processes in high energy density/short time domain.

II- Most Attractive Features

a. Simplicity - design and operation
b. Low cost
c. Reliability
d. Good for training students to do research
e. There is much more to learn than what is now known regarding mechanisms of ion desorption.
f. Interdisciplinary - organic, inorganic, physical analytical chemistry, biotechnology, physics-surface, plasma, materials research.

III- My views of the Optimal Features of a new PDMS system

a. Do not miniaturize. Provide plenty of room for adding and modifying the system, keeping wires and high voltage surfaces isolated. Fringe Fields".
b. Get an ORSAY TDC + Associated hardware and software. PC compatible.

c. Think carefully about transmission vs. reflection geometry for ^{252}Cf source. I prefer transmission because:

(1) Less intense sources can be used ≤ 5 μCi
(2) More flexibility introduced by varying target/source distance
(3) Order ^{252}Cf source with 0.2 μ Ni foils rather than 1 μ to provide higher effective energies.

A Modern Research / Applications PDMS System with Possibilities for Important New Studies must be able to:

1. Obtain high quality mass spectra for applications

a. Good mass accuracy (± 0,2μ, up to 10,000μ)
(1) Attention to alignment of parallel surfaces that influence TOF. Laser alignment ~ 1 m length
(2) At least 15 KV acceleration, stable power supplies
(3) Use of "standing" design reflectron to give an effective 1 m length

2. Measure initial kinetic energy distributions in the simplest most straight forward manner

a. Axial velocity - TOF spectrum at low V
b. Radial velocity - Iris collimators, beam stops

3. Correlations with fission fragment directions

a. Variable target/source distance
b. Beam stop method

4. Study Metastable Decays

a. Ion deflection - pulsed, continuous
b. Energetic neutrals spectra
c. Coincidence studies between neutrals and reflected ions (Reflectron)

5. Study Secondary Ion/Surface Interactions

THE HOT TOPICS OF 1991 (my group)

1. Developing a more efficient ion detector for large (> 5000 u) molecular ions.
2. Studying the role of the substrate in determining the yields of molecular ions of different species adsorbed on surfaces.
EXAMPLE: Nitrocellulose is excellent for proteins but poor for fatty acids; nylon is excellent for fatty acids but poor for proteins.
3. Metastable decay of ions using reflectron.
4. Looking for better adsorption/desorption matrices based on principles of ion desorption mechanisms.
5. Making use of fragmentation ions in spectrum to understand molecular excitation by fission fragments.

6. Study of Solution Properties and Reactions of Proteins by PDMS
   EXAMPLE: Pd^{+2} + Insulin ⟶ Insulin . Pd Complex (I. Pd)

   I. Pd (solution) ⟶ I. Pd (nitrocellulose)

   I. Pd (n.c.) $\xrightarrow{\text{PDMS}}$ Presence and location Pd^{+2} in insulin

SUGGESTIONS FOR NEW EXPERIMENTS

Yvon Le Beyec

I would like to encourage students and scientists to think about what has been said during the School. This is an open field in many ways.

I would like to say also that PDMS has in a sense opened this field.

Of PDMS TOF mass spectrometry is very useful.

Related experiments on basic physics and chemistry can be done with simple apparatus.

Laser impacts with matrix is also very interesting and offer many facets in applications and basic phenomena.

Cluster ion impacts with large ions and molecules is also something new.

citation from Le Beyec's message

I. <u>MeV Ion Projectiles</u>

- Measurement of neutrals, instead of ions secondary (for comparison with models)
- Other types of multilayer targets (depth of emission)
- Angular distributions of S.I.
- ^{252}Cf double TOF

II. <u>Charge States</u> (variations of charge states "q" of fast ions in solids

- q = f (x) for various projectiles and targets in collaboration with plasma physics people
- <u>H^+ emission</u>
  Measurement of H^+ velocity distribution as function of incident charge with KeV and MeV ions
  Measurement of yield of D^+ to compare with theory

III. <u>Impacts of Clusters</u>

1. **Double TOF experiments**. Use of several kinds of targets as "ion sources" - molecules and clusters
   a) Studies on electron emission and ion emission (event by event)
   b) Studies on time correlation of secondary ions ejected by KeV and MeV ions.
   c) Studies on non linear effects in cluster impacts.

2. **Ion Sources (for large cluster ion intensity)**
   - Electrospray ion sources (beams of large molecules or large clusters)
   - Any kind of cluster ion source

3. **Experiments with beams from accelerator**
   - Use of 4 MeV Van de Graaff
   - Use of tandem
   - Open field (cluster ions accelerated to MeV energy)

IV. Impact of Photons

With the development of new small Lasers ⟶ simple experiments on desorption phenomena.

- Time of emission
- Metastable decays
- Studies of large molecule impacts
- Applications in biology: a large expanding field

V. Models and Mechanisms

- d Molecular dynamic calculation
- Energy dissipation, fast collisions...

VI. Instrumentation

BO SUNDQVIST

THEORY OF SPUTTERING OF LARGE ORGANIC MOLECULES

1. Microscopic models for transfer of electronic energy to atomic and molecular motion.
2. Models for ionization mechanisms.
3. Microscopic theory for H^+ ejection.

MOLECULAR DYNAMICS SIMULATIONS

1*. Matrix assisted laser desorption
2. Desorption from adsorbed states
3*. Include ionization mechanisms
4*. Energetic macromolecular ion impact

*Under way !

EXPERIMENTS ON SPUTTERING OR LARGE ORGANIC MOLECULES

1. Sputtering yields of neutrals
2. Exploit PDMS for studies of macromolecular-surface interaction studies
3. Energetic molecular ion impact studies
4. Matrix assisted laser description

PDMS FOR M < 1000 u

Raul Baragiola

FUNDAMENTAL ASPECTS

. Electronic sputtering by individual excitation of adsorbed particles on metal surfaces
. role of the nuclear track ?
. simple frozen gases
. influence of nuclear stopping
. total yields !!

APPLICATIONS

. Mixtures of simple compounds
. Geological samples (Schweikert's correlation method)
. Corrosion... phase transitions ?

INSTRUMENTATION

. Clean and annealed surfaces
. Temperature variation
. High mass resolution at low masses!
. Knowledge of mechanisms leads to applications

UNKNOWNS

. Primary vs Total (stopping power)
. Energy and charge flow
. Lifetimes of excited states
. Charge transfer H^+ vs H^-
. Final-state interactions ion-ion, ion-surface
. Internal energies and breakup in flight
. Relation to SIMS
. Clusters to dust grains
. Initial/Final state of target - Hot chemistry

SYSTEMS AND METHODS

Things should be as simple as possible, but not simpler.

. Minimize uncontrolled variables ⟶ reproducibility
  - Clean vacuum - trap mechanical pump , dry N_2/Ar
  - Clean target - remove oxide
. Combine with surface analysis ions + total yields
. Combine with gas-phase studies

THEORY

. Microscopic models
. Use known σ
. Model "hot" region/track